Pathology in Marine Science

Proceedings of the Third International Colloquium on Pathology in Marine Aquaculture held in Gloucester Point, Virginia, October 2–6, 1988.

Academic Press Rapid Manuscript Reproduction

Pathology in Marine Science

Edited by

Frank O. Perkins
School of Marine Science
College of William & Mary
Gloucester Point, Virginia

Thomas C. Cheng
Marine Biomedical Research Program
Medical University of South Carolina
Charleston, South Carolina

ACADEMIC PRESS, INC.
Harcourt Brace Jovanovich, Publishers
San Diego New York Berkeley Boston
London Sydney Tokyo Toronto

This book is printed on acid-free paper. ∞

Academic Press, Inc.
San Diego, California 92101

United Kingdom Edition published by
Academic Press Limited
24–28 Oval Road, London NW1 7DX

Library of Congress Cataloging-in-Publication Data

International Colloquium on Pathology in Marine Aquacultue (3rd : 1988 : Gloucester Point, VA.)
Pathology in marine science : proceedings of the Third International Colloquium on Pathology in Marine Aquaculture held in Gloucester Point, Virginia (USA), 2-6 October 1988 / edited by Frank O. Perkins, Thomas C. Cheng.
p. cm.
ISBN 0-12-550755-0 (alk. paper)
1. Marine fishes--Diseases--Congresses. 2. Shellfish--Diseases--Congresses. I. Perkins, Frank O. II. Cheng, Thomas C. (Thomas Clement) III. Title.
SH171.I56 1988
639.3'01'5912--dc20 89-18108
CIP

Printed in the United States of America
89 90 91 92 9 8 7 6 5 4 3 2 1

CONTENTS

EPIDEMIOLOGY/EPIZOOTIOLOGY SECTION

NUTRITIONAL PATHOLOGY SECTION

IMMUNOLOGY SECTION

Contributors

Numbers in parentheses indicate the pages on which the authors' contributions begin.

Vicki A. Alberts (65), Department of Biology, University of South Florida, Tampa, Florida 33620

M. R. Alvarez (489, 501), Department of Biology, University of South Florida, Tampa, Florida 33620

R. Andral (33), Laboratoire Départemental des Services Vétérinaires de l'Hérault, 306 rue de la Croix de las Cazes, 34000 Montpellier, France

Carlos Azevedo (175, 181), Department of Cell Biology, Institute of Biomedical Science, University of Oporto, Porto, Portugal

Isabel Bandín (101, 123), Departmento de Microbiología y Parasitología, Universidad de Santiago, Santiago de Compostela 15707, Spain

Bruce J. Barber (377), Virginia Institute of Marine Science, Gloucester Point, Virginia, 23062

Juan L. Barja (77, 91), Departmento de Microbiología y Parasitología, Universidad de Santiago, Santiago de Compostela 15707, Spain

Ana Baya (91), Department of Microbiology, University of Maryland, College Park, Maryland 20742

Teresa Kraxberger Beatty (65), Department of Biology, University of South Florida, Tampa, Florida 33620

Jorge Bolinches (77), Departmento de Microbiología y Parasitología, Universidad de Santiago, Santiago de Compostela 15707, Spain

J. A. Boutillier (267), Department of Fisheries and Oceans, Biological Sciences Branch, Pacific Biological Station, Nanaimo, British Columbia, Canada

Susan M. Bower (267), Department of Fisheries and Oceans, Biological Sciences Branch, Pacific Biological Station, Nanaimo, British Columbia, Canada

Thomas Braunbeck (347), Department of Zoology, University of Heidelberg, D-6900, Heidelberg, Federal Republic of Germany

D. W. Bruno (101), Department of Agriculture and Fisheries for Scotland, Marine Laboratory, Aberdeen AB9 8DB, Scotland

Jacques Brusle (441, 461), Labortoire de Biologie Marine, Université de Perpignan, Avenue de Villeneuve, Perpignan Cedex, France

R. Cachola (181), National Institute of Fisheries Research, Faro, Portugal

Thomas C. Cheng (439, 513), Marine Biolomedical Research Program, Medical University of South Carolina, Charleston, South Carolina 29410

Fu Lin E. Chu (533), Virginia Institute of Marine Science, Gloucester Point, Virginia 23062

Claude Combes (291, 331), Centre de Biologie et d'Ecologie Tropicale at Méditerranéenne, Université de Perpignan, Perpignan, France

Ramón F. Conchas (77, 123), Departmento de Microbiología y Parasitología, Universidad de Santiago, Santiago de Compostela 15707, Spain

E. L. Cooper (471), Department of Anatomy and Cell Biology, School of Medicine, University of California at Los Angeles, Los Angeles, California 90024

L. Corral (175), Department of Cell Biology, Institute of Biomedical Science, University of Oporto, 40000 Porto, Portugal

John A. Couch (343), Environmental Research Laboratory, U. S. Environmental Protection Agency, Gulf Breeze, Florida 32561

C. Coustau (291), Laboratoire de Parasitologie Comparée, Université de Montpellier, 34060 Montpellier Cedex, France

B. Delay (291), Institut des Sciences de l'évolution, Université de Montpellier, 34000 Montpellier Cedex, France

Ariel Diamant (7), Israel Oceanographic and Limnological Research Institute, Eilat, Israel 88112

M. M. Dougherty (387), Marine Biomedical Research Program, Medical University of South Carolina, Charleston, South Carolina 29412

William J. Dougherty (387), Marine Biomedical Research Program, Medical University of South Carolina, Charleston, South Carolina 29412

W. Eaton (407), Alaska Department of Fish and Game, Fisheries Rehabilitation Enhancement and Development Division, Juneau, Alaska 99802

A. E. Ellis (101), Department of Agriculture and Fisheries for Scotland, Marine Laboratory, Aberdeen AB9 8DB, Scotland

Ralph Elston (163), Battelle Marine Sciences Lab, Sequim, Washington 98382

P. Elziere-Papayanni (33), Laboratoire Departmental des Services Vétérinaires de l'Hérault, 34060 Montpellier Cedex, France

Mohamed Faisal[1] (283, 471), Department of Avian and Aquatic Animal Medicine, Faculty of Veterinary Medicine, University of Alexandria, Alexandria, Egypt

Elisabeth Faliex (299), Marine Biology Laboratory, Université de Perpignan, Avenue de Villeneuve, 66025 Perpignan Cedex, France

Susan E. Ford (229), Shellfish Research Laboratory, Rutgers University, Port Norris, New Jersey 08349

[1]Present address: Virginia Institute of Marine Science, School of Marine Science, College of William and Mary, Gloucester Point, Virginia 23062

John W. Forsythe (51), Marine Biomedical Institute, University of Texas Medical Branch, Galveston, Texas 77550

Craig K. Foster (215), Salmon Enterprises of Tasmania, Dover, Tansmania 7117, Australia

Belén Fouz (77), Departmento de Microbiología y Parasitología, Universidad de Santiago, Santiago de Compostela 15707, Spain

F. E. Friedl (489, 501), Department of Biology, University of South Florida, Tampa, Florida 33620

H. Fukuda (15), Department of Aquatic Biosciences, Tokyo University of Fisheries, Minato-ku, Tokyo 108 Japan

C. L. Goggin (189), Department of Parasitology, University of Queensland, Brisbane Q4067, Australia

S. Gusmão (175), Department of Cell Biology, Center of Biological Sciences, Federal University of Pará, 66000 Belém, Brazil

O. L. M. Haenen (317), Netherlands Institute for Fishery Investigations, 1970 AB, Ijmuiden, The Netherlands

Sonja Hall-Mendelin (421), Graduate School of Tropical Veterinary Science, James Cook University of North Queensland, Townsville, Queensland 4811, Australia

Roger T. Hanlon (51), Marine Biomedical Institute, University of Texas Medical Branch, Galveston, Texas 77550-2772

Frank M. Hetrick (91), University of Maryland, College Park, Maryland 20742

B. D. Hicks (305), Pacific Aqua Foods, Vancouver, British Columbia VGB 1G1, Canada

Gertrude W. Hinsch (479), Department of Biology, University of South Florida, Tampa, Florida 33620

Michele Hunte (479), Department of Biology, University of South Florida, Tampa, Florida 33620

C. Hurard (33), Laboratoire Départemental des Services Vétérinaires de l'Hérault, 306 rue de la Croix de las Cazes, 34000 Montpellier, France

E. A. Imam (283), Faculty of Veterinary Medicine, Cairo University, Cairo, Egypt

J. N. Kraeuter (277), Shellfish Research Laboratory, Rutgers University, Port Norris, New Jersey 08349

Jerome F. La Peyre (533), Virginia Institute of Marine Science, Gloucester Point, Virginia 23062

Phillip G. Lee (51), Marine Biomedical Institute, University of Texas Medical Branch, Galveston, Texas 77550-2772

Manuel L. Lemos (123), Departmento de Microbiología y Parasitología, Universidad de Santiago, Santiago de Compostela 15707, Spain

R. J. G. Lester (189, 215), Department of Parasitology, University of Queensland, Brisbane Q4067, Australia

Donald V. Lightner (3), Department of Veterinary Science, University of Arizona, Tucson, Arizona 85721

Daniel V. Lim (65), Department of Biology, University of South Florida, Tampa, Florida 33620

C. Maillard (291), Laboratoire de Parasitologie Comparée, Université de Montpellier, 34000 Montpellier Cedex, France

V. M. Marryatt (165), Halifax Fisheries Research Laboratory, Halifax, Nova Scotia, Canada B3J 2S7

E. S. Mathews (369), Division of Chemistry and Toxicology, Virginia Institute of Marine Science, Gloucester Point, Virginia, 23062

E. Matos (175), Department of Cell Biology, Center of Biological Sciences, Federal University of Pará, 66000 Belém, Brazil

M. A. McCarthy (277), Baltimore Gas and Electric, Crane Aquaculture Facility, Baltimore, Maryland 21203

Donald J. McGarey (65), Department of Biology, University of South Florida, Tampa, Florida 33620

Sharon E. McGladdery (305), Department of Fisheries and Oceans, Gulf Fisheries Centre, Moncton, New Brunswick, Canada

Alasdair H. McVicar (397), Department of Agriculture and Fisheries for Scotland Marine Laboratory, Aberdeen AB98DB, Scotland

Theodore R. Meyers (407), Alaska Department of Fish and Game, Fisheries Rehabilitation Enhancement and Development Division, Juneau, Alaska 99802

Jamie Montes (223), Conselleria de Pesca, Vilagarcia de Arousa, Spain

J. Frank Morado (201), National Marine Fisheries Service, Seattle, Washington 98115

Katsuyoshi Mori (465, 525), National Research Institute of Aquaculture, Fisheries Agency, Nansei, Mie 516-01, Japan

Carol M. Morrison (165), Halifax Fisheries Research Laboratory, Halifax, Nova Scotia, Canada B3J 2S7

Nadine Mosconi-Bac (455), Marine Biology Laboratory, Université de Perpignan, Perpignan, France

B. L. Munday (215), Tasmanian State Institute of Technology, Newnham, Tasmania 7250, Australia

L. Murphy (305), Department of Fisheries and Oceans, Charlottetown, Prince Edward Island C1A 7M8, Canada

Soledad Núñez (91), Departmento de Microbiología y Parasitología, Universidad de Santiago, Santiago de Compostela 15707, Spain

Motoichi Nakamura (525), Faculty of Agriculture, Tohoku University, Sendai 980, Japan

Teresa P. Nieto (101), Departmento de Microbiología y Parasitología, Universidad de Santiago, Santiago de Compostela 15707, Spain

Edward J. Noga (143), Department of Companion Animal and Special Species Medicine, College of Veterinary Medicine, North Carolina State University, Raleigh, North Carolina 27606

Tadashi Nomura (525), Faculty of Agriculture, Tohoku University, Sendai 980, Japan

Robin M. Overstreet (261), Gulf Coast Research Laboratory, Ocean Springs, Mississippi 39564

Leigh Owens (421), Graduate School of Tropical Veterinary Science, James Cook University of North Queensland, Townsville, Queensland 4811, Australia

Frank O. Perkins (181), Virginia Institute of Marine Science, College of William and Mary, Gloucester Point, Virginia 23062

P. S. Perkins (471), Department of Anatomy and Cell Biology, School of Medicine, University of California at Los Angeles, Los Angeles, California 90024

F. Renaud (291), Laboratoire de Parasitologie Comparée, Université de Montpellier, 34000 Montpellier Cedex, France

Jesús L. Romalde (77, 123), Departmento de Microbiología y Parasitología, Universidad de Santiago, Santiago de Compostela 15707, Spain

Joshua Rosenthal (7), Rectortown, Virginia 22140

F. R. Roubal (215), Department of Parasitology, University of Queensland, Brisbane, Queensland 4067, Australia

D. R. Sager (277), Texas Parks and Wildlife, Austin, Texas 78744

M. Sano (15), Department of Aquatic Biosciences, Tokyo University of Fisheries, Minato-ku, Tokyo 108 Japan

Tokuo Sano (15, 49), Department of Aquatic Biosciences, Tokyo University of Fisheries, Minato-ku, Tokyo 108 Japan

Ysabel Santos (101), Departmento de Microbiología y Parasitología, Universidad de Santiago, Santiago de Compostela 15707, Spain

Robert A. Scro (229), New Jersey Department of Environmental Protection, Trenton, New Jersey 08625

Helmut Segner (347), Zoologie II (Physiologie), Universität Karlsruhe, Karlsruhe, Federal Republic of Germany

K. B. Sewell (189), Department of Parasitology, University of Queensland, Brisbane Q4067, Australia

S. Short (407), Alaska Department of Fish and Game, Fisheries Rehabilitation Enhancement and Development Division, Juneau, Alaska 99802

A. K. Sparks (201), National Marine Fisheries Service, Seattle, Washington 98115

Diane Te Strake (65), Department of Biology, University of South Florida, Tampa, Florida 33620

Alicia E. Toranzo (77, 91, 101, 123), Departmento de Microbiología y Parasitología, Universidad de Santiago, Santiago de Compostela, Spain 15706

P. E. Vestergaard Jørgensen (43), Statens Veterinaere, Serumlaboratorium, Arhus, N, Denmark

Martine Vigneulle (117), Laboratoire de Pathologie des Animaux Aquatiques, Plouzané, France

Christian P. Vivarès (33, 255), Laboratoire de Pathologie Comparée, Université des Sciences et Techniques du Languedoc, 34060 Montpellier Cédex, France

S. K. Wagner (305), Department of Fisheries and Oceans, Gulf Fisheries Centre, Moncton, New Brunswick E1C 9B6, Canada

J. E. Warinner (369), Division of Chemistry and Toxicology, Virginia Institute of Marine Science, Gloucester Point, Virginia, 23062

B. Anne Weeks (369), Division of Chemistry and Toxicology, Virginia Institute of Marine Science, Gloucester Point, Virginia, 23062

A. Wishkovsky (369), Division of Chemistry and Toxicology, Virginia Institute of Marine Science, Gloucester Point, Virginia, 23062

L. Curry Woods III (277), Baltimore Gas and Electric, Crane Aquaculture Facility, Baltimore, Maryland 21203

Paul van Banning (317, 433), Netherlands Institute for Fishery Investigations, 1970 AB, Ijmuiden, The Netherlands

PREFACE

The vastness of aquaculture ventures worldwide has taken impressive leaps during the past two decades, and a large portion of this pertains to mariculture. Significant strides have been made in the design and implementation of more efficient and cost effective growing areas, feed, and harvesting techniques. The economic impact of aquaculture will certainly increase in the years ahead.

As the development of mariculture has advanced, it has become increasingly apparent that one essentially unsolved problem is the prevention of mass mortalities and cure of mass morbidities. It is a well-established principle that animals maintained under crowded conditions are more susceptible to infections and other categories of diseases. This is true of human as well as fish and shellfish populations. In the case of infectious diseases, the frequent introduction of economically important animals from other geographic areas with the intent of establishing new products has resulted in the introduction of devastating pathogens that have caused mass mortalities of native stocks. The accidental introduction of *Bonamia* into western European oyster grounds is a case in point.

As the recognition of disease problems in mariculture became more acute, pathobiologists of a variety of persuasions, e.g., virologists, microbiologists, parasitologists, immunologists, and epidemiologists have been enlisted to help solve such crises.

Because one of the major areas of the world where diseases have taken their toll among animals maintained in mariculture is France and adjacent coastal areas, the idea of originating an international pathology in marine aquaculture (PAMAQ) was formulated by Drs. Jean-Michel Arcier, Noel Boemare, Jean-Robert Bonami, Michel Brehélin, Jocelyne Mari, and Christian Vivarès of the University of Montpellier, France. These pathobiologists were encouraged and sponsored by Professor C. Vago, a member of the Academy of Sciences of France. As a result, the first International Colloquium on Pathology in Marine Aquaculture (PAMAQ I) was held in Montpellier in 1984. Based on the decision that a similar colloquium should be called every other year, PAMAQ II was organized by Professor Carlos Azevedo in 1986 in Oporto, Portugal.

At the Oporto meeting, the International Scientific Committee[1] decided that the 1988 Colloquim should be convened in the United States; specifically, at the Virginia Institute of

[1]Members of the International Scientific Committee: C. Azevedo (Portugal), T. C. Cheng (U. S. A.), E. Egidius (Norway), B. J. Hill (United Kingdom), I. Paperna (Israel), F. O. Perkins (U. S. A.), T. Sano (Japan), P. van Banning (the Netherlands), and C. P. Vivarès (France: Executive Secretary).

Marine Science, School of Marine Science, College of William and Mary, Gloucester Point, Virginia. Thus, PAMAQ III was held in Virginia on October 2–6 1988, with the editors of this volume as the co-hosts. The meeting attracted over 140 scientists worldwide. Most of the arrangements were made by Ms. Sarah Hamrick, of the Virginia Institute of Marine Science, to whom all who participated owe an immense debt of gratitude. We would also like to thank Mr. Dan Sved for his help in editing, Ms. Janet Walker for typography, and Ms. Kay Stubblefield for preparing the layout of the contributions that follow. Their diligence is greatly appreciated.

The following pages represent the majority of papers presented at PAMAQ III. Each of the included contributions has been evaluated and found to include information that should be shared with fellow scientists worldwide who were unable to attend. It is our hope that these proceedings will not only serve as an additional tangible landmark in the progress of concerted research in marine pathobiology but also as a document that will be a useful reference. Reflective of the subdisciplines of the biological sciences that are of immediate concern to investigators interested in pathology in marine aquaculture, included herein are contributions pertaining to viroses, bacterioses, mycoses, protozoan diseases, metazoan parasitic diseases, toxicological syndromes, teratological and neoplastic diseases, epidemiology/ epizootiology, nutritional pathology, and immunology. May the readers find this volume a worthwhile record.

Frank O. Perkins
Thomas C. Cheng

DEDICATION

This volume is dedicated to the memory of Dr. Emmy Egidius, a member of the International Scientific Committee of the Colloquium on Pathology in Marine Aquaculture (PAMAQ). Emmy passed away on February 3, 1989. Emmy devoted much of her career to investigating bacterial diseases of fish and shellfish. During the 1960s she studied gaffkemia in lobsters at the Institute of Marine Biology of the University of Bergen, Norway. This was followed by studies on piscian vibriosis. She won wide acclaim in the early 1980s for her work on Hitra disease, which she helped to reveal as being caused by a bacterium. She named it Vibrio salmonicida. During ensuing years, Emmy was a member of a team that developed a new vaccine for Hitra disease (cold water vibriosis).

Recognizing the economic problem caused by salmon lice, she tested various compounds and was instrumental in gaining wide usage of Neguvon against this parasite.

Emmy was also actively involved in fisheries administration. She was an advisor to the Norwegian Directorate of Fisheries, a central figure in the International Council for the Exploration of the Seas (ICES). In 1987 she was elected President of the European Association of Fish Pathologists. As stated, she was intimately involved in the development of PAMAQ. It was unfortunate that because of her other duties, she was unable to be with us at Gloucester Point, Virginia, in 1988.

Emmy will long be remembered by her former students and colleagues worldwide as a friendly, dedicated scientist and will be greatly missed.

VIROSES SECTION

VIROSES SECTION: INTRODUCTORY REMARKS

D.V. Lightner

Environmental Research Laboratory, University of Arizona,
2601 East Airport Drive, Tucson, Arizona 85706, U.S.A.

There are nearly as many viruses known in marine invertebrates and vertebrates of interest to aquaculturists as there are species cultured. With only a few exceptions (for example, with lobsters) at least one virus caused disease is recognized in each significant marine species now being cultured. A perusal of some recent reference books on the subject of disease and pathology of these animals shows this to be true (Table 1).

Table 1. A partial list (from three recent reviews) of cultured marine animals that are affected by virus caused diseases.

	Amos (1985)	Sindermann & Lightner (1988)	Sparks (1985)
Finfish			
Salmonids	7	1	-
Others	1	2	-
Crustacea			
Penaeid shrimp	-	6	4
Macrobrachium	-	0	0
Crabs	-	3	13
Lobsters	-	0	0
Mollusks			
Oysters	-	2-3	5
Others	-	0	2
Turtles	-	1	-
Totals	8	15-16	24

The Fish Health Blue Book (published by the Fish Health Section of the American Fisheries Society; Amos, 1985) lists eight types of virus caused diseases, six of which are of concern to the marine aquaculture industry. Another reference, "Disease Diagnosis and Control in North American Marine Aquaculture" that Carl Sindermann and I recently edited (1988) lists 16 virus diseases in cultured marine crustaceans, mollusks, finfish, and turtles. Sparks (1985) in his "Synopsis of Invertebrate Pathology Exclusive of Insects" lists 24 virus caused diseases of cultured and commercially important invertebrate animals. In the current literature we find that several more new viruses, new hosts, or geographic records for previously known viruses, are reported each year. Viruses are important pathogens of cultured marine species.

Many of these virus diseases are of only minor apparent significance to the aquaculture industry. However, others of considerable significance cause catastrophic losses whenever host or environmental conditions favor their development. Of the viruses that we now recognize, several have significantly affected the development of commercial marine aquaculture. Some important examples include IHN, IPN, and VEN of salmonids; and IHHNV, BP, BMN, and MBV of penaeid shrimp. This year's crash of Taiwan's crop of *Penaeus monodon* may cost that country as much as 500 million dollars in lost export revenue, and one or more viruses (MBV and HPV) are believed to have caused or contributed to the epizootic (Rosenberry, 1988).

A dilemma that has faced the shrimp culture industry as it has developed over the past two decades has been our desire as an industry composed of scientists and businessmen to obtain and test a variety of local and exotic penaeid shrimp. The reasons for these transfers are sometimes justifiable, sometimes not, but shrimp growers have made them seeking shrimp that grow well and provide more profit from their culture systems. Some desirable characteristics used to justify importing exotic species have included a larger harvest size, faster growth rates, disease resistance, a higher market price, easier reproduction and larval rearing, and growth at colder water temperatures. Hence, larvae, postlarvae, and broodstock from shrimp farms, from experimental facilities, or from wild stocks collected by commercial fishermen, have been transferred countless times from one geographic location to another for aquaculture purposes. Adding to this dilemma was our paucity of knowledge of shrimp pathogens. Less than a decade ago the shrimp culture industry knew of only one virus in penaeids (Couch, 1974). We now recognize at least six different virus diseases in these animals, and we suspect several more (Lightner, 1988; Lightner and Redman, 1989).

The examples shown in Figure 1 of published and unpublished records of live shrimp transfers are typical of transfers made beginning nearly two decades ago and continuing in today's shrimp culture industry. When combined into the single map shown in Figure 2, the plotted shrimp transfers illustrate "an exotic shrimp transfer network." From this web of shrimp culture facilities that have exported and imported penaeid shrimp, it is apparent that if an unrecognized pathogen entered any network facility, the mechanism exists for it to be rapidly transferred to several other facilities, and if it remains undetected, to all of the facilities in the transfer network. We have named this pathogen transfer mechanism "the jumbo jet vector of virus disease".

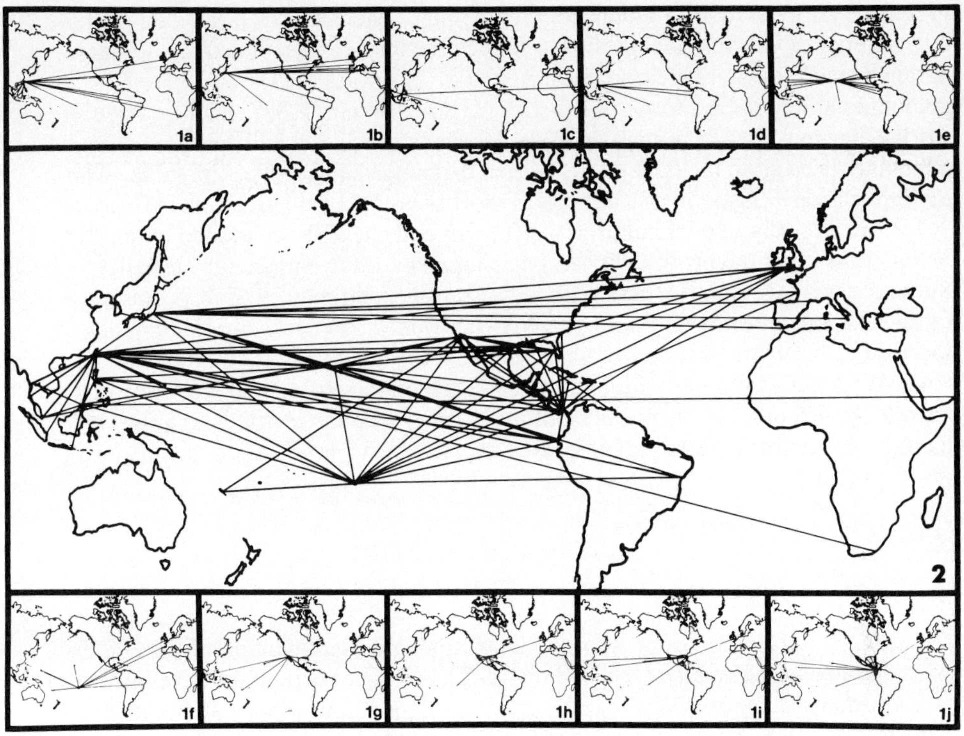

Fig. 1. Examples of published and unpublished records of live shrimp transfers that are typical of transfers made beginning nearly two decades ago and continuing in today's shrimp culture industry. Specific examples show exports or imports of penaeids from or to: 1a. Taiwan; 1b. Japan; 1c. Malaysia; 1d. Philippines; 1e. Hawaii; 1f. Tahiti; 1g. Puerto Penasco, Mexico; 1h. Texas; 1i. Florida; 1j. Panama.

Fig. 2. A hypothetical "exotic shrimp transfer network", that results from the combination of Fig. 1 shrimp transfer maps into a single map. It is apparent that if an unrecognized pathogen entered any facility in the transfer network, the mechanism exists for it to be rapidly transferred to several other facilities, and if the pathogen remained undetected, it could be potentially introduced into all of the facilities in the transfer network.

IHHNV of penaeid shrimp was introduced into this transfer network from an unknown source at least a decade ago. IHHNV disease was first recognized in 1981 in imported shrimp in Hawaii that had been imported from at least five different sources (Lightner et al., 1983a,b). None of the suppliers knew of the existence of IHHNV, nor did the shrimp they supplied (*P. vannamei*) show any external signs of disease or of infection by this virus. However, when the imported *P. vannamei* stocks were grown adjacent to stocks of *P. stylirostris*, most of the latter species were soon lost to IHHNV. An interesting coincidence is that all of the sources that provided IHHNV-contaminated *P. vannamei* had previously abandoned most of their attempts at culture of the larger, faster growing (and hence more valuable) *P. stylirostris* because of its poor survival in their culture systems.

As I will point out in a contribution included in this volume, other shrimp viruses have entered this transfer network, and they too are being introduced to farms in regions far from their original geographic range. Mechanisms have been proposed by a number of international groups to reduce the importation of exotic pathogens and pests with transfers of aquatic species. Two examples are the FAO (1977) guidelines and the ICES Code of Practice (in Sindermann, 1988). Both provide a workable mechanism to reduce this risk. However, for these guidelines to work, adequate quarantine facilities and qualified diagnosticians must be available. The shrimp culture industry to this day is still short of both, but the situation is improving.

The virus diseases of cultured marine fish and shellfish are indeed important factors that affect the profitability and development of the aquaculture industry. Successful management of these pathogens translates directly into jobs, commerce, foreign exchange revenues, and most importantly into new sources of high quality foodstuffs. New information like what will be presented by the participants of PAMAQ 3 will add to our understanding of these pathogens, and collectively that knowledge will provide the marine culture industry with the methods to avoid and manage those virus disease problems that now or will adversely affect its further development.

LITERATURE CITED

Amos, K.H. 1985. "Procedures for the Detection and Identification of Certain Fish Pathogens" (The Fish Health Blue Book), 3rd ed. Special Publ. Am. Fisher. Soc., Washington, D.C.

Couch, J.A. 1974. Free and occluded virus, similar to *Baculovirus*, in hepatopancreas of pink shrimp. Nature, 247, 229-231.

FAO. 1977. "Control of the Spread of Major Communicable Fish Diseases." Report of the FAO/OIE Government Consultation on an International Convention for the Control of the Spread of Major Communicable Fish Diseases. FAO Fisher. Repts., No. 192. FID/R192 (EN).

Lightner, D.V. 1988. Diseases of cultured penaeid shrimp and prawns. In: "Disease Diagnosis and Control in North American Marine Aquaculture" (C.J. Sindermann and D.V. Lightner, eds.), pp. 8-127. Elsevier, Amsterdam.

Lightner, D.V. and R.M. Redman. 1989. Host, geographic range and diagnostic procedures for the penaeid virus diseases of concern to shrimp culturists in the Americas. In: "Frontiers of Shrimp Research" (W. Dougherty, ed.), in press. Elsevier, New York.

Lightner, D.V., R.M. Redman and T.A. Bell. 1983a. Infectious hypodermal and hemotopoietic necrosis (IHHN), a newly recognized virus disease of penaeid shrimp. J. Invert. Pathol., 42, 62-70.

Lightner, D.V., R.M. Redman, T.A. Bell and J.A. Brock. 1983b. Detection of IHHN virus in *Penaeus stylirostris* and *P. vannamei* imported into Hawaii. J. World Maricult. Soc., 14, 212-225.

Rosenberry, B. 1988. Crash in Taiwan. Aquacult. Dig., 13.9.1.

Sindermann, C.J. 1988. Disease problems created by introduced species. In: "Disease Diagnosis and Control in North American Marine Aquaculture" (C.J. Sindermann and D.V. Lightner, eds.), pp. 394-398. Elsevier, Amsterdam.

Sindermann, C.J. and D.V. Lightner, eds., 1988. "Disease Diagnosis and Control in North American Marine Aquaculture." Elsevier, Amsterdam.

Sparks, A.K. 1985. "Synopsis of Invertebrate Pathology Exclusive of Insects". Elsevier, Amsterdam.

IN VITRO PRIMARY CELL CULTURES FROM *PENAEUS SEMISULCATUS*

Joshua Rosenthal and Ariel Diamant

National Center for Mariculture, Israel Oceanographic and Limnological Research Institute, P.O. Box 1212, Eilat 88112, Israel

ABSTRACT: Establishment of primary cell cultures from *Penaeus semisulcatus* intended for *in vitro* virus replication is described. Cultures were initiated from hematopoietic tissue/lymphoid organ, hepatopancreas, and ovary. The best results were obtained using cell culture medium M-199 supplemented with 15% FBS and 5% heat inactivated shrimp hemolymph. Cells originating from the hematopoietic/ lymphoid tissue explants formed fairly dense, though not confluent, monolayers. The cultured cells were maintained for 3 weeks, but attempts to sub-culture them were unsuccessful.

INTRODUCTION

Mortalities of cultured shrimp associated with viral infection represent an emerging worldwide problem as a result of the constant spreading of viruses with newly introduced brood stock (Lightner, 1988). In Israel, penaeid shrimp culture has similarly been subject to persistent infections, mainly with MBV (Monodon-type Baculovirus), causing mass mortality breakouts in post-larval and young animals (Colorni et al., 1987). Although useful when intranuclear inclusion bodies become detectable (at medium or high intensity infections), diagnostics of MBV through histology, malachite green staining, and acridine orange fluorescence are inadequate when screening for early or carrier state infections (Diamant and Colorni, 1987; Nash et al., 1988). Development of sensitive detection methods for latent viral shrimp infections depends largely on the availability of purified virus material. At present, the state of crustacean *in vitro* cell culture that could support *in vitro* virus replication is still at a very early experimental stage. However, some promising results have recently been achieved with shrimp (Chen et al., 1986; Ellender et al., 1988; Chen and Kou, 1989). This paper reports on the results obtained during some preliminary trials to initiate cell cultures from Mediterranean stock of *Penaeus semisulcatus*.

MATERIALS AND METHODS

Adult *P. semisulcatus* were obtained from the Shrimp Culture Unit at the National Center for Mariculture, Eilat, Israel, during October, 1987 and April, 1988. The animals had been caught several months earlier off Haifa Bay (Mediterranean coast of Israel) and transferred to Eilat (Red Sea) as intended brood stock.

Hematopoietic tissue typically possesses high generative potential, and was therefore selected as a cell source. The location of hematopoietic tissue centers in *P. semisulcatus* was determined by serial paraffin sections stained with hematoxylin and eosin (H&E), which readily demonstrated diffuse hematopoietic centers on the anterio-dorsal surface of the hepatopancreas (midgut gland), overlying the cardiac stomach. Lymphoid organ tubuli were found to be located adjacently. Since neither hematopoietic nor the lymphoid organ are plainly discernable in gross examination during dissection of live shrimp, the entire membraneous anterio-dorsal region overlying the hepatopancreas, including both tissue types, was removed and utilized. Excision of ovarian and hepatopancreatic tissue for explants was straightforward.

Shrimp were placed on ice for 10 min and then dipped briefly (5-10 sec) in 0.2% benzalkonium chloride (Roccal), rinsed in sterile sea water, and dissected under aseptic conditions. Initially, we attempted to disperse bits of minced tissue enzymatically. However, both trypsin (0.25%) and collagenase (0.15%) (Sigma) treatments proved ineffective in liberating cells from tissue fragments. Subsequently, the tissue bits were minced with two scalpel blades to pieces approximately 1 mm in diameter, and explants were plated directly after repeated washes with osmotically-adjusted sterile phosphate buffered saline.

The cell cultures rapidly displayed bacterial contamination and sterility of the hemolymph was questioned. In an attempt to test the presumptive source of contamination, the primary cultures, aseptically collected *P. semisulcatus* hemolymph, were plated on tryptic soy agar (TSA) and thiosulfate citrate bile sucrose (TCBS) agar prepared with 25% sea water and monitored for development of bacterial colonies. The susceptibility of subsequently isolated bacteria to various antibiotics was determined by the method of Colorni (1985).

The animal cell culture basal media used were the following: Eagle's minimum essential medium (Earle's salt base, with nonessential amino acids); nutrient mixture Ham F-12; Leibovitz L-15 medium; and tissue culture medium 199 (Hank's salt base). All basal media were supplemented with 1% L-glutamine. Osmolality values were adjusted to match that of Eilat pond-cultured *P. semisulcatus* hemolymph, which was measured at 923 $\pm$ 10 mOsm/kg using a Wescor 5100C vapor pressure osmometer (osmolality values are considerably elevated in Eilat-reared shrimp owing to the high salinity of the ambient sea water). The pH values of all media and solutions were adjusted to 7.3, as measured in *P. semisulcatus* hemolymph, and incubations were carried out at 18°C. The basic components used are given in Table 1. Hemolymph was extracted from *P. semisulcatus* using the following routine: a heparin-rinsed sterile hypodermic needle was carefully inserted ventrally into the pericardial cavity (via the abdomen-cephalothorax junction) and hemolymph was drawn out at a slow, steady pace. To avoid coagulation, this procedure was carried out on ice under an atmosphere of pure nitrogen. The hemolymph was then spun at 500*g* for 10 min to remove hemocytes, and heat inactivated at 65°C for 10 min in a water bath. It was then respun at 10,000*g* for 10 min. The resulting pellet of denatured proteins was discarded. This procedure yielded approximately 1 ml of treated hemolymph (supernatant fluid) per 20 g body weight.

Table 1. Components used and their concentrations in the primary cell cultures from *P. semisulcatus*.

Component	Final concentration
basal medium*	1X
non-essential amino acid mix*	1X
essential amino acid mix*	1X
vitamin solution*	1X
fetal bovine serum*	5-20%
shrimp hemolymph	5-20%
HEPES pH 7.3	20 mM
sodium bicarbonate	0.36%
penicillin	100 u/ml
streptomycin	100 ug/ml
amphotericin B	12.5 ug/ml
osmolality	923 mOSM/kg

*Biological Industries, Kibbutz Beit HaEmek, Israel.

For quantification of initial cell migration from the explants, the number of cells is expressed per 1 cm^2 area of culture dish (at 48 hr).

RESULTS

Results of cell growth with the various culture media combinations are presented in Figure 1. The best results by far were obtained with M-199. Addition of fetal bovine serum (FBS) and shrimp hemolymph yielded variable results (Fig. 2). When each of these was used separately, little cell proliferation was evident. Considerably better results were obtained when the two were used in conjunction. When variable concentrations of hemolymph were used while keeping FBS at a constant 5% level, we found that cell proliferation increased steadily up to a level of 5% hemolymph, after which it decreased. Conversely, when FBS concentration was increased (while hemolymph kept at a constant 5%), cell proliferation was significantly enhanced, and the trend persisted up to a concentration of approximately 15% FBS, after which a plateau was reached where further increase made no detectable difference.

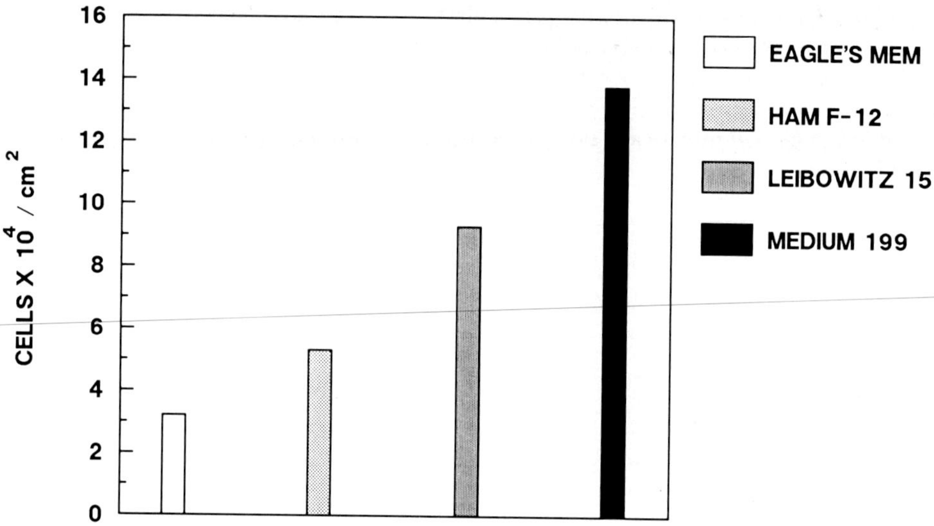

Fig. 1. Cell number per cm^2 generated by fragments of hematopoietic/lymphoid organ tissue in various culture media.

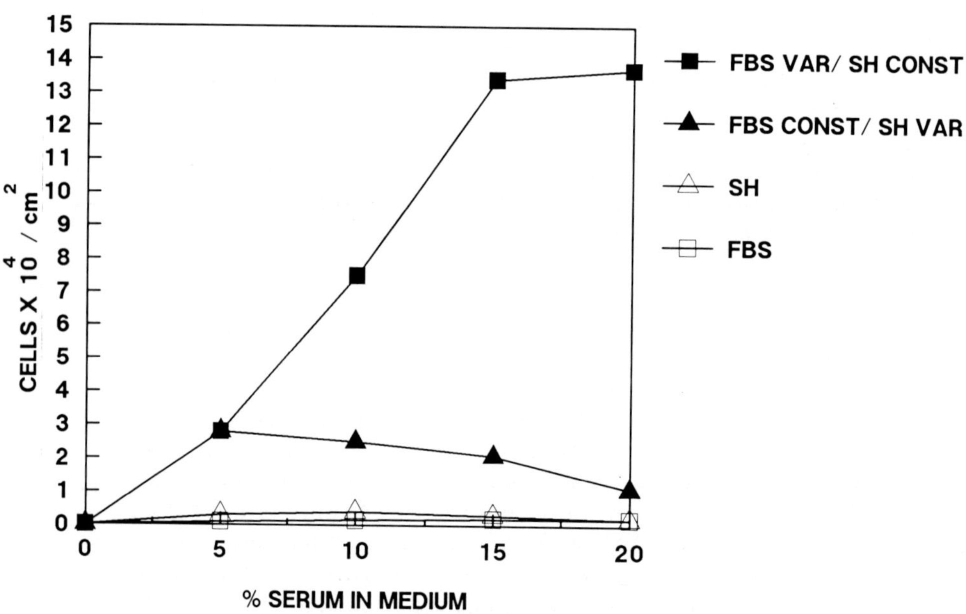

Fig. 2. Effects of fetal bovine serum (FBS) and shrimp hemolymph (SH) concentration (VAR = variable; CONST = constant) on the number of cells originating from plated hematopoietic/lymphoid tissue fragments.

Three distinct bacterial strains were isolated from the shrimp hemolymph and evaluated for sensitivity to various antibiotics (Table 2). The results indicated that both kanamycin and streptomycin were effective, but only the latter was subsequently used since both have very similar bactericidal spectra. Penicillin and amphotericin B were added to help eliminate any possible airborne bacterial and fungal contamination.

Table 2. Antibiotic susceptibility (+) of bacterial strains isolated from the hemolymph of *P. semisulcatus*.

Antibiotic	Bacterial strain		
	A	B	C
Tetracycline	+	+	–
Kanamycin	+	+	+
Streptomycin	+	+	+
Ampicillin	–	+	+
Penicillin	–	+	+

Cell monolayers generated by hematopoietic tissue/lymphoid organ fragments were maintained successfully for 3 weeks, at which time a density of 1.34 x 10^6 cells/cm was reached, equalling roughly 40% of a confluent monolayer. Attempts to sub-culture the cells following trypsinization failed.

Attachment of hepatopancreatic and ovary-derived fibroblast-like cells in primary culture in M-199 medium occurred within 24 hr following plating of explants of both tissue types. The cells in both cases displayed no changes for 6 days, after which they began to deteriorate, display cytoplasmic vacuolization, and disintegrate.

No mitotic figures could be discerned in any of the attached cells of either the hematopoietic/lymphoid organ or tissue of ovarian origin.

DISCUSSION

Of the two tissue sources used in this work, i.e., hematopoietic/lymphoid and ovarian tissue, our results suggest that the former is by far more suitable for *in vitro* culture. The results indicate that cell culture basal media used are variable in their suitability for *P. semisulcatus* cell culture. M-199 yields roughly 3-4 times better growth than Eagle's MEM, L-15, and Ham F-12. It is likely that the results obtained with M-199 were due to the fact that it is the richest and most complex medium of all those tested. The supplement trials show that a combination of shrimp hemolymph and FBS yield significantly better results than each of these separately, suggesting a form of synergistic effect. An additional trial that yielded similar results was carried out by supplementing with 5% heat inactivated hemolymph taken from spiny lobster,

Panulirus penicillatus (a local Red Sea species) (Diamant and Rosenthal, unpubl.). Shrimp and lobster hemolymph have been used in the past for *in vitro* maintenance of crustacean cells (Paterson and Stewart, 1974; Chen et al., 1986). In the present study, although hemolymph was demonstrated as essential, cell proliferation declined when a baseline concentration (5%) was exceeded. Crustacean hemolymph evidently contains essential trace elements; however, it also apparently contains heat-stable inhibitory factors that are not eliminated by the heat inactivation process employed here. Considering FBS, its rich nutritional content is clearly required. However, the results demonstrate that concentrations above 15% FBS are not necessary.

Bacteremias are common in crustaceans (Sizemore et al., 1975; Scott and Thune, 1986). Tareen (1982) found that *P. semisulcatus* hemolymph contains *Vibrio* spp. In the present study, no effort was made to identify the isolates. However, since they grew readily on TCBS agar, it seems likely that they also were *Vibrio* spp., which are very common aquatic bacteria in the Eilat culture facilities (A. Colorni, pers. comm.). There was little doubt, in any case, that the bacteria originated from the shrimp hemolymph and were not of external contamination.

Recent experience with *in vitro* invertebrate cell culture in the U.S. have shown that occasionally, proliferation of fungal cells that closely mimic hemocytes in the cultures may occur (L. Ellis, pers. comm.). Although such a possibility exists in the present case, we believe it is unlikely. First, many fungi are sensitive to amphotericin B, which was a standard medium component in all trials. Secondly, fungi characteristically possess extraordinary generative potential, a fact which is inconsistent with our unsuccessful attempts to subculture the cells of hematopoietic/lymphoid origin. Finally, cells were observed to develop from certain tissue fragments (presumed to contain functional hematopoietic or lymphoid tissue), and were never observed to develop freely in the nutrient-rich medium.

Further work is concentrating on (1) attempts to subculture cells of hematopoietic/lymphoid origin, and (2) *in vitro* MBV infection of primary cell cultures.

LITERATURE CITED

Chen, S.N. and G.H. Kou. 1989. Infection of cultured cells from lymphoid organ of *Penaeus monodon* by *monodon*-type baculovirus (MBV). J. Fish Dis., 12, 73-76.

Chen, S.N., S.C. Chi, G.H. Kou and I.C. Liao. 1986. Cell culture from tissues of grass prawn *Penaeus monodon*. Fish Pathol., 21, 161-166.

Colorni, A. 1985. A practical method for determining the susceptibility of aquatic *Vibrio* and *Aeromonas* spp. to antibacterial agents. Aquaculture, 46, 263-266.

Colorni, A., T. Samocha and B. Colorni. 1987. Pathogenic viruses introduced into Israeli mariculture systems by imported penaeid shrimp. Bamidgeh, 39, 21-28.

Diamant, A. and A. Colorni. 1987. Viral infections of penaeid shrimp in Israel, and use of acridine orange fluorescence as a rapid diagnostic method. Poster presented at the EAFP Third International Conference, 31 Aug.-3 Sept. 1987, Bergen, Norway.

Ellender, R.D., B.L. Middlebrooks, S.L. McGuire and R.M. Overstreet. 1988. Evaluation of various growth enhancement factors, media formulations, and support matrices for the development of primary and established cell lines from *Penaeus* hepatopancreas. World Aquaculture Society 19th Annual Conference & Exposition, Hawaii, Abstract No. 80.

Lightner, D.V. 1988. New host and geographic records for the penaeid shrimp viruses BP, MBV, IHHN and HPV. PAMAQ 3rd International Conference, 2-6 October 1988, Gloucester Point, Va., U.S.A., Abstract No. 5.

Nash, G., A. Poernomo and M.B. Nash. 1988. Baculovirus infection in brackishwater pond cultured *Penaeus monodon* Fabricius in Indonesia. Aquaculture, 73, 1-6.

Paterson, W.D. and J.E. Stewart. 1974. *In vitro* phagocytosis by hemocytes of American lobster (*Homarus americanus*). J. Fish Res. Bd. Can., 31, 1051-1056.

Scott, J.R. and R.L. Thune. 1986. Bacterial flora of hemolymph from red swamp crawfish, *Procambarus clarkii* (Girard), from commercial ponds. Aquaculture, 58, 161-165.

Sizemore, R.K., R.R. Colwell, H.S. Tubiash and T.E. Lovelace. 1975. Bacterial flora of the hemolymph of the blue crab, *Callinectes sapidus*: numerical taxonomy. Appl. Microbiol., 29, 393-399.

Tareen, I.U. 1982. Control of diseases in the cultured population of penaeid shrimp, *Penaeus semisulcatus* (de Haan). Proc. World Maricult. Soc., 13, 157-161.

ISOLATION AND CHARACTERIZATION OF A NEW HERPESVIRUS FROM EEL

M. Sano, H. Fukuda, and T. Sano

Department of Aquatic Biosciences, Tokyo University of Fisheries, 4-5-7 Konan, Minato-ku, Tokyo, 108, Japan

ABSTRACT: A new virus was isolated from both the Japanese eel (*Anguilla japonica*) and the European eel (*A. anguilla)* cultivated in an intensive culture system in Japan. The virus grew less well in BF-2, CEA, FHM, CHH-1, RTG-2 than in eel cell lines (EK-1, EO-2). It could not grow in amphibian and mammalian cell lines. The cytopathic effect observed in an eel kidney (EK-1) cell line consisted of typical syncytium formation and rounded cells in which Cowdry type A intranuclear inclusions were observed. The optimal growth temperature was 20-25°C. This virus differed serologically from Ictalurid herpesvirus 1 (Channel catfish virus; CCV), Salmonid herpesvirus 2 (Nerka virus of Towada lake, Akita, and Aomori; NeVTA), and Cyprinid herpesvirus 1 (*Herpesvirus cyprini*; CHV). There were 25 virion polypeptides found in this virus, ranging from 19.5K to 320K as determined by SDS-polyacrylamide gel electrophoresis. Viral DNA, whose molecular weight was calculated to be about 100 x 10^6, was infectious. We have designated this virus as Anguillid herpesvirus 1, with *Herpesvirus anguillae* as the latinized name and Eel herpesvirus as the vernacular name.

INTRODUCTION

Five herpesviruses, including Ictalurid herpesvirus 1 (Channel catfish virus; CCV) (Wolf and Darlington, 1971), Salmonid herpesvirus 1 (*Herpesvirus salmonis*) (Wolf et al., 1978), Salmonid herpesvirus 2 (Nerka virus of Towada lake, Akita, and Aomori; NeVTA [Sano, 1976], *Onchorynchus masou* virus; OMV [Kimura et al., 1981], Yamame tumor virus; YTV [Sano et al., 1983]), Percinid herpesvirus 1 (*Herpesvirus vitreum*) (Kelly et al., 1983), and Cyprinid herpesvirus 1 (*Herpesvirus cyprini*; CHV) (Sano et al., 1985), have been isolated from fish and others have been observed by electron microscopy (Wolf, 1984).

We recently isolated a herpesvirus from Japanese eel and European eel cultured in two different nurseries in Japan. This virus possesses the general properties of the herpesvirus group, which include a DNA genome and sensitivity to ether, acid, and heat. The virus grows well only in cells derived from eel.

We describe here a more detailed characterization of this herpesvirus; the first one isolated from eel.

MATERIALS AND METHODS

Histopathological examination. In August, 1985, undefined mortalities were observed in Japanese eels and European eels, at rates of 1% and 6.8%, respectively, cultured in circulating water systems at about 30°C in Aichi Prefecture, Japan. Paraffin-embedded sections (5-6 μm) of tissues from naturally infected eel were prepared and stained with hematoxylin-eosin for histopathological examination.

Virus isolations. A virus was isolated from moribund eels using the eel kidney cell line (EK-1) developed by Chen et al. (1982). This virus, designated as 85-8, was cloned three times in EK-1 by limiting dilution and then passed two additional times in EK-1. Independently, in May, 1987, we isolated a serologically similar virus, named 87-5, in EK-1 from Japanese eels cultured in another nursery in Kanagawa Prefecture, Japan.

CCV was kindly provided from Dr. J.A. Plumb (Auburn University, Alabama, USA).

Cell lines. The CHSE-214 (Lannan et al., 1984), CHH-1 (Lannan et al., 1984), RTG-2 (Wolf and Quimby, 1962), BB (Wolf and Quimby, 1969), BF-2 (Wolf et al., 1966), FHM (Gravell and Malsberger, 1965), and EPC (Fijan et al., 1983) were grown with Eagle's minimal essential medium containing 2 or 5% fetal bovine serum (FBS), (MEM-2 or MEM-5). The EK-1 and EO-2 lines (Chen and Kou, 1981) were grown with Leibovitz medium L-15 with 5% FBS (L-15-5). The amphibian cell line, A-6, was grown with 2/3 strength L-15 with 5% FBS. The mammalian cell lines, L-929, VERO, and FL, were grown with MEM-5.

Virus titers were determined by endpoint dilution using microtiter plates. The following cells were routinely used: EK-1 for 85-8 isolate, BB for CCV, CHSE-214 for NeVTA, and FHM for CHV.

Electron microscopy. Ultra-thin sections of virus-infected cells were fixed, embedded, and stained for electron microscopy as described by Sano et al. (1983). For negative staining, 2% uranyl acetate or 2% phosphotungstic acid (pH 7.0) was used. Specimens were examined with a Hitachi H-700 electron microscope.

Cross neutralization tests. Cross neutralization tests among the viruses, 85-8 isolate, CCV, CHV, and NeVTA, were carried out using specific rabbit antisera prepared against all viruses except CCV.

Virus purification. EK-1 cells were inoculated with the 85-8 isolate at a multiplicity of infection (MOI) of 0.01 $TCID_{50}$/cell and then incubated at 25°C for 4 days. The virus was concentrated by polyethylene glycol precipitation, followed by centrifugation at 18,500*g* for 30 min. The virus pellet was suspended in TN buffer

(10mM Tris-HCl, pH 7.2, 0.85% NaCl), and clarified by centrifugation at 1,000*g* for 10 min, and layered onto a 20-50% linear sucrose gradient made up in Tris saline buffer (0.2 M Tris-Hcl, pH 7.2, 0.85% NaCl). After centrifugation at 77,000*g* for 30 min with a MSE 6 x 14 ml rotor, virions were found in a diffuse light scattering, thick band in the middle of the tube. This band, containing partially purified, mostly enveloped virus, was removed and dialyzed against TN buffer.

For viral structural proteins analysis, two more purification steps were employed. Briefly, the partially purified virus was banded in isopycnic CsCl (175,000*g* for 20 hr with a MSE 3 x 5 ml rotor). The virus band was removed, and after dialysis against TN buffer, the virus was rebanded in sucrose as described above. The collected band was diluted with TN buffer and pelleted by centrifugation at 100,000*g* for 30 min. The pelleted virus was resuspended in electrophoretic buffer (Laemmli, 1970), boiled for 2 min, and stored at -80°C prior to use.

SDS-polyacrylamide gel electrophoresis (SDS-PAGE). SDS-PAGE was carried out as described by Laemmli (1970) on three concentrations of acrylamide: 7.5, 10, and 12.5%. After 20 hr of electrophoresis at 4 mA, proteins in the gel were stained with silver following the procedure of Wray et al. (1981).

Viral DNA isolation and transfection. The partially purified virus solution was mixed with an equal volume of step buffer (0.1 M Tris-HCl, pH 7.5, at 37°C, 2% SDS, 10mM EDTA) and proteinase K (Sigma) was added to a final concentration of 1 mg/ml, and then incubated at 37°C for 1 hr. The digested solution was extracted three times with saturated phenol and two times with chloroform-isoamyl alcohol. After ethanol precipitation, nucleic acid was suspended in TE buffer (10mM Tris-HCl, 1mM EDTA, pH 7.8) and stored at 4°C prior to use. This nucleic acid was DNA because it was sensitive to DNase I (Sigma, type II) but not to RNase A (Sigma, type I-AS).

DNA transfection to EK-1 was carried out using DEAE-dextran (200 μg/ml) as described by Sheldrick et al. (1973).

DNA restriction endonuclease digestion and fragments electrophoresis. The isolated viral DNA was digested with Eco RI, Bam HI, Hind III, Bgl II, and Pst I with buffers described by the supplier (Boehringer Mannheim). After digestion, electrophoresis of the fragments of the DNA in horizontal agarose gel containing 1 μg/ml of ethidium bromide was carried out at 5 mA for 18 hr with Tris-acetate buffer (40mM Tris, 20mM sodium acetate, and 2mM EDTA, pH 7.8). As molecular weight markers, digested λ phage DNA (Boehringer Mannheim, DNA molecular weight maker III) and Hind III digested CCV DNA were used. CCV DNA was prepared by the same procedures used for the 85-8 isolate DNA, and the molecular weights of the fragments of CCV DNA compared to those reported by Chousterman et al. (1979).

RESULTS

Histopathological examination. Affected eels showed varying degrees of erythema on the skin and gills. In the epidermal tissue of the lesion, club cells were affected, indicating eosinophilic granulations that were irregular in size (Fig. 1). This followed the necrosis and partial desquamation of the epidermis. In the dermal tissue of the

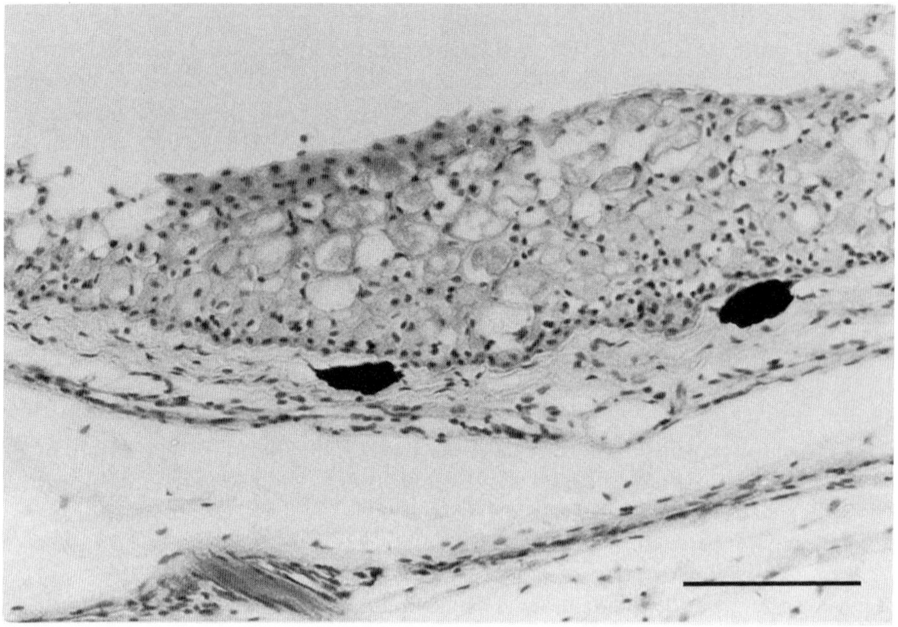

Fig. 1. Histological section of epidermal tissue showing irregular eosinophilic granulation and necrosis of club cells. H&E stain. Bar = 100 μm.

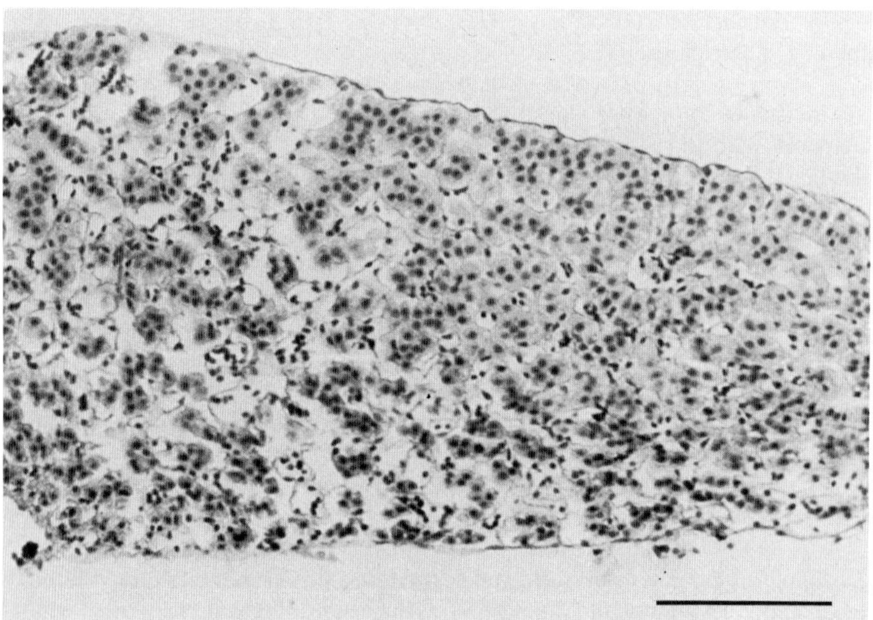

Fig. 2. Histological section of liver showing mild necrosis of hepatic cells. H&E stain. Bar = 100 μm.

lesion, blood capillaries increased and melanophores contracted. The gills showed partial fusion of the branchial lamellae, resulting in mild necrosis. The liver was dotted with necrotic hepatic cells (Fig. 2).

Cell susceptibility and virus growth. The BF-2, CHH-1, RTG-2, FHM, EK-1, and EO-2 cell lines were susceptible to 85-8 isolate (Table 1). The highest virus titers were obtained with the eel lines EK-1 and EO-2. Syncytium formation was most evident in EK-1, CHH-1, and RTG-2 cells and intranuclear acidophilic inclusions, Cowdry type A, were observed in EK-1, EO-2, and CHH-1 stained by the May-Grünwald-Giemsa method (Fig. 3). Cyncytium formation was enhanced at MOI less than 1 $TCID_{50}$/cell.

High virus yields occurred only in EK-1 and EO-2 among the susceptible cell lines (Table 1). Virus in EK-1 or EO-2 occurred over the 15-37°C range. Virus titers reached approximately 10^8-10^9 $TCID_{50}$/ml at 15-30°C. The maximum titer was reached in less time at the higher temperatures (Fig. 4), but virus titer rapidly decreased at 30°C because of virus inactivation (Fig. 5). We concluded that the optimal growth temperature *in vitro* was 20-25°C.

The one-step growth curve at 25°C on EK-1 is shown in Figure 6. The logarithmic growth phase began at 9 hr after infection, and the plateau was reached by 21 hr. There was always about 10-fold more released virus than cell-associated virus.

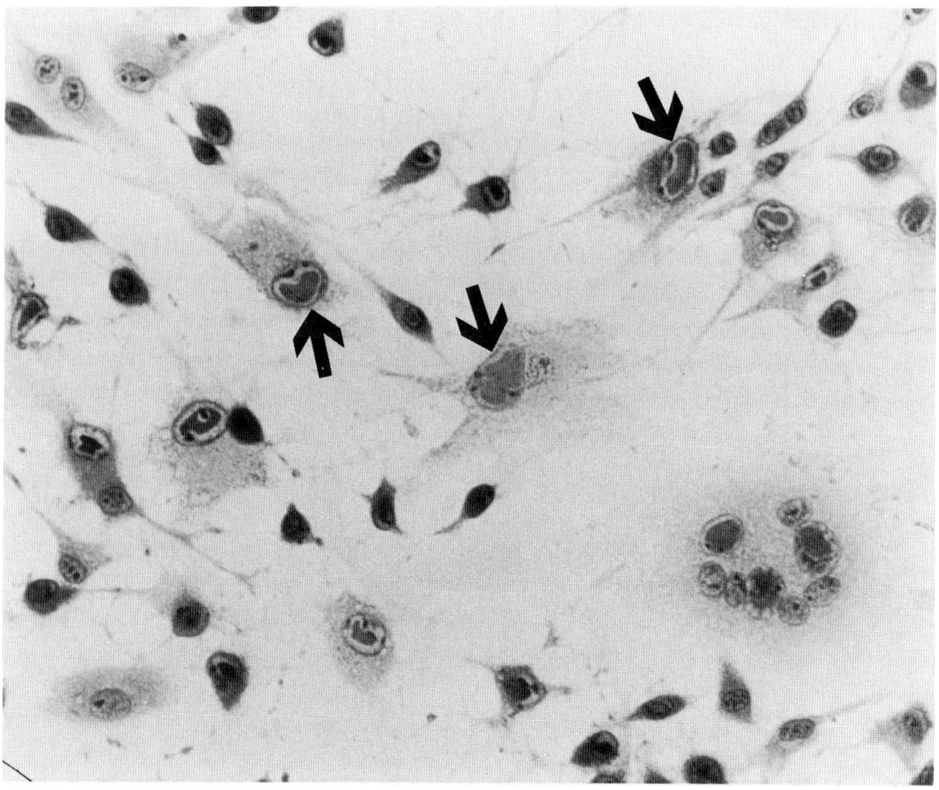

Fig. 3. Syncytia formed and Cowdry type A intranuclear inclusions (arrows) in EK-1 cells infected with 85-8 isolate. May-Grünwald-Giemsa stain.

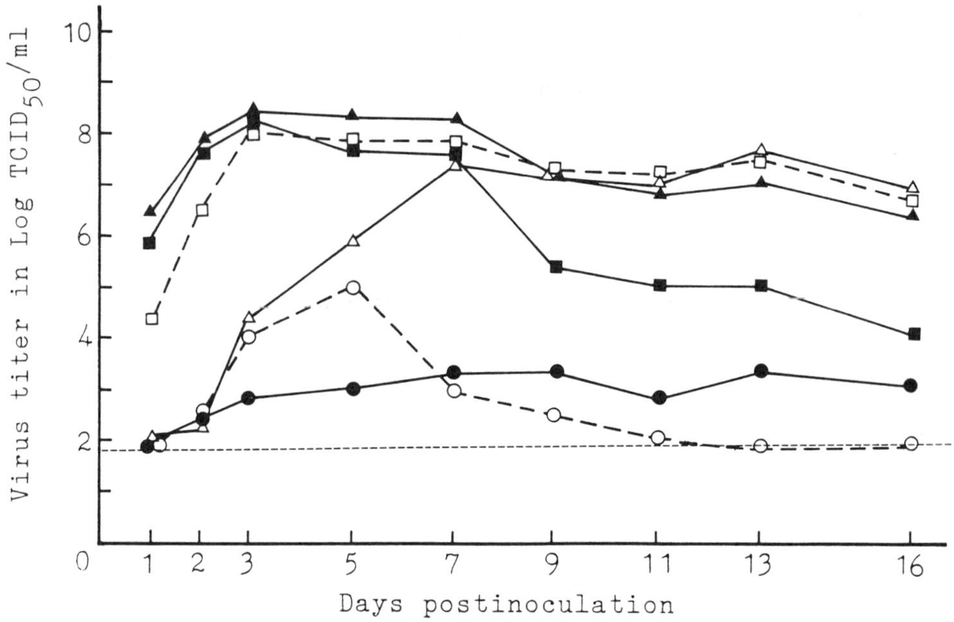

Fig. 4. Effect of incubation temperature on replication of 85-8 isolate in EK-1 cells. The virus was inoculated at MOI of 0.01 $TCID_{50}$/cell. (●) 10°C, (△) 15°C, (□) 20°C, (▲) 25°C, (■) 30°C, (○) 37°C, (----) detection limit.

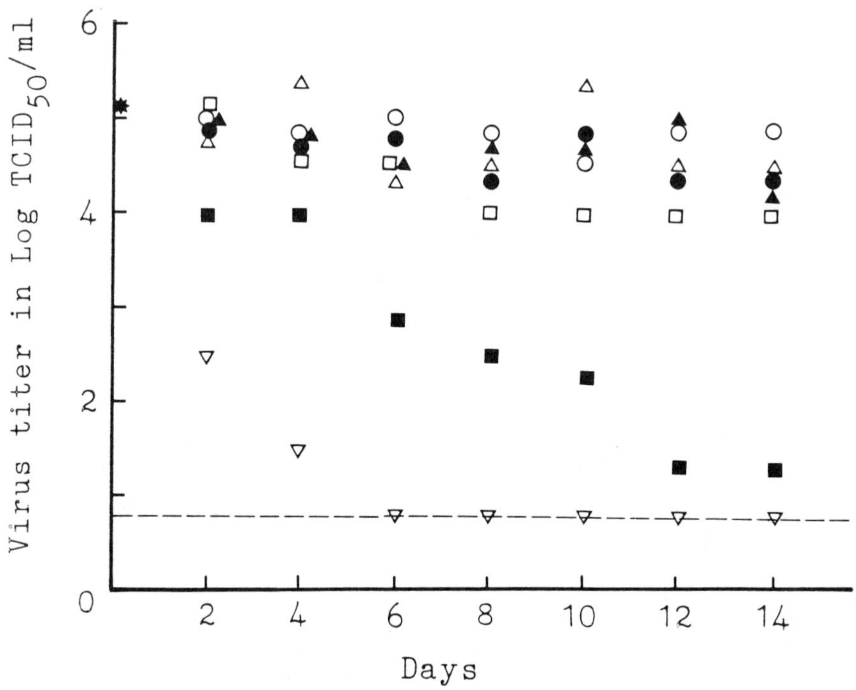

Fig. 5. Stability of 85-8 isolate at several temperatures in L-15-5. (○) 4°C, (●) 10°C, (△) 15°C, (▲) 20°C, (□) 25°C, (■) 30°C, (▽) 37°C, (----) detection limit.

Table 1. Cell culture spectrum of 85-8 isolate.

Cell lines	CPE	Virus yield[a] Log $TCID_{50}$/ml)
Fish		
BB	-	
BF-2	+	ND
EPC	-	
FHM	+	<2.3
CHH-1	+	<2.3
CHSE-214	-	
RTG-2	+	3.0
EK-1	+	8.0
EO-2	+	7.5
Amphibia		
A-6	-	
Mammal		
L-929	-	
VERO	-	
FL	-	

[a] Cells were inoculated with 85-8 isolate of $10^4$$TCID_{50}$/ml, and incubated at 20°C for 14 days. Then virus titer was determined using EK-1 cell line. ND = not done.

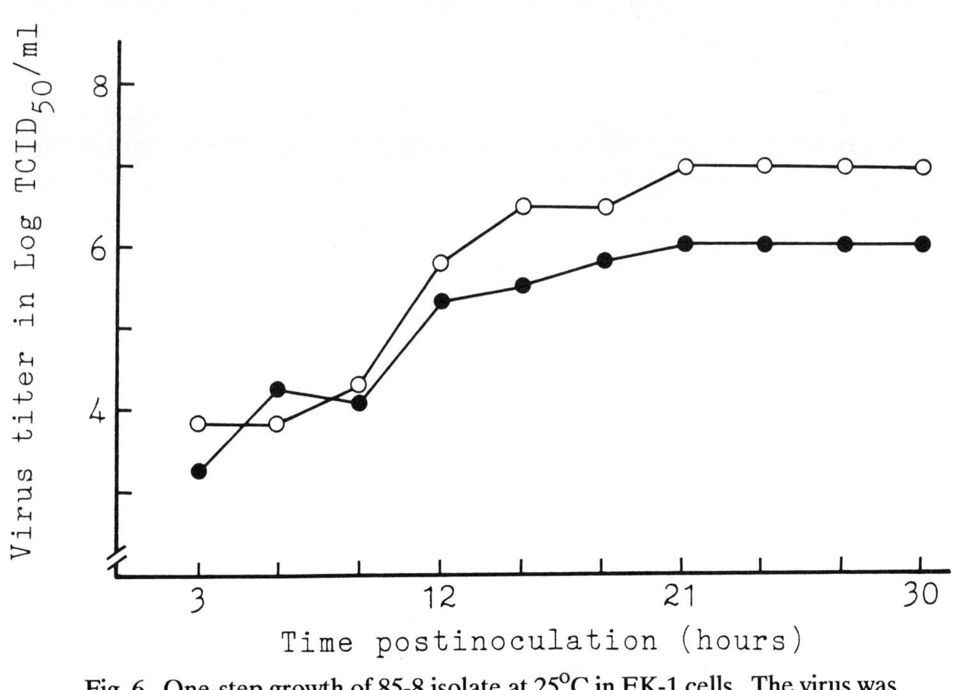

Fig. 6. One-step growth of 85-8 isolate at 25°C in EK-1 cells. The virus was inoculated at MOI of 10 $TCID_{50}$/cell. (○) released virus, (●) cell associated virus.

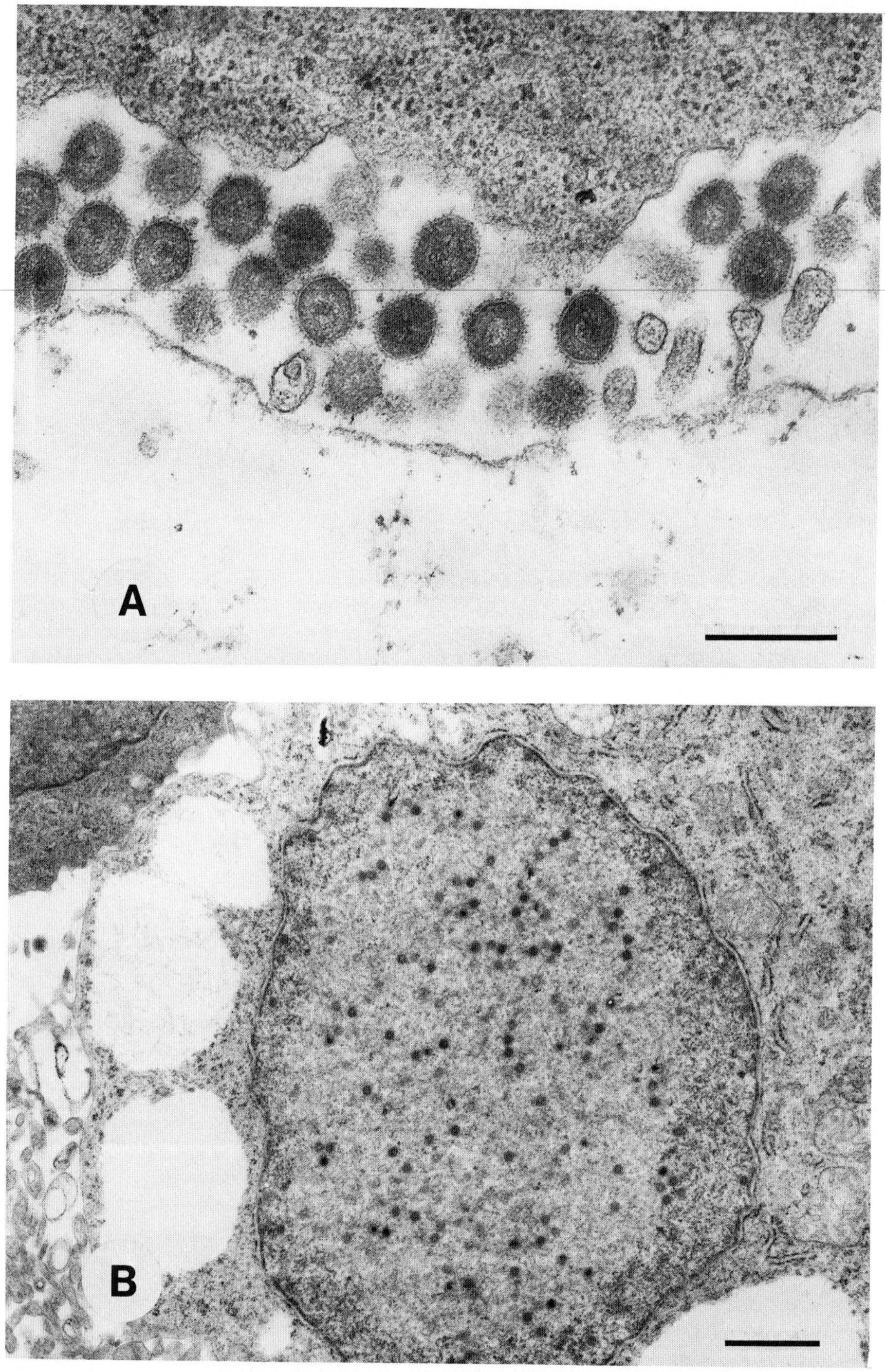

Fig. 7. (A) Electron micrograph of a thin sectioned virion of 85-8 isolate consisting of an envelope and a capsid surrounded by a tegmentum. Bar = 300nm. (B) Electron micrograph of a thin sectioned EK-1 cell infected with 85-8 isolate showing a maturation of viral capsid occur in the nucleus. Bar = 1 μm.

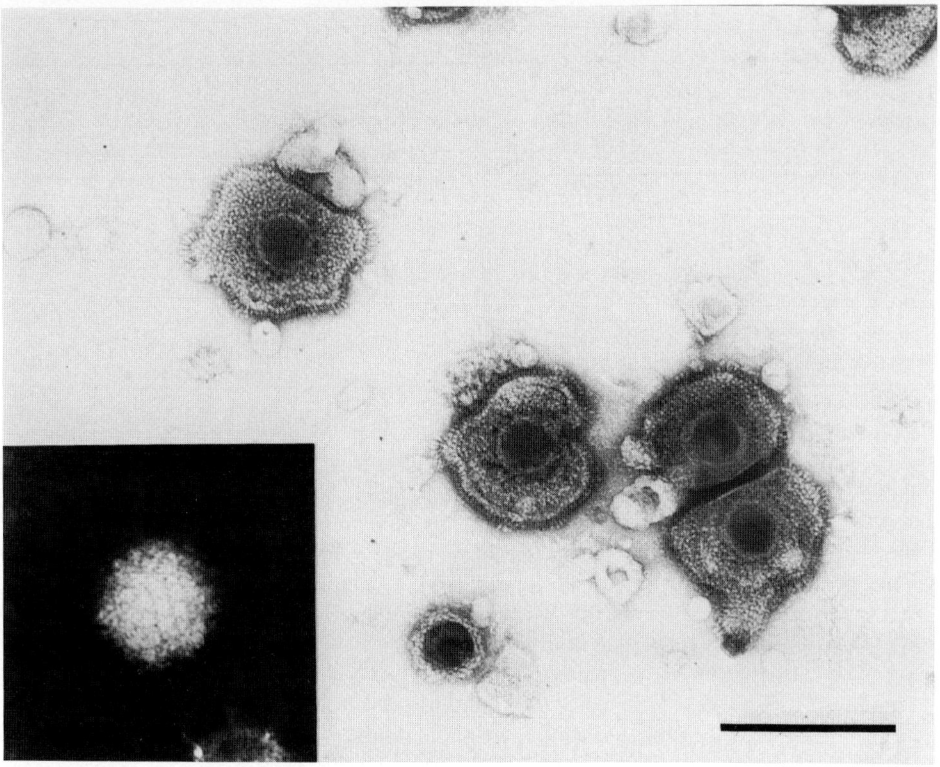

Fig. 8. Electron micrograph of a negatively stained 85-8 isolate showing the profile of a herpesvirus. Insert indicates the nucleocapsid. Bar = 300nm.

Virus morphology. Thin sections of infected EK-1 cells and the negatively stained virion showed that the virion consisted of an envelope and capsid surrounded by a tegumentum, and the nucleocapsid, with a mean diameter of 110 nm, was made up of hollow capsomeres and was icosahedral (Figs. 7, 8). The virion had a mean diameter of 200 nm, ranging from 185 to 210 nm including spikes. Because five capsomeres could be seen along the edge, the nucleocapsid included 162 capsomeres. These morphological characteristics are similar to that of the herpesvirus group. Although many stages of capsid development were observed in the nucleus, there were few enveloped particles in the cytoplasm, and many enveloped particles outside of the cell (Fig. 7).

Cross neutralization tests. The 85-8 isolate was not neutralized by anti-CHV and anti-NeVTA serum. Conversely, anti-85-8 serum did not demonstrate neutralization activities against CHV, CCV, and NeVTA (Table 2). These results indicate that 85-8 isolate is serologically different from CCV, CHV, and NeVTA.

Virus purification. The first sucrose gradient centrifugation gave partially purified virus that contained few contaminants when observed by electron microscopy. When cycled through an isopycnic CsCl gradient, two bands were observed. The upper band, at a buoyant density of 1.24g/cm^3, contained primarily enveloped particles, and a lower band consisted mainly of nucleocapsids. When the complete virions were again run in a sucrose gradient, only one band was evident. Virus purified by this three step process was used for analysis of the viral structural proteins.

Table 2. Cross neutralization test results using four fish herpesviruses and specific antisera.

Virus	Antisera (ND_{50})		
	85-8	CHV	NeVTA
85-8	340	-	-
CHV	-	40	ND
NeVTA	-	ND	360
CCV	-	ND	ND

- = not neutralized ND = not determined

Table 3. Estimation of molecular weights of the viral structural proteins from SDS-PAGE.

Viral protein	molecular weight ($x 10^{-3}$)[a]
vp 1	320
2	242
3	234
4	202
5	189
6	180
7	162
8	148
9	123
10	106
11	97
12	90.5
13	83.5
14	67
15	51
16	49
17	42
18	39.5
19	32.5
20	30.5
21	27
22	26
23	24.5
24	22
25	19.5
Total	2607.5

[a] Average molecular weight calculated from three concentrations of polyacrylamide gels.

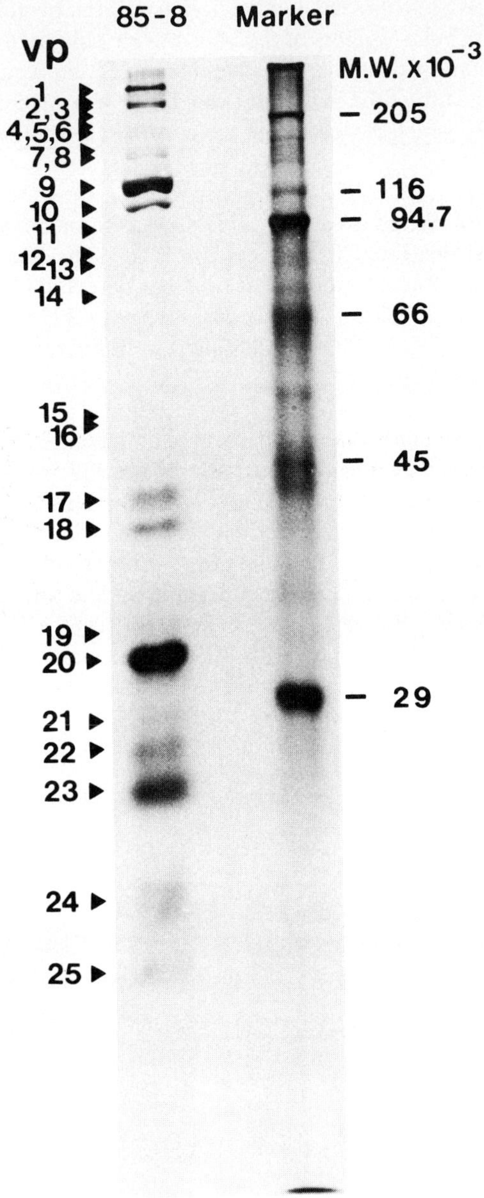

Fig. 9. SDS-PAGE analysis of total virion polypeptides of 85-8 isolate. Photograph of a 12.5% acryamide gel stained with silver after electrophoresis of the purified viral polypeptides.

Viral proteins. Analysis of purified viral structural proteins revealed the presence of 25 discernible bands on SDS-PAGE, and we designated the proteins as vp1 through vp25 (Fig. 9). The molecular weights of the bands were estimated by relative migration rates with marker proteins. The average molecular weights ranged from 19.5K to 320K, and the sum of the molecular weights of viral structural proteins was approximately 2607.5K, which is well within the cording capacity of herpesvirus group DNA (Table 3).

Viral DNA transfection. When isolated viral DNA was transfected to EK-1 cells in the presence of DEAE-dextran, CPE was evident in the cells 4 or 5 days postinoculation at 25°C. Furthermore, virus titer was detected in the culture medium. Viral DNA pretreated with DNase I did not produce any CPE on the cells. These results indicated that the DNA of 85-8 isolate possessed infectivity as has been reported for other herpesviruses.

Restriction endonuclease digestion. Restriction endonuclease cleavage patterns in agarose gels are shown in Figure 10. These patterns were characteristic (compared with that of CCV DNA), and were not changed when the DNA further purified by centrifugation in 5-20% linear sucrose gradient was used (data not shown).

The molecular weights of the fragments were estimated by relative migration rate with marker DNA and are summarized in Table 4. The total molecular weights of 115.81 x 10^6 with Eco RI, 104.31 x 10^6 with Hind III, 108.69 x 10^6 with Bam HI, 90.25 x 10^6 with Bgl II, and 103.31 x 10^6 with Pst I, giving a mean of 104.47 x 10^6. This value is consistent with the molecular weights reported for the DNA of members of the herpesvirus group, which range from 80-150 x 10^6.

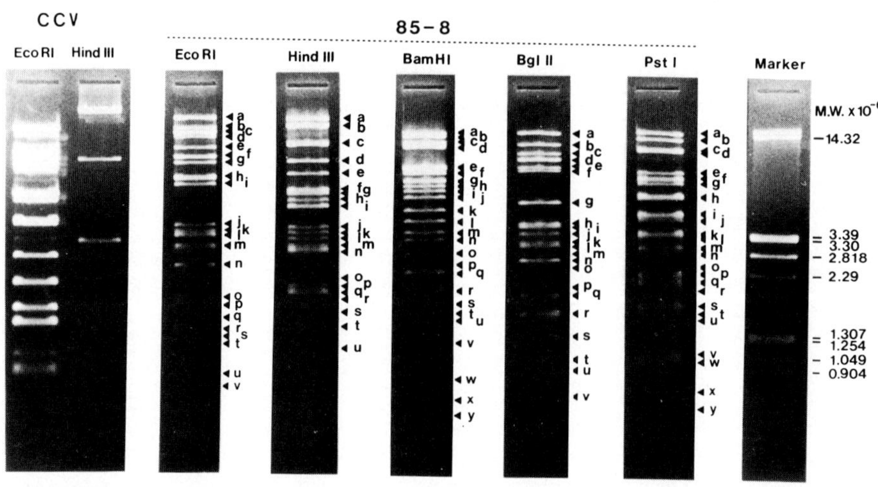

Fig. 10. Restriction endonuclease profiles of 85-8 isolate DNA. Viral DNA was cleaved with the restriction endonucleases and separated by electrophoresis on 0.5% agarose gel. Gel was stained with ethidium bromide and photographed under UV illumination.

Table 4. Molecular weights of the fragments produced by digestion of the viral DNA with restriction endonucleases.

Fragment	Molecular weight ($x 10^{-6}$) of the fragment produced by				
	Eco RI	Hind III	Bam HI	Bgl II	Pst I
a	20.2	20.7	11.9	12.0	12.6
b	13.1	14.1	10.8	9.7	11.1
c	11.6	9.7	10.4	8.8	10.0
d	10.5	7.6	9.8	7.9	9.4
e	9.2	6.1	7.0	6.9	6.5
f	7.8	5.2	6.3	6.6	6.0
g	7.1	5.0	5.8	4.7	5.6
h	5.9	4.6	5.4	3.76	5.0
i	5.4	4.4	5.1	3.59	4.1
j	3.51	3.51	4.9	3.39	3.89
k	3.35	3.35	4.2	3.16	3.43
l	3.20	3.13	3.85	3.02	3.39
m	2.82	2.88	3.39	2.75	2.99
n	2.45	2.73	3.13	2.60	2.82
o	1.68	2.11	2.79	2.37	2.45
p	1.62	1.95	2.45	1.95	2.29
q	1.41	1.84	2.20	1.82	2.15
r	1.27	1.57	1.91	1.53	1.86
s	1.19	1.48	1.68	1.22	1.56
t	1.10	1.31	1.41	0.96	1.51
u	0.73	1.05	1.34	0.88	1.38
v	0.68	-	1.09	0.65	1.05
w	-	-	0.71	-	0.95
x	-	-	0.61	-	0.71
y	-	-	0.53	-	0.58
Total	115.81	104.31	108.69	90.25	103.31

DISCUSSION

The virus described here belongs to the family Herpesviridae as defined by the International Committee for the Taxonomy of Viruses (Roizman et al., 1981) because of the following: viral infection of cell culture resulted in the formation of syncytia and Cowdry type A intranuclear inclusions; maturation of the capsid occurred in the nucleus and the morphological features of the virus were characteristic of herpesvirus, including size (200 nm of enveloped particle, 110 nm of capsid) and number of capsomeres (162); the virus was sensitive to acid, ether, and heat treatment (56°C for 30 min); buoyant density of the virion in CsCl was 1.24 g/cm^3; nucleic acid of the virus was DNA with a molecular weight of approximately 100×10^6; one-step growth curves showed the logarithmic growth phase began at 9 hr postinoculation, the plateau reached by 21 hr, and there was always about 10-fold more released virus than cell-associated virus. This replication pattern is characteristic of a group α herpesvirus. Although other viruses including EV-1, EV-2, EVE, EVA, and EVEX have been isolated from eel and, moreover, recently Békési et al. (1986) reported herpesvirus-like particles in skin lesions of European eel, this is the first isolation of herpesvirus from both the Japanese eel and the European eel. We designated our virus Anguillid herpesvirus 1, with *Herpesvirus anguillae* (HVA) as the latinized name and Eel herpesvirus as the vernacular name in accordance with a revision of the nomenclature of the Herpesviridae proposed by Herpesvirus Study Group appointed by the International Committee on Taxonomy of Viruses (Roizman et al., 1981).

In determining the type of nucleic acid in a virus, we routinely titrate by employing the microtiter system with culture medium containing 10^{-4}M of IUdR (5-iodo-2'-deoxyuridine). In the case of Nerka virus, with NeVTA as a positive control, the titer was greatly decreased. Although HVA is a DNA virus, the titer was not considerably decreased. The reason is not clear. However, when sensitivity of HVA to IUdR was measured by plaque reduction using an overlay containing 7% methylcellulose, the IC_{50} (50% inhibitory concentration) was estimated at 8 μg/ml ($10^{-4.6}$M). This value resembles those of the Herpes simplex virus type-1 and OMV (Coen et al., 1985, Suzuki et al., 1987).

HVA showed CPE in several fish cell lines, including BF-2, CHH-1, RTG-2, FHM, EK-1, and EO-2, but viral replication occurred only in EK-1 and EO-2 derived from eel. This parallels the situation with herpesviruses from fish in that there is the tendency to grow only in cell lines derived from the same family of affected fish. The results of cell susceptibility and cross neutralization studies indicate that HVA is clearly different from CCV, CHV, *H. vitreum,* and salmonid herpesviruses.

Although viral replication occurred over the temperature range of 10-37°C, the optimal growth temperature was 20-25°C *in vitro*, which reflects the temperature of eel-rearing water. Herpesviruses isolated from fish do not replicate at 37°C; however, HVA does and should therefore be handled carefully. We are planning to investigate the relationships among HVA and mammalian herpesviruses.

As reported herein, sucrose and CsCl equilibrium density gradient centrifugation was used to purify HVA preparations. Herpesvirus purification is very difficult with the main contaminants being cellular membranes (Spear and Roizman, 1972). In the case of our preparations, we cannot at present exclude this possibility, but contaminants were not seen by electron microscopy. We also cannot exclude the possibility that minor bands may be masked by the major bands.

In the virion, we have detected 25 polypeptides that are presumably virus-specific structural proteins in spite of above mentioned possibilities. The molecular weights ranged from 19.5K to 320K. The sum of the molecular weights of the 25 virus-specific proteins is approximately 2600K. This value is similar to that reported for HSV-1 (Spear and Roizman, 1972), Bovine herpesvirus type 1 (Misra et al., 1981), and Varicella-zoster virus (Shemer et al., 1980).

It is thought that the vp-9 (123K) polypeptide may be the major component of the nucleocapsid. In other herpesviruses, molecular weight of the major capsid protein was determined to be: 155K for HSV-1 (Heine et al., 1974), 150K for Human cytomegalovirus, 160-165K for Epstein-Barr virus and Herpes simplex virus type 2 (Killington et al., 1977), 180K for VZV (Shemer et al., 1980), 146K for Equine herpesvirus type 1 (Kemp et al., 1974), and 155K and 122K for CCV (Dixon and Farber, 1980). It is interesting that the molecular weight of the major capsid protein of HVA is about 30K-40K less than those reported for other herpesviruses, except the 122K polypeptide of CCV.

Viral DNA isolated from the virion possessed infectivity against EK-1. As far as we are aware, this is the first report of viral nucleic acid from a fish virus that has been shown to be infectious.

HVA DNA cleaved by several restriction endonucleases revealed characteristic electrophoresis patterns on agarose gel. These cleavage patterns were distinguishable from those of CCV DNA. The cleavage patterns of the 87-5 isolate were homologous to those of the 85-8 isolate, which was mainly used in this study. Therefore, if additional new herpesviruses are isolated from eel, these patterns will be useful for comparing similarities between HVA and the new isolates.

In the pursuing examinations, the virus could be isolated 2 weeks after initial isolation, but not 1 month after, when mortality had ceased.

In Japan, eels are almost always cultured under conditions of high density and high water temperature. It is quite possible that these conditions could cause stress in eels accruing in immunodepression. So, we need to conduct HVA infectivity trials against eels under unusual physical conditions, and examine its pathogenicity.

LITERATURE CITED

Békési, L., I. Horváth, E. Kovács-Gayer and G. Csaba. 1986. Demonstration of herpesvirus like particles in skin lesions of European eel (*Anguilla anguilla*). J. Appl. Ichthyol., 4, 190-192.

Chen, S.N. and G.H. Kou. 1981. A cell line derived from Japanese eel (*Anguilla japonica*) ovary. Fish Pathol., 16, 129-137.

Chen, S.N., Y. Ueno and G.H. Kou. 1982. A cell line derived from Japanese eel (*Anguilla japonica*) kidney. Proc. Natl. Sci. Council, B, ROC, 6, 93-100.

Chousterman, S., M. Lasaca and P. Sheldrick. 1979. Physical map of the channel catfish virus genome: Location of sites for restriction endonucleases Eco RI, Hind III, Hpa I, and Xba I. J. Virol., 31, 73-85.

Coen, D., H. Edward Fleming, Jr., L. Leslie and M. Retondo. 1985. Sensitivity of arabinosyladenine-resistant mutants of herpes simplex virus to other antiviral drugs and mapping of drug hypersensitivity mutants to the DNA polymerase locus. J. Virol., 53, 477-488.

Dixon, R.A.F. and F.E. Farber. 1980. Channel catfish virus: Physicochemical properties of the viral genome and identification of viral polypeptides. Virology, 103, 267-278.

Fijan, N., D. Sulimanovic, M. Bearzotti, D. Muzinic, L.O. Zwillenberg, S. Chilmonczyk, J.F. Vautherot and P. de Kinkelin. 1983. Some properties of the epitheliosum cyprini (EPC) cell line from carp, *Cyprinus carpio*. Ann. Virol., 1134E, 202-207.

Gravell, M. and R.G. Malsberger. 1965. A permanent cell line from the fathead minnow (*Pimephales promelas*). Ann. N.Y. Acad. Sci., 126, 555-565.

Heine, J.W., R.W. Honess, E. Cassal and B. Roizman. 1974. Proteins specified by herpes simplex virus. XII. The virion polypeptides of type 1 strains. J. Virol., 14, 640-651.

Kelly, R.K., O. Nielsen, S.C. Mitchell and T. Yamamoto. 1983. Characterization of Herpesvirus vitreum isolated from hyperplastic epidermal tissue of walleye, *Stizostedion vitreum vitreum* (Mitchell). J. Fish Dis., 6, 249-260.

Kemp, M., M. Perdue, H. Rogers, D. O'Callaghan and C. Randall. 1974. Structural polypeptides of the hamster stain of Equine herpes virus type 1: Products associated with purification. Virology, 61, 361-375.

Killington, R.A., J. Yeo, R.W. Honess, D.H. Watson, B.E. Duncan, I.W. Halliburton and J. Muford. 1977. Comparative analyses of the proteins and antigens of five herpes viruses. J. Gen. Virol., 37, 297-310.

Kimura, T., M. Yoshimizu, M. Tanaka and H. Sannohe. 1981. Studies on a new virus (OMV) from *Oncorhynchus masou*. I. Characterization and pathogenicity. Fish Pathol., 15, 143-147.

Laemmli, U.K. 1970. Cleavage of the structural proteins during the assembly of the head of bacteriophage T4. Nature, 227, 680-685.

Lannan, C.N., J.R. Winton and J.L. Fryer. 1984. Fish cell lines: Establishments and characterization of nine cell lines from salmonids. In Vitro, 20, 671-676.

Misra, V., R. Blumenthal and L. Babiuk. 1981. Proteins specified by bovine herpes virus 1 (infectious bovine rhinotracheitis virus). J. Viorol., 40, 367-378.

Roizman, B., L. Carmichael, F. Deinhardt, G. de-The, A. Nahmias, W. Plowright, F. Rapp, P. Sheldrick, M. Takahashi and K. Wolf. 1981. Herpesviridae: Definition, provisional nomenclature, and taxonomy. Intervirology, 16, 201- 217.

Sano, T. 1976. Viral diseases of cultured fish in Japan. Fish Pathol., 10, 221-226.

Sano, T., H. Fukuda, N. Okamoto and F. Kaneko. 1983. Yamame tumor virus: Lethality and oncogenicity. Bull. Japan. Soc. Sci. Fish., 49, 1159-1163.

Sano, T., H. Fukuda, M. Furukawa, H. Hosoya and Y. Moriya. 1985. A herpesvirus isolated from carp papilloma in Japan. In: "Fish and Shellfish Pathology" (A.E. Ellis, ed.), pp. 307-311. Academic Press, New York and London.

Sheldrick, P., M. Liaither, D. Lando and M. Ryhiner. 1973. Infectious DNA from herpes simplex virus: Infectivity of double-stranded and single-stranded molecules. Proc. Natl. Acad. Sci. U.S.A., 70, 3621-3625.

Shemer, Y., S. Leventon-Kriss and I. Sarov. 1980. Isolation and polypeptide characterization of varicella-zoster virus. Virology, 106, 133-140.

Spear, G. and B. Roizman. 1972. Proteins specified by herpes simplexvirus V. Purification and structural proteins of the herpesvirion. J. Virol., 9, 143- 159.

Suzuki, S., H. Machida and M. Saneyoshi. 1987. Antiviral activity of various 1-β-D-arabinofuranosyl-E-5-halogenovinyluracils and E-5-bromovinyl-2'-deoxyuridine against salmon herpes virus, Oncorhyncus masou virus (OMV). Antiviral Res., 7, 79-86.

Wolf, K. 1984. Fish viruses: Their biology, classification, hosts, pathology, and control. In: "Control of Virus Diseases" (E. Kirstak, ed.), pp. 197-215. Marcel Dekker, New York and Basel.

Wolf, K. and R.W. Darlington. 1971. Channel catfish virus: a new herpesvirus of Ictalurid fish. J. Virol., 8, 525-533.

Wolf, K. and M.C. Quimby. 1962. Established eurythermic line of fish cell in vitro. Science, 135, 1065-1066.

Wolf, K. and M.C. Quimby. 1969. Fish cell and tissue culture. In: "Fish Physiology" (W.S. Hoar and D.J. Randall, eds.), pp. 253-305. Academic Press, New York and London.

Wolf, K., M. Gravell and R.G. Malsberger. 1966. Lymphocystis virus: isolation and propagation in centrachid fish cell lines. Science, 151, 1004- 1005.

Wolf, K., R.W. Darlington, W. Taylor, M. Quimby and T. Nagabayashi. 1978. Herpesvirus salmonis: Characterization of a new pathogen of rainbow trout. J. Virol., 27, 659-666.

Wray, W., T. Boulikas, V.P. Wray and R. Hancock. 1981. Silver staining of proteins in polyacrylamide gels. Anal. Biochem., 118, 197-203.

ESTABLISHMENT AND CHARACTERIZATION OF A RAINBOW TROUT KIDNEY CELL-LINE, RTK MONTPELLIER

R. Andral[1], C. Hurard[1], P. Elziere-Papayanni[1], and C.P. Vivarès[2]

[1]Laboratoire Départemental des Services Vétérinaires de l'Hérault, 306 rue de la Croix de las Cazes, 34 000 Montpellier, France
[2]Laboratoire de Pathologie Comparée (URA INRA-CNRS no. 11840), Université des Sciences et Techniques du Languedoc, 34 060 Montpellier Cédex, France

ABSTRACT: Trout (*Salmo gairdneri*) primary kidney cells were cultivated and serial subcultures were realized. Thus, a fibroblastic cell-line was obtained (more than 80 subcultures). The main characteristics of this cell-line are described.

Cell growth occurred at 20°C. We observed a fast decrease of the culture density at 29°C and the cells could be kept at 4°C. At -80°C, as already mentioned with other fish cell-lines, we found it difficult to cultivate the cells after thawing. After freezing the cells at -180°C under liquid nitrogen, we obtained a monolayer cell culture after one week.

Of the foetal calf serum provided, 2% was used for maintenance and 10-15% for growth. First investigations on karyotype showed 2n = 64 chromosomes, with several cells having around 90 chromosomes.

We have tried to use this cell-line for cultivation of fish viruses: VHSV at 14°C, IPNV at 14°C, and IPNV-like viruses from marine origin (LNPAA Brest) all showed cytopathic effect, while SVCV at 20°C was unable to provoke any cytopathic effect.

Characterization of this cell-line (RTK Montpellier) is in progress, especially its use in viral diagnosis and for the study of biological cycles of kidney parasites as the PKD agent.

INTRODUCTION

Recent progress in diagnostic and screening of viral ichthyopathological diseases were related to the use of cell cultures (Wolf and Mann, 1980; Lannan et al., 1984). Thus, cell-lines of salmonids are available. However, the only one available to study the virology of fresh water and marine salmonids and also *in vitro* intracellular cycle of the PKD agent was obtained from an organ mixture (Nicholson and Byrne, 1973). We decided to obtain a new line from kidney cells. The aim of this article is to characterize such a line from *Salmo gairdneri* kidney and to study it when infected by fresh or marine waters viruses.

MATERIALS

Rainbow trout (*Salmo gairdneri*) were a generous gift of the Fédération des Associations Agréées de Pêche de l'Hérault. The young 5-cm long trout were kept in pisciculture in medium mountains during January to May 1987.

Culture medium was Eagle's minimal essential medium (Glasgow modification) supplemented with the following: foetal calf serum (FCS) from Boehringer (different concentrations from 2%-15% were used in order to check the tolerance of the cell-line; 15% was finally chosen); trypto-casein soja from Pasteur IPP (10%); L-glutamine from Boehringer (200mM, 50 times diluted); and penicillin (300 Ul/ml), streptomycine (300 μg/ml), and amphotericine B (1 μg/ml) from Boehringer.

The medium was adjusted to pH 7.3-7.4 with 0.16 M Tris-HCl. This medium was used for the primary culture and also the subcultures. Penicillin (100 Ul/ml) and streptomycine (100 ug/ml) were used for the subcultures only.

Establishment of primary culture and subcultures. Kidneys were first treated with 2.5% trypsin from Boehringer without any calcium or magnesium. The final concentration used was 0.125%. EDTA (0.04%), FCS (2.5%), penicillin (200 Ul/ml), and streptomycine (200 μg/ml) were added and the final mixture adjusted to pH 7.3-7.4.

For the subcultures, the same medium was used without FCS and without antibiotics. After 20 subcultures, confluent cell sheets were further treated with trypsin-EDTA as usual for fish cell-lines (Lannan et al., 1984; Nicholson and Byrne, 1973).

Washings. Kidneys were washed with pH 7.3-7.4 PBS with 100 Ul/ml penicillin and 100 μg/ml streptomycine. For the confluent cell sheets, the same washing solution was used without antibiotics.

Viruses. IPN virus, strain 31/75; VHS virus, strain 23/75; and SVC virus, strain 86 were inoculated to the confluent cell sheets. Fresh water viruses of salmonids were kindly provided by Dr. P. de Kinkelin (INRA, Thiverval-Grignon, France). Viruses of marine fishes like Birnavirus IPN-like, strain I-8 B, from sea bass (*Dicentrarchus labrax*), and I-8 D, from young gilt head (*Sparus auratus*) were also used. They were kindly provided by Dr. J. Castric from Laboratoire National de Pathologie des Animaux Aquatiques, Brest, France. I-8 B and I-8 D viruses were isolated from asymptomatic carriers. I-8 B virus was isolated three times from HPC lines before being inoculated in the RTK cells at dilutions of 1/10, 1/100, and 1/1,000 as described below. I-8 D virus was isolated three times from SBL lines. Both viruses were consequently inoculated two times to the RTK cells.

METHODS

Primary cultures. All experiments with living material were performed on ice (Wolf and Dunbar, 1957; Wolf et al., 1960). Aseptic removal of trout kidneys was performed after washing the fish skin with 95% ethanol. After extraction, organs were immediately transferred to cold washing solution and then washed several times. In order to eliminate blood cells, 1-mm diameter pieces were washed several times; kidney fragments were then digested with 5-10 ml of enzymatic solution (see above). Trypsinization occurred at 4°C for 5 hr under slow stirring (Wolf et al., 1960; Gravell and Malsberger, 1965; Wolf and Quimby, 1969). The mixture was centrifuged at 4°C for 30 min at 200g, the pellet was resuspended in the culture medium and filtered to eliminate undigested fragments and cell clusters. Counting of living cells was realized with trypan blue. Another control for viability was the observation of ciliated cells in a cell suspension (Wolf and Quimby, 1969).

Cells were inoculated into 25 cm^2 flasks at a concentration of about 1-2 x 10^7 cells/10 ml. Inoculation occurred at 20°C, which has been reported to be the optimal temperature for salmonid cells (Wolf et al., 1960). The culture medium was changed 48 hr after inoculation. During three weeks, change of the medium was realized irregularly depending on the growth rate of the cells. As soon as cell aggregates started to occur, half of the medium needed to be changed every four days (Hayashi et al., 1976).

Subcultures. Trypsinization was performed at 4°C until the 50th subculture, then at room temperature.

Growing rate. At each step, an aliquot was frozen at -80°C and also at -180°C for subcultures 17 to 21, 23, 24, 34, 35, 40, 54, and 55. The cells were frozen as usual (10% DMSO) at a concentration between 5 x 10^5 to 1 x 10^6 cells per ml. After thawing, cells were kept at 20°C in the culture medium for 48 hr. Then the medium was changed to eliminate the DMSO.

Thermosensitivity. The effect of temperature was studied at different stages (subcultures 5, 25, and 40). The reference temperature was 20°C. The experiments were performed at 4°C, 14°C, 29°C, and 37°C. Cell growth was followed by counting the cells every 24 hr until the third day and then every 48 hr until the 10th day. The number of cells was the average of two experiments.

Infection. Cells were infected with IPN or VHS virus at different stages (subcultures 5, 40, and 58). With VPC virus, only cells at the 61st step were infected. With viruses from a marine source, RTK cells were infected at subcultures 62 and 63. The confluent cell sheet growing for 24-48 hr was filtered, inoculated with a 1/1,000 diluted SHV virus suspension, a 1/1,000 and 1/10,000 NPI virus suspension or a 1/100 and 1/1,000 VPC virus suspension, all in MEMG supplemented with 2% SVF and 10% tryptose-casein soja. Virus suspensions were obtained from the supernatant fluids of cell cultures (RTG for the IPN and VHS viruses and CPE for SVC virus). After the infection, the cultures were kept at 14°C for the VHS, IPN, and IPN-like viruses, and at 20°C for the SVC virus. The control cultures, growing with limiting FCS concentration, were kept at both temperatures.

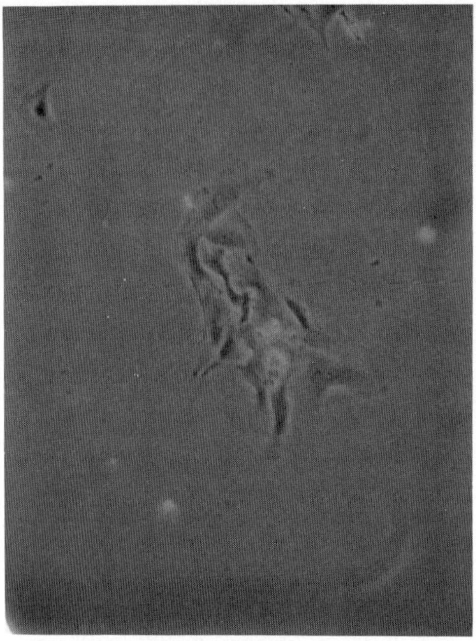

Fig. 1. RTK primary culture at day 8: cluster of cells.

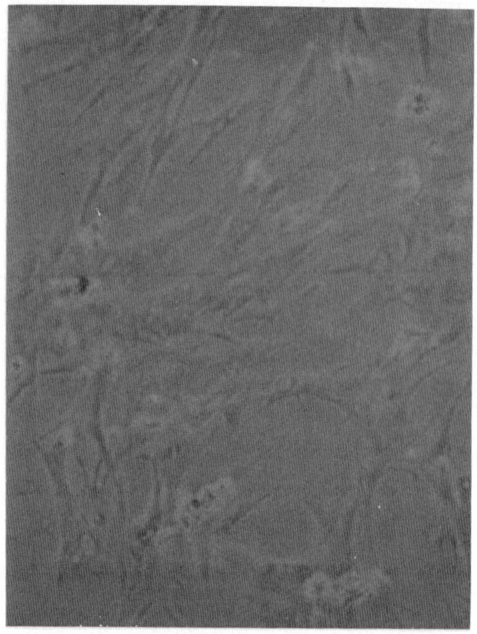

Fig. 2. RTK primary culture at day 21: cluster of cells.

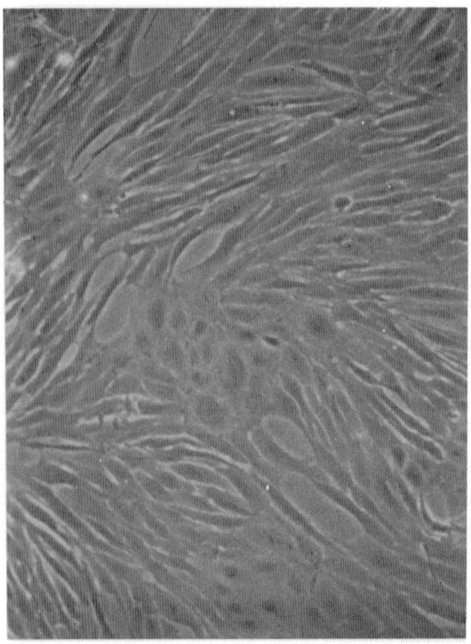

Fig. 3. RTK primary culture at day 21: cluster of cells.

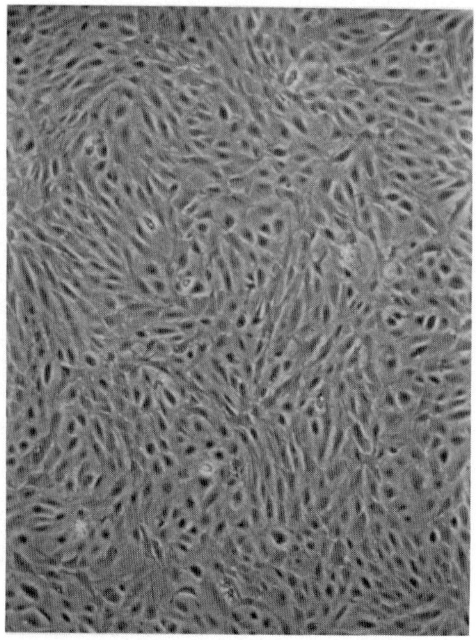

Fig. 4. RTK confluent cells sheet after 5 subcultures.

RESULTS

Establishment of culture-subcultures. After 15 days, aggregates of 2-4 cells were observed (Fig. 1). One week later (Figs. 2 and 3), the number of cells per aggregate was significantly increased. The fifth week from the beginning of the culture, a confluent cell sheet (CCS) was observed. Nontheless, total confluence was not obtained, even after two months.

At the beginning of the fourth month, cell degeneration was observed and then the first subculture was trypsinized in order to realize the second subculture. The new CCS appeared after 7 days. Subsequently, new subcultures were started every 4-8 days depending on the cell density. In one year, 61 subcultures of young rainbow trout kidney were obtained.

Morphology. The shape of the cells was clearly fibroblastic (see Figs. 2 and 4 showing their evolution with time).

Thermostability and growth rates with temperature. At 4°C, the growth rate is slow (Fig. 5), but a CCS was stable at such a temperature for 6 months and gave us the

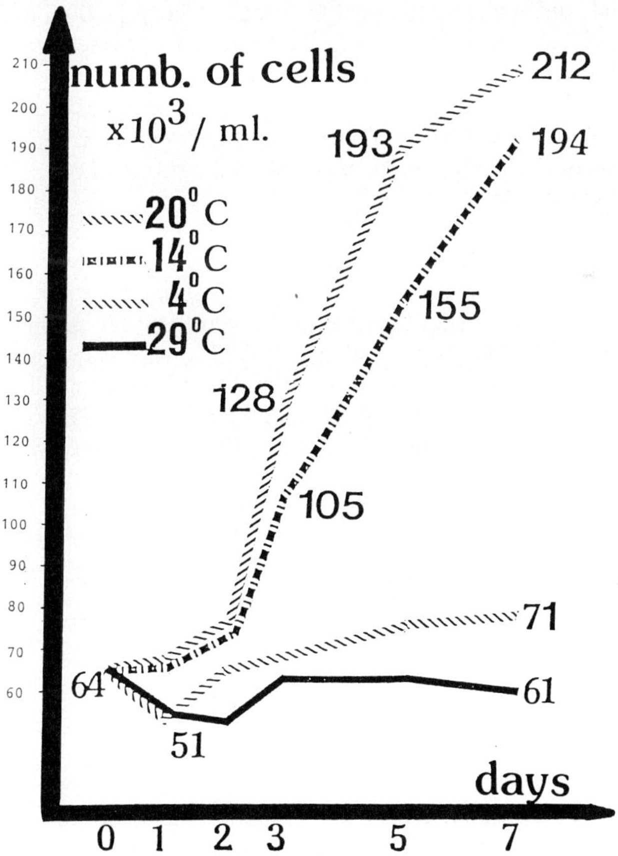

Fig. 5. Curves of growth of RTK cells versus incubation time.

opportunity to start a new culture at 20°C after trypsinization of the CCS. At 14°C, the delay to obtain a new subculture was reduced to 3 months. The optimal temperature of the growth was 20°C. At 29°C, the formation of the CCS stopped and then disappeared after 2 weeks. Finally, at 37°C there was no anchoring of the cells to the flask bottom.

Conservation. We tried to keep the RTK cell line at -80°C. The thawing of the cells before 3 months led to different times of the appearance of the CCS (one week to one month). After 3 months, cell growth was not observed.

After conservation in liquid nitrogen, a new CCS was obtained one week after thawing.

Growth factors. Three different concentrations of FCS were used. A CCS was obtained in 24 hr with 1 x 10^5 cells/ml in medium supplemented with at least 10% SVF. It was necessary to change the medium 8 days after the beginning of the subculture; while with 15% SVF, the change must take place after 10 days. With 2% SVF, CCS already obtained were maintained for a period of 3 months, at 14°C.

Viral replication and sensitivity.

VHS virus. RTK cell line was sensitive to the presence of the VHS virus and cytopathological effect (CPE) was observed (Figs. 6 and 7). The infected cells were round, refringent, and agglomerates of cells detached from the bottom of the flask. Such CPE started 3 days after infection of the CCS with 1/1,000 diluted viral suspension and was achieved after 6 days, at 14°C.

IPN virus. The delay of the infection varied with the concentrations of the virus (Figs. 8 and 9). With 1,000 dilution, a partial destruction of the CCS occurred at day 3 post-inoculation, and 5 days later CPE was obtained. With 1/10,000 viral solution the same result was obtained but with a delay of 24 hr, the CPE being maximal 10 days after the infection.

SVC virus. With 1/100 and 1/1,000 viral dilution, no CPE was observed even 15 days after the infection.

IPN-like I-8 B virus. After the first infection of the RTK cell line, CPE began with the 1/20 and 1/100 viral dilutions after 2 days. The CPE was generalized after 3 days with the same viral dilutions (Fig. 10). With the 1/1,000 viral dilution, after 3 days the CPE started. After the second infection, the delay of the CPE was 2 days in comparison with the first infection. The biggest CPE was obtained with 1/1,000 viral dilution.

IPN-like I-8 D virus. After the first infection, we obtained the same pattern as the one obtained in the IPN-like I-8 B virus. On the other hand, after the second infection CPE was important after 4 days (Fig. 11). A maximal destruction of the CCS occurred after 5 days with 1/100 viral dilution and after 6 days with 1/1,000 viral dilution.

DISCUSSION AND CONCLUSION

Until now, the only salmonid cell line available originated from a mixture of organs (Nicholson and Byrne, 1973). The main interest of this work was to obtain a RTK cell line from a single organ, namely the kidney.

Fig. 6. RTK cell line: cytopathogenic effect (CPE) 4 days after infection with VHS virus.

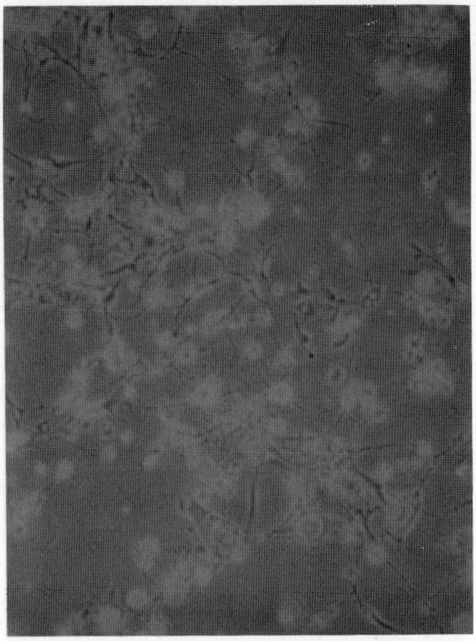

Fig. 7. RTK cell line: CPE 6 days after infection with VHS virus.

Fig. 8. RTK cell line: CPE 3 days after infection with IPN virus.

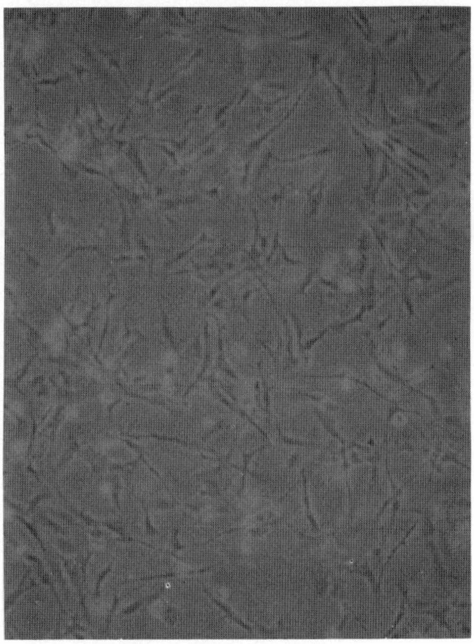

Fig. 9. RTK cell line: CPE 5 days after infection with IPN virus.

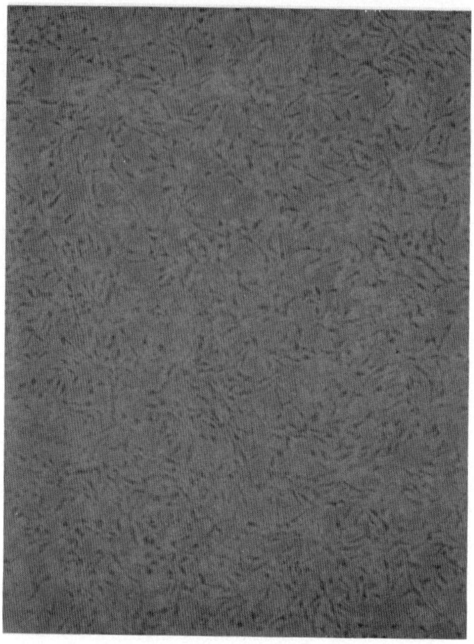

Fig. 10. RTK cell line: CPE 3 days after infection with IPN-like virus, strain I-8 B isolated from sea bass (*Dicentrarchus labrax*).

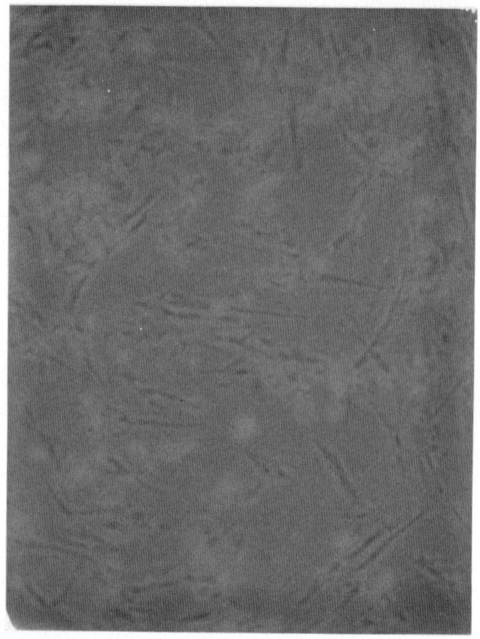

Fig. 11. RTK cell line: CPE 4 days after infection with IPN-like virus, strain I-8 D isolated from gilthead (*Sparus auratus*).

Such a line allowed the same studies as with the one previously available to be performed, especially the study of viruses from freshwater salmonids (Wolf and Mann, 1980; Nicholson and Byrne, 1973; De Kinkelin et al., 1986). Sensitivity for ichthyopathogenic viruses was similar compared to the RTG cell line (De Kinkelin et al., 1986). Sensitivity was identical for VHS and IPN viruses with a delay of CPE for the RTK cell line. Furthermore, the RTK cell line seemed less sensitive to the SVC virus. A transmission electron microscopic study will be necessary in order to check the replication of the VPC virus in the RTK cells.

Also, viruses isolated from marine fishes not from the salmonid group were able to infect RTK cell line with a CPE. I-8 B and I-8 D virus strains had different rates of visible CPE. The time between the second infection and the beginning of the CPE varied depending on the marine viruses used. The time was shorter for I-8 D virus compared to I-8 B. For I-8 D virus the CPE started more rapidly and was more important after the infection with the highest concentration of the virus. On the contrary, for I-8 B, the CPE was more important and started more rapidly with the lowest concentration of the virus. Such a paradoxal effect could be due to viral interference leading to the presence of defective viral particles at the highest concentration (Nicholson and Dunn, 1974).

The characteristics of the cell-line, the number of subcultures (91), the type of the cell growth curves, and the shape of the confluent cells sheet after one month without trypsinization were in agreement with the ones of a typical cell line. Further characterization of karyotype will be necessary; the first results showed that the presence of triploid cells (2n = 64, but a lot of cells had nearly 90 chromosomes; Fig. 12).

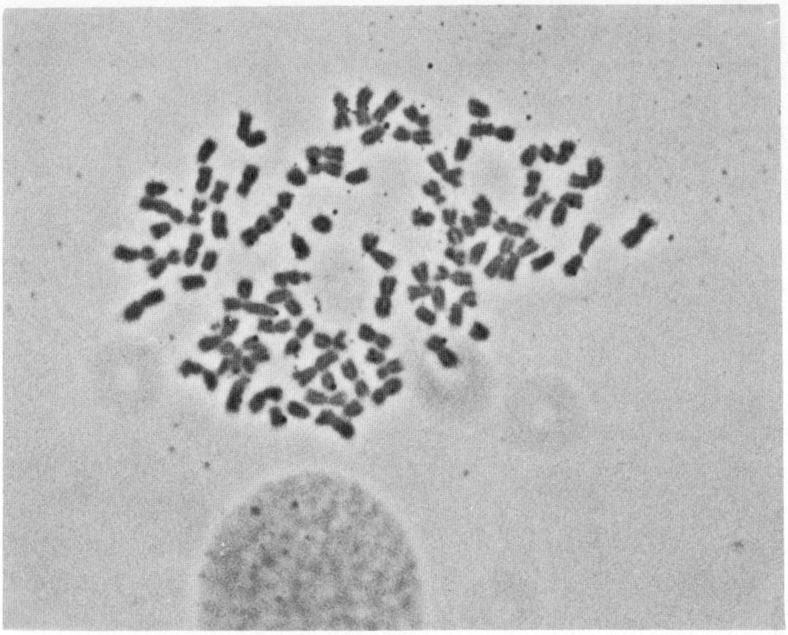

Fig. 12. Karyotype of RTK Montpellier showing 88 chromosomes.

In conclusion, our cell line, which is stable, could be used for study of salmonid virology. We are now checking the sensitivity of the RTK Montpellier cell line to different ichthyopathogenic pathogens, particularly parasites with a life cycle showing intracellular stages in the kidney.

LITERATURE CITED

De Kinkelin, P., C. Michel and P. Ghittino. 1986. Précis de pathologie des poissons, OIE, INRA, Paris, 247-249, 256.

Gravell, M. and R.G. Malsberger. 1965. A permanent cell-line from the fathead minnow (*Pimephales promelas*). Ann. N.Y. Acad. Sci., 126, 555-565.

Hayashi, M., Y. Ojima and N. Asano. 1976. A cell-line from teleost fish: establishment and cytogenetical characterization of the cells. Japan J. Genetics, 51, 65-68.

Lannan, C.N., J.R. Winton and J.L. Fryer. 1984. Fish cell-lines; establishment and characterization of nine cell-lines from salmonids. In Vitro, 20, 671-676.

Nicholson, B.L. and C. Byrne. 1973. An established cell-line from Atlantic salmon (*Salmo salar*). J. Fish Res. Board Can., 30, 913-916.

Nicholson, B.L. and J. Dunn. 1974. Homologous viral interference in trout and Atlantic salmon cell cultures infected with infectious pancreatic necrosis virus. J. Virol., 14, 180-182.

Wolf, K. and C.E. Dunbar. 1957. Cultivation of adult teleost tissues *in vitro*. Proc. Soc. Exptl. Biol. Med., 95, 455-458.

Wolf, K. and J.A. Mann. 1980. Poïkilotherm vertebrate cell-lines and viruses: a current listing for fishes. In Vitro, 16, 168-179.

Wolf, K. and M.C. Quimby. 1969. Fish cell and tissue culture. In: "Fish Physiology" (W.S. Hoar and D.J. Randall, eds.), Vol. 3, 253-305. Academic Press, New York and London.

Wolf, K., M.C. Quimby, E.A. Pile and R.P. Dexter. 1960. Preparation of monolayer cell cultures from tissues of some lower vertebrates. Science, N.Y., 132, 1890-1891.

VIROSES SECTION: CONCLUDING REMARKS

P.E. Vestergaard Jørgensen

National Veterinary Laboratory, Hangøvej 2,
8200 Arhus N, Denmark

This section contained a comprehensive description of a new herpes virus found in eel. The study included use of serological as well as advanced biochemical methods such as restriction endonuclease cleavage analysis, a method that will probably become very important in the future. Another paper dealt with the spread of shrimp viruses.

Another presentation dealt with diagnostic procedures related to IPN virus, the first virus to be isolated from an aquatic organism, and interestingly enough, the authors demonstrated how centrifugation had different effects on virus recovery from different sexual fluids. Such information is very relevant for diagnostic procedures and particularly for virus carrier detection.

Permit me to share some thoughts relative to new areas of future research. If additional research capacity could be made available, the following new research areas should be considered: (1) production and utilization of immunological reagents against all viruses encountered in connection with mariculture; and (2) studies dealing with epizootiology of viral diseases with the aim of control by SPF methods or vaccination.

Immunological reagents. The ideal virus antigen for production of an antiserum is a gradient purified virus suspension of a high titer. In practice, however, it is not always possible to obtain such a preparation; for instance, if the virus does not multiply in cell culture, or multiplies only to a low titer, or if an ultracentrifuge is not at hand. In such situations it should be attempted to produce antisera on the basis of crude antigens since even these can lead to useful reagents, particularly if some of the techniques outlined briefly below are applied. Crude antigen in this connection means pelleted virus from cell cultures, untreated infectious cell culture fluid, or homogenized "victim" tissue with a high concentration of virus.

Production of monoclonal antibodies. Crude virus antigen is injected into mice. Mice spleen cells are fused with myeloma cells and hybrid cells secreting antibodies of the desired specificity are selected by means of an appropriate screening procedure. This may be some immunochemical method (e.g., immunofluorescence, immunoperoxidase, enzyme linked immunosorbent assay, immunoblotting) or some biological method such as neutralization tests.

Screening procedures are of extreme importance in this type of work, whereas purity of antigen is less critical.

Production of second-generation antibodies by immunization with immune precipitates from crossed immunoelectrophoresis. Antiserum is produced by immunizing rabbits with crude antigen (first-generation antiserum). In a two-dimensional (crossed) immunoelectrophoresis, the original crude antigen is reacted against the first-generation antiserum incorporated into a top-gel. Depending on the number of antigen-antibody combinations present in the system, a smaller or larger number of precipitation lines will develop in the gel. Each of the lines (or some of them) are cut out of the gel, washed in saline, homogenized, and injected into rabbits.

The resulting second-generation antiserum frequently is of extremely high quality and well fitted for use in immunochemistry. Some of the antisera will be directed against non-viral antigens, so again, a certain selection work has to be carried out. High quality antisera to VHS (N-protein) have been produced this way (N.J. Olesen and P.E. Vestergaard Jørgensen, unpubl.).

In-vivo **absorption**. Antiserum from rabbits resulting from immunization with crude virus antigen frequently contains so many tissue reactive antibodies that its utilization in immunochemistry is not possible. Such an antiserum can be injected intraperitoneally into an animal of the species that gave rise to the cells from which the virus originated and the rabbit antibodies recovered by repeated bleedings between 24 and 72 hr later. If the virus for immunization of rabbits was cultivated in a fish cell line, the antiserum would have to be injected into a fish. The fish then by itself would remove fish-reactive antibodies from the rabbit-serum inoculum before delivering the virus specific antibodies via its blood. The technique has allowed purification of VHS-antibody (P.E. Vestergaard Jørgensen, unpubl.), IHN-antibody (S. LaPatra, pers. commun.), CCV-antibody (R. Hedrick, pers. commun.).

If necessary, the virus specific rabbit antibody can be "extracted" from the fish serum through protein-A sepharose filtration. The protein A will bind IgG of rabbit origin but none of the fish serum components.

Absorption of antisera with cell culture cells or tissue homogenates. Although useful for removing traces of tissue reactive antibodies from viral antisera, the methods are much less efficient than the ones mentioned before.

Epizootiology. Although SPF conditions may be obtained for juvenile stages of certain marine species in land-based hatcheries supplied with sterilized sea water, it probably has to be realized that SPF methods cannot be broadly applied in mariculture. This is quite unlike the situation in fresh water aquaculture.

Another approach at disease prophylaxis is vaccination, and very likely this approach will be widely applied in the future. This is true at least for fish, but later may become realistic also for other marine organisms with sufficiently developed immune systems.

As an example of a modern protocol for virus vaccine production, current experiments in our group concerning development of a VHS vaccine shall be briefly outlined. Monoclonal antibodies are produced against each of the structural proteins of Egtved virus. Through passive immunization of fish with monoclonal antibodies, it is clarified which of the proteins induce protective antibodies. Subsequently, the identity of the protective antigen is further documented through active immunization of fish with antigens obtained in pure form by affinity chromatography with monoclonal antibodies.

Once the identity of the protective antigen has been established, work can proceed in the identification of the RNA that induces the synthesis of that particular protein (or an immunogenic part of it). This is done by screening a cDNA library constructed from mRNA from infected cell cultures.

Vaccine production subsequently is attempted in appropriate microorganisms into which the relevant DNA has been incorporated.

Vaccine development along the above mentioned lines as a spin-off effect leads to products that can be used as DNA probes in diagnostic and epizootiological work.

It should be added that virus vaccines also may be produced along more conventional lines, such as cultivation of attenuated or virulent virus in cell cultures.

BACTERIOSES SECTION

BACTERIOSES SECTION: INTRODUCTORY REMARKS

T. Sano

Department of Aquatic Biosciences, Tokyo University of Fisheries
4-5-7 Konan, Minato-ku, Tokyo 108, Japan

Included herein are several papers that pertain to bacterial diseases. I would like to point out some matters brought out by these papers that I feel are crucial to the issues surrounding marine fish and shellfish diseases, further study of these diseases, and prevention and regulation of such diseases.

Bacterial causative agents are divided into two categories: facultative pathogens and obligatory pathogens. Fish bacterial facultative agents represent the overwhelming majority, but obligatory pathogens are limited to the following species: *Aeromonas salmonicida, Haemophilus piscium*[1], and *Renibacterium salmoninarum*. These are the only known salmonid pathogens. Although *A. salmonicida* and *R. salmoninarum* were described from freshwater salmonids, these pathogens have been introduced into marine salmonid culture by carrier fish, spreading from the wild population, or intensive salmonid culture with development of marine aquaculture. Also, *Yersinia ruckeri*, indigenous to freshwater salmonids, seems to have invaded such marine fish as the turbot. In the future, the obligatory pathogens specific to marine fish and shellfish, other than *R. salmoninarum* and *A. salmonicida*, must be isolated from these animals.

There is high possibility that a rickettsia is an obligatory pathogen occurring in the sea scallop, *Pecten maximus*, and a second bivalve, *Chlamys opercularis*. This pathogen causes mass mortalities among both hosts where it occurs as a cytoplasmic parasite.

Vibrio spp., facultative pathogens in fish, cause general septicemia following systemic infection. What pathological changes do such bacteria eventually bring to cephalopods? To answer this question, one needs to understand the comparative pathology between fish and cephalopods. Takahashi et al., (1985) have reported

---

[1]Austin, B. and D. Allen-Austin (1985) state that it would appear that *Haemophilus piscium* is synonymous with *Aeromonas salmonicida*, possibly comprising a new subspecies in their review.

vibriosis with severe mortalities of *Penaeus japonicus,* including hatchery-cultured larvae and adult shrimp in Japan, which affects the lymphoid organ and results in the development of brown pigmentation, hypertrophy, and eventually intensive necrosis. The isolated *Vibrio* sp. has been identified as the causative agent and Koch's postulates have been satisfied.

Here, I am citing the classical and presently still accepted components of Koch's postulates: (1) the microorganism must be obtained in every case of the disease; (2) it must be isolated and grown in a pure culture; (3) the pure culture must, when inoculated into a susceptive animal, reproduce the disease; and (4) the microorganism must be observed in and recovered from the experimentally diseased animal.

For artificial infection or transmission, particularly in the case of facultative pathogens, in order to fulfill Koch's postulates, the most important aspect is to perform the experiments under conditions as close as possible to the natural environment in which the epizootic occurs.

It is extremely difficult to completely cure fish and shellfish diseases once these occur in mass production aquacultural environments; even if successful, it would be monetarily expensive. Therefore, prophylactic treatment is the most important and crucial practice in the aquaculture industry. This methodology includes: (1) improvements in the aquatic environment including nutrition and habitat; (2) introduction of animals genetically resistant to disease; (3) introduction of suitably immunized animals; and (4) extermination of pathogens present. Among these techniques, exterminating or preventing invasion of the pathogen is crucially significant and practical. However, this needs the establishment of quarantine and regulation systems for transporting fish and shellfish under domestic and international legislation, in order to prevent any epizootic in aquatic animals. FAO (1977) proposed a "Draft Convention for the Control of Communicable Fish Diseases" for the prevention, control, and eradication of the major communicable fish diseases for both exporting and importing countries. This convention has yet to be adopted. Jasiorowski et al. (1987), in the FAO-WHO-OIE 1987 Animal Health Yearbook, designated 12 communicable fish diseases as being important in international commerce. No disease of marine fish and shellfish is included in both international documents mentioned. This is a crucial and timely matter. PAMAQ must endeavor to designate major communicable diseases of marine fish and shellfish into the draft now.

LITERATURE CITED

Austin, B. and D. Allen-Austin. 1985. A review of bacterial pathogens of fish. J. Appl. Bacteriol., 58, 483-506.

FAO. 1977. "Draft Convention for the Control of Communicable Fish Diseases and Amended Technical Annexes." FAO, Rome, Italy.

Jasiorowski, H.A., G. Torrigiani and K. Bogel. 1987. "FAO-WHO-OIE 1987 Animal Health Yearbook" (V. Kouba and L. Velloso, eds.). Animal Production and Health Division, FAO, Rome, Italy.

Takahashi, Y., Y. Shimoyama and K. Momoyama. 1985. Pathogenicity and characterizations of *Vibrio* sp. isolated from cultured Kuruma prown, *Penaeus japonicus* Bate. Bull. Jpn. Soc. Sci. Fish., 51, 721-730.

A FORMULARY FOR TREATING CEPHALOPOD MOLLUSC DISEASES

John W. Forsythe, Roger T. Hanlon, and Phillip G. Lee

Marine Biomedical Institute, University of Texas Medical Branch, League Hall H63, Galveston, Texas 77550-2772, U.S.A.

ABSTRACT: During extensive mariculture experiments, we encountered various cases of external skin lesioning or internal systemic infections caused by bacteria and one epidemic of ectoparasitic protozoans responsible for extensive skin lesioning. In attempting to formulate disease treatment protocols, it became clear that no information existed on appropriate therapeutic agents, safe concentrations, or methods of application. For a wide range of antibacterial and antiprotozoal compounds, we present information on concentrations, duration and frequency of exposure, efficacy of the treatment, and any effects on the cephalopod species treated. Most work was on octopuses, with additional data pertaining to cuttlefishes and squids.

The antibacterial agents included: penicillin, piperacillin, neomycin, kanamycin, amikacin, gentamycin, cefotaxime, chloramphenicol, oxytetracycline, minocycline, furazolidone, nitrofurazone, nifurpirinol, and piromidic acid. The antiprotozoacides were: acetic acid, acriflavin, calcium hypochlorite, formaldehyde, malachite green, metronidazole, and quinacrine. These compounds were administered by: (1) short-term baths or dips (minutes to hours); (2) continuous baths (days); (3) intramuscular injection; or (4) orally via injection of food items. Nitrofurazone (1.6 - 3.2 mg/l in two 3-day baths) and nifurpirinol (1 - 2 mg/l in daily short-term baths) were effective in treating skin infections of octopuses, but exposure of squids to 0.1 mg/l nifurpirinol for 16 hr resulted in 100% mortality within 2 days. Systemic infections in octopuses were treated successfully with initial intramuscular injections of chloramphenicol (100 mg/kg octopus for 1-3 days) followed by feeding the octopuses for 7 days on shrimp that had been injected with the antibiotic. Experiments evaluating the effectiveness of several antibiotics against *Vibrio* spp. cultured in liquid T-soy broth made with sterile artificial sea water were conducted to better predict antibiotic performance in bath application. Preliminary findings showed nitrofurazone and choramphenicol to have the lowest minimum inhibitory concentration (MIC), nifurpirinol was intermediate in effectiveness, and oxytetracycline and kanamycin were ineffective. No effective treatment method for protozoans was found. Octopus eggs can be safely dipped in malachite green (0.05

Pathology in Marine Science

mg/l) to treat fungal growth, while squid eggs may be dipped in iodine (100 mg/l) to reduce epibiont populations.

INTRODUCTION

Secondary bacterial infection of abraded skin is the principal cause of disease-related mortality in cephalopods (Summers et al., 1974; Leibowitz et al., 1977; Hulet et al., 1979; Hanlon et al., 1983). Abraded skin results from inter-animal contact in octopuses or abrasions caused by collisions with tank walls in squids and cuttlefishes. The infections may be acute (death within 8-10 days) but are more commonly chronic (lasting weeks or months before an animal dies). Occasionally, acute internal infections causing muscular necrosis or eye damage have been identified. As we began to encounter disease problems, it quickly became apparent that no published information existed on appropriate antibiotics, safe concentrations, or methods of application for cephalopods, or other marine invertebrates for that matter.

Since 1976 we have conducted research on the capture, maintenance, and culture of cephalopod species with potential or demonstrated value in biomedical research (Forsythe and Hanlon, 1980; Hanlon et al., 1983, 1989; Hanlon and Forsythe, 1985; Hanlon, 1987). We have maintained, reared, or cultured 14 octopus species, eight squid species, and two species of cuttlefish. Animal densities and seawater system crowding have increased as our research emphasis has shifted from capture to culture, sometimes resulting in an increased occurrence of disease. Bacterial infections, usually *Vibrio* spp., have been observed in every species. In addition, we have observed several new problems that have not been previously described. An epidemic infestation of ectoparasitic protozoans caused extensive skin damage resulting in acute secondary bacterial infections in several species of octopuses (Forsythe et al., 1988). The observed deterioration of octopus and particularly squid egg capsules prior to hatching is due to bacteria and epibionts.

The obvious lack of information on disease therapy and the impact of disease-related mortalities on our culture programs acted as the impetus to begin systematic evaluations of drug effectiveness and treatment methods. Several reports (Hanlon et al., 1983; Ford et al., 1986) have described our initial therapeutic methods. The objective of this "formulary" is to provide a compilation of all our experiences over the past 8 years, successful and unsuccessful, in exposing cephalopods to a wide range of therapeutic agents. This information should provide cephalopod investigators a starting point for treating a pathogen believed susceptible to one of the listed antibiotics.

MATERIALS AND METHODS

The therapeutic agents were obtained either from the hospital pharmacy or in bulk from aquaculture supply houses. Drugs from the pharmacy typically came in vial or capsule form. Antibiotics in vials were reconstituted with 10 ml of sterile saline to form a stock solution. Capsules were opened and their contents dissolved in a fixed amount of sterile saline to form stock solutions. Drugs obtained from aquaculture supply houses in bulk quantities were weighed and dissolved in sterile saline to make stock solutions.

Infected animals or eggs were typically removed from their culture systems and placed in aerated, static baths containing a known concentration of a drug or mixture of drugs. Baths were made up fresh daily with UV sterilized sea water from the culture system. Two to four different drugs were independently administered over the same time period. Small groups (5-10 animals) of comparably sick octopuses were placed in each bath and their condition followed for 7-14 days to monitor their condition. Some drugs were applied directly to the culture system. In those cases, activated carbon was removed from the filtration circuit during treatment (up to 72 hr) and then replaced to adsorb the drug within a few hours. Intramuscular injections were made using 26 gauge, 1 ml, tuberculin syringes. For oral administration, live food organisms such as crabs, shrimps, or fishes were injected with an antibiotic and immediately fed to infected animals. Frozen shrimp could be thawed, injected, and fed to older octopuses.

In a separate experiment, liquid T-soy broth tubes were used to evaluate the effectiveness of antibiotics in bath applications. To more realistically evaluate the antibiotics, we used a liquid broth method where the broth was made with sterile artificial sea water and then buffered to at least pH 7.5 or higher. This technique exposed the drug to the highly ionic environment and pH of sea water while the bacteria were also grown in a more optimal environment. Results in this broth were compared to broth made with deionized water at pH 6.0. Increasing concentrations of antibiotics were added to serial tubes of each type of broth. The tubes were then innoculated with 0.5 McFarlands of bacteria and allowed to incubate at 25°C for 24 hr. *Vibrio damsela* and *Vibrio alginolyticus*, originally isolated from infected octopuses, were the bacterial species used. The minimum inhibitory concentration (MIC) was the first tube dilution with no visible growth. The minimum bactericidal concentration (MBC) was the first tube dilution in which no bacterial colonies grew after replating on T-soy agar.

RESULTS AND DISCUSSION

Most agents listed in Table 1 were evaluated in 1984 during an epidemic outbreak of an ectoparasitic protozoan (Forsythe et al., 1988). The first visible signs of the disease were several white spots with small lesions at the center. These were initially mistaken as typical *Vibrio* skin infections and treated as such. After many antibiotics were tried, it was discovered that a protozoan parasite was the primary cause of the disease. As a result, we learned a great deal about drug tolerances of octopuses but rather little about their effectiveness against disease. A similar outbreak occurred again in 1985 and a large number of antiprotozoan agents were evaluated. These trials largely make up Table 2.

Table 1 lists the antibacterial drugs we have used in bath application for cephalopod disease problems along with maximal concentrations used for a given exposure sequence. Unless noted in the comments, the dosages in Table 1 should be sub-lethal for octopuses in early stages of infections (i.e., animals still behave normally and feed). Lower dosages or shorter exposure times should be quite safe. As

Table 1. Antibacterial drugs used in treating disease problems in laboratory cultured cephalopods.

DRUG Family Generic/(Trade)	Species Treated*	Concentration (mg/l)	Duration (hours)	Frequency (x/day)	Total Therapy Duration (days)	Comments
Penicillins						
potassium penicillin G	D	500	1	2	6	Was ineffective against *Vibrio* infections.
piperacillin sodium (Pipracil)	A	400	1	2	7	Became toxic beyond 4 days.
Aminoglycocides						
neomycin sulfate (Mycifradin)	A	20	20 min	2	7	
kanamycin sulfate	A,C	100	1	2	7	
(Kantrex)	A,C	150	30 min	2	14	
amikacin sulfate (Amikin)	A	25	1	2	7	Poor survival in bath solution.
gentamycin sulfate	A	20	15 min	2	7	
(Garamycin)		20	24	--	1	
Cephalosporins						
cefotaxime (Claforan)	A	200	1	2	7	

*Species treated: A = *Octopus bimaculoides*, B = *Octopus briareus*, C = *Octopus digueti*, D = *Octopus joubini*, E = *Octopus maya*, F = *Sepia lattimanus*, G = *Sepia officinalis*, H = *Loligo forbesi*, I = *Lolliguncula brevis*.

(continued)

Table 1 (continued)

DRUG Family Generic/(Trade)	Species Treated*	Concentration (mg/l)	Duration (hours)	Frequency (x/day)	Total Therapy Duration (days)	Comments
Chloramphenicols						
chloramphenicol sodium succinate (Chloromycetin)	D	100	1	2	6	
Tetracyclines						
oxytetracycline (Terramycin)	A,C	50	8	1	7	
minocycline (Minocin)	B,D	0.53	24	--	20	
	A,C	20	24	1	1	
Nitrofurans						
furazolidone (Furoxone)	A,C,E,F,G	0.5	72	--	7	Always used in conjunction with 1.5 mg/l nitrofurazone.
nitrofurazone (Furacin)	A,C,E	50	10 min	2	7	
	A,C,E	25	1	2	7	Can be stressful to very sick octopuses.
	A,C,E,F,G	1.5	72	--	7	Well tolerated by cuttlefish and octopuses.

*Species treated: A = *Octopus bimaculoides*, B = *Octopus briareus*, C = *Octopus digueti*, D = *Octopus joubini*, E = *Octopus maya*, F = *Sepia lattimanus*, G = *Sepia officinalis*, H = *Loligo forbesi*, I = *Lolliguncula brevis*.

(continued)

Table 1 (continued)

DRUG Family Generic/(Trade)	Species Treated*	Concentration (mg/l)	Duration (hours)	Frequency (x/day)	Total Therapy Duration (days)	Comments
nifurpirinol (Furanace, Prefuran)	A,B,C,D,E	1.5-2.0	3-10 min	1	74	An apparent toxic reaction to this by *O. briareus* after 32 days.
	A,B,C,D,E	1.5	30	1	14	
	A,B,C,D,E	2.0	30	1	7	
	H,I	0.2	24	--	1	100% mortality with *Loligo forbesi.*
Others						
piromidic acid (Panos)	A	30.0	8	1	1	Low solubility in sea water. Lethal within 8 hr.
	A	3.0	16	1	1	This dosage was lethal by 16 hr.

*Species treated: A = *Octopus bimaculoides*, B = *Octopus briareus*, C = *Octopus digueti*, D = *Octopus joubini*, E = *Octopus maya*, F = *Sepia lattimanus*, G = *Sepia officinalis*, H = *Loligo forbesi*, I = *Lolliguncula brevis*.

Table 2. Antiprotozoal compounds used in treating ectoparasitic protozoans (morphologically similar to *Ichthyobodo necator*) on *Octopus bimaculoides* and *O. maya*.

Generic Name (Trade)	Concentration	Duration	Frequency (x/day)	Total Therapy Duration (days)	Tolerance	Comments
Acetic Acid	1 ml/l	30 sec	1	2	Good	Lethal if continued beyond 2d.
Acriflavin	0.25 mg/l	16 hr	1	4	Good	Usually in conjunction with 50-100ppm Formaldehyde.
	0.50 mg/l	16 hr	1	1	Poor	Lethal within 24 hr. Turns gills yellow.
Calcium Hypochlorite	10 mg/l	2 min	1	1	Poor	50% mortality at 48 hr post-dip.
Formaldehyde, 37%	100 µl/l	1 hr	1		Fair	Can expect some mortality among weaker octopuses.
	75 µl/l	30 min	1/48 hr	8	Good	Was usually concurrently given with Acriflavin (0.25 mg/l) or Malachite green (0.05 mg/l).
	25 µl/l	20 min	2	7	Good	Given with 0.05 mg/l Malachite Green Oxylate.

(continued)

Table 2 (continued)

Generic Name (Trade)	Concentration	Duration	Frequency (x/day)	Total Therapy Duration (days)	Tolerance	Comments
Malachite	0.075 mg/l	30 min	1/48 hr	6	Good	Given with 75 μl/l formaldehyde.
Green/Chloride	0.100 mg/l	24 hr	1	1	Good	
	0.050 mg/l	16 hr	1/72 hr	18	Good	
Malachite	0.030 mg/l	16 hr	1	6	Poor	
Green/Oxylate	0.050 mg/l	7 hr	1	6	Poor	
Metronidazole	100.0 mg/l	16 hr	1	1	Excellent	
(Flagyl)	150.0 mg/l	1 hr	1	7	Good	
	200.0 mg/l	1 hr	1	4	Good	
Quinacrine HCl	5.0 mg/l	5-8 min	1	1	Fair	
(Atabrine)	5.0 mg/l	30 min	1	1	Poor	Tends to turn gills yellow.
	1.0 mg/l	4-5 hr	1	3	Fair	
	0.5 mg/l	48 hr	1	2	Poor	Mortalities observed 24 hr after treatment completed.
	0.5 mg/l	16 hr	1	4	Fair	

octopuses become lethargic and stop feeding, some of the more toxic antibiotics and extended exposures may cause mortality. Only piromidic acid at all dosages tried with octopuses and nifurpirinol at 0.2 mg/l with the squid *Loligo forbesi* were outright lethal. No definable differences in drug tolerance among the five octopus species were noted, suggesting that these dosages may be broadly applicable to most common littoral octopods. The two cuttlefish species were also tolerant to the antibiotics. Our experiences, as well as that of Ford et al. (1986), has shown loliginid squids to be much more sensitive to antibiotics than cuttlefishes and octopuses.

Among the bath treatments, only nifurpirinol and nitrofurazone with furazolidone have been beneficial in treating octopus bacterial infections. Our initial successes with nifurpirinol are outlined in Hanlon et al. (1984), where we used the original formulation called Furanace[R] made by Dainippon Pharmaceutical in Japan. Shortly after this publication (Hanlon et al., 1984), production of Furanace ceased completely. Currently, a product called Prefuran[R] manufactured in the U.S. is the only nifurpirinol available, but this current formulation has seemed less effective against *Vibrio* spp. than the original Furanace. This lowered effectiveness of nifurpirinol along with that of other common antibiotics in bath application has led us to look for better methods to evaluate the sensitivity of pathogens to drugs dissolved in sea water.

Traditional disc diffusion techniques (see Ford et al., 1986) have not been reliable since drugs that inhibited bacterial growth on culture plates have not performed well in practice. Alternatively, we evaluated antibiotic effectiveness against known bacterial pathogens grown in liquid broth media made with either sterile artificial sea water or de-ionized water. Table 3 displays the preliminary results of this research. The bacteria grew well in both types of broth, although slightly better in the saline broth. Of the five antibiotics listed in Table 3, only nitrofurazone was more effective in the saline broth than in the deionized water broth; the other four were less effective; oxytetracycline dramatically so. Nitrofurazone was effective at the lowest dosages followed closely by chloramphenicol, whereas nifurpirinol, kanamycin, and oxytetracycline required much higher dosages. The nifurpirinol was clearly not effective at the manufacturer's suggested maximal dosages of 1-2 mg/l. We are continuing this line of research using additional antibiotics and additional strains of bacteria in an effort to pinpoint the best drug for a given pathogen.

Intramuscular injections of chloramphenicol have been highly effective in arresting acute systemic infections of *Vibrio carchariae* in *Octopus bimaculoides* and *Octopus maya* as well as some chronic mantle tip lesions on cuttlefish (Table 4). Daily injections were typically made for 1-3 days and thereafter the same dosage was put into live food organisms for another 7-10 days. In octopuses and cuttlefishes, injections were made in the highly vascularized muscular arm bases near the mantle. The total dose was usually split between 2-4 injection sites on different arms. Before injection, octopuses were anesthetized in a 3% solution of pure ethanol in sea water (30 ml/l) and cuttlefishes in a 2% solution. Infected animals usually recovered when this regimen was followed.

Our current treatment protocols for octopus or cuttlefish infections are based upon the results of Table 3; nitrofurazone being the most effective antibiotic in bath application. At the first sign of infections, we dose the entire closed culture system with a combination of nitrofurazone and furazolidone (see Table 1). Activated carbon is removed from the filtration circuit of the system for 72 hr. Carbon is replaced to remove the antibiotics, a 25% water change is made and the system

Table 3. The effect of sea water on antibiotic activity against marine bacteria. These experiments utilized T-soy broth (TSB) tubes prepared with sterile deionized water (DW) and sterile-filtered artificial seawater (SW). The antibiotics were added and appropriate dilutions made. Bacteria (0.5 McFarland) were added and allowed to incubate 24 hr at 25°C. The MIC was the tube in which no visible growth could be recognized. The MBC was the tube from which no bacterial colonies grew after replating on T-soy agar.

Antibiotic Bacteria	Minimum Inhibitory Concentration (μg/ml)		Minimum Bactericidal Concentration (μg/ml)	
	DW	SW	DW	SW
Nitrofurazone				
Vibrio damsela	12.5	3.12	50	12.5
Vibrio alginolyticus	6.25	1.6	25	25
Nifurpirinol				
V. damsela	25	50	25	100
V. alginolyticus	12.5	100	12.5	100
Chloramphenicol				
V. damsela	6.25	12.5	6.25	12.5
Kanamycin sulfate				
V. damsela	200	200	200	>400
V. alginolyticus	100	>400	100	>400
Oxytetracycline				
V. damsela	0.8	200	6.25	100
V. alginolyticus	0.8	100	12.5	100

redosed for 72 more hours without activated carbon. After the second treatment, the antibiotics are removed with carbon and a final 25% water change is made. The nitrifying bacteria in the filter bed are not killed by this procedure; however, they cease their ammonia (NH_4) and nitrite (NO_2) breakdown processes during the treatment. Ammonia and nitrite levels drop rapidly in the 3-5 days post-treatment, indicating that these beneficial bacteria have resumed their activities. If the infections progress despite these measures or if the infection is found to be systemic, intramuscular injections and injected food are administered. No effective treatment protocols have been developed for squids at this time.

Table 4. Antibiotic injections and oral dosages given cephalopods. Animals were typically given a one-time intramuscular injection under anesthesia, then fed live food that had been injected with the drug. These daily oral doses were provided for at least a week.

Drug	Dose (mg/kg BW**)	Frequency (x/day)
Intramuscular injection		
Octopus (species A,C,E)*		
Chloramphenicol	100-150	1
Cuttlefish (species F,G)		
Chloramphenicol	75-100	1
Oral Dose		
Octopus (species A,C,E)		
Chloramphenicol	100-150	1 (for 7-10 days)
Cuttlefish (species F,G)		
Chloramphenicol	75-100	1 (for 7-10 days)

* See species listed in Table 1.
** BW = Body weight of animal.

Antiprotozoan drugs used in combating an epidemic infestation of an ectoparasitic protozoan flagellate morphologically very similar to the freshwater fish parasite *Ichthyobodo necator* (see Becker, 1977) are listed in Table 2. For the most part, effective drugs and dosages used for fish infestations of *I. necator* were lethal to octopuses. The microvilli-bearing epidermis of octopods appears to make them more sensitive to the noxious chemicals traditionally used in treating protozoan parasites on fishes. No effective treatment was discovered for this particular host-parasite relationship. It remains problematical treating an invertebrate parasite on an invertebrate host.

The eggs of *Octopus bimaculoides* have been dipped in malachite green (oxylate and chloride salts) to retard growth of fungus on the external surfaces of the egg. Daily dips of 30-60 min for 3 days in 5 mg/l of both types of malachite green were well tolerated by developing embryos. Slight reduction in fungus proliferation was seen with the oxylate salt. The egg covering or chorion layer of normal intact eggs was impervious to the malachite green, while eggs with chorion damage or minute holes became tinted green internally, resulting in death of the embryo.

The eggs of the squids *Loligo forbesi* and *Sepioteuthis lessoniana* have been dipped in 100 mg/l baths of iodine on alternate days for 10 min during the first 3 weeks of development to retard breakdown of the outer gelatinous tunic. Results from surface swabs indicate that bacterial populations were not reduced significantly, but light microscope surveys of the surface indicate that the number of epibionts may be reduced. In two trials, hatching rate of iodine-dipped versus undipped eggs was 100% versus 73% and 31% versus 0%, respectively. The effectiveness of elemental iodine in sea water is complicated further by its conversion to iodate. Research on the efficacy of iodine and its residence time in sea water is underway. Although the results of these treatments are not conclusive, they have been sufficiently positive to make this treatment routine procedure for laboratory-laid eggs. Embryos within egg strands tolerate the iodine without problem but hatchlings invariably die upon direct iodine exposure.

In summary, the ineffectiveness of most antibiotics when dissolved in sea water severely limits their usefulness for mariculture. A strong need exists for palatable artificial ration that can serve as a vehicle for oral administration of therapeutic compounds to large numbers of animals. Intramuscular injections of sick individuals and their food items are effective, but logistically impractical on a large scale. Finally, though several effective treatments have been established for octopuses and cuttlefishes, none yet exist for loliginid squids.

ACKNOWLEDGMENTS

The excellent technical efforts of Randall DeRusha, Karen Gilligan, and Anthony DiNuzzo on the majority of this work are gratefully acknowledged. Philip Turk, Keith Johnson, and Joanne Hollyfield also assisted in various ways. Numerous conversations with Nelson Herwig (Houston Aquarium), Steve Robertson (Dallas Aquarium), and Dr. Michael Stoskopf (National Aquarium in Baltimore) were invaluable. We thank Laura Koppe for the typing. We are grateful for funding from DHHS grant RR01279 and for continued funding on DHHS grant RR01024 from the Division of Research Resources of NIH and from the Marine Medicine Account of the Marine Biomedical Institute.

LITERATURE CITED

Becker, C.D. 1977. Flagellate parasites of fish. In: "Parasitic Protozoa. Vol. I: Taxonomy, Kinetoplastids, and Flagellates of Fish" (J.P. Kreier, ed.), pp. 357-416. Academic Press, New York.

Ford, L.A., S.K. Alexander, K.M. Cooper and R.T. Hanlon. 1986. Bacterial populations of normal and ulcerated mantle tissue of the squid, *Lolliguncula brevis*. J. Invert. Pathol., 48, 13-26.

Forsythe, J.W. and R.T. Hanlon. 1980. A closed marine culture system for rearing *Octopus joubini* and other large-egged benthic octopods. Lab. Anim., 14, 137-142.

Forsythe, J., R. Hanlon and R. DeRusha. 1988. First observation of an ectoparasitic bodonid flagellate on a marine invertebrate host. Absts. of 3rd Internat. Colloq. Pathol. Marine Aquacul., 145-146.

Hanlon, R.T. 1987. Mariculture. In: "Cephalopod Life Cycles, Vol. II: Comparative Reviews" (P.R. Boyle, ed.), pp. 291-305. Academic Press, London.

Hanlon, R.T. and J.W. Forsythe. 1985. Advances in the laboratory culture of octopuses for biomedical research. Lab. Anim. Sci., 35, 33-40.

Hanlon, R.T., R.F. Hixon and W.H. Hulet. 1983. Survival, growth, and behavior of the loliginid squids *Loligo plei, Loligo pealei*, and *Lolliguncula brevis* (Mollusca: Cephalopoda) in closed sea water systems. Biol. Bull., 165, 637-685.

Hanlon, R.T., J.W. Forsythe, K.M. Cooper, A.R. DiNuzzo, D.S. Folse and M.T. Kelly. 1984. Fatal penetrating skin ulcers in laboratory-reared octopuses. J. Invert. Pathol., 44, 67-83.

Hanlon, R.T., W.T. Yang, P.E. Turk, P.G. Lee and R.F. Hixon. 1989. Laboratory culture and estimated life span of the Eastern Atlantic squid *Loligo forbesi* (Mollusca: Cephalopoda). Aquacul. Fish. Manag., 20, 15-34.

Hulet, W.H., M.R. Villoch, R.F. Hixon and R.T. Hanlon. 1979. Fin damage in captured and reared squids. Lab. Anim. Sci., 29, 528-533.

Leibowitz, L., T.R. Meyers, R. Elston and P. Chanley. 1977. Necrotic exfoliative dermatitis of captive squid (*Loligo pealei*). J. Invert. Pathol., 30, 369-376.

Summers, W.C., J.J. McMahon and G.N.P.A. Ruppert. 1974. Studies on the maintenance of adult squid (*Loligo pealei*). II. Empirical extensions. Biol. Bull., 146, 291-301.

INVESTIGATIONS OF POTENTIAL MICROBIAL PATHOGENS ASSOCIATED WITH ULCERATIVE DISEASE SYNDROME (UDS) OF FLORIDA FISH

Donald J. McGarey, Teresa Kraxberger Beatty, Vicki A. Alberts, Diane Te Strake, and Daniel V. Lim

Department of Biology, University of South Florida, Tampa, Florida 33620, USA

ABSTRACT: Reports from fishermen reviewed by Grier and Quintero (1987) in Florida, as early as 1979, revealed that several types of fresh and salt water fish developed lesions not previously characterized. Subsequent to these local reports, Noga and Dykstra (1986) described oomycete fungi specifically associated with ulcerative mycosis in menhaden from North Carolina. UDS, occurring annually in recent years in the St. Johns River during the months of November to March, was characterized by severe hemorrhagic or open dermal ulcers on the body of the fish, usually the head, mandible, or midbody regions. A variety of fish including gizzard shad (*Dorosoma cepedonium*), blue gill (*Lepomis macrochirus*), weakfish (*Cynoscion regalis*), Atlantic menhaden (*Brevoortia tyrannus*), and ocellated flounder (*Ancylopsetta quadrocellata*) have been predominately affected by UDS. Reports have indicated that over 50% of commercial fish catches are lesioned and unmarketable during the months of occurrence.

Results of our bacterial and mycological examinations of the Florida fish revealed that their lesions often contained bacterial species of *Aeromonas* and *Vibrio* but do not always contain oomycete fungi. The distribution of the Oomycetes in the St. Johns River was monitored monthly at 5 stations ranging in salinity from 0 ‰ to 9 ‰. Both carbohydrate and protein baits were used to collect an array of watermolds. Isolates of these fungi were obtained from fish and the river system. These fungal isolates were physiologically surveyed for extracellular production of proteases, hemolysins, and DNAse activity.

Ulcerative Disease Syndrome does not appear to be a localized problem. In addition to reports from the southeastern coast of the United States, similar disease outbreaks affecting mariculture and fisheries operations have been reported from Australia, Philippines, Thailand, Malaysia, New Guinea, Laos, Burma, Indonesia, and Kampuchea. Whether or not these disease outbreaks contain a common causitive agent remains to be reported. Future studies on UDS will focus on the detection and

Pathology in Marine Science

enumeration of potential pathogens with endemic areas of the St. Johns River system as well as the environmental monitoring of temperature, salinity, pH, and rainfall.

Noga, E.J. and M.J. Dykstra. 1986. J. Fish. Dis., 9, 47-53.
Grier, H and I. Quintro. 1987. Florida Department of Environmental Regulation Report WM-164.

INTRODUCTION

The St. Johns River estuary located in northeast Florida between Duval, Clay, and St. Johns counties is an area of extreme productivity for the regional fisheries. As early as 1979, epizootic outbreaks of fish disease affecting both fresh and salt water fish were reported (Grier and Quintro, 1987) in which the fish had developed dermal lesions not previously described. The term ulcerative disease syndrome (UDS) has been assigned as the descriptive title of this disease (Hargis, 1986), which was reported by fishermen to affect as much as 50% of commercial fish catches during the peak of occurrence, usually in February or March. UDS is a degenerative disease characterized by deep, ulcerative lesions located mainly on the head, mandible, or middorsal regions of the fish. In the St. Johns River estuary a variety of marine fish including weakfish (*Cynoscion regalis*), Atlantic menhaden (*Brevoorita tyrannus*), and ocellated flounder (*Ancyloosetta quadrocellata*) have been predominately affected by UDS.

Similar outbreaks of disease have been reported to occur in estuaries located on the Eastern coast of the United States. In North Carolina, Noga and Dykstra (1986) have described ulcerative mycosis (UM) associated with oomycete fungi in Atlantic menhaden in which the fish exhibit similar lesions as in UDS. A review of fish mortalities by Hargis (1986) indicates the occurrence of similar cases in Virginia and Maryland coastal tributaries associated with Chesapeake Bay.

Although the appearances of UDS and UM fish are clinically similar, lower recovery rates of fungi from the lesions of the Florida fish than those to the north is evident (Colwell and Grimes, 1984; Te Strake and Lim, 1986). The results of bacterial and mycological examinations of the Florida fish revealed that their lesions often contained species of the bacteria *Aeromonas* and *Vibrio* but did not always contain oomycete fungi, which will be discussed in further detail later.

Several biological agents including viruses, fungi, bacteria, and parasites have been implicated or shown to cause lesion syndromes in wild and cultured marine fish populations (Alderman, 1982; Austin and Austin, 1987; Colwell and Grimes, 1984; Dykstra et al., 1986; Munro, 1982; Noga and Dykstra, 1986; Schulz, 1980; Wolke, 1975). Our intentions were to examine the occurrence of bacterial and fungal organisms in UDS fish and to understand their possible roles in this disease. The distribution of bacteria and fungi at varying temperatures and salinities was also examined. Specific microbial populations may be controlled by these environmental parameters that may also be conducive to the development of UDS.

MATERIALS AND METHODS

Fish collection and processing. Monthly collection was carried out by trawling (Continental Shelf Associates, Jupiter, Florida) the St. Johns and Ortega (interconnecting river) rivers. Lesioned and non-lesioned fish were either kept alive in a holding tank or wrapped in a sterile bag and placed on ice for transport to the dock or laboratory. A cardiac puncture was performed on the living fish and the blood was then placed in a blood culture bottle containing tryptic soy broth (GIBCO Labs., Madison, Wisconsin) and incubated at 37°C for up to 72 hr. Tissue samples from lesions and skin were taken, homogenized into a slurry, and plated on two each of the following bacteriological media: thiosulfate citrate bile salt sucrose agar (TCBS), MacConkey agar, tryptic soy agar, blood agar (5% sheep blood) (GIBCO Labs., Madison, Wisconsin), marine agar (Difco Labs., Detroit, Michigan), and Flexibacter agar (Alderman, 1982). All inoculated media were incubated at 37°C and 22°C for 24 hr. Isolated colonies were subcultured and identified using standard bacteriological tests and references (Austin and Austin, 1987; Baumann and Schubert, 1984; Palleroni, 1984).

Tissue samples were also examined for the presence of fungal hyphae by microscopical examination using wet mounts and lacto-phenol preparations. Small samples were taken from lesions suspected of containing a fungus and were isolated following the method of Dykstra et al. (1986). River water was examined for the presence of Saprolegniales by collecting 50-ml water samples in sterile vials "baited" with hempseed halves, rice, or boiled egg white that were incubated and subcultured until pure culture was achieved. Identification was accomplished by microscopical examination of morphological characteristics of the fungal isolate (Lim and Te Strake, 1988).

Water conductivity and temperature were determined at the time of collection and recorded. The results of the bacterial and mycologic studies will be reported separately.

RESULTS

Potential bacterial pathogens. The results of bacteria isolated from various types of fresh and salt water fish (Table 1) indicate that a number of different bacterial species are found to be associated with normal and lesioned fish. Of these organisms, members of *Aeromonas, Vibrio, Plesiomonas,* and *Pseudomonas* are particularly interesting since they represent known piscine pathogens of primary and opportunistic infectious nature (Austin and Austin, 1987; Colwell and Grimes, 1984; Schulz, 1980; Wolke, 1975). Extreme concern about *Vibrio anguillarum* and *Aeromonas hydrophila* in fisheries and aquaculture systems is due to the pathogenicity of these organisms that results in heavy fish mortality and subsequent high economic loss. In addition, *Plesiomonas shigelloides, Aeromonas sobria, A. hydrophila, Vibrio vulnificus, Vibrio parahaemolyticus, Vibrio cholerae* non-01, and *Vibrio damsela* have, in recent years, become increasingly important as human pathogens (Baumann and Schubert, 1984; Farmer et al., 1985).

Table 1 demonstrates bacteria isolated from various fresh and salt water fish but does not show any evidence of bacterial specificity for each type of fish. The highest recovery was of *A. hydrophila* and *A. sobria*, bacteria normally associated with fresh water environments. *Pseudomonas putrefaciens* was also frequently recovered and was previously reported as a possible infectious agent during an ulcerative disease outbreak in carp fish farms in Europe (Schulz, 1980). *V. anguillarum, V. cholerae* non-Ol, *V. damsela*, and *V. vulnificus* were also isolated from many marine fish.

A. hydrophila, A. sobria, and *P. putrefaciens* were generally recovered from fish collected in waters below 5 ‰ salinity, although some were found at salinities as high as 22 ‰ (Table 2). On the other hand, the *Vibrio* spp. were more frequently recovered when salinities were greater than 2 ‰ and only *V. anguillarum* and *V. cholerae* non-Ol were found at lower salinities. The association of marine bacteria (vibrios) with fish caught in fresh water and the fresh water bacteria (motile aeromonads and *Plesiomonas*) found in marine species of fish suggests that these fish were migrating between the St. Johns River estuary and its fresh water tributaries.

Water temperatures ranged from 12° to 33°C (Table 3). Lesioned fish were more commonly reported when water temperatures ranged between 12° to 22°C. The recovery of bacteria from fish at different temperatures was slightly higher below 14°C for the aeromonads, *P. putrefaciens*, and *V. anguillarum* than for other bacterial species.

Subsurface water samples were taken from a river dock and at mid-river during two separate fish collections within a week and were cultured for bacteria using the media and incubation conditions described earlier for fish tissue samples. Although direct enumeration of bacterial species was not performed, the results did show that *A. hydrophila* and *P. putrefaciens* were the most abundant bacterial species when water conditions in the estuary were 14°C and 2 ‰ salinity. *Vibrio alginolyticus* and *V. anguillarum* were also present but in significantly lower abundance within the same sample. At the second sampling, water conditions were 13.5°C and 16 ‰ salinity. *A. hydrophila* and *A. sobria* were moderately abundant, but *V. anguillarum*, *V. vulnificus*, and other unidentified vibrios were more populous. These findings have a direct correlation to the bacterial species recovered from the skin and lesions of fish collected at these times.

There is no indication that temperature and salinity are direct factors in the debilitation of fish; instead these conditions influence the types and numbers of bacterial flora in the water column at a given site and time within the estuary. Consequently, this may reflect the etiology of UDS organisms by the exposure of the healthy or compromised fish to the microbes present. The ability of these organisms to survive in adverse environments may be important in the advancement of UDS.

By comparing the data of bacteria recovered from UDS versus healthy fish (Table 4), it was found that *A. hydrophila, A. sobria, V. anquillarum,* and *V. cholerae* non-Ol were isolated at a significantly higher frequency in lesioned fish than in healthy fish. Although these rates may signify a potential pathogenic role for the bacteria, a single causative agent cannot be shown by this evidence. Instead, the bacteria may act as opportunists or secondary invaders to a compromised fish. This would explain the recovery of different pathogenic and opportunist bacteria within a single lesion. The ability to form ulcerative lesions in fish is a common characteristic for some of these bacteria, which makes it even more difficult to elucidate the strain(s) involved with UDS. Further examination of the internal organs may clarify each species' position as a pathogen or as part of the normal flora.

Table 1. Isolation of bacteria from various marine* and fresh water# fish.[a]

Bacterium	Fish												
	Bass#	Perch*	Menhaden*	G. shad#	Weakfish*	Catfish*#	Croaker*	Spot*	Fundulus*	Flounder*	Sunfish#	Mullet#	Other
Aeromonas hydrophila			5	9	9	1	3	1	2	4	1		3
Aeromonas salmonicida			2										
Aeromonas sobria			8	5	9	3	1	3		3	2		3
Corynebacterium sp.					2	1							
Pl. shigelloides			3	1		1						1	
Ps. fluorescens							1						
Ps. putida	1										1		
Ps. putrefaciens			5	2	9	1	2	1		1	1	1	1
Pseudomonas sp.			1		1					2	1		
Vibrio alginolyticus			1		1								
Vibrio anguillarum			2	1					4				1
Vibrio cholerae non01			3	2	3	1				2			2
Vibrio damsela					1	2	1		4				
Vibrio parahaemolyticus					2								
Vibrio spendidus					1								
Vibrio sp.			4		5	1	2			3			2

[a]The number of bacterial isolates from each type of fish is indicated for the period 5/7/86 to 4/28/88.

Table 2. Isolation of bacteria at different salinities.[a]

Bacterium	Salinity (‰)														
	22	19	16	14	13.5	8	5	4	3	2	1	0.8	0.7	0.05	0
Aeromonas hydrophila		1	2	1		1	4	1	1	9		2			12
Aeromonas salmonicida										2					
Aeromonas sobria	1		2	1			4	2	1	8		1	1	1	14
Corynebacterium sp.							2								
Pl. shigelloides															2
Ps. fluorescens							1								
Ps. putida															1
Ps. putrefaciens	1						3	1		8	1				7
Pseudomonas sp.			3							1					1
Vibrio alginolyticus			1				1								
Vibrio anguillarum		1	4							2	1				1
Vibrio cholerae non01			1			1		3		2				2	
Vibrio damsela		2	2	3				1							
Vibrio parahaemolyticus			1				1								
Vibrio splendidus										1					
Vibrio sp.			4			1	1	5		4					1

[a]Salinity was determined with a refractometer at fish collection site. The number of fish bacterial isolates at that salinity are indicated during the period 5/7/86 to 4/28/88.

Table 3. Isolation of bacteria at different temperatures.[a]

Bacterium	Temperature (°C)									
	33	32	29	28.5	28	27	25	24	23.5	22
Aeromonas hydrophila				2	1	1	1	1	3	1
Aeromonas sobria	1	4	1	1	1	1	1	3	1	
Corynebacterium sp.								1		
Pl. shigelloides								1		
Ps. fluorescens										
Ps. putida										
Ps. putrefaciens	1							1		
Pseudomonas sp.										
Vibrio alginolyticus										
Vibrio anguillarum						1				
Vibrio cholerae non01		2	2							1
Vibrio damsela			1			2	3			
Vibrio parahaemolyticus										
Vibrio splendidus		1								
Vibrio sp.	3	1								1

[a]Temperatures represent measurements of water at fish collection site. Number of fish bacterial isolates are indicated at that temperature during the period 5/7/86 to 4/28/88.

Table 3 (continued)

Bacterium	Temperature (°C)									
	21.5	21	19	18	17	16.8	14	13.5	13	12
Aeromonas hydrophila	1		1	4		2	8	1	4	2
Aeromonas sobria	1	1	2	3	1	1	5	2	2	4
Corynebacterium sp.										
Pl. shigelloides							1			
Ps. fluorescens				1						
Ps. putida								1		
Ps. putrefaciens	2	1	1	2			6	1	3	3
Pseudomonas sp.							1	3		1
Vibrio alginolyticus				1				1		
Vibrio anguillarum	1						2	1	1	
Vibrio cholerae non01			3							
Vibrio damsela										
Vibrio parahaemolyticus										
Vibrio splendidus										
Vibrio sp.			4	1			1	4	1	

[a]Temperatures represent measurements of water at fish collection site. Number of fish bacterial isolates are indicated at that temperature during the period 5/7/86 to 4/28/88.

Table 4. Bacterial isolates from UDS and healthy fish.[a]

Bacterium	Number isolates/Number fish (%)	
	UDS fish	Healthy fish
Acinetobacter calcoaceticus	1/42 (2%)	9/59 (15%)
Aeromonas hydrophila	19/42 (45%)	17/59 (29%)
Aeromonas sobria	20/42 (48%)	21/59 (36%)
Enterobacter sp.	1/42 (2%)	3/59 (5%)
Flavobacterium sp.	0/42 (0%)	7/59 (12%)
Plesiomonas shigelloides	4/42 (10%)	1/59 (2%)
Pseudomonas putrefaciens	11/42 (26%)	12/59 (20%)
Vibrio anguillarum	6/42 (14%)	3/59 (5%)
Vibrio cholerae non01	9/42 (21%)	2/59 (3%)
Vibrio damsela	5/42 (12%)	3/59 (5%)
Vibrio splendidus	0/42 (0%)	1/59 (2%)
Moraxella sp.	0/42 (0%)	4/59 (7%)
Gram positive cocci	6/42 (14%)	15/59 (25%)

[a]Representative bacteria isolated from fish collected from the St. Johns River, Lake George, and Tampa Bay from 5/7/86 to 4/28/88. Only fish with suspected UDS and healthy fish with no visible lesions are included. Fish with minor or moderate lesions are excluded.

Results from the blood samples provide additional evidence that the role of bacteria in UDS is not as the primary infectious agent. *Aeromonas sobria, A. hydrophila, P. shigelloides,* and *P. putrefaciens* were sometimes found in pure culture in the blood of UDS and non-ulcerated fish. It is not known whether the bacteria found in the blood of normal fish were due to transient infection or the early stages of UDS. Since lesioned fish did not always exhibit positive cultures, the existence of bacteria in the blood is probably secondary to the onset of disease or negative cultures may represent immune clearance by the host.

Oomycete fungi and UDS. A total of 43 fish were collected, with 39 of them having one or more lesions. Six of these lesions were positive for the presence of fungi. Of these six, three of them produced fungi on media plates. In all cases potential bacterial pathogens were isolated whereas the fungi were not isolated. The absence of fungi in ulcerative lesions strongly indicates that these microbes also played secondary or opportunistic roles in UDS. The fungi involved with UDS fish belonged to the Saprolegniales in the genera *Saprolegnia, Achlya,* and *Aphanomyces.* These fungi are normally found in fresh water systems but Dykstra et a1. (1986) described a salt tolerant *Aphanomyces* associated with UM in Atlantic menhaden. Various genera, *Aphanomyces, Pythium, Achlya,* and *Saprolegnia*, were found in the St. Johns River at salinities up to 12 ‰ (Lim and Te Strake, 1988). Stratification of these genera according to salinity may play a role in their involvement in UDS.

Enzymatic studies done by Alberts et al. (1989) using oomycetes isolated from the river and the lesions of fish showed that some of these strains elaborate hemolysin, DNAse, and proteinase that may be factors aiding the pathogenicity of these fungi.

DISCUSSION

The results of our study indicated that bacteria and fungi act as secondary or opportunistic invaders to the piscine host with UDS. The involvement of other biological primary pathogens, such as parasites and viruses, in UDS remains a question. Several other groups investigated various parameters of the river including heavy metal and hydrocarbon pollutants, nutrient loadings, sewage effluent from urbanized areas, and retrospective studies on temperature, salinity, rainfall, and past disease outbreaks.

The lower St. Johns River is a complex system that is subjected to the activities and products of a U.S. Navy station, pulp and paper mills, seafood processors, residential developments, agriculture, waste water and sewage plants, and boat marinas. Studies on the impact that these products may exert on the river's flora and fauna at these sites were only recently begun. Early reports (unpubl. data) indicate mercury and cadmium contamination in some of the tributaries and polynuclear aromatic hydrocarbons and domestic sewage contamination in other parts of the river. The environmental impact of these findings has not been fully explored, but the possibility of inducing a compromised immunity in fish at these sites is certainly possible.

LITERATURE CITED

Alberts, V.A., S.S. Kahn, D.V. Lim and D. Te Strake. 1989. Extracellular enzyme activity of some Saprolegniales from a Florida estuary. Mycologia, 81, 460-463.

Alderman, D.J. 1982. Fungal diseases of aquatic animals. In: "Microbial Diseases of Fish" (R.J. Roberts, ed.), pp. 189-242. Academic Press, New York, N.Y.

Austin, B. and D.A. Austin. 1987. "Bacterial Fish Pathogens: Diseases in Farmed and Wild Fish." Ellis Horwood Ltd, West Sussex, England.

Baumann, P. and R.H.W. Schubert. 1984. Family II. Vibrionaceae. In: "Bergey's Manual of Systemic Bacteriology", Vol. 1 (N.R. Krieg and J.G. Holt, eds.), pp. 516-550. Williams and Wilkins, Baltimore, Maryland.

Colwell, R.R. and D.J. Grimes. 1984. *Vibrio* diseases of marine fish populations. Helgoland. Meeresunt., 37, 265-287.

Dykstra, M.J., E.J. Noga, J.F. Levine, D.W. Moye and J.H. Hawkins. 1986. Characterization of the *Aphanomyces* species involved with ulcerative mycosis (UM) in menhaden. Mycologia, 78, 664-672.

Farmer, J.J., III, F.W. Hickman-Brenner and M.T. Kelley. 1985. *Vibrio*. In: "Manual of Clinical Microbiology," 4th ed. (E.H. Lennette, A. Balows, W.J. Hausler, Jr. and H.J. Shadomt, eds.). Amer. Soc. Microbiol., Washington, D.C.

Grier, H. and I. Quintro. 1987. Florida Department of Environmental Regulation Report, WM-164.

Hargis, W.J., Jr. 1986. Quantitative effects of marine diseases on fish and shellfish populations. In: "NOAA Technical Memorandum NMFS F/NWR - 16 Transactions of the Fiftieth North American Wildlife and Natural Resources Conference," pp. 627-631. U.S. Department of Commerce, Washington, D.C.

Lim, D.V. and D. Te Strake. 1988. Studies of pathogens causing ulcerative disease syndrome (UDS) in fish in the St. Johns River estuary. Florida Department of Environmental Regulation Final Report, SP 129.

Munro, A.L.S. 1982. The pathogenesis of bacterial diseases of fishes. In: "Microbial Diseases of Fish" (R.J. Roberts, ed.), pp. 131-148. Academic Press, New York.

Noga, E.J. and M.J. Dykstra. 1986. Oomycete fungi associated with ulcerative mycosis in menhaden, *Brevoorita tyrannus* Latrobe. J. Fish Dis., 9, 47-53.

Palleroni, N.J. 1984. Family I. Pseudomonadaceae. In: "Bergey's Manual of Systemic Bacteriology", Vol. 1. (N.R. Krieg and J.G. Holt, eds.), pp. 141-210. Williams and Wilkins, Baltimore, Maryland.

Schulz, D. 1980. Erythrodermatitis of carp: studies of the mode of infection. In: "Fish Diseases, Third COPRAQ Session" (W. Ahne, ed.), pp. 137-144. Springer-Verlag, Berlin, Heidelberg, and New York.

Te Strake, D. and D.V. Lim. 1986. Bacterial and fungal studies of ulcerative fish in the St. Johns River. Florida Department of Environmental Regulation Report, WM 138.

Wolke, R.E. 1975. Pathology of bacterial and fungal diseases affecting fish. In: "The Pathology of Fishes" (W.E. Ribelin and G. Migaki, eds.), pp. 33-116. University of Wisconsin Press, Madison, Wisconsin.

RELATIONSHIP AMONG PATHOGENIC *VIBRIO ANGUILLARUM* AND *VIBRIO TUBIASHII* WITH ENVIRONMENTAL VIBRIOS[1]

Belén Fouz, Ramón F. Conchas, Jorge Bolinches, Jesús L. Romalde, Juan L. Barja, and Alicia E. Toranzo

Departamento de Microbiología y Parasitología, Facultad de Biología, Universidad de Santiago. Santiago de Compostela 15706, Spain

ABSTRACT: Galicia (Northwestern Spain) is a very important environment for marine aquaculture. However, vibriosis caused by *V. anguillarum* and *V. tubiashii* is the most threatening bacterial disease limiting the production of marine fish and shellfish in our area (Toranzo et al., 1987; Lodeiros et al., 1987). The present work was carried out to establish the taxonomic, serological, and ecological relationship among isolates from diseased fish and oysters with similar estuarine vibrios.

In this study, we used 341 strains isolated from water, sediment, and phyto- and zooplankton, as well as diseased fish and oysters. The isolates were grouped, by numerical taxonomy, in 22 phena according with their relatedness (Fig. 2). Phena were classified by means of their characteristics and by the clustering with reference strains and data of Bryant et al. (1986). The serological assays were performed using the dot blot method as described by Cipriano et al. (1985).

The taxonomical and serological results indicate that although pathogenic *V. anguillarum* and *V. tubiashii* are differentiated homogeneous groups associated to disease of fish and oysters respectively, there are some very similar bacteria in the estuarine environment that may represent a phylogenetically related group. It is noteworthy that using only the classical biochemical tests, these strains, grouped in phena 3 to 14 (Fig. 2), are indistinguisable from the pathogenic *V. anguillarum* or *V. tubiashii* isolates. However, although taxonomically and serologically similar, the

[1]In memory of J. Bolinches, who died as a result of a car accident during the elaboration of the present work.

Pathology in Marine Science

strains isolated from diseased oysters and fish appeared very ecologically differentiated from the environmental vibrios (Fig. 3).

The results of this work raise the question about whether *V. anguillarum* and *V. tubiashii* are obligate or opportunistic pathogens. It may be possible that these species are saprophitic normal flora of the estuarine environments being capable to infect oysters and fish by some environmental or genetic changes.

Bryant, T.N., J.V. Lee, P.A. West and R.R. Colwell. 1986. J. Appl. Bacteriol., 61, 469-480.

Cipriano, R.C., J.B. Pyle, C.E. Starliper and S.W. Pyle. 1985. J. Wildlife Diseases, 21, 211-218.

Lodeiros, C., J. Bolinches, C.P. Dopazo and A.E. Toranzo. 1987. Aquaculture, 65, 15-29.

Toranzo, A.E., Y. Santos, M.L. Lemos, A. Ledo and J. Bolinches. 1987. Aquaculture, 67, 41-52.

INTRODUCTION

The genus *Vibrio* includes a great diversity of species that are common inhabitants of the aquatic environment where they are usually closely associated with many kinds of marine organisms. In addition, some of these *Vibrio* spp. are considered typical opportunistic pathogens for poikilothermic and homoiothermic aquatic animals. Galicia (Northwestern Spain) is a very important environment for marine aquaculture. However, vibriosis caused by *V. anguillarum* and *V. tubiashii* are the most threatening bacterial disease limiting the production of marine fish and shellfish in our area (Toranzo et al., 1987; Lodeiros et al., 1987).

Sørensen and Larsen (1986), by comparing a great number of strains of *V. anguillarum* isolated from diseased fish with environmental strains, established ten distinct serotypes (01 through 010). Serotypes 01 and 02 were the most associated with outbreaks of vibriosis (in cultured fish and feral fish, respectively), whereas the environmental strains belonged to the remaining serotypes or were not typable. However, data about the pathogenicity of those environmental isolates of *V. anguillarum* have not been reported.

Although until now most of the epizootics occurred in trout, salmon, and turbot reared in the northwest of Spain have been associated with the 01 and 02 serotypes of *V. anguillarum* (Toranzo et al., 1987), recently we have reported (Lupiani et al., 1989) that other members of the genus *Vibrio*, such as *V. splendidus* and *V. pelagius*, closely related (taxonomically and serologically) with *V. anguillarum* and *V. tubiashii*, have been implicated directly or indirectly with mortalities of turbot. Nevertheless, their importance as pathogens in marine aquaculture has yet to be determined.

At present, the majority of studies dealing with environmental vibrios have been focussed only in taxonomical (West et al., 1984, 1986; Bryant et al., 1986a, b) or in serologic (Sørensen and Larsen, 1986) aspects. Therefore, in order to clarify the pathobiological properties of these "*V. anguillarum*-like" organisms, in this work we establish the taxonomic, serological, and ecological relationship among isolates from diseased fish and oysters with estuarine vibrios.

MATERIALS AND METHODS

Bacterial strains. All the bacterial strains included in this study were isolated from "Ria de Arosa" (north west of Spain) (Fig. 1). Samples of subsurface water (5 m depth), bottom sediment, and phyto- and zooplankton, as well as from gills, skin, and gut of wild fish were collected in four stations of this estuary at the different seasons of the year. Some strains were also isolated from liver and kidney of diseased fish and from oyster larvae cultured in seafarms located in our area. The water distribution system of these rearing facilities was also analyzed. Samples were streaked on marine agar (MA) Z 2216 (Difco) plates and incubated at 22°C for a 3-day period.

For comparative purposes, several species from culture collections were included in the taxonomical and serological studies: *V. parahaemolyticus* ATCC 27969, *V. vulnificus* ATCC 27562, *V. alginolyticus* ATCC 17749, *V. proteolyticus* ATCC 15338, *V. aestuarianus* ATCC 5048, *V. costicola* CCM 2811, *V. harveyi* KN-96, *V. fischeri* NCMB 1274 and 1281, *V. cholerae* ATCC 14035, *V. anguillarum* ATCC 19264 and 14181, *V. tubiashii* ATCC 19106 and 19109, *V. diazotrophicus* ATCC 33466, *V. natriegens* ATCC 14048, *V. fluvialis* ATCC 33812, *V. nigrapulchritudo* ATCC 27043, *V. ordalii* ATCC 33509, *V. damsela* NCMB 2184, *V. logei* NCMB 1143, and *V. nereis*

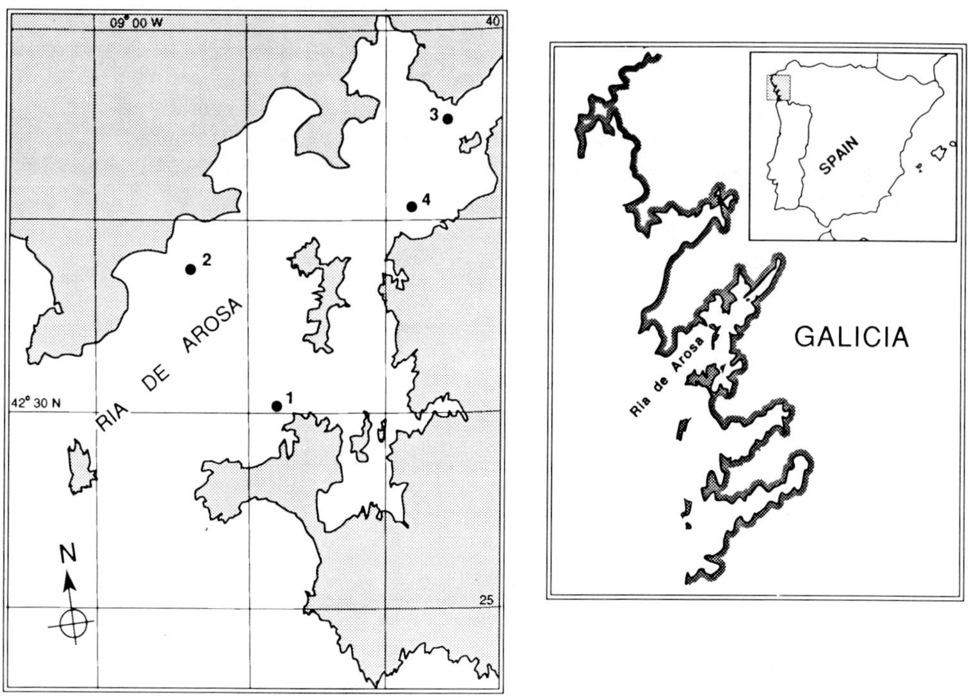

Fig. 1. Location of the sampling stations in the Ria de Arosa (NW Spain).

ATCC 25917. We have also included some non-*Vibrio* strains: *Aeromonas salmonicida* ATCC 14174, *Aeromonas hydrophila* ATCC 7966, *Photobacterium angustum* ATCC 25915, *Plesiomonas shigelloides* ATCC 14029, *Pseudomonas fluorescens* TIS, *Escherichia coli* ATCC 10536, *Yersinia ruckeri* NCMB 1316, *Deleya pacifica* ATCC 27122, *Deleya marina* ATCC 25374, *Salmonella typhimurium* LT-2, and *Staphylococcus aureus* ATCC 25923.

Characterization tests. In preliminary screening we selected the presumptive *Vibrio* spp. on the basis of Gram-negative stain, colony and cell morphology, motility, fermentation of glucose, growth on thiosulphate citrate bile sucrose (TCBS) agar (Oxoid), and sensitivity to the vibriostatic agent 0/129. The strains were routinely maintained on MA slants at 15°C and kept frozen at -70°C in marine broth (MB) with 15% glycerol (v/v).

All the *Vibrio* isolates were characterized using a total of 135 physiological and biochemical taxonomic tests. All of them, except the range of salinities and the growth on selective media, were carried out using a final concentration of 2% NaCl. The characterization tests, except motility, indol, Voges-Proskauer, nitrate, and methyl red, were performed by replica-plating using a Denley multipoint inoculator (Watt, 1979).

Twenty-one strains were inoculated in each plate; the swarming strains were tested separately. Plates were incubated at 22°C until reading.

Extracellular enzymes. Following the procedures of Colwell and Wiebe (1970), Furniss et al. (1978), Hsu et al. (1981), West et al. (1984, 1986), and Bryant et al. (1986a, b), the bacterial strains were tested for ability to hydrolyze esculin, gelatin, casein, starch, agar, Tween 20, Tween 80, lecitin, DNA, alginate, elastin, chitin, phenanthrene, tyrosine, xanthine, cellulose, and urea. The presence of sulfatase, phosphatase, and β-galactosidase (ONPG) (Burman and Ostensson, 1978) was also determined.

Fermentation of carbohydrates. According to the procedures of Burman and Ostensson (1978), strains were tested for acid production from D-melibiose, D-cellobiose, sucrose, lactose, maltose, D-trehalose, D-glucose, L-arabinose, D-mannose, D-galactose, D-fructose, D-mannitol, inositol, D-sorbitol, arbutin, and salicin.

Nutritional screening. Strains were screened for their ability to utilize the following compounds as sole carbon source (West et al., 1984; 1986): D-alanine, L-threonine, L-leucine, L-arginine, L-hydroxyproline, glycine, L-citrulline, putrescine, ethanolamine, D-glucosamine, arbutin, propionate, laurate, DL-3-hydroxybutyrate, valerate, heptanoate, acetate, DL-glycerate, malonate, succinate, D-gluconate, D-glucuronate, γ-aminobutyrate, α-ketoglutarate, D-galacturonate, D-mannitol, D-sorbitol, m-inositol, m-erythritol, glycerol, ethanol, D-glucose, D-xylose, D-fructose, L-arabinose, D-mannose, D-galactose, D-trehalose, D-cellobiose, sucrose, lactose, maltose, D-melibiose, adenine, guanine, cytosine, thymine, uracil, and vitamin assay casamino acids (Difco).

Growth on selective media and range of salinity and temperature. Strains were tested for the ability to grow on Simmons citrate agar (Difco), Agar CLED (Hispanagar), SS Agar (Difco), MacConkey Agar (Difco), TCBS (Oxoid), sodium dodecyl sulfate-polymyxin β-sucrose (SPS) (Kitaura et al., 1983), and trypticase soy agar-sucrose-bile-triphenyltetrazolium (TSAT) (Kourany, 1983). Ability to grow on iron limiting conditions was investigated by growth on M9 supplemented with ethylenediamine-di (o-hydroxyphenyl acetic acid) (EDDA) 20mM (M9-EDDA 20) and with EDDA 200mM (M9-EDDA 200) (Lemos et al., 1988). The salt tolerance and range of temperature of the Vibrio isolates was assayed on MA containing 0, 0.5, 2, 6, 8, and 10% (w/v) NaCl and at 4, 25, and 37°C.

Resistance to inhibitors and chemotherapeutants. Organisms were tested on MA to which were added a variety of antibacterial compounds: brilliant green (20 μg/ml), crystal violet (20 μg/ml), tellurite (5 μg/ml), methylene blue (100 μg/ml), methyl violet (2 μg/ml).

The drug sensitivity pattern of the *Vibrio* strains was also evaluated on MA plates containing the following antibiotics and chemotherapeutant agents: ampicillin (25 μg/ml), novobiocin (5 μg/ml), polymyxin B (12 μg/ml), tetracycline (12 μg/ml), penicillin (25 μg/ml), chloramphenicol (25 μg/ml), nitrofurantoin (100 μg/ml), streptomycin (12 μg/ml) and O/129 (150 μg/ml).

Additional examinations. As described by Bryant et al. (1986a, b), all the *Vibrio* strains were subjected to the following additional characterization tests: bioluminescence, haemolysis of sheep and human erytrocytes, Kanagawa phenomenum, swarming, production of indol, nitrate reduction, oxidase, H_2S production, methyl red, and Voges-Proskauer reactions together with the production of gas from glucose, Moeller L-arginine dihydrolase (ADH), L-lysine (LDC) and L-ornithine decarboxylase (ODC), and phenylalanine deaminase (Oxoid).

Physiological indices. Following Van Wanbeke and Bianchi (1985), we established 21 indices (Table 1), which were calculated by the formula:

$$\text{Index(i)} = (\Sigma + \text{ x } 100) / \text{n x } \Sigma\text{i}$$

where $\Sigma +$ indicates the positive responses in the i family for all the strains of the sample; n is the number of strains of the sample; and Σi is the number of characters tested in the family.

Numerical analysis. The strains isolated from all the samples together with the species from culture collections were subjected to numerical taxonomy. Similarities among them were calculated using the simple matching (S_{SM}) coefficient; dendogram was constructed by means of the unweighted pair-group method with arithmetic averages (UPGMA) (Sneath, 1972). A correspondence analysis was performed to show the association among phena using as numerical data the pattern of physiological indices of each phena.

The numerical analysis was performed in a UNIVAC 1100 computer after developing the appropriate programs (Bolinches et al., 1988).

Serological procedures.

Source of antigens and antisera. We have used as antigen for the immunization 11 strains that were selected from different phena after the numerical taxonomy studies.

The antisera were obtained as described by Sørensen and Larsen (1986). Briefly, rabbits were injected intravenously with formalin-killed cells twice weekly in consecutive doses of 0.2, 0.4, 0.8, and 1 ml (10^9 cells/ml). One week after the last injection, the rabbits were bled from the ear vein. The blood was allowed to clot and the sera was separated and stored at -30°C until used.

Dot blot assay (DBA). This serological method was performed basically as described by Cipriano et al. (1985). Live whole cells obtained from each strain were

Table 1. Physiological indices used in the present study.

	INDEX	TESTS COMPRISED
Haemolysis	HEM	Sheep, Human.
Resistance to:		
Antibiotics and Chemotherapeutants	ANT	Ampicillin, Streptomycin, Novobiocin, Polymyxin, Tetracycline, Penicillin, Chloramphenicol, Nitrofurantoin.
Inhibitors	INH	Brilliant Green, Tellurite, Methylene blue, Methyl Violet, O/129.
Growth on:		
Selective media	SEL	TSAT, TCBS, CLED, SS, MacConkey, SPS.
Iron limitation	LIM	M9, EDDA 20, EDDA 200.
Range of salinity	SAL	0, 0.5, 2, 6, 8, and 10% NaCl.
Range of temperature	TEM	4, 37, and 42°C.
Hydrolysis of:		
Proteins	PRO	Casein, Gelatin, Elastin.
Lipids	LIP	Tween 20, Lecitin.
Polysaccharides	POL	Starch, Esculin, Alginate, Agar, Chitin.
Other compounds	OTH	Xanthine, Tyrosine, Urea, Phosphate, Sulphate, Galactosidase.
Acid production from:		
Disaccharides	ADI	Sucrose, Lactose, Maltose, Trehalose.
Monosaccharides	AMO	Glucose, Arabinose, Mannose, Galactose, Fructose.
Organic alcohols	AAL	Inositol, Sorbitol, Mannitol.
Utilization of:		
Aminoacids	UAA	Alanine, Leucine, Hydroxyproline, Citrulline.
Organic acids	UOA	Malonate, Gluconate, Glucuronate, Galacturonate.
Inorganic acids	UIA	Propionate, Laurate.
Alcohols	UAL	Sorbitol, Glycerol, Inositol, Ethanol.
Monosaccharides	UMO	Fructose, Galactose.
Disaccharides	UDI	Cellobiose, Maltose, Melibiose.

dotted onto nitrocellulose. Subsequently, it was blocked in Tris-buffered saline plus 0.05% (v/v) Tween 20 and exposed to the antisera at a routine dilution (one 10-fold dilution under the titer in DBA). Goat antirabbit IgG-horseradish peroxidase (Sigma) was used as conjugate and H_2O_2 together with 4-chloro-1-napthol were used to develop the reaction. This is a sensitive serological method that allows processing a great number of samples while using very low quantities of antisera (Bolinches et al., 1989).

RESULTS AND DISCUSSION

Grouping of strains by numerical taxonomy. We have selected 341 vibrios that include representative strains of the *Vibrio* community of the northwestern Spain estuary, as well as some isolates from diseased fish and shellfish in our area. All the strains were subjected to numerical analysis and grouped in 22 phena (Fig. 2).

According to the characteristics of the phena and by clustering with reference strains (see Materials and Methods), we can establish five defined species and other unidentified clusters: Phena I and II correspond with *V. anguillarum* and *V. tubiashii*, respectively. Phena III to XIV are homogeneous groups composed of related vibrios. Phenon XV contains strains of the halophilic *V. costicola* and phena XVI and XVII were composed by *V. fisheri*.

The most separate phenon was *V. alginolyticus* (phenon XXII), followed by four heterogeneous unidentified phena (XVIII, XIX, XX, and XXI).

In the most recent numerical study of vibrios performed by Bryant et al. (1986a, b), it was found that *V. anguillarum* and *V. tubiashii* had an internal similarity of 90% and, in addition, they clustered together at a level of 81% with the phena of related organisms composed of "*V. anguillarum*-like", *V. pelagius*, and *V. splendidus* biotypes I and II. The phenon of "*V. anguillarum*-like", proposed by these authors, differed from *V. anguillarum* in the Voges-Proskauer reaction, hydrolysis of elastin, and production of acid from arabinose, and differed from *V. tubiashii* in the inability to degrade xanthine, to utilize melibiose, and to grow at 4°C.

Our results are very similar, since we have obtained defined phena of environmental *Vibrios* spp. (III to XIV) differing from *V. anguillarum* and *V. tubiashii* in the characteristics cited above. Therefore, these phena can be considered formed by the same species of vibrios described by Bryant et al. (1986a). Interestingly, using only the classical biochemical tests, the strains grouped in phena III to XIV are indistinguishable from the pathogenic *V. anguillarum* and *V. tubiashii* isolates, which support the similarity of 83% and 77%, respectively, found in the numerical taxonomy studies (Fig. 2). All these findings allow us to consider that these environmental vibrios can represent phenotypic variants of a major and unique group that should be named "*V. anguillarum*-like".

Serological comparisons. Because of the high biochemical similarity detected among *V. anguillarum, V. tubiashi,* and the group of vibrios designated under the term "*V. anguillarum*-like", only the use of a serological method, such as DBA, allows us to determine the relationship and the distribution of the serotypes in the aquatic environment.

We have previously demonstrated (Bolinches et al., 1989) that whereas the serotype 01 of *V. anguillarum* constitutes a homogeneous group, within the serotype

02 different patterns of serological reactions were detected. The serological heterogeneity observed in this 02 serotype was also detected by Rasmussen (1987a).

In the present study we have found strong cross-reactions between the pathogenic *V. tubiashii* and *V. anguillarum* serotype 02 but not with the serotype 01 (Table 2). Interestingly, the antisera raised against "*V. anguillarum*-like" strains from each of the phena III to XIV displayed cross-reactions with all the strains included in these clusters of related vibrios as well as with *V. anguillarum* 02 and *V. tubiashii*. The *V. fischeri* antisera was less specific, because it cross-reacted with all the strains.

It is known that strains with different thermostable somatic O antigens can harbor related thermolabile capsular and flagellar antigens (and vice versa). In fact, it has been reported that the serotype 02 of *V. anguillarum* possess at least two capsular antigens, some of which are shared by some other *Vibrio* and *Aeromonas* species (Rasmussen, 1987b; Tajima et al., 1987). Therefore, these findings can explain the serological relationship detected in our study among *V. anguillarum* 02, *V. tubiashii*, and "*V. anguillarum*-like" organisms, which include mainly *V. pelagius* and *V. spendidus*.

Ecophysiological studies. Correspondence analysis was carried out to determine the possible ecophysiological relationship among the pathogenic *V. anguillarum* and *V. tubiashii* with the phena of environmental vibrios. Figure 3 shows a simplified spatial relation between phena and physiological indices where only two axes are represented: axis I is related with the level of resistance to environmental fluctuations such as temperature and salinity, while axis II can be associated with nutritional versatility. The situation of the phena could indicate some difference or affinity in the colonization of inanimated surfaces and host-specificity.

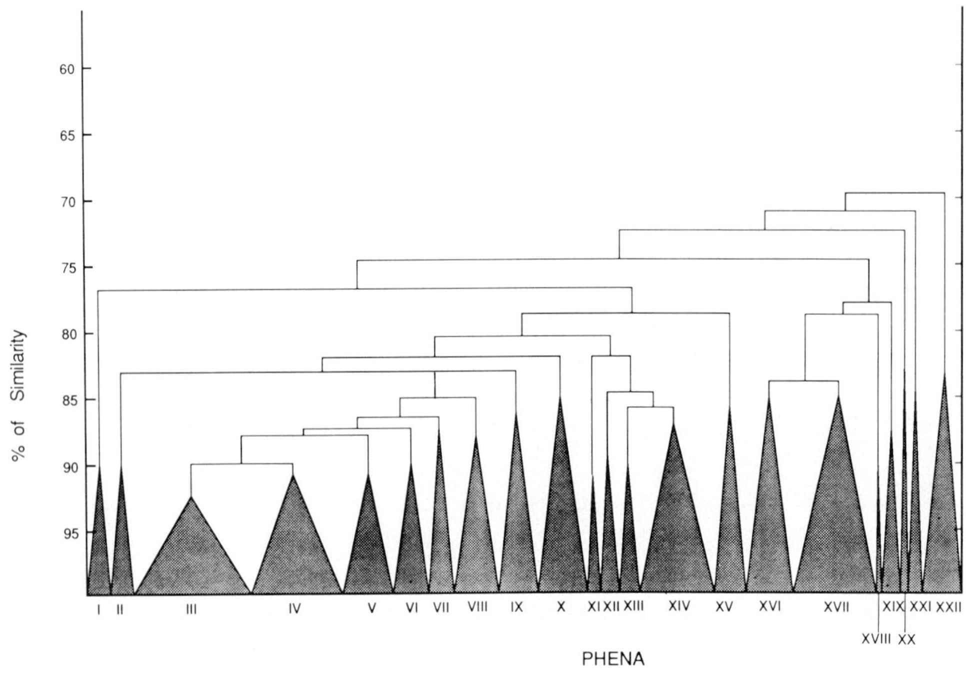

Fig. 2. Simplified dendogram of Vibrio strains based on simple matching coefficients and unweighted pair-group method with arithmetic averages (UPGMA).

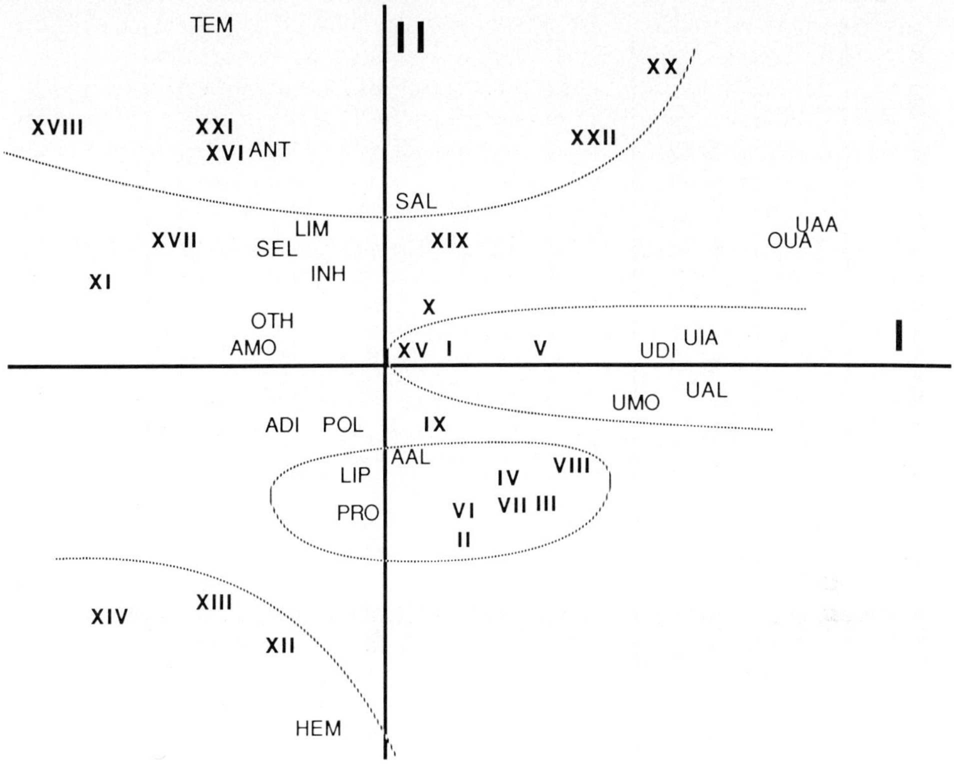

Fig. 3. Correspondence analysis showing the correlation between different phena and the physiological and nutritional characteristics.

We find that *V. anguillarum* phenon I appears closely associated to phenon V and with the utilization as sole carbon source of a variety of compounds like: monosaccharides, disaccharides, alcohols, and inorganic acids.

On the other hand, phenon II of *V. tubiashii* formed a strongly related group with phena III, IV, VI, VII, and VIII of "*V. anguillarum*-like", all associated with the ability to hydrolyze proteins and lipids and with acid production from polyalcohols.

Interestingly, phena I of *V. anguillarum* and XV of *V. costicola* were also related. This relation can be due to the halophilic characteristics of both species.

In addition, bacterial isolates from samples taken in winter and summer were, in general, more versatile than those isolated in spring and autumn (data not shown), which are periods of phytoplanktonic blooms and production of organic matter. These results are in agreement with the studies of Bell (1984) who found that bacteria adapted to low nutrient environments present a major nutritional versatility. Nevertheless, within the *Vibrio* community of each season coexist phena with differences in physiological and nutritional versatility promoting, surely, the niche segregation.

Table 2. Strong reactions (●) and cross-reactions (+) displayed by strains from the different phena against eleven antisera by the Dot Blot Assay.

Bacterial Antigens		Source of antisera										
STRAIN	PHENA	TM14 $(I)^{a}$	RV22 (I)	EX1 (II)	12F4 (III)	14A2 (III)	4IR2 (IV)	34Z2 (V)	33Z1 (V)	34A5 (XI)	4IF1 (XVI)	41S5 (XXII)
V. anguillarum												
TM14(01)[b]	I	●	–	–	–	–	–	–	–	–	+	–
V. anguillarum												
RV22(02)	I	–	●	+	+	+	+	+	+	+	+	–
V. tubiashii EX1	II	–	+	●	+	+	+	+	+	+	+	–
V. anguillarum-Like												
12F4	III	–	–	–	●	●	+	+	+	+	+	–
4IR2	IV	–	–	–	●	●	●	+	+	●	+	–
34Z2	V	–	–	–	+	+	+	●	●	+	+	–
34A3	VI	–	–	–	+	●	+	+	+	●	+	–
31A4	VII	–	–	–	+	+	+	+	+	+	+	–
33A1	VIII	–	–	–	+	+	+	+	+	+	+	–
52Z1	IX	–	–	–	+	●	+	+	●	+	+	–
42Z5	X	–	–	–	+	●	+	+	+	+	+	–

[a] Phena
[b] Serotype

(continued)

Table 2 (continued)

Bacterial Antigens		Source of antisera										
STRAIN	PHENA	TM14 (I)[a]	RV22 (I)	EX1 (II)	12F4 (III)	14A2 (III)	4IR2 (IV)	34Z2 (V)	33Z1 (V)	34A5 (XI)	4IF1 (XVI)	41S5 (XXII)
34A5	XI	–	–	–	+	+	+	+	+	●	+	–
32A1	XII	–	–	–	+	+	+	+	+	+	+	–
32Z4	XIII	–	–	–	+	+	+	+	+	+	+	–
32S1	XIV	–	–	–	+	●	+	●	+	+	+	–
V. costicola 11F1	XV	–	–	–	–	–	–	–	–	–	+	–
V. fisheri 4IF1	XVI	–	–	–	+	+	+	+	+	+	●	–
V. fisheri 31Z2	XVII	–	–	–	+	+	+	+	+	+	+	–
Unidentified *Vibrio* spp.												
43A5	XVIII	–	–	–	–	–	–	–	–	–	+	–
51Z2	XIX	–	–	–	–	–	–	–	–	–	+	–
33A7	XX	–	–	–	–	–	–	–	–	–	+	–
54Z4	XXI	–	–	–	–	–	–	–	–	–	+	–
V. alginolyticus 41S5	XXII	–	–	–	–	–	–	–	–	–	+	●

[a] Phena
[b] Serotype

CONCLUSIONS

The taxonomical, serological, and ecophysiological relationships among *Vibrio anguillarum, V. tubiashii,* and the group of environmental strains designated as "*V. anguillarum*-like" indicate that although *V. anguillarum* and *V. tubiashii* are differentiated homogeneous groups pathogenic for fish and oysters, respectively, there are some bacteria in the estuarine environment that may represent a phylogenetically closely related group.

The findings of this study raise the question about whether *V. anguillarum* and *V. tubiashii* are obligate or opportunistic pathogens and if the environmental vibrios represent a potential risk in marine aquaculture. It may be possible that these species are normal saprophytic microflora of the estuarine environments capable of infecting oysters and fish in special conditions derived from environmental or genetic changes.

ACKNOWLEDGMENTS

This work was supported by grant PB87-1027 from the Dirección General de Investigación Científica y Técnica (DGICYT), and grant AQ-0018/84 from the Comisión Asesora de Investigación Científica y Técnica (CAICYT), Ministerio de Educación y Ciencia, Spain.

J.L. Romalde thanks the Ministerio de Educación y Ciencia, Spain, for a Research Fellowship.

The authors acknowledge Dr. Javier Barja for his contribution in the presentation of the data.

LITERATURE CITED

Bell, W.H. 1984. Bacterial adaptation to low nutrient conditions as studied with algal extracellular products. Microb. Ecol., 10, 217-230.

Bolinches, J., M.L. Lemos and J.L. Barja. 1988. Population dynamics of heterotrophic bacterial communities associated with *Fucus vesiculosus* and *Ulva rigida* in an estuary. Microb. Ecol., 15, 345-357.

Bolinches, J., B. Fouz, M. Cambra, L.J. Larsen and A.E. Toranzo. 1989. Serological relationship within *V. anguillarum* strains. J. Aquat. Anim. Health, 1, in press.

Bryant, T.N., J.V. Lee, P.A. West and R.R. Colwell. 1986a. Numerical classification of species of *Vibrio* and related genera. J. Appl. Bacteriol., 61, 437-467.

Bryant, T.N., J.V. Lee, P.A. West and R.R. Colwell. 1986b. A probability matrix for the identification of species of *Vibrio* and related genera. J. Appl. Bacteriol., 61, 469-480.

Burman, L.G. and R. Ostenson. 1978. Time- and media-saving testing and identification of microorganisms by multipoint inoculation on undivided agar plates. J. Clin. Microbiol., 8, 219-227.

Cipriano, R.C., J.B. Pyle, C.E. Starliper and J.B. Pyle. 1985. Detection of *Vibrio anguillarum* antigen by the dot blot assay. J. Wild. Dis., 21, 211- 218.

Colwell, R.R. and W.S. Wiebe. 1970. "Core" characteristics for use in classifying aerobic, heterotrophic bacteria by numerical taxonomy. Bull. G. Acad. Sci., 28, 165-185.

Furniss, A.L., J.V. Lee and T.J. Donovan. 1978. In: "The Vibrios." Public Health Laboratory Service Monograph Series No. 11. Public Health Laboratory Service, London.

Hsu, T.C., W.D. Waltman and E.B. Shotts. 1981. Correlation of extracellular enzymatic activity and biochemical characteristics with regard to virulence of *Aeromonas hydrophila*. Develop. Biol. Standard., 49, 101-111.

Kitaura, T., S.Doke, I. Azuma, M. Imaida, K. Miyano, K. Harada and E. Yabuuchi. 1983. Halo production by sulfatase activity of *V. vulnificus* and *V. cholerae* 01 on a new selective sodium dodecyl sulfate-containing agar medium: a screening marker in environmental surveillance. FEMS Microbiol. Lett., 17, 205-209.

Kourany, M. 1983. Medium for isolation and differentiation of *Vibrio parahaemolyticus* and *Vibrio alginolyticus*. Appl. Environ. Microbiol., 45, 310-312.

Lemos, M.L., P. Salinas, A.E. Toranzo, J.L. Barja and J.H. Crosa. 1988. A chromosome-mediated iron uptake system in pathogenic strains of *Vibrio anguillarum*. J. Bacteriol., 170, 1920-1924.

Lodeiros, C., J. Bolinches, C.P. Dopazo and A.E. Toranzo. 1987. Bacillary necrosis in hatcheries of *Ostrea edulis* in Spain. Aquaculture, 65, 15-29.

Lupiani, B., C.P. Dopazo, A. Ledo, B. Fouz, J.L. Barja, F.M. Hetrick and A.E. Toranzo. 1989. A new syndrome of mixed bacterial and viral etiology in cultured turbot *Scophthalmus maximus*, l. J. Aquat. Anim. Health, 1, in press.

Rasmussen, H.B. 1987a. Subgrouping of lipopolysaccharide antigens from *Vibrio anguillarum* serogroup 02 by immunoelectrophoretic analyses. Curr. Microbiol., 16, 39-42.

Rasmussen, H.B. 1987b. Evidence for two new *Vibrio anguillarum* K antigen. Curr. Microbiol., 16, 105-107.

Sneath, P.H.A. 1972. Computer taxonomy. In: "Methods in Microbiology," Vol. 7A (J.R. Norris and D.W. Ribbons, eds.), pp. 29-98. Academic Press, London.

Sørensen, U.B. and J.L. Larsen. 1986. Serotyping of *Vibrio anguillarum*. Appl. Environ. Microbiol., 51, 593-597.

Tajima, K., Y. Ezura and T. Kimura. 1987. The possible use of a thermolabile antigen in detection of *Vibrio anguillarum*. Fish Pathol., 22, 237-242.

Toranzo, A.E., Y. Santos, M.L. Lemos, A. Ledo and J. Bolinches. 1987. Homology of *Vibrio anguillarum* strains causing epizootics in turbot, salmon and trout reared on Atlantic coast of Spain. Aquaculture, 67, 41-52.

Van Wanbeke, F. and M.A. Bianchi. 1985. Bacterial biomass production and ammonium regeneration in mediterranean seawater supplemented with aminoacids. 2. Nitrogens flux through heterotrophic microplankton food chain. Mar. Ecol. Prog. Ser., 23, 117-128.

Watt, P.R. 1979. An improved automatic multipoint inoculator. In: "Identification methods for microbiologists," 2nd ed. (F.A. Skinner and D.W. Lovelock, eds.), pp. 297-304. Academic Press, London.

West, P.A., G.C. Okpokwasili, P.R. Brayton, D.J. Grimes and R.R. Colwell. 1984. Numerical taxonomy of phenanthrene-degrading bacteria isolated from the Chesapeake Bay. Appl. Environ. Microbiol., 48, 988-993.

West, P.A., P.R. Brayton, T.N. Bryant and R.R. Colwell. 1986. Numerical taxonomy of vibrios isolated from aquatic environments. Int. J. Syst. Bacteriol., 36, 531-543.

ASSOCIATION OF A *MORAXELLA* SP. AND A REO-LIKE VIRUS WITH MORTALITIES OF STRIPED BASS, *MORONE SAXATILIS*

Ana Baya[1], Alicia E. Toranzo[2], Soledad Núñez[2], Juan L. Barja[2], and Frank M. Hetrick[1]

[1]Department of Microbiology, University of Maryland, College Park, Maryland 20742, USA
[2]Departamento de Microbioligía y Parasitología, Facultad de Biología, Universidad de Santiago de Compostela, 15706, Spain

ABSTRACT: During winter 1987, mortalities occurred in juvenile striped bass (*Morone saxatilis*) in the discharge channel of a power plant located on the Potomac River, Maryland, USA. Affected fish showed large hemorrhagic lesions along the dorso-lateral portions of the body where scales were missing. Hemorrhages were found in the swim bladder and the liver was pale, enlarged, and mottled. Large amounts of membranous material were seen connecting the liver and body wall. The gills of all fish examined were invaded by the parasites *Trichodina* and *Ergasilus*.

From some, but not all of the moribund striped bass, a bacterium was recovered in pure culture from all the organs and lesions. The isolate was a Gram-negative, non-motile short rod (0.8-1 x 1-1.3 μm) giving negative reactions in the majority of biochemical tests conducted. On the basis of a positive oxidase test and sensitivity to penicillin, it most likely belongs to the Moraxella-Acinetobacter group of organisms. These organisms are common inhabitants of fresh and marine waters and are part of the normal flora of various fish tissues and can invade and cause pathological effects in fish when stressed or when subjected to an environmental insult.

In addition, a viral agent was isolated from several fish, which from its characteristics appears to be a reo-like virus. This is but another of an increasing list of such agents that have been isolated from aquatic animals in recent years. Because of their potential importance for the aquaculture of striped bass, transmissions studies need to be conducted with both agents to determine if they are primary pathogens or whether they act synergistically.

INTRODUCTION

Previous epizootics in cultured striped bass (*Morone saxatilis*) have incriminated *Vibrio anguillarum* (Toranzo et al., 1983) or Birnaviruses (Schutz et al., 1984) as

Pathology in Marine Science

potential causes. In the present report we describe a new pathological syndrome in adult striped bass from which both a *Moraxella* sp. and a Reo-like virus have been isolated as possible etiological agents.

To date, the major aerobic Gram-negative pathogens with demonstrated implications in aquaculture belong to the genera *Cytophaga, Flexibacter, Pseudomonas, Vibrio,* and *Aeromonas* (Austin and Austin, 1987). However, unreactive oxidative rods resembling *Acinetobacter* or *Moraxella* species have been referred to as pathogens for Atlantic salmon in Norway (Roald and Håstein, 1980).

On the other hand, several Reo-like viruses have been isolated from fish and studied with some detail: golden shiner Reovirus, GSV (Plumb et al., 1979); chum salmon Reovirus, CSV (Winton et al., 1981); channel catfish Reovirus, CRV (Amend et al., 1984); grass carp Reovirus, GRV (Chen and Jiang, 1984); tench and chub Reovirus (Ahne and Kolbl, 1987); and more recently, the turbot Reovirus, TRV (Lupiani et al., 1989). In general, low mortality rates were associated with these infections and few experimental studies have been conducted to assess the viral roles in the disease process.

All the above described oxidative bacterial groups have a widespread occurrence in fresh and marine waters (Austin, 1982) and, in addition, they form part of the normal microflora of fish (Horsley, 1973; Toranzo et al., 1985). Therefore, any biological or environmental stress may lead to the start of a disease process where, in some occasions, different microorganisms are involved. Thus, it is sometimes difficult to determine which is the primary invader.

In this study we carried out a characterization of the bacterial and viral agents associated with mortalities of wild striped bass.

Description of the pathologic syndrome.

Location and conditions of the epizootic. Mortalities occurred during winter of 1987-88 in adult wild striped bass (*Morone saxatilis*) (average size 30 cm) in the discharge channel of a power plant located on the Potomac River, Maryland, U.S.A. The water temperature ranged between 0°C (during a plant shut down) and 8°C. Of 73 fish caught, 11 showing external hemorrhagic lesions were processed for microbiological analysis.

Clinical symptoms of diseased fish. Affected fish had large hemorrhagic lesions along the dorso-lateral portions of the body where the scales were missing. The gills of all fish examined were heavily infected with the parasites *Trichodina* sp. (more than 100 per gill lamella) and *Ergasilus* sp. (more than 50 per gill lamella). Internal examination showed a pale liver, which was enlarged and mottled. Hemorrhages were found in the swim bladder, and large amounts of membraneous material were observed connecting the liver and body wall.

MATERIALS AND METHODS

Bacteriological analysis. For bacterial isolation, samples taken from kidney, liver, and pancreas of moribund striped bass were inoculated simultaneously into different media: trypticase soy agar (TSA) and broth (TSB); Cytophaga agar and broth; and brain heart infusion agar and broth. Duplicate cultures were incubated at 25°C under aerobic and anaerobic conditions.

Pure cultures derived from isolated colonies were subjected to taxonomic analysis using a battery of morphological, biochemical, and physiological conventional plate and tube tests (Toranzo et al., 1985) as well as miniaturized API systems (API 20E, API 20B, and API -20EC) (Analytab Products, Planview, New York).

Bacterial identification was made using the criteria of Amos (1985), Austin and Austin (1987), and Bergey's Manual of Systematic Bacteriology (Bøvre, 1984).

Drug sensitivity of the bacterial isolates was assayed by the disc diffusion method on Mueller-Hinton Agar (Oxoid) using the following drugs (μg/disc): penicillin (10), erythromycin (15), chloramphenicol (30), kanamycin (30), tetracycline (30), novobiocin (5), nalidixic acid (30), oxolinic acid (2), nitrofurantoin (300), trimethroprim sulphamethoxazole (25), and pteridine (O/129) (150).

Slide agglutination tests were conducted to examine the possible serological relationship of our isolate with other oxidative fish pathogenic strains (*Acinetobacter, Flexibacter,* and *Pseudomonas* spp.). The reactions were performed as previously described (Toranzo et al., 1987) using whole cells and the thermostable somatic "O" antigens, which were obtained by heating the respective cell suspensions (10% v/v in PBS) at 100°C for 1 hr.

Virological assays. Spleen, kidney, liver, and skin tissues were removed from the fish, pooled and processed following standard virological procedures (Amos, 1985). From some fish, internal organs were analyzed separately.

Chinnok salmon embryo (CHSE-214) cells incubated at 15°C, fathead minnow (FHM), and epithelioma papillosum cyprini (EPC) cell lines maintained at 25°C, were utilized for virus isolation attempts. All cell types were routinely cultivated with Eagle's minimal essential medium (MEM) with Earle's salts supplemented with 10% newborn calf serum (NCS). Samples were inoculated onto the different cell monolayers, which were incubated at 15°C and examined daily for the appearance of cytopathic effects (CPE).

To determine if the virus was enveloped and the nature of its genome, the effects of chloroform and 5-bromo-2-deoxyuridine (BDU) were determined as previously described (Lupiani et al., 1989). Neutralization tests were conducted using infectious pancreatic necrosis virus (IPNV) antisera.

Virus morphology was determined by electron microscopy of negatively stained (phosphotungstic acid) preparations of partially purified virus. The micrographs were taken with a JEOL CXII electron microscope.

Virulence tests. The bacterial isolates were tested for pathogenicity in fingerling rainbow trout and striped bass (8-10 cm) maintained in fresh water with aeration. Fish were inoculated intraperitoneally with bacterial doses ranging from 10^3 to 10^7 cells as previously described (Toranzo et al., 1983). Mortalities were recorded daily during a 10-day period, and the degree of virulence was expressed as 50% lethal dose (LD_{50}) calculated by the Reed and Muench (1938) method.

RESULTS

Bacterial isolation and identification. From some, but not all of the moribund striped bass, a bacterium was recovered in pure culture from all of the organs and lesions. The organism produced round, raised, translucent, and mucoid colonies of about 1.5 mm in diameter within 48 hr at 22°C on TSA. The isolate was a nonmotile, Gram-negative, short rod (0.8-1 x 1.3 μm) often in pairs and exhibiting bipolar bodies (Fig. 1).

The bacterium was oxidase and catalase positive, non-fermentative, and displayed inability to produce acid from the majority of the carbohydrates tested. Similarly, the bacteria did not exhibit proteolytic activities (Table 1). All these results allowed us to place the bacterial isolate from striped bass (strain SB-9) into the Neisseriaceae family.

Our isolate is closely related to the *Acinetobacter*-like bacterium described by Roald and Håstein (1980) in Atlantic salmon, which differed from the SB-9 strain only in its capacity to produce acid from maltose. In addition, both isolates failed to lyse salmon and striped bass erythrocytes and displayed the same drug sensitivity pattern, being resistant only to the vibriostatic agent O/129 (Table 1). Taking as important phenotypic traits, the sensitivity to penicillin and the positive oxidase reaction, both strains can be classified as belonging to the genus *Moraxella*.

Agglutination reactions using whole cells and the somatic O antigens revealed that the antiserum raised against the striped bass isolate failed to react with the Atlantic salmon strain.

Virulence assays demonstrated that our *Moraxella* sp. strain was pathogenic for striped bass and rainbow trout with a mean LD_{50} of 10^5-10^6 organisms. The inoculated strain was recovered as pure culture from the internal organs of all dead fish.

Virus isolation and characterization. A viral agent was isolated in CHSE-214 monolayers from only two of the fish sampled. The agent appears to be a member of the *Reovirus-Rotavirus* group, based mainly on its cythopathology (formation of giant syncytia), its size, morphology by electron microscopy (Fig. 2) (icosahedral particles of 70-75 nm in diameter), and its insensitivity to chloroform (no envelope). The exposure of the cells to bromo-deoxyuridine (BDU) did not affect viral replication, indicating the presence of an RNA genome. Moreover, the virus from striped bass was not neutralized by the antisera against IPNV. The agent is tentatively named striped bass reovirus (SRV).

DISCUSSION

Two different microorganisms were isolated as possible etiological agents causing mortalities in striped bass: an unreactive and aerobic bacterial strain (SB-9) tentatively assigned to the genus *Moraxella*, and a virus (SRV) belonging to the Reoviridae family. Interestingly, a similar bacterial and viral association was recently detected in diseased turbot in northwestern Spain where *Vibrio spendidus* and a Reo-like virus were implicated in the pathological process (Lupiani et al., 1989).

Although the bacterial isolate from striped bass has a high degree of resemblance with the strain isolated from Atlantic salmon by Roald and Håstein (1980) in Norway (Table 1), the agglutination assays indicated the absence of a serological relationship between both strains. In addition, the Atlantic salmon isolate exhibited a higher psycrophilic nature and a lower growth rate than the striped bass strain (data not shown).

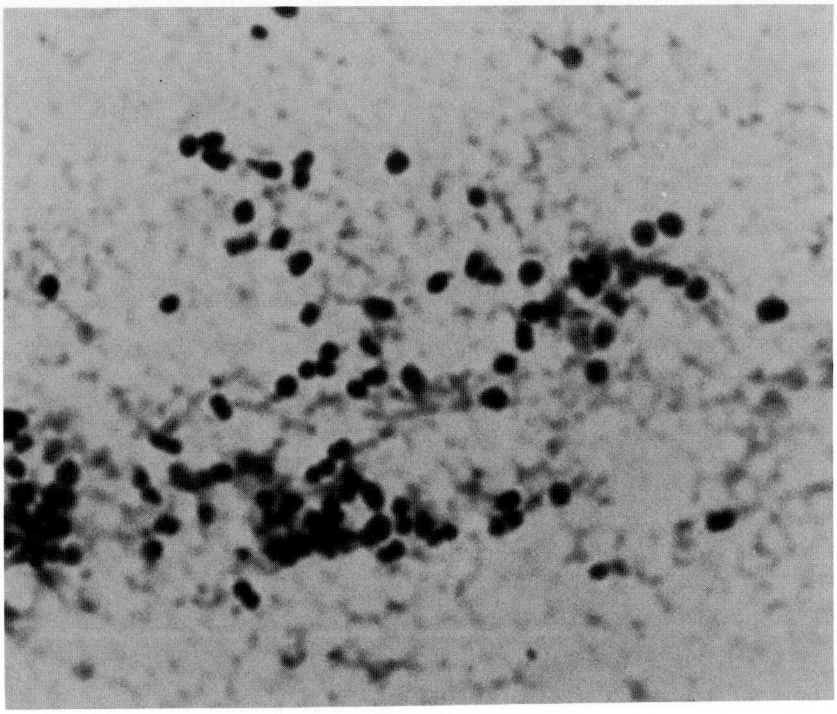

Fig. 1. Gram stain of the Moraxella sp (strain SB-9) isolated from striped bass. Short rods often in pairs with a mean size of 0.8-1 x 1.3 μm.

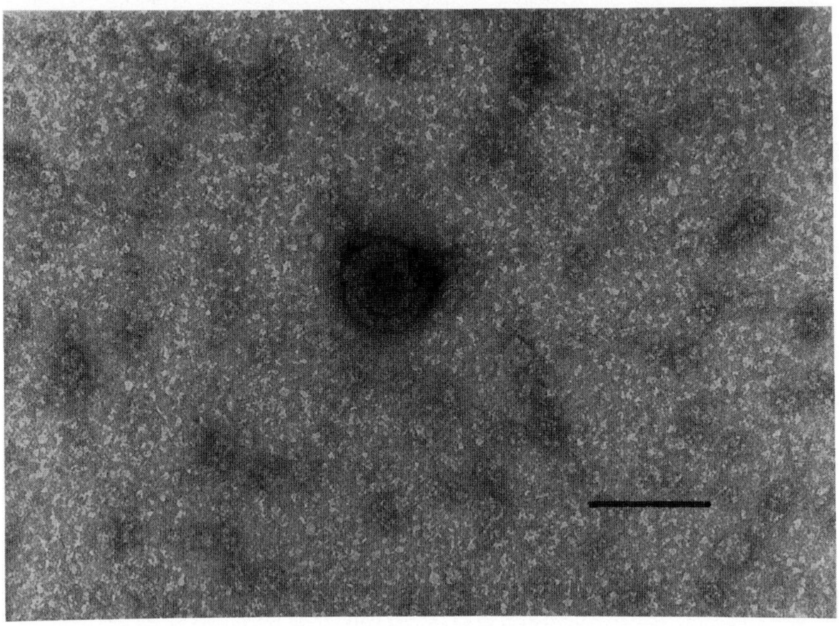

Fig. 2. Electron micrograph of the striped bass reovirus (SRV) showing icosahedral particles of 70-75 nm diameter with the typical double capsid. Bar = 100 nm.

Table 1.Comparison of the characteristics of the presumptive *Moraxella* sp. (strain SB-9) isolated from striped bass with those exhibited by the isolate from Atlantic salmon.

CHARACTER	Strains isolated from	
	Striped bass (present study)	Atlantic salmon (Roald and Håstein, 1980)
Gram	–	–
Oxidase	+	+
Catalase	+	+
Motility	–	–
Indole	–	–
Voges-Proskauer	–	–
Citrate (Simmons)	–	–
Nitrate reduction	–	–
H_2S production	–	–
Arginine dihydrolase	–	–
Lysine decarboxiylase	–	–
Ornithine decarboxylase	–	–
Acid from		
Glucose	–	–
Maltose	–	+
Lactose	–	–
Mannose	(+)	+
Galactose	(+)	+
Sucrose	–	–
Arabinose	–	–
Amygdaline	–	–
Rhamnose	–	–
Salicin	–	–
Mannitol	–	–
Sorbitol	–	–
Inositol	–	–
β-galactosidase	–	–
Urease	–	–
Gelatinase	–	–
Hemolysis of erythrocytes from		
sheep	+	(+)
salmon	–	–
striped bass	–	–

(+), weak and delayed positive reaction
R, resistant strain; S, sensitive strain; NT, Not Tested

(continued)

Table 1 (continued)

CHARACTER	Strains isolated from	
	Striped bass (present study)	Atlantic salmon (Roald and Håstein, 1980)
Resistance to		
Penicillin	S	S
Chloramphenicol	S	S
Tetracyclin	S	S
Nitrofurantoin	S	NT
Oxolinic acid	S	NT
O/129	R	R

(+), weak and delayed positive reaction
R, resistant strain; S, sensitive strain; NT, Not Tested

Roald and Håstein (1980) reported that the pathogen corresponded to an unnamed species of *Acinetobacter.* However, according to Juni (1984), *Acinetobacter* spp. should not produce oxidase. Therefore, although the Atlantic salmon and striped bass isolates should be placed in the area loosely bound by *Moraxella, Acinetobacter*, and *Neisseria* genera; on the basis of the positive oxidase reaction and sensitivity to penicillin, both unreactive strains can be presumptively considered as *Moraxella* spp. Future DNA hybridization studies will better determine their taxonomic position.

The viral agent isolated in the present study exhibited the characteristics of the other members of the novel group of aquatic reoviruses detected in fish such as: GSV (Plumb et al., 1979); CSV (Winton et al., 1981); CRV (Amend et al., 1984); GRV (Chen and Jiang, 1984); tench and chub reoviruses (Ahne and Kolbl, 1987); and the TRV (Lupiani et al., 1989). Although these aquatic reoviruses do not fit within any of the six established genera of the Reoviridae (i.e., *Reovirus, Rotavirus, Orbivirus, Fijivirus, Phytovirus,* and *Cypovirus*), whether these viruses represent a new genus of the family or whether they are atypical rotaviruses is still not clear (Winton et al., 1987). Interestingly, similar viruses were isolated from oysters (Meyers, 1979; Nagabayashi and Mori, 1983) and clams (Hsu et al., 1988). The primary hosts for these shellfish viruses are not known with certainty, since their isolation could be simply a result of the efficient filter-feeding mechanism of bivalves.

Except for the GRV (Chen and Jiang, 1984), other reoviruses isolated from fish did not show specific pathogenicity to their hosts in experimental trials. Nevertheless, we must take in consideration that clinically inapparent infections with any virus may weaken the resistance to other infectious agents or stress caused by the environment that can result in further fish mortalities. Therefore, although our *Moraxella* strain proved to be pathogenic for striped bass, additional challenge studies with the Reo-like virus need to be conducted to determine if the bacterium is a primary or secondary invader.

ACKNOWLEDGEMENTS

This work was supported by contract # F139-88-008 from the Maryland Department of Natural Resources, and by Grant AQ-018/84 from the Comisioón Asesora de Investigación Científica y Técnica (CAICYT) of Spain.

LITERATURE CITED

Ahne, W. and O. Kolbl. 1987. Occurrence of reoviruses in European cyprinid fishes (*Tinca tinca*, Lin.) and chub (*Leuciscus cephalus* Lin.). J. Appl. Ichthyol., 3, 139-141.

Amend, D.F., T. McDowell and R.P. Hedrick. 1984. Characteristics of a previously unidentified virus from channel catfish (*Ictalurus punctatus*). Can. J. Fish Aquat. Sci., 41, 807-811.

Amos, A.H. (ed.). 1985. "Procedures for the Detection and Identification of Certain Fish Pathogens," 3rd ed. Fish Health Section, American Fisheries Society. Corvallis, Oregon.

Austin, B. 1982. Taxonomy of bacteria isolated from a coastal marine fish- rearing unit. J. Appl. Bacteriol., 53, 253-268.

Austin, B. and D.A. Austin. 1987. "Bacterial Fish Pathogens: Disease in Farmed and Wild Fish." John Wiley and Sons, New York.

Bøvre, K. 1984. Genus II. *Moraxella*. In: "Bergey's Manual of Systematic Bacteriology," Vol. 1 (N.K. Krieg, ed.), pp. 296-303. Williams & Wilkins, Baltimore.

Chen, Y. and Y. Jiang. 1984. Morphological and physico-chemical characterization of the hemorrhagic virus of grass carp. Kexue Tongbao, 29, 832-835.

Horsley, R.W. 1973. The bacterial flora of the Atlantic salmon (*Salmo salar*, L.) in relation to its environment. J. Appl. Bacteriol., 36, 337-386.

Hsu, Y.L., B.S. Chen and J.L. Wu. 1988. Biological and biochemical characteristics of a reovirus isolated from hard clams (*Meretrix lusoria*). In: "Abstr. Intern. Fish Health Conf.", Am. Fish. Soc. Vancouver, Canada.

Juni, E. 1984. Genus III. *Acinetobacter*. In: "Bergey's Manual of Systematic Bacteriology," Vol. 1 (N.K. Krieg, ed.), pp. 303-307. Williams & Wilkins, Baltimore.

Lupiani, B., C.P. Dopazo, A. Ledo, B. Fouz, J.L. Barja, F.M. Hetrick and A.E. Toranzo. 1989. A new syndrome of suspected viral etiology in cultured turbot (*Scophthalmus maximus*, L.). J. Aquat. Anim. Health., 1, in press.

Meyers, T.R. 1979. A reo-like virus isolated from juvenile American oysters (*Crassostrea virginica*). J. Gen. Virol., 43, 203-212.

Nagabayashi, T. and S. Mori. 1983. Characterization of a virus (JOV-1) isolated from the Japanese oysters (*Crassostrea gigas*). In: "Abstr. XVI Ann. Meet. Soc. Invert. Pathol.", Cornell University, New York.

Plumb, J.A., P.R. Bowser, J.M. Grizzle and A.J. Mitchell. 1979. Fish viruses: a double-stranded RNA icosahedral virus from a North American cyprinid. J. Fish. Res. Bd. Canad., 36, 1390-1394.

Reed, L.J. and H. Muench. 1938. A simple method of estimating fifty percent endpoints. Am. J. Hyg., 27, 493-497.

Roald, S.O. and T. Håstein. 1980. Infection with an *Acinetobacter*-like bacterium in Atlantic salmon (*Salmo salar*). In: "Fish Diseases" (W. Ahne, ed.), pp. 154-156. Third COPRAQ-Session, Springer-Verlag, Berlin.

Schutz, M., E.B. May, J.L. Kraeuter and F.M. Hetrick. 1984. Isolation of Infectious Pancreatic Necrosis virus from an epizootic occurring in cultured striped bass, *Morone saxatilis* (Walbaum). J. Fish Dis., 10, 29-34.

Toranzo, A.E., J.L. Barja, S.A. Potter, R.R. Colwell, F.M. Hetrick and J.H. Crosa. 1983. Molecular factors associated with virulence of marine vibrios isolated from striped bass in Chesapeake Bay. Infect. Immun., 39, 1220-1227.

Toranzo, A.E., P. Combarro, Y. Conde and J.L. Barja. 1985. Bacteria isolated from rainbow trout reared in freshwater in Galicia (Northwestern Spain): Taxonomic analysis and drug resistance patterns. In: "Fish and Shellfish Pathology" (A.E. Ellis, ed.), pp. 141-152. Academic Press, London.

Toranzo, A.E., A.M. Baya, B.S. Roberson, J.L. Barja, D.J. Grimes and F.M. Hetrick. 1987. Specificity of Slide Agglutination test for detecting Bacterial Fish Pathogens. Aquaculture, 61, 81-97.

Winton, J.R., C.N. Lannan, J.L. Fryer and T. Kimura. 1981. Isolation of a new reovirus from chum salmon in Japan. Fish Pathol., 15, 155-162.

Winton, J.R., C.N. Lannan, J.L. Fryer, R.P. Hedrick, T.R. Meyers, J.A. Plumb and Y. Yamamoto. 1987. Morphological and biochemical properties of four members of a novel group of reoviruses isolated from aquatic animals. J. Gen. Virol., 68, 353-364.

COMPARISON OF THE CELL SURFACE HYDROPHOBICITY OF BACTERIAL FISH PATHOGENS BY DIFFERENT PROCEDURES

Ysabel Santos[1], Isabel Bandin[1], Teresa P. Nieto[1],
David W. Bruno[2], Anthony E. Ellis[2], and Alicia E. Toranzo[1]

[1]Departamento de Microbiología y Parasitología, Facultad de Biología,
Universidad de Santiago. Santiago de Compostela 15706, Spain.
[2]Department of Agriculture and Fisheries for Scotland, Marine Laboratory,
Aberdeen AB9 8DB, Scotland.

ABSTRACT: Recent studies have demonstrated that hydrophobic interactions play a role in the adherence of pathogenic bacteria to eukaryotic cells. It was also suggested that the possession of a hydrophobic cell surface enables the microorganism to avoid host defense mechanisms and contributes to its intracellular survival (Beachey et al., 1980; Daly and Stevenson, 1987). In the present report, we compare the cell surface hydrophobicity of some isolates of *Renibacterium, Yersinia, Aeromonas,* and *Vibrio* using three different methods: salt aggregation test (SAT), phase partitioning with hydrocarbon solvents, and adherence to nitrocellulose filters (NCF). Influence of growth medium, salinity, as well as the effects of sugars, enzymes, and heat treatment on hydrophobicity were examined. In addition, autoagglutinating and hemagglutinating capacities of our strains were evaluated in order to know their possible relationship with the hydrophobic character.

Tables in the text indicate the results obtained with a representative number of the strains tested. Most of the isolates were relatively hydrophobic regardless of the method used to evaluate this capacity. In general, *Renibacterium salmoninarum* and *Aeromonas* spp. seem to have increased hydrophobic abilities in comparison with *Vibrio anguillarum* strains. A lack of hydrophobicity was observed in *Y. ruckeri* isolates with the only exception of NCMB 1316 strains, which also presented a high hemagglutinating activity for trout erythrocytes. The variable results obtained in the phase partitioning assay depending on the hydrocarbon employed were noteworthy. Interestingly, only the hydrophobic capacities of *R. salmoninarum* and *Y. ruckeri* cell suspensions were affected by the heat treatment. In the present study no relationship was found among hydrophobicity and hemagglutinating or autoagglutinating activities.

From our results we must conclude that the standardization of growth conditions, initial concentration of cell suspensions, and type and amount of hydrocarbon is necessary in order to compare data obtained by different workers.

Beachey, E.H., W.A. Simpson and I. Ofek. 1980. In: "Microbial Adhesion to Surfaces" (R.C.W. Berkeley, J.M. Lynch, J. Melling, P.R. Rutter and B. Vincent, eds.). Soc. Chem. Industry, London.

Daly, J.G. and R.M.W. Stevenson. 1987. J. Gen. Microbiol., 133, 3575-3580.

INTRODUCTION

It has been reported that hydrophobic interactions play a role in the nonspecific attachment of bacteria to inanimate surfaces as well as in the mechanism of adhesion to eukaryotic cells (Magnusson et al., 1980; Trust et al., 1983; Fattom and Shilo, 1984). Recent studies suggest that the possession of a hydrophobic cell wall would be advantageous to pathogenic bacteria, allowing their intracellular survival and multiplication in the phagocytic cells of diseased fish (Beachey et al., 1980; Daly and Stevenson, 1987; Bruno, 1988). In addition, conflicting reports exist about the possible relationship among the ability of virulent strains to agglutinate red blood cells, the possession of an autoagglutinating phenotype, and the relative hydrophobic nature of the bacterial cell surfaces (Trust et al., 1980; Daly and Stevenson, 1987; Larsen et al., 1988). In this contribution, we present data on the hydrophobic nature of the cell surface of fish pathogenic strains belonging to the genera *Renibacterium, Aeromonas, Vibrio,* and *Yersinia*. The usefulness of the salt aggregation test, phase partitioning with hydrocarbon solvents, and adherence to nitrocellulose filters as methods for discriminating strains with respect to their hydrophobic properties is critically evaluated. The influence of growth medium, salinity, as well as the effect of sugars, enzymes, and heat treatment on hydrophobicity was also examined. In addition, the adhesiveness to fish cell lines and the autoagglutinating and hemagglutinating capacities of our strains were determined to investigate any possible relationship with the hydrophobic character.

MATERIALS AND METHODS

Bacterial strains. In the present study we have used a group of 48 bacterial fish pathogens belonging to the species *Renibacterium salmoninarum* (5 strains), *Aeromonas hydrophila* (9 strains), *A. sobria* (5 strains), *A. salmonicida* (10 strains), *Vibrio anguillarum* (11 strains), and *Yersinia ruckeri* (8 strains). The strains were obtained from ATCC (American Type Culture Collection, Rockville, Maryland, USA), NCMB (National Collection of Marine Bacteria, Aberdeen, Great Britain), laboratories in different countries, or isolated as causal agents of epizootics in trout, salmon, and turbot reared in the northwest of Spain or Scotland. Bacteria were routinely cultured at 22°C on trypticase soy agar (TSA) or broth (TSB) (Difco) with the appropriate salt concentration (Gram-negative bacteria) and at 15°C on Mueller Hinton agar or broth (Difco) supplemented with 0.1% L-cysteine hydrochloride (MHA-C or MHB-C) (*R. salmoninarum* strains). Stock cultures were kept on TSA or MHA-C slants at 15°C under mineral oil and frozen at -70°C in medium with 15% (v/v) glycerol.

Measurement of bacterial hydrophobicity. The relative cell surface hydrophobicity of the strains were comparatively assayed using three standard methods: salt aggregation test (SAT), phase partitioning with hydrocarbon solvents (BATH), and adherence to nitrocellulose filters (NCF). The influence of medium and salt concentration on the hydrophobic character of *R. salmoninarum* was simultaneously evaluated, growing the strains in MHA-C (Wolf and Dunbar, 1959; Bruno, 1988), KDM-2 (Evelyn, 1977) and KDM-C (Daly and Stevenson, 1985) without or with 1% and 2% NaCl.

Salt aggregation test (SAT). The assay was performed basically as described by Lindhal et al. (1981). Bacterial cultures on TSA or MHA-C (*Renibacterium* strains) were removed, washed twice, and suspended to an absorbance at 420 nm (A_{420}) of 1.0 in 0.002 M sodium phosphate buffer (pH 6.8). These suspensions (30 µl) were mixed with an equal volume of various molarities of buffered ammonium sulphate solutions that ranged from 0.05 to 4.0 M. Bacterial suspensions, to which were added a 4 M $(NH_4)_2SO_4$ solution or with only sodium phosphate buffer, were used as positive and negative controls, respectively.

BATH test. Partitioning in liquid hydrocarbons was determined essentially according to the procedures of Rosenberg et al. (1980). Briefly, cultures grown as described above were harvested, washed twice, and suspended to give an absorbance of 0.16 at 600 nm (A_{600}) in phosphate buffered saline (PBS), pH 7.3. The bacterial suspensions were overlaid with various volumes on n-hexadecane or n-octane (Sigma, St. Louis, Missouri) in 10-mm diameter glass tubes. After 2 min agitation, the mixtures were allowed to separate for 15 min. The percent of partitioning in the hydrocarbon phase was calculated using the following formula:

$$\{[A_{600}\text{ (original bacterial suspension)} - A_{600}\text{ (aqueous phase)}] / A_{600}\text{ (original bacterial suspension)}\} \times 100$$

NCF test. The ability to bind to nitrocellulose filters was evaluated by the procedure of Lachica and Zink (1984). Overnight cultures of bacteria in the above described media were harvested, washed twice, and suspended to an A_{600} of 1.0 in saline solution (0.9% NaCl, pH 7.2). In the case of *R. salmoninarum*, 7-day-old cultures were used. Four ml aliquots of these suspensions were passed dropwise through an 8 µm pore size nitrocellolose filter (Type SC, Millipore, Bedford, Massachusetts) and the absorbance of the resultant filtrate was compared with that of the initial bacterial suspension (control). The relative NCF binding efficiency (%) of each strain was calculated as follows:

$$\{[A_{600}\text{ (control)} - A_{600}\text{ (NCF treated)}] / A_{600}\text{ (control)}\} \times 100$$

Autoagglutination. Autoagglutination of strains was determined basically as described by Janda et al. (1987). Briefly, a loopful of bacteria was inoculated in 10-mm diameter glass tubes containing 5 ml of TSB with 1% NaCl (Gram-negative strains) or MHB-cysteine (*R. salmoninarum* strains) and incubated at the appropriate temperature for 18 hr without shaking. Autoagglutination was assessed by the absence of growth in the broth and the appearance of a cellular pellet in the bottom of the tube. Strains that spontaneously pelleted in static cultures were designated as SP^+.

Hemagglutination assays. The hemagglutinating activity of strains was determined using human and rainbow trout erythrocytes according to Toranzo et al. (1983). To avoid possible interferences with autoagglutination, hemagglutinating activity by SP^+ strains was determined using suspensions of red blood cells and bacteria prepared in a 0.2% saline solution, pH 7.2. Briefly, bacterial suspensions adjusted by optical density to yield 10^{10} cells/ml were mixed with equal amounts (20 μl) of erythrocyte suspensions. Strains were considered negative if visible hemagglutination had not occurred within 10 min.

Inhibition of hydrophobic and hemagglutinating activities. The effect of heating, enzymes, and sugars on the hemagglutinating capacity and hydrophobic character (measured by SAT) was evaluated in a selected group of strains. For heat treatment, the strains were suspended in the above described buffers and heated at 80°C for 10 min. Buffered solutions of trypsin and proteinase K (Sigma) at a concentration of 10 mg ml^{-1} were added to the bacterial suspensions and incubated at 22°C for 1 hr. The effect of sugars was evaluated after mixing equal volumes of bacterial suspensions and solutions of 1% (w/v) D-mannose or L-fucose (Sigma) in PBS or sodium phosphate buffer. Untreated control cell suspensions were tested simultaneously with the treated samples.

Adherence to fish cell lines. The cell adherence capacity of the strains was assayed on EPC (epithelioma papillosum cyprini) fish cell line. Semiconfluent cell monolayers were prepared in 24-well culture plates (Costar) using Minimal Essential Medium (MEM; Flow) supplemented with 10% fetal calf serum and 50 μg/ml of gentamycin. Two hr before performing adhesion assays, the medium was removed and replaced by fresh medium without antibiotic. Cells were then rinsed with PBS and fixed in Carnoy's for 20 min. For adhesion studies, 1 ml of bacterial cell suspensions prepared in PBS were inoculated on the fixed tissue cultures and incubated at 20°C for 1 hr. Monolayers inoculated with 1 ml of PBS were used as controls. After the incubation period, plates were washed several times with PBS, fixed, stained with Giemsa, and examined for evidence of bacterial adherence using an inverted microscope.

RESULTS AND DISCUSSION

The overall results of autoagglutination, hydrophobicity, hemagglutination, and cell adherence of 48 fish pathogenic strains are shown in Tables 1-5. The majority of the strains were relatively hydrophobic, but clear differences in the hydrophobicity index were detected according to the bacterial group analyzed and the method and substratum employed in the assays. In addition, whereas a notable number of strains displayed hemagglutinating activities against trout or human erythrocytes and adhesive abilities on fish cell lines, only 8 of 43 strains tested pelleted in static broth cultures (SP^+ phenotype).

Cell surface properties of *Renibacterium salmoninarum*. The *R. salmoninarum* isolates were strongly hydrophobic in SAT, BATH, and NCF assays (Table 1). All the strains agglutinated in $(NH_4)_2SO_4$ in a range of 0.05 to 0.1 M, displayed a 75-100% binding to nitrocellulose filters, and a 30-52% adherence to hydrocarbon solvents. Since *Renibacterium* is a very fastidious organism in its nutritional requirements (Evelyn, 1977) and causes disease in salmonids reared in fresh and marine waters (Fryer and Sanders, 1981), we studied the role of medium and salinity on its hydrophobicity. Whereas the growth in presence of 2% NaCl seems to slightly decrease the adherence to hydrocarbons, there was no observed influence of media on the hydrophobic character measured by the three techniques (data not shown).

Daly and Stevenson (1987) reported a correlation between hydrophobicity of *R. salmoninarum* and hemagglutination on homeotherm erythrocytes. Our studies have demonstrated that such correlation cannot be established with poikilotherm erythrocytes because all our isolates failed to agglutinate trout erythrocytes. It is noteworthy that only one *R. salmoninarum* strain displayed adherence to EPC fish cell lines.

Cell surface properties of motile and non-motile *Aeromonas* strains. The motile and non-motile *Aeromonas* strains exibited a heterogeneous pattern of response in the three different methods used to determine the cell surface hydrophobicity, as well as in cell adhesion, hemagglutination, and autoagglutination tests (Tables 2, 3).

The strains of *A. hydrophila* and *A. sobria* aggregated, in general, with a lower concentration of ammonium sulphate than *A. salmonicida* strains. The majority of *Aeromonas* strains bound strongly to nitrocellulose filters. Although heterogeneous indices of hydrophobicity were found in the BATH assay, the *Aeromonas* isolates displayed a slightly higher affinity towards octane than hexadecane.

With respect to hemagglutination assays, more isolates agglutinated human than trout erythrocytes. Only one strain, P-351, displayed specific hemagglutination of trout red blood cells. Similar results have been previously reported in pathogenic *A. hydrophila* and *V. anguillarum* (Santos et al., 1987, 1988; Toranzo et al., 1983). Regardless of their autoagglutinating, hemagglutinating, and hydrophobic capacities, all the *A. hydrophila* strains tested showed adherence ability to fish cell lines. However, variable responses were found within *A. sobria* and *A. salmonicida*.

Cell surface properties of *V. anguillarum*. *V. anguillarum* isolates had similar values in the phase partitioning test (Table 4) to motile and non-motile *Aeromonas* (Tables 2, 3), exhibiting clearly a higher affinity towards octane than hexadecane. However, in the adherence to nitrocellulose filters the *Vibrio* strains displayed low hydrophobic values.

Comparing the surface hydrophobicity of serotypes 01 and 02, the highest values of hydrophobicity in all three methods used were displayed by some strains of serotype 02. This finding contrasts with the results reported by Larsen et al. (1988) where the 01 strains exhibited more hydrophobic tendencies than 02. This discrepancy may be attributed to the different origin of the isolates.

On the other hand, differences in adherence to fish cell lines among strains belonging to 01 and 02 serotypes were not detected. We have previously reported

Table 1. Hydrophobicity and adherence properties of *R. salmoninarum*.

STRAINS	BATH (% change of absorbance)		NCF (% binding to filters)	SAT[a]	SP[b]	Hemagglutination of erythrocytes		Adherence to EPC-cells
	hexadecane	octane				Human	Trout	
R. salmoninarum								
ATCC 33209	52	48	85	0.05-0.1	+	+	–	ND
NCMB 1113	50	30	99	0.05-0.1	–	+	–	–
419	41	40	75	0.05-0.1	(+)[c]	+	–	+
420	50	44	100	0.05-0.1	–	+	–	–
421	45	30	80	0.05-0.1	+	+	–	–

[a]Values represent the lowest molarity of ammonium sulfate yielding a strong aggregation of bacteria.
[b]SP, self-pelleting
[c](+), weak response
ND, Not determined

Table 2. Hydrophobicity and adherence properties of motile *Aeromonas* strains.

STRAINS	BATH (% change of absorbance)		NCF (% binding to filters)	SAT[a]	SP[b]	Hemagglutination of erythrocytes		Adherence to EPC-cells
	hexadecane	octane				Human	Trout	
A. hydrophila								
ATCC 7966	ND	0	96	3.5	–	+	–	ND
Y-62	13	14	ND	1.0-1.5	–	+	–	ND
1.54	13	21	100	1.0-1.5	–	+	+	+
67-P-24	35	32	96	0.5-1.0	–	+	+	+
81-83	11	19	ND	2.0-2.5	–	–	–	+
80-A1	20	38	71	1.0-1.5	–	–	–	+
80-A2	11	0	ND	2.0-2.5	–	+	–	+
B-51	16	48	78	1.5-2.0	–	+	+	+
B-32	17	48	91	1.0-1.5	–	–	–	+
A. sobria								
P-33	25	58	45	1.5-2.0	+	–	–	ND
P-261	17	56	83	0.5-1.0	–	+	+	–
P-281	25	ND	96	2.5-3.0	ND	–	–	–
P-334	29	37	58	1.0-1.5	ND	–	–	ND
P-351	ND	35	83	1.0-1.5	ND	–	+	ND

[a]Values represent the lowest molarity of ammonium sulfate yielding a strong aggregation of bacteria.
[b]SP, self-pelleting
ND, Not determined

Table 3. Hydrophobicity and adherence properties of *A. salmonicida*.

STRAINS	BATH (% change of absorbance)		NCF (% binding to filters)	SAT[a]	SP[b]	Hemagglutination of erythrocytes		Adherence to EPC-cells
	hexadecane	octane				Human	Trout	
A. salmonicida								
3.101	21	0	92	2.0-2.5	+	+	–	ND
E-8101	ND	22	88	0.5-1.0	–	+	–	–
V5/80	29	16	39	2.0-2.5	–	+	–	ND
H22/77	0	0	100	2.0-2.5	(+)[c]	(+)	–	+
ATCC 33658	0	25	25	25-3.0	ND	–	–	ND
ATCC 14174	11	11	ND	3.5	–	+	–	–
3.93	7	20	33	3.5	–	+	(+)	ND
2261	30	ND	ND	0.5-1.0	ND	(+)	–	ND
004	28	36	64	1.5-2.0	–	+	–	ND
416	39	63	96	1.5-2.0	+	+	–	+

[a]Values represent the lowest molarity of ammonium sulfate yielding a strong aggregation of bacteria.
[b]SP, self-pelleting
[c](+), weak response
ND, Not determined

Table 4. Hydrophobicity and adherence properties of *V. anguillarum*.

STRAINS	BATH (% change of absorbance)		NCF (% binding to filters)	SAT[a]	SP[b]	Hemagglutination of erythrocytes		Adherence to EPC-cells
	hexadecane	octane				Human	Trout	
V. anguillarum								
serotype 01								
775	18	25	ND	2.0-2.5	–	–	–	+
96-F	13	26	ND	2.5-3.0	–	+	+	+
R-82	7	23	30	2.0-2.5	–	–	–	–
TM-14	9	28	35	2.0-2.5	–	–	–	+
RT-32	0	ND	ND	2.5-3.0	–	–	–	–
serotype 02								
NCMB6	ND	33	34	0.5-1.0	–	+	+	–
ATCC 14181	38	29	84	1.5-2.0	–	–	ND	–
43-F	18	62	68	1.5-2.0	–	+	+	+
RV-22	13	27	38	2.0-2.5	–	+	+	–
91079	25	52	59	ND	+	–	ND	+
RG-111	3	38	ND	ND	–	–	ND	(+)[c]

[a]Values represent the lowest molarity of ammonium sulfate yielding a strong aggregation of bacteria.
[b]SP, self-pelleting
[c](+), weak response
ND, Not determined

Table 5. Hydrophobicity and adherence properties of *Y. ruckeri*.

STRAINS	BATH (% change of absorbance)		NCF (% binding to filters)	SAT[a]	SP[b]	Hemagglutination of erythrocytes		Adherence to EPC-cells
	hexadecane	octane				Human	Trout	
Y. ruckeri								
serotype 01								
11.4	25	0	36	3.0-3.5	–	–	(+)[c]	–
8R	29	0	ND	3.5	–	–	–	ND
PP-31	20	0	20	>4	–	–	–	–
AG-41	0	0	24	>4	–	–	–	–
FP-11	15	0	18	>4	–	–	–	–
serotype 02								
11.29	11	4	41	3.0	–	–	–	ND
1622	0	0	47	3.5	–	–	–	–
1638	0	0	86	3.0	–	–	–	ND

[a]Values represent the lowest molarity of ammonium sulfate yielding a strong aggregation of bacteria.
[b]SP, self-pelleting
[c](+), weak response
ND, Not determined

similar variable responses between both serotypes in extracellular virulence factors of *V. anguillarum* (Toranzo et al., 1987).

Cell surface properties of *Yersinia ruckeri.* None of the *Y. ruckeri* strains were strongly hydrophobic regardless of the technique employed (Table 5). Interestingly, in contrast to *Aeromonas* and *Vibrio* isolates, *Y. ruckeri* showed a greater ability to partition in hexadecane in the BATH assay. None of the strains adhered to EPC cell line and only one strain (11.4) showed a weak hemagglutinating activity with trout erythrocytes. At difference with *V. anguillarum*, the different serotypes of *Y. ruckeri* all had similar cell surface associated characteristics.

Criteria of hydrophobicity. The results of this study together with the data reported by others (Janda et al., 1987; Paula et al., 1988; Larsen et al., 1988; Bruno, 1988) allow us to propose the criteria of hydrophobicity presented in Table 6. According to these indices we have determined the number of strains showing positive hydrophobicity in each test (Table 7).

A better correlation seems to exist between NCF and SAT assays since we can establish the same decreasing order of hydrophobicity from *Renibacterium*, (the strongest hydrophobic group) to *Y. ruckeri,* considered as not hydrophobic, with motile *Aeromonas, A. salmonicida,* and *V. anguillarum* intermediate. In addition, Table 7 clearly demonstrates the influence of the hydrocarbon used in the BATH test with *V. anguillarum* and *Y. ruckeri* (see Tables 4, 5).

Table 6. Proposed criteria of hydrophobicity.

TEST	VALUES		RESULTS
SAT	0.0-1.0 M	STRONG	POSITIVE TEST
	1.0-2.0 M	Moderate	
	2.0-4.0 M	Weak	
	> 4.0 M	No hydrophobic	
BATH	> 50%	STRONG	POSITIVE TEST
	20-50%	Moderate	
	< 20%	No hydrophobic	
NCF	> 75%	STRONG	POSITIVE TEST
	50-75%	Moderate	
	< 50%	No hydrophobic	

Table 7. Percentage of strains exhibiting hydrophobic properties in the different assays according to the criteria proposed.

STRAINS	BATH				NCF		SAT	
	hexadecane		Octane					
R. salmoninarum (5 strains)	5/5[a]	(100)[b]	3/3	(100)	5/5	(100)	5/5	(100)
Motile *Aeromonas* (14 strains)	5/12	(41.6)	9/13	(69.2)	10/11	(90.9)	10/14	(71.4)
A. salmonicida (10 strains)	5/9	(55.5)	5.9	(55.5)	5/8	(62.5)	4/10	(40.0)
V. anguillarum (11 strains)	2/10	(20.0)	10/10	(100)	3/7	(42.8)	3/9	(33.3)
Y. ruckeri (8 strains)	3/8	(37.5)	0/8	(0.0)	1/7	(14.3)	0/8	(0.0)

[a]Number of positive strains/Number of strains tested
[b]Percentage of positive strains

We must take into consideration that on the basis of our criteria, only two *A. hydrophila*, one *A. sobria*, two *A. salmonicida*, one *V. anguillarum*, and zero *Y. ruckeri* were hydrophobic simultaneously in all the tests employed. Therefore, in agreement with Dillon et al. (1986) our findings support that the results of a single assay cannot be a reliable representation of the cell surface hydrohobicity of different isolates.

Effect of heat, enzymes, and sugars on hydrophobicity and hemagglutination. The effects of the different treatments on the cell surface hydrophobicity and hemagglutinating activities of a selected group of fish pathogens were evaluated. The hydrophobicity of the strains were not affected by trypsin or proteinase K (10 mg ml^{-1}). However, both enzyme sensitive and resistant hemagglutinations were observed within the pathogenic strains. In addition, the agglutinating activity of some *Aeromonas* strains was sensitive to proteinase K but not affected by trypsin. Similar findings were reported by Adams et al. (1983) in clinical *A. hydrophila* isolates. Analogously, different responses were found in both hemagglutination and hydrophobicity assays after heating or treatment with mannose or fucose. All these results support the notion that the cell wall of fish pathogenic bacteria possess a wide variety of adhesions with distinct natures and structures that recognize specific receptors on the host cells.

CONCLUSIONS

The findings reported in this study can be summarized in the following conclusions:

(1) Results of the different assays to assess the cell surface associated properties of bacterial fish pathogens are variable depending on the method and bacterial group. Only in *R. salmoninarum* was a good correlation found among the different methods used to test hydrophobic characteristics.

(2) Regardless of the method used, we can conclude that *R. salmoninarum* have the highest hydrophobicity, followed by motile *Aeromonas, A. salmonicida,* and *V. anguillarum*. A lack of hydrophobicity was displayed by *Yersinia ruckeri* strains.

(3) Among the assays used to measure the cell surface hydrophobicity, the salt aggregation test gave the most uniform results within each bacterial group.

(4) No relationship was found between hydrophobicity and hemagglutination or adherence to fish cell lines.

(5) The effect of inhibitors on the hydrophobicity and hemagglutinating properties provide evidence of the different nature of the adhesive components on the cell surface and, hence, the existence of separate attachment mechanisms among the bacterial strains.

(6) Because the methods used in this study can measure different types of interactions, it is necessary to standardize the assay(s) and conditions (e.g., growth media, temperature, initial cell concentration, hydrocarbon, buffer, etc.) to determine *in vitro* the adherence properties of fish pathogens. In addition, we recommend the establishment of adequate criteria of hydrophobicity in each test to facilitate comparison of results from different laboratories.

ACKNOWLEDGEMENTS

We gratefully thank Dr. Javier Barja for his contribution to the presentation of the data. This work was supported by Grant AQ-018/84 from the Comisión Asesora de Investigación Científica y Técnica (CAICYT) and by Grant PB87-1027 from the Dirección General de Investigación Científica y Técnica (DGICYT), Spain.

Y. Santos and I. Bandín acknowledge the Ministerio de Educación y Ciencia and the Universidad de Santiago de Compostela (Spain) for research fellowships.

LITERATURE CITED

Adams, D., H.M. Atkinson and W.H. Woods. 1983. The differing effect of proteases on the adhesions of *Aeromonas hydrophila*. FEMS Microbiol. Lett., 20, 197-200.

Beachey, E.H., W.A. Simpson and W.H. Ofek. 1980. Interaction of surface polymers of *Streptococcus pyogenes* with animal cells. In: "Microbial Adhesion to Surfaces" (R.C.W. Berkeley, J.M. Lynch, J. Melling, P.R. Rutter and V. Vincent, eds.), pp. 389-405. Ellis Horwood Ltd, Chichester, England.

Bruno, D.W. 1988. The relationship between auto-agglutination, cell surface hydrophobicity and virulence of the fish pathogen *Renibacterium salmoninarum*. FEMS Microbiol Lett., 51, 135-140.

Daly, J.G. and R.M.W. Stevenson. 1985. Charcoal agar, a new growth medium for the fish disease bacterium *Renibacterium salmoninarum*. Appl. Environ. Microbiol., 50, 868-871.

Daly, J.G. and R.M.W. Stevenson. 1987. Hydrophobic and Haemagglutinating properties of *Renibacterium salmoninarum*. J. Gen. Microbiol., 133, 3575- 3580.

Dillon, J.K., J.A. Fuerst, A.C. Hayward and G.H.G. Davis. 1986. A comparison of five methods for assaying bacterial hydrophobicity. J. Microbiol. Meth., 6, 13-19.

Evelyn, T.P.T. 1977. An improved growth medium for the Kidney Disease bacterium and some notes on using the medium. Bull. Off. Int. Epiz., 87, 511-513.

Fattom, A. and M. Shilo. 1984. Hydrophobicity as an adhesion mechanism of benthic cyanobacteria. Appl. Environ. Microbiol., 47, 135-143.

Fryer, J.L. and J.E. Sanders. 1981. Bacterial kidney disease of salmonid fish. Ann. Rev. Microbiol., 35, 273-298.

Janda, M.J., L.S. Oshiro, S.L. Abbott and P.S. Duffey. 1987. Virulence markers of mesophilic aeromonads: Association on the autoagglutination phenomenon with mouse pathogenicity and the presence of a peripheral cell-associated layer. Infect. Immun., 55, 3070-3077.

Lachica, R.V. and D.L. Zink. 1984. Plasmid-associated cell surface charge and hydrophobicity of *Yersinia enterocolitica*. Infect. Immun., 44, 140-543.

Larsen, J.L., H.B. Rasmussen and I. Dalsgaard. 1988. Study of *Vibrio anguillarum* strains from different sources with emphasis on ecological and pathobiological properties. Appl. Environ. Microbiol., 54, 2264-2267.

Lindhal, M., A. Faris, T. Wadstrom and S. Hjerten. 1981. A new test based on "salting out" to measure relative surface hydrophobicity of bacterial cells. Biochim. Biophys. Acta, 677, 471-476.

Magnusson, K.E., J. Davies, T. Grundstrom, E. Kihltrom and S. Normak. 1980. Surface charge and hydrophobicity of Salmonellae, *E. coli* and *Gonococi* in relation to their tendency to associate with animal cells. Scand. J. Infect. Dis., 24, 130-140.

Paula, S.J., P.S. Duffey, S.L. Abbott, R.P. Kokka, L.S. Oshiro, J.M. Janda, T. Shimada and R. Sakazaki. 1988. Surface properties of mesophilic Aeromonads. Infect. Immun., 56, 2658-2665.

Rosenberg, M., D. Gutnick and E. Rosenberg. 1980. Adherence of bacteria to hydrocarbons: A simple method for measuring cell-surface hydrophobicity. FEMS Microbiol. Lett., 9, 29-33.

Santos, Y., A.E. Toranzo, C.P. Dopazo, T.P. Nieto and J.L. Barja. 1987. Relationships among virulence for fish, enterotoxigenicity, and phenotypic characteristics of motile *Aeromonas*. Aquaculture, 67, 29-39.

Santos, Y., A.E. Toranzo, J.L. Barja, T.P. Nieto and T.G. Villa. 1988. Virulence properties and enterotoxin production of *Aeromonas* strains isolated from fish. Infect. Immun., 56, 3285-3293.

Toranzo, A.E., J.L. Barja, R.R. Colwell, F.M. Hetrick and J.H. Crosa. 1983. Haemagglutinating, haemolytic and cytotoxic activities of *Vibrio anguillarum* and related vibrios isolated from striped bass on the Atlantic coast. FEMS Microbiol. Lett., 18, 257-262.

Toranzo, A.E., Y. Santos, M.L. Lemos, A. Ledo and J. Bolinches. 1987. Homology of *Vibrio anguillarum* strains causing epizootics in turbot, salmon and trout reared on the Atlantic coast of Spain. Aquaculture, 67, 41-52.

Trust, T.J., P.S. Howard, J.B. Chamberlain, E.E. Ishiguro and J.T. Buckley. 1980. Additional surface protein in autoaggregating strains of atypical *Aeromonas salmonicida*. FEMS Microbiol. Lett., 9, 35-387.

Trust, T.J., W.W. Kay and E.E. Ishiguro. 1983. Cell surface hydrophobicity and macrophage association of *Aeromonas salmonicida*. Curr. Microbiol., 9, 315-318.

Wolf, K. and C.E. Dunbar. 1959. Test of 34 therapeutic agents for control of kidney disease in trout. Trans. Amer. Fish. Soc., 88, 117-124.

SUSCEPTIBILITY OF SEA BASS, *DICENTRARCHUS LABRAX*, AND TURBOT, *SCOPHTHALMUS MAXIMUS*, TO *YERSINIA RUCKERI*

M. Vigneulle

Centre National d'Etudes Vétérinaires et Alimentaires, Laboratoire de Pathologie des Animaux Aquatiques (CNEVA - PAA), B.P. 70, 29263 Plouzané, France

ABSTRACT: As a result of experimental infection of juvenile sea bass (*Dicentrarchus labrax*) and turbot (*Scophthalmus maximus*) to the bacterium *Yersinia ruckeri* by inoculation and bathing in contaminated water, it is now known that both species of fish are susceptible to this pathogen. In addition, the data indicate that the turbot is more susceptible to *Y. ruckeri* than sea bass.

INTRODUCTION

Yersinia ruckeri is the bacterial etiological agent of the enteric redmouth disease (ERMD), an hemorrhagic enterosepticemia of fish. This bacterium was first recognized as a cause of mortalities in farmed rainbow trout (*Salmo gairdneri*) in the Hagerman Valley of Idaho, USA, in the early 1950's (Ross et al., 1966). Following this first record, the endemic range of *Y. ruckeri* increased with time in North America. In 1982, Lesel et al. (1983) reported its introduction into France and, at the same time, the disease spread in nearly all European countries.

It has been considered to be a severe problem in salmonid farms, especially rainbow trout. A few wild and reared nonsalmonid fish species have been reported to be susceptible to yersiniosis, including carp, *Cyprinus carpio* (Fuhrman et al., 1984), goldfish, *Carassius auratus* (McArdle and Dooley-Martyn, 1985), burbot, *Lota lota* (Dwilow et al., 1987), minnows, *Pimephales promelas* (Michel et al., 1986), sturgeon, *Acipenser baeri* (Vuillaume et al., 1987), tench, *Tinca tinca*, and eel, *Anguilla anguilla* (Fuhrman et al., 1984).

The contaminated effluents from freshwater salmonid farms or from the wild river populations and the transfer in sea water of asymptomatic salmonid carriers of *Y. ruckeri* constitute a hazard for marine farms situated along the coast or in estuaries. The purpose of this study was to investigate the susceptibility of sea bass

(*Dicentrarchus labrax*) and turbot (*Scophthalmus maximus*), two marine fish species intensively cultured in France, to *Y. ruckeri*.

MATERIALS AND METHODS

Fish. Juvenile fish used in these experiments were free of *Y. ruckeri* infection. They were reared in our laboratory on a commercial dry diet and when used in these experiments, they weighed 2-9 g. The fish were randomly distributed in aquaria holding 10 liters of aerated sea water in which the water temperature was regulated to within ± 1°C of the desired temperature.

Bacterium. The strain, recently isolated from an epizootic in rainbow trout, was grown in tryptic soy agar (TSA) for 24 hr at 20°C. This strain belongs to serotype I, the most commonly encountered, most pathogenic, and is associated with clinical cases of ERMD (Busch, 1982). All the strains isolated in our laboratory since 1982 belong to this serotype.

Infection of fish. Cultures of *Y. ruckeri* were grown in sterile tryptic soy broth (TSB) at 20°C for 24 hr. Cell concentrations were titrated by serial 10-fold dilutions in sterile physiological saline using four plates of each dilution in TSA. The plates were incubated at 20°C for 48 hr before counting. Experimental contaminations of fish were performed by inoculation and by bath.

Inoculation. Sea bass were anaesthetized with 2-phenoxyethanol at a concentration of 200 mg l^{-1} and then injected intraperitoneally (ip) with 0.1 ml of serial dilutions in sterile physiological saline of a 24-hr culture of *Y. ruckeri* in TSB.

Turbot were injected (ip) with 0.05 ml of the same dilutions.

For each experiment, control fish were injected, respectively, with 0.1 ml and 0.05 ml of a sterile dilution of TSB in physiological saline.

Bath. The water flow was turned off and the water volume of the aquarium was reduced to 2 l. A sample (200 ml) of the 24-hr culture of *Y. ruckeri* in TSB was added. Contact was continued for 1 hr and the water flow was resumed. Control fish were similarly immersed in a sterile dilution of TSB. Mortalities were recorded daily from each aquarium.

Necropsy. Samples of kidneys were removed and transferred to TSA plates. The plates were incubated at 20°C for 24 hr to 48 hr. Typical *Y. ruckeri* colonies were confirmed by slide agglutination with an anti-*Y. ruckeri* serum (CNEVA - PAA). Only dead fish that yielded cultures serologically identical to *Y. ruckeri* were counted as being caused by ERMD.

RESULTS

Several trials were performed with sea bass and turbot with serial doses (22 to 10^7 bacteria/fish) and temperatures (12.5°C to 20°C). The results are summarized in Tables 1 and 2.

Table 1. Experimental transmission of *Yersinia ruckeri* to sea bass.

Transmission methods	Trial	No. of fish*	Weight (g)	Dose	Temperature °C	Mortality %	Duration of mortality (days)
Inoculation	no. 1	20	2.7	6.0×10^{7}**	20.0	100	2
	no. 2	25	6.0	9.0×10^{6}		28	2
		25	6.0	9.0×10^{5}	14.3	32	1
		25	6.3	9.0×10^{4}		20	9
		17	7.0	9.0×10^{3}		21	15
	no. 3	10	9.4	6.0×10^{6}		100	2
		10	8.8	6.0×10^{5}		100	2
		10	7.8	6.0×10^{4}	12.8	60	1
		10	8.2	6.0×10^{3}		50	7
		10	8.0	1.2×10^{3}		0	
		10	8.5	1.2×10^{2}		0	
Bath	no. 1	20	2.1	1.2×10^{8}***	20.0	10	30
	no. 2	10	3.9	5.0×10^{7}	17.0	0	
	no. 3	25	6.8	1.2×10^{8}	14.3	8	15
	no. 4	10	8.5	8.0×10^{7}	12.5	20	6

* for each trial a same number of control fish were injected or immersed in a sterile solution.
** inoculation: number of bacteria injected/fish.
*** bath: final concentration of bath in number of bacteria ml^{-1}.

Table 2. Experimental transmission of *Yersinia ruckeri* to turbot.

Transmission methods	Trial	No. of fish*	Weight (g)	Dose	Temperature °C	Mortality %	Duration of mortality (days)
Inoculation	no. 1	15	2.2	6.0×10^{7}**	20.0	100	2
	no. 2	10	3.0	3.5×10^{6}	20.5	100	2
		10	3.0	3.5×10^{5}		100	6
	no. 3	7	5.0	0.2×10^{5}	20.0	100	6
		7	3.9	0.2×10^{4}		100	6
	no. 4	5	6.8	220	19.0	100	8
		5	4.9	22		80	6
	no. 5	14	3.1	350	18.0	58	20
		17	3.0	35		65	20
	no. 6	10	3.8	460	14.3	80	4
		10	4.4	46		60	4
	no. 7	8	3.9	3.3×10^{3}	13.0	100	8
		8	3.5	300		100	3
		10	3.3	30		100	5
Bath	no. 1	15	2.6	1.2×10^{8}***	20.0	13	60
	no. 2	14	3.2	5.0×10^{7}	17.0	28	20
	no. 3	10	4.0	1.2×10^{8}	14.3	100	6
	no. 4	10	5.0	8.0×10^{7}	12.5	20	6

* for each trial a same number of control fish were injected or immersed in a sterile solution.

** inoculation: number of bacteria injected/fish.

*** bath: final concentration of bath in number of bacteria ml^{-1}.

The high mortality induced by injection ranged from 20 to 100% and never exceeded a period of 20 days (0% in the control). The percent mortality was a function of the infectious dose. Fish were readily infected in 2 to 3 days. Nevertheless, significant mortality was obtained in injected sea bass only with a dose from 10^3 to 10^5 bacteria/fish, whereas injected turbot died in a repetitive way with tens of bacteria per fish.

Losses in bath-infected fish never exceeded 20% (0% in control). This lower mortality spread over a long period, up to 2 months. Nevertheless, one trial with turbot with the same experimental conditions induced a loss of 100%.

The clinical signs in ERMD-infected fish were similar to those reported for rainbow trout.

In sea bass, the external signs of the disease were subcutaneous hemorrhages into and around the mouth and oral cavity, at the base of the fins, and around the vent. These lesions were often observed in moribund injected sea bass, and in few cases after the bath.

Moribund turbot displayed variable external signs of the disease. Only some petechial hemorrhages in the jaws and the eyes were observed.

At autopsy, the fish showed hemorrhages in the lower intestine and congestion of the abdominal cavity. In addition, turbot displayed petechial hemorrhages in the liver and muscles and abundant reddish ascitic fluid.

DISCUSSION AND CONCLUSION

From our results it appears that juveniles of sea bass and turbot are susceptible to ERMD following inoculation and bath infection at 18° to 20°C, the normal temperature of fish farming, and at 12.8° to 14.3°C. The same results were observed in susceptibility of sea bass and turbot to viral hemorrhagic septicemia by Castric and De Kinkelin (1984). Except for trial No. 3 in the case of the turbot where the 100% loss remains unexplained, the mortalities recorded after water-borne challenges were lower than those usually obtained in salmonids. As demonstrated by Bullock et al. (1981), water-borne challenge with 10^8 to 10^9 bacteria/ml^{-1} in 10-g rainbow trout induced a loss of 65 to 95%. The injected infectious dose in turbot, about 10 to 50 bacteria per fish, lower than this in sea bass, agreed with the LD_{50} dose, 30 bacteria per fish, defined in rainbow trout by Busch (1982).

Turbot appears to be more susceptible to *Y. ruckeri* than sea bass under experimental conditions. This suggests that this species would also be susceptible to natural epizootics of ERMD when reared in sea farms and exposed to effluents from diseased salmonids.

Two natural outbreaks of yersiniosis have been diagnosed recently in turbot hatcheries in France – the first from juveniles (Michel et al., 1986) with high mortality, the second from 100-g fish (CNEVA - PAA). In this group of turbot reared at 17°C the mortality was low, about 1% during a week. This suggests a chronic form of the ERMD. However, considering the expensive cost of such fish, this mortality was an important economical loss.

LITERATURE CITED

Bullock, G.L., E.B. Shotts and C. Starliper. 1981. Biochemical, serological, and virulence studies with *Yersinia ruckeri*. Proc. 5th Ann. FHS/AFS Workshop. Starkville, Mississippi.

Busch, R.A. 1982. Enteric redmouth disease (*Yersinia ruckeri*). In: "Antigens of Fish Pathogens: Development and Production for Vaccines and Serodiagnostics," Symposium International de Talloires, 10, 11, 12 May, 1982 (D.P. Anderson, M. Dorson, and Ph. Dubourget, eds.), pp. 201-223.

Castric, J. and P. De Kinkelin. 1984. Experimental study of the susceptibility of two marine fish species, sea bass (*Dicentrarchus labrax*) and turbot (*Scophthalmus maximus*), to viral haemorrhagic septicaemia. Aquaculture, 41, 203-212.

Dwilow, A.G., B.W. Souter and K. Knight. 1987. Isolation of *Yersinia ruckeri* from burbot, *Lota lota* (L.) from the Mackenzie River, Canada. J. Fish Dis., 10, 315-317.

Fuhrman, H., K.H. Böhm and H.-J. Schlotfeldt. 1984. On the importance of enteric bacteria in the bacteriology of freshwater fish. Bull. Eur. Assoc. Fish Path., 4, 42-46.

Lesel, R., M. Lesel, F. Gavini and A. Vuillaume. 1983. Outbreak of enteric redmouth disease in rainbow trout, *Salmo gairdneri*-Richardson, in France. J. Fish Dis., 6, 385-387.

McArdle, J.F. and C. Dooley-Martin. 1985. Isolation of *Yersinia ruckeri* type I (Hagerman strain) from goldfish *Carassius auratus* (L.). Bull. Eru. Assoc. Fish. Pathol., 5, 10-11.

Michel, C., B. Faibre and P. De Kinkelin. 1986. A clinical case of enteric redmouth in minnows (*Pimephales promelas*) imported in Europe as bait-fish. Bull. Eur. Assoc. Fish. Pathol., 6, 97-99.

Ross, A.J., R.R. Rucker and W.H. Ewing. 1966. Description of a bacterium associated with redmouth disease of rainbow trout (*Salmo gairdneri*). Can. J. Microbiol., 12, 763-770.

Vuillaume, A., R. Brun, P. Chene, E. Sochon and R. Lesel. 1987. First isolation of *Yersinia ruckeri* from sturgeon, *Acipenser baeri* Brandt, in south west of France. Bull. Eur. Assoc. Fish. Pathol., 7, 18-19.

ADHESIVE PROPERTIES AND OTHER VIRULENCE FACTORS IN *YERSINIA RUCKERI*

Jesús L. Romalde, Manuel L. Lemos, Ramón F. Conchas,
Isabel Bandín, and Alicia E. Toranzo

Departamento de Microbiología y Parasitología, Facultad de Biología,
Universidad de Santiago. Santiago de Compostela 15706, Spain

ABSTRACT: *Yersinia ruckeri* is the causative agent of enteric redmouth disease (ERMD) of several salmonid and non-salmonid species reared in both fresh and marine waters (Dalsgaard and Rasmussen, 1987). Although a great number of reports about the epidemiological and serological aspects of *Y. ruckeri* exist (Stevenson and Airdrie, 1984), few studies have been performed to understand the virulence mechanism of this fish pathogen.

In the present study, the virulence attributes of 24 strains of *Y. ruckeri* from different hosts and geographic origin were analyzed with the aim of establishing the possible relationship among some of these phenotypic characters and pathogenicity. Several environmental and clinical isolates belonging to other species of the genus *Yersinia* were included for comparative purposes.

Autoagglutination (AA), hydrophobicity, and hemagglutination were evaluated to determine the cell surface properties of our strains. Hemolytic and cytotoxic activities were determined as previously described (Toranzo et al., 1983) and siderophore production was performed on CAS medium (Schwyn and Neilands, 1987) and by cross-feeding assays (Lemos et al., 1988). The plasmid profile of the *Y. ruckeri* isolates was also analyzed.

Regardless of their serotype, the majority of the *Y. ruckeri* strains grew at 37°C and in general were negative in all the tests conducted. Only the type strain NCMB 1316 was hemagglutinating, hydrophobic, and displayed cytotoxic response in both poikilotherm and homeotherm cell lines. All of the isolates compared in this study showed ability to grow under iron limiting conditions and were positive on CAS medium, but failed to cross-feed the indicator strains for production of enterobactin and aerobactin siderophores.

Pathology in Marine Science

The overall results indicate that these virulence determinants analyzed are not adequate criteria of pathogenicity in *Y. ruckeri*.

Dalsgaard, I. and H.B. Rasmussen. 1987. 3rd Int. Conf. Fish Pathol. (EAFP) pp. 53.
Lemos, M.L., P. Salinas, A.E. Toranzo, J.L. Barja and J.H. Crosa. 1988. J. Bacteriol., 170.
Schwyn, B. and J.B. Neilands. 1987. Anal. Biochem., 160, 47-56.
Stevenson, R.M.W. and D.W. Airdrie. 1984. J. Fish Diseases, 7, 247-254.
Toranzo, A.E., J.L. Barja, R.R. Colwell, F.M. Hetrick and J.H. Crosa. 1983. FEMS Microbiol. Lett., 18, 257-262.

INTRODUCTION

Yersinia ruckeri is the causative agent of enteric redmouth disease (ERMD) affecting salmonids and non-salmonids reared in both fresh and marine waters (Stevenson and Daly, 1982; Dalsgaard and Rasmussen, 1987; Villaume et al., 1987).

ERMD was initially reported among hatchery-reared rainbow trout (*Salmo gairdneri*) from the Hagerman Valley in Idaho, USA (Rucker, 1966). No descriptions of ERMD in Europe were reported until 1978 when *Y. ruckeri* was isolated in a fish farm in Italy (Busch, 1978). The disease has since been observed in Great Britain (Roberts, 1983), France (Lesel et al., 1983), Germany (Fuhrman et al., 1983), Denmark (Dalsgaard et al., 1984), Spain (Cruz et al., 1986), and Australia (Bullock et al., 1978). The disemination of the disease was attributed to the transport of asymptomatic carriers of the bacterium (Bullock et al., 1978).

Y. ruckeri was classically divided into two serotypes that correlated biochemically with the ability of some strains to ferment sorbitol (O'Leary et al., 1979). Strains that did not ferment sorbitol and were strongly virulent for fish were considered as serotype 1 (Bullock et al., 1978), whereas the fermenting sorbitol strains were included in the second serotype. However, additional investigations demonstrated a great antigenic diversity among the strains belonging to serotype 2 (Stevenson and Daly, 1982; Stevenson and Airdrie, 1984). Then, two biotypes were established: biotype 1 corresponds to serotype 1 (sorbitol-negative strains) and biotype 2 with a total of five different "O" serotypes. Although field observations indicated that the majority of the isolates of biotype 2 were obtained from asymptomatic fish, recent studies have demonstrated the virulence of strains of this biotype of *Y. ruckeri* (Cipriano et al., 1986; Raymond et al., 1988).

In contrast to the great number of reports concerning the epidemiological and serological aspects of *Y. ruckeri*, few studies have been performed to understand the virulence mechanism in comparison with other bacterial fish pathogens such as *Vibrio* and *Aeromonas* (Toranzo et al., 1983b,c, 1987; Lemos et al., 1988; Santos et al., 1987; Janda et al., 1987; Ellis et al., 1988). In this work, we analyzed cell surface properties, putative extracellular virulence factors, and several phenotypic traits in *Y. ruckeri* strains isolated from fish culture systems with the aim to establish their possible relationship with pathogenicity.

MATERIAL AND METHODS

Bacterial strains. In this study, we used 24 strains of *Yersinia ruckeri* from different hosts and geographic origin (Table 1). Several environmental and clinical isolates of other species of the genus *Yersinia, Y. kristesenii, Y. intermedia, Y. fredericksenii,* and *Y. enterocolitica*, were included for comparison.

The taxonomic positions of these strains were confirmed using conventional tube and plate tests and the miniaturized 20E Api system (Analytab Products, Plainview, New York) and their serotypes were determined by the slide agglutination method as previously described (Toranzo et al., 1987).

All the strains were routinely cultured on tryptic soy agar (TSA) or broth (TSB) (Difco) at 25°C for 24 hr. The stock cultures were maintained on TSA slants at 15°C under mineral oil and frozen at -70°C in TSB with 15% (v/v) glycerol.

Characterization of autoagglutinating strains. The isolates were evaluated for the ability to autoagglutinate (AA) in brain heart infusion broth (BHIB) (Oxoid). To know the influence of the growth temperature on autoagglutination, two tubes containing 6 ml of sterilized BHIB were inoculated with each strain and were incubated for 18 hr, one at 25°C and the other at 37°C, in static culture. After the incubation period, strains displaying self-pelleting were considered positive for autoagglutination (SP^+).

After this determination, cultures were divided in two equal fractions. One of these fractions was held at room temperature for 1 hr, whereas the other was placed in a boiling water bath for the same period. Strains that showed a reduction in the turbidity of heated tubes were considered positive for precipitation after boiling (PAB^+).

Hydrophobicity assays. *Yersinia* isolates were evaluated for their relative cell surface hydrophobicity by the salt aggregation test (SAT) and adherence to nitrocellulose filters (NCF).

The SAT was performed following the procedures of Lindhal et al. (1981). The bacteria were washed and suspended in 0.002 M sodium phosphate buffer (pH 6.8). The optical density of the suspension was adjusted to 1.0 at 420 nm. An aliquot of 30 µl of this suspension was mixed with an equal volume of various concentrations of ammonium sulphate $(NH_4)_2SO_4$ (from 0.05-4 M) on glass slides and rotated for 2 min at room temperature. The SAT value was defined as the lowest molarity of $(NH_4)_2SO_4$ that caused a visible agglutination of the test organism.

The adherence to nitrocellulose filters was determined basically as described by Lachica and Zink (1984). Strains were suspended in saline solution (0.9% NaCl, pH 7.2) and adjusted to an absorbance of 1.0 at 600 nm (A_{600}). Five ml of this suspension were passed through an 8.0 µm pore size NCF (type SC; Millipore Corp.), and the turbidity of the resultant filtrate was compared with that of the initial bacterial suspension (control). The relative NCF binding efficiency (%) of each strain was calculated as follows:

$$\{[A_{600}\text{ (control)} - A_{600}\text{ (NCF treated)}] / A_{600}\text{ (control)}\} \times 100$$

Table 1. Bacterial strains used in this study.

STRAIN	SOURCE	DONOR[a]	COUNTRY	SEROTYPE
Yersinia ruckeri				
NCMB 1316	*Salmo gairdneri*	NCMB	U.K.	1
NCMB 2194	*Salmo gairdneri*	NCMB	U.K.	1
860627-1/1	*Salmo garidneri*	I. Dalsgaard	Denmark	1
11.4	*Salmo gairdneri*	D.P. Anderson	U.S.A.	1
11.29	*Oncorhychus tshawytscha*	D.P. Anderson	U.S.A.	2
11.40	*Salmo gairdneri*	G.L. Bullock	U.S.A.	1
11.47	*Salmo gairdneri*	G.L. Bullock	U.S.A.	4
PP-11	*Salmo gairdneri*	J.L. Romalde	Spain	1
PP-12	*Salmo gairdneri*	J.L. Romalde	Spain	1
PP-31	*Salmo gairdneri*	J.L. Romalde	Spain	1
FP-11	*Salmo gairdneri*	J.L. Romalde	Spain	1
FP-13	*Salmo gairdneri*	J.L. Romalde	Spain	1
FP-21	*Salmo gairdneri*	J.L. Romalde	Spain	1
FP-32	*Salmo gairdneri*	J.L. Romalde	Spain	1
AG-52	*Salmo gairdneri*	J.L. Romalde	Spain	1
AG-71	*Salmo gairdneri*	J.L. Romalde	Spain	1

[a]Donor address: NCMB (National Collection of Marine Bacteria), Aberdeen, Scotland (Great Britain); I. Dalsgaard, Royal Veterinary & Agricultural University, Copenhagen (Denmark); D.P. Anderson & G.L. Bullock, National Fisheries Health Laboratory, Leetown, W. Va. (U.S.A.); T. Håstein, National Veterinary Institute, Oslo (Norway); I. Márquez, Laboratorio de Sanidad Animal. Consejería de Agricultura de Asturias (Spain); E. Carniel, Unité d'Ecologie Bacterienne. Instut Pasteur. Paris (France).

(continued)

Table 1 (continued)

STRAIN	SOURCE	DONOR[a]	COUNTRY	SEROTYPE
1533/87	*Salmo salar*	T. Håstein	Norway	1
1621/87	*Salmo salar*	T. Håstein	Norway	1
1622/87	*Salmo salar*	T. Håstein	Norway	2
1638/87	*Salmo salar*	T. Håstein	Norway	2
1	*Salmo gairdneri*	I. Márquez	Spain	1
8R	*Salmo gairdneri*	I. Márquez	Spain	1
15R	*Salmo gairdneri*	I. Márquez	Spain	1
14B	*Salmo gairdneri*	I. Márquez	Spain	1
Yersinia kristesenii				
IP 16832	Sangone River	E. Carniel	Italy	–
Yersinia intermedia				
IP 16835	Po River	E. Carniel	Italy	–
Yersinia fredericksenii				
IP 16840	Po River	E. Carniel	Italy	–
Yersinia enterocolitica				
IP 383	Human	E. Carniel	France	0:9
Ye 8081	Human	E. Carniel	France	0:8
IP 16849	Mud	E. Carniel	France	0:39

[a]Donor address: NCMB (National Collection of Marine Bacteria), Aberdeen, Scotland (Great Britain); I. Dalsgaard, Royal Veterinary & Agricultural University, Copenhagen (Denmark); D.P. Anderson & G.L. Bullock, National Fisheries Health Laboratory, Leetown, W. Va. (U.S.A.); T. Håstein, National Veterinary Institute, Oslo (Norway); I. Márquez, Laboratorio de Sanidad Animal. Consejería de Agricultura de Asturias (Spain); E. Carniel, Unité d'Ecologie Bacterienne. Instut Pasteur. Paris (France).

Hemagglutination assays and inhibition of hemagglutination. Blood from rainbow trout was collected directly into Alsever's solution from the caudal peduncle of killed fish. Human group 0 blood was obtained by venipuncture and also immediately mixed with Alsever's solution.

According to Toranzo et al. (1983c), blood cells were washed three times in PBS (pH 7.4) and a 3% (v/v) suspension was prepared in this saline solution. Suspensions of *Yersinia* strains were prepared in PBS and adjusted by absorbance to 10^{10} cells/ml. Slide agglutinations were performed at room temperature by mixing 20 μl of a blood suspension with an equal volume of bacterial suspensions and gently rocking the slide by hand. A PBS-blood control was always included. Strains were considered positive hemagglutinators if visible agglutination occurred within 10 min. Inhibition of hemagglutination was performed essentially as the agglutination test using 1% (w/v) solution in PBS of D-mannose (Sigma).

Hemolytic activity. Hemolysin production was detected on TSA plates containing 5% final concentration of defibrinated blood from rainbow trout, coho salmon (*Oncorhynchus kisutch*), striped bass (*Morone saxatilis*), rabbit, sheep, and human. Results were recorded after 24 hr and 48 hr of incubation at 25°C.

Calcium dependence. The test was performed using magnesium oxalate agar plates (MGOX) (BHI-agar supplemented with 8% (v/v) of 0.25 M $MgCl_2$ and 8% (v/v) of 0.25 M sodium oxalate) (De Grandis and Stevenson, 1982) containing oxalate ions to chelate calcium ions.

Strains were streaked on two MGOX plates and on two plates without magnesium and oxalate ions. One plate of each pair was incubated at 37°C and the other at 25°C. A calcium dependent strain does not grow on MGOX plates incubated at 37°C or grows very slowly forming pinpoint colonies; however, it will form normal colonies in 24-48 hr on all other plates.

Cytotoxicity assays. Cultures in TSB from all the isolates were incubated at 25°C for 24 hr in a rotary shaker and centrifuged at 9,000*g* for 10 min at 4°C. Supernatant fluids were filtered and the cytotoxicity of extracellular products was measured as previously described (Toranzo et al., 1983c) using four cell lines: Chinook salmon embryo (CHSE-214), fathead minnow peduncle (FHM), epithelioma papillosum of carp (EPC) and mouse lung fibroblast (L-929).

For the toxicity test, the different cell lines, grown as monolayers in 24-well culture plates, were inoculated with 0.1 ml of each culture filtrate. Non-inoculated monolayers as well as cells added to fresh TSB were used as controls. Plates were incubated at 18°C (fish cell lines) and 37°C (homoeothermic cell line) and wells showing totally or partially destroyed monolayers within a period of 3 days were scored as positive cytotoxic responses.

Plasmid contents. The extraction of plasmid DNA from the *Y. ruckeri* isolates was conducted as previously described (Toranzo et al., 1983a) using small volumes of bacterial cultures (3 ml). The presence of extrachromosomal DNA was analyzed by electrophoresis through 0.7% agarose (Type I, Sigma) in tris-acetate buffer (40 mM Tris, 2 mM Na_2-EDTA); adjusted to pH 7.9 with glacial acetic acid) at 150 V, 110 mA, for 4 hr on a horizontal apparatus. Plasmids from *Escherichia coli* V517 ranging from 35.8 to 1.4 megadaltons (Md) and *Yersinia enterocolitica* IP 383 with a plasmid of

47 Md was included to estimate the molecular weight of the DNA plasmids in our *Y. ruckeri* strains.

RESULTS

Autoagglutinating activity. Of 30 *Yersinia* strains (24 *Y. ruckeri* and 6 of the other species) evaluated for the ability to autoagglutinate before or after boiling, only four strains belonging to *Y. kristesenii, Y. intermedia, Y. fredericksenii,* and *Y. enterocolitica* were AA^+ (Table 2). Two of these four AA^+ strains were negative to self pelleting and positive to precipitation after boiling (SP^-, PAB^+), and the other two isolates were positive in the two methods tested (SP^+, PAB^+). None of the *Y. ruckeri* strains displayed this cell surface property.

Hydrophobicity of *Yersinia* strains. In general, *Y. ruckeri* isolates showed to be a low hydrophobic species regardless of their serotype. However, differences could be detected depending on the method used to determine this property (Table 2).

According to the criteria of hydrophobicity proposed in our laboratory (Santos et al., 1989), a great number of strains (15) were shown to be nonhydrophobic by SAT because these agglutinated in a concentration of $(NH_4)_2SO_4$ higher than 4 M. Eight strains displayed values in the range from 2.0-4.0 M being considered weakly hydrophobic, and only the reference strain *Y. ruckeri* NCMB 1316 exhibited a strong hydrophobicity since it had an agglutination value of 0.2 M ammonium sulphate.

The adherence to nitrocellulose filters displayed more heterogeneous results within the isolates of *Y. ruckeri* than the SAT method. Although the majority of *Y. ruckeri* tested displayed a low capacity to adhere to nitrocellulose filters (less than 50%), strains showing a high value of hydrophobicity were observed (Table 2).

Growth at 37°C and calcium dependence. The existence in *Y. ruckeri* of temperature-calcium dependence that is present in some clinical species of *Yersinia* was assessed by growth tests on magnesium oxalate agar at 37° and 25°C. The majority of our *Y. ruckeri* isolates were able to grow on either the control or magnesium oxalate agar plates at both temperatures, which indicates that this fish pathogen does not possess the calcium dependence property (Table 2).

Hemagglutinating and hemolytic activities. We studied the hemagglutinating activity for rainbow trout and human erythrocytes in our isolates. None of the tested strains showed a positive agglutination for human red blood cells. Only two strains belonging to *Y. ruckeri* species produced hemagglutinins for fish erythrocytes (Table 3). These hemagglutinations were sensitive to inhibition by D-mannose, suggesting that this carbohydrate, or an analog, is part of a receptor on the erythrocytes for a bacterial adhesion.

In the hemolytic assays, only one isolate of *Y. ruckeri* (FP-13) and the *Y. fredericksenii* strain displayed lytic activity against human erythrocytes. All the other strains showed no lytic activities in the other poikilothermic and homoeothermic blood cells used (Table 3).

Table 2. Autoagglutination, calcium-dependence and hydrophobicity of *Yersinia* strains.

STRAIN	SP[a]		PAB[b]	GROWTH AT 37°C	CALCIUM DEPENDENCE	HYDROPHOBICITY	
	25°C	37°C				SAT[c]	NCF[d]
Y. ruckeri							
NCMB 1316	–	–	–	–	–	0.2	87.3
NCMB 2194	–	–	–	–	–	>4	35.7
860627-1/1	–	–	–	+	–	>4	5.2
11.4	–	–	–	+	–	3.5	20
11.29	–	–	–	+	–	2	80
11.40	–	–	–	+	–	2	0
11.47	–	–	–	+	–	2	24
PP-11	–	–	–	+	–	>4	NT[e]
PP-12	–	–	–	+	–	>4	20
PP-31	–	–	–	+	–	>4	15
FP-11	–	–	–	+	–	>4	0
FP-13	–	–	–	+	–	>4	20
FP-21	–	–	–	+	–	>4	NT
FP-32	–	–	–	+	–	>4	0

[a]Self pelleting.
[b]Precipitation after boiling.
cSalt aggregation test, values represent the lowest molarity of ammonium sulphate yielding a strong aggregation of bacteria.
[d]Adherence to Nitrocellulose filters, values represent the % of binding efficiency.
[e]Not Tested.

(continued)

Table 2 (continued)

STRAIN	SP[a] 25°C	SP[a] 37°C	PAB[b]	GROWTH AT 37°C	CALCIUM DEPENDENCE	HYDROPHOBICITY SAT[c]	HYDROPHOBICITY NCF[d]
AG-52	–	–	–	+	–	>4	63.6
AG-71	–	–	–	+	–	>4	55.5
1533/87	–	–	–	+	–	2	10
1621/87	–	–	–	+	–	2	93
1622/87	–	–	–	+	–	3	40
1638/87	–	–	–	+	–	2	88.2
1	–	–	–	+	–	>4	22.9
8R	–	–	–	+	–	>4	26.3
15R	–	–	–	+	–	>4	46.3
14B	–	–	–	+	–	>4	15
Y. kristesenii							
IP 16832	–	–	+	+	–	2.5	23.5
Y. intermedia							
IP 16835	(+)	(+)	+	+	–	2.5	40.9

[a]Self pelleting.
[b]Precipitation after boiling.
cSalt aggregation test, values represent the lowest molarity of ammonium sulphate yielding a strong aggregation of bacteria.
[d]Adherence to Nitrocellulose filters, values represent the % of binding efficiency.
[e]Not Tested.

(continued)

Table 2 (continued)

STRAIN	SP[a]		PAB[b]	GROWTH AT 37°C	CALCIUM DEPENDENCE	HYDROPHOBICITY	
	25°C	37°C				SAT[c]	NCF[d]
Y. fredericksenii							
IP 16840	–	–	–	+	–	1.5	49
Y. enterocolitica							
IP 383	–	+	(+)	+	–	3	NT
Ye 8081	–	–	–	+	–	2	100
IP 16849	–	–	+	+	–	3	NT

[a]Self pelleting.
[b]Precipitation after boiling.
cSalt aggregation test, values represent the lowest molarity of ammonium sulphate yielding a strong aggregation of bacteria.
[d]Adherence to Nitrocellulose filters, values represent the % of binding efficiency.
[e]Not Tested.

Table 3. Hemagglutinating and hemolytic activities of the *Yersinia* strains used in this study.

STRAINS	Hemagglutination		Hemolysis of erythrocytes from					
	Trout erythrocytes	Human erythrocytes	Trout	Salmon	Striped Bass	Rabbit	Sheep	Human
Y. ruckeri								
NCMB 1316	+MSHA[a]	–	–	–	–	–	–	–
NCMB 2194	–	–	–	–	–	–	–	–
860627-1/1	–	–	–	–	–	–	–	–
11.4	(+)MSHA	–	–	–	–	–	–	–
11.29	–	–	–	–	–	–	–	–
11.40	–	–	–	–	–	–	–	–
11.47	–	–	–	–	–	–	–	–
PP-11	–	–	–	–	–	–	–	–
PP-12	–	–	–	–	–	–	–	–
PP-31	–	–	–	–	–	–	–	–
FP-11	–	–	–	–	–	–	–	–
FP-13	–	–	–	–	–	–	–	+
FP-21	–	–	–	–	–	–	–	–
FP-32	–	–	–	–	–	–	–	–
AG-52	–	–	–	–	–	–	–	–
AG-71	–	–	–	–	–	–	–	–
1533/87	–	–	–	–	–	–	–	–
1621/87	–	–	–	–	–	–	–	–

[a]Hemagglutination sensitive to inhibition by D-mannose.

(continued)

Table 3 (continued)

STRAINS	Hemagglutination		Hemolysis of erythrocytes from					
	Trout erythrocytes	Human erythrocytes	Trout	Salmon	Striped Bass	Rabbit	Sheep	Human
1622/87	–	–	–	–	–	–	–	–
1638/87	–	–	–	–	–	–	–	–
1	–	–	–	–	–	–	–	–
8R	–	–	–	–	–	–	–	–
15R	–	–	–	–	–	–	–	–
14B	–	–	–	–	–	–	–	–
Y. kristesenii								
IP 16832	–	–	–	–	–	–	–	–
Y. intermedia								
IP 16835	–	–	–	–	–	–	–	–
Y. fredericksenii								
IP 16840	–	–	–	–	–	–	–	+
Y. enterocolitica								
IP 383	–	–	–	–	–	–	–	–
Ye 8081	–	–	–	–	–	–	–	–
IP 16849	–	–	–	–	–	–	–	–

[a]Hemagglutination sensitive to inhibition by D-mannose.

Cytotoxic activity on cell lines. The cytotoxic assays conducted with both pathogenic *Y. ruckeri* strains and nonpathogenic environmental isolates revealed that only eight strains displayed cytotoxic activity on some of the cell lines tested. Degenerative changes started within 3 hr or 6 hr after inoculation of the extracellular products (ECP). In general, all the positive samples produced rounding, shrinking, and vacuolization of the cells. Finally, all the cells detached from the wells.

We found three different patterns of cytotoxicity (Table 4). One pattern produced only by the ECP of *Y. ruckeri* NCMB 1316 strain with toxic effects in all the cell lines; a second pattern (three strains) displaying cytotoxicity only in FHM; and the third pattern, represented by four strains, with cytotoxic activity in FHM and EPC cells.

Heat treatment of the culture filtrates resulted in a loss of cytotoxicity, indicating the thermolabile nature of the toxins.

Table 4. Cytotoxic activity of *Yersinia ruckeri* strains in different poikilotherm and homoeotherm cell lines.

STRAINS	FISH CELL LINES			HOMEOTHERMIC CELLS
	CHSE-214	FHM	EPC	L-929
Yersinia ruckeri				
NCMB 1316	+	+	+	+
NCMB 2194	–	+	+	–
860627-1/1	–	–	–	–
11.4	–	–	–	–
11.29	–	+	–	–
11.40	–	–	–	–
11.47	–	–	–	–
PP-11	–	–	–	–
PP-12	–	–	–	–
PP-31	–	+	+	–
FP-11	–	–	–	–
FP-13	–	–	–	–
FP-21	–	–	–	–
FP-32	–	–	–	–
AG-52	–	–	–	–
AG-71	–	–	–	–
1533/87	–	–	–	–
1621/87	–	+	+	–
1622/87	–	–	–	–
1638/87	–	–	–	–

(continued)

Table 4 (continued)

STRAINS	FISH CELL LINES			HOMEOTHERMIC CELLS
	CHSE-214	FHM	EPC	L-929
1	–	–	–	–
8R	–	+	–	–
15R	–	+	–	–
14B	–	–	–	–
Yersinia kristesenii				
IP 16832	–	+	+	–
Yersinia intermedia				
IP 16835	–	–	–	–
Yersinia fredericksenii				
IP 16840	–	–	–	–
Yersinia enterocolitca				
IP 383	–	–	–	–
Ye 8081	–	–	–	–
IP 16849	–	–	–	–

Plasmid profiles. Analysis of the plasmid content revealed that the isolates of *Y. ruckeri* in our area belonging to serotype 1 contain two high molecular weight plasmid bands of about 40-50 Md and 25-30 Md, respectively. The low molecular weight plasmid band may represent open circular forms of the 40-50 Md plasmid. In the *Y. ruckeri* strains assigned to the serotype 2, no extrachromosomal DNA or only small plasmids were detected.

DISCUSSION AND CONCLUSIONS

It is known that some biochemical, enzymatic, and cell surface properties are potential indicators of virulence in other species of bacterial fish pathogens as *Aeromonas hydrophila, Aeromonas salmonicida,* or *Vibrio anguillarum* (Crosa, 1980; Toranzo et al., 1983b,c, 1987; Janda et al., 1987; Santos et al., 1987; Lemos et al., 1988; Ellis et al., 1988).

In general, the negative results found in the hemagglutinating and hemolytic capacities of the *Y. ruckeri* isolates indicate that the ability to recognize and attach to the receptors on the host cell surface does not appear to be a prerequisite for successful invasion in *Y. ruckeri*. In addition, no correlation was found between virulence and the cytotoxic activity of the *Yersinia* strains on the fish cell lines (Tables 3, 4). We have reported similar conclusions in pathogenic *Vibrio anguillarum* and motile *Aeromonas* (Toranzo et al., 1983c, 1987; Santos et al., 1987) since, although these fish pathogens are stronger producers of extracellular products than *Y. ruckeri*, none of these putative virulence factors can be used as an absolute criterion of pathogenicity for all the isolates.

The properties of autoagglutination, calcium dependence, and hydrophobicity studied in this work are recognized as possible indicators of pathogenicity in other species of the genus *Yersinia*, like *Y. enterocolitica* or *Y. pseudotuberculosis*. Interestingly, these properties are, in these species, associated with virulence plasmids, with a size ranging between 40 and 50 Md (Skurnik, 1984; Prpic et al., 1985). In agreement with De Grandis and Stevenson (1982), all the *Y. ruckeri* strains belonging to the serotype 1 strains carry, at least, a 40-50 Md plasmid band but, at difference with the findings of these authors, our plasmid-carrying strains were able to grow at 37°C on both control and magnesium oxalate agar plates. At present, the nature or function of the high molecular weight of *Y. ruckeri* plasmid remains cryptic and curing experiments are necessary to assess its possible contribution with the virulence of this fish pathogen.

The negative results in the virulence factors studied in *Y. ruckeri* support the hypothesis that a minimal percent of hybridization occurred between the chromosomal DNAs of *Y. ruckeri* and *Y. enterocolitica* (De Grandis et al., 1988) and suggest that *Y. ruckeri* and the other species of the genus *Yersinia* are very scantily related. Then, the taxonomic classification of this fish pathogen needs to be modified and *Y. ruckeri* should be considered as an independent genus.

All the findings reported here indicate that the virulence determinants analyzed are not adequate criteria of pathogenicity in *Y. ruckeri*. Therefore, it is necessary to look for other pathogenic mechanism(s) that can contribute to the successful invasion and proliferation of these bacteria in the host. In this sense, preliminary data obtained in our laboratory about the production of siderophores indicate that *Y. ruckeri* strains possess an iron uptake system. The relationship between this system and pathogenicity has been demonstrated for other species of fish pathogens, like *Vibrio anguillarum* (Crosa, 1980; Lemos et al., 1988). The characterization of this iron uptake system will allow us to assess its possible role in the virulence of *Y. ruckeri*.

ACKNOWLEDGMENTS

This work was supported by Grant AQ-0018/84 from the Comisión Asesora de Investigación Científica y Técnica (CAICYT), Ministerio de Educación y Ciencia, Spain. J.L. Romalde and I. Bandín thank the Ministerio de Educación y Ciencia and Universidad de Santiago (Spain) for research fellowships.

LITERATURE CITED

Bullock, G.L., H.M. Stuckey and E.B. Shotts. 1978. Enteric redmouth bacterium: comparison of isolates from different geographic areas. J. Fish. Dis., 1, 351,356.

Busch, R.A. 1978. Enteric redmouth disease (Hagerman strain). Mar. Fish. Rev., 40, 42-51.

Cipriano, R.C., W.B. Schill, S.W. Pyle and R. Horner. 1986. An epizootic in chinook salmon (*Oncorhynchus* Tshawytscha) caused by a sorbitol-positive serovar 2 strain of *Yersinia ruckeri*. J. Wild. Dis., 22, 488-492.

Crosa, J.H. 1980. A plasmid associated with virulence in the marine fish pathogen *Vibrio anguillarum* specifies an iron sequestering system. Nature (London), 283, 566-568.

Cruz, J.A. de la, A. Rodriguez, C. Tejedor, E. de Lucas and L.R. Orozco. 1986. Isolation and identification of *Yersinia ruckeri*, causal agent of enteric redmouth disease (ERM), for the first time in Spain. Bull. Eur. Assoc. Fish. Pathol., 6, 43-44.

Dalsgaard, I., J. From and V. Horlyck. 1984. First observation of *Yersinia ruckeri* in Denmark. Bull. Eur. Assoc. Fish. Pathol., 4, 10.

Dalsgaard, I. and H.B. Rasmussen. 1987. Biochemical and serological characteristics and antibiotic sensitivity of *Yersinia ruckeri*. 3rd Intl. Conf. Fish Pathol. (EAFP) p. 53.

De Grandis, S.A. and R.M.W. Stevenson. 1982. Variations in plasmid profiles and growth characteristics of *Yersinia ruckeri* strains. FEMS Microbiol. Lett., 15, 199-202.

De Grandis, S.A., P.J. Krell, D.E. Flett and R.M.W. Stevenson. 1988. Deoxyribonucleic acid relatedness of serovars of *Yersinia ruckeri*, the enteric redmouth bacterium. Int. J. Syst. Bacteriol, 38, 49-55.

Ellis, A.E., A.S. Burrows and K.J. Stapleton. 1988. Lack of relationship between virulence of *Aeromonas salmonicida* and the putative virulence factors: A-layer, extracellular proteases and extracellular haemolysins. J. Fish. Dis., 11, 309-323.

Fuhrman, H., K.H. Böhm and H.J. Schlotfeldt. 1983. An outbreak of enteric redmouth disease in West Germany. J. Fish. Dis., 6, 309-311.

Janda, J.M., L.S. Oshiro, S.L. Abbott and P.S. Duffey. 1987. Virulence markers of mesophilic Aeromonads: Association of the autoagglutination phenomenon with mouse pathogenicity and the presence of a peripheral cell- associated layer. Infect. Immun., 55, 3070-3077.

Lachica, R.V. and D.L. Zink. 1984. Plasmid-associated cell-surface charge and hydrophobicity of *Yersinia enterocolitica*. Infect. Immun., 44, 540-543.

Lemos, M.L., P. Salinas, A.E. Toranzo, J.L. Barja and J.H. Crosa. 1988. Chromosome-mediated iron uptake system in pathogenic strains of *Vibrio anguillarum*. J. Bacteriol., 170, 1920-1925.

Lesel, R., M. Lesel, F. Gavini and A. Villaume. 1983. Outbreak of enteric redmouth disease in rainbow trout, *Salmo gairdneri* Richardson, in France. J. Fish. Dis., 6, 385-387.

Lindhal, M., A. Faris, T. Wadstrom and S. Hjerten. 1981. A new test based on "salting out" to measure relative surface hydrophobicity of bacterial cells. Biochim. Biophys. Acta, 677, 471-476.

O'Leary, P.J., J.S. Rohovec and J.L. Fryer. 1979. A further characterization of *Yersinia ruckeri* (Enteric redmouth bacterium). Fish. Pathol., 14, 71-78.

Prpic, J.K., R.M. Robins-Browne and R.B. Davey. 1985. In vitro assessment of virulence in *Yersinia enterocolitica* and related species. J. Clin. Microbiol., 22, 105-110.

Raymond, B.T., D. Flett and R.M.W. Stevenson. 1988. Characteristics and virulence of sub-groups of *Yersinia ruckeri,* the enteric redmouth bacterium. Intl. Fish. Health Conf., Vancouver, Canada, 1988, (Abstract).

Roberts, M.S. 1983. A report of an epizootic in hatchery rainbow trout, *Salmo gairdneri* Richardson, at an English trout farm, caused by *Yersinia ruckeri*. J. Fish. Dis., 6, 551-552.

Rucker, R.R. 1966. Redmouth disease of rainbow trout (*Salmo gairdneri*). Bull. Off. Int. Epizoot., 65, 825-830.

Santos, Y., A.E. Toranzo, C.P. Dopazo, T.P. Nieto and J.L. Barja. 1987. Relationship among virulence for fish, enterotoxigenicity, and phenotypic characteristics of motile *Aeromonas*. Aquaculture, 67, 29-39.

Santos, Y., I. Bandín, T.P. Nieto, D.W. Bruno, A.E. Ellis and A.E. Toranzo. 1989. Comparison of the cell surface hydrophobicity of bacterial fish pathogens by different procedures. In: "Pathology in Marine Aquaculture" (F.O. Perkins and T.C. Cheng, eds.). Academic Press, San Diego, (This Volume).

Skurnik, M. 1984. Lack of correlation between the presence of plasmids and fimbriae in *Yersinia enterocolitica* and *Yersinia pseudotuberculosis*. J. Appl. Bacteriol., 56, 355-363.

Stevenson, R.M.W. and D.W. Airdrie. 1984. Serological variation among *Yersinia ruckeri* strains. J. Fish. Dis., 7, 247-254.

Stevenson, R.M.W. and J.G. Daly. 1982. Biochemical and serological characteristics of Ontario isolates of *Yersinia ruckeri*. Can. J. Fish. Aqua. Scien., 39, 870-876.

Toranzo, A.E., J.L. Barja, R.R. Colwell and F.M. Hetrick. 1983a. Characterization of plasmids in bacterial fish pathogens. Infect. Immun., 39, 184-192.

Toranzo, A.E., J.L. Barja, S.A. Potter, R.R. Colwell, F.M. Hetrick and J.H. Crosa. 1983b. Molecular factors associated with virulence of marine vibrios isolated from striped bass in Chesapeake Bay. Infect. Immun., 39, 1220-1227.

Toranzo, A.E., J.L. Barja, R.R. Colwell, F.M. Hetrick and J.H. Crosa. 1983c. Haemagglutinating, haemolytic and cytotoxic activities of *Vibrio anguillarum* and related vibrios isolated from striped bass on the Atlantic coast. FEMS Microbiol. Lett., 18, 257-262.

Toranzo, A.E., Y. Santos, M.L. Lemos, A. Ledo and J. Bolinches. 1987. Homology of *Vibrio anguillarum* strains causing epizootics in turbot, salmon and trout reared on the Atlantic coast of Spain. Aquaculture, 67, 41-52.

Villaume, A., R. Brun, P. Chene, E. Sochon and R. Lesel. 1987. First isolation of *Yersinia ruckeri* from sturgeon, *Acipenser baeri* Brandt, in south west of France. Bull. Eur. Assoc. Fish. Pathol., 7, 18-19.

MYCOSES SECTION

A SYNOPSIS OF MYCOTIC DISEASES OF MARINE FISHES AND INVERTEBRATES

Edward J. Noga

Department of Companion Animal and Special Species Medicine,
College of Veterinary Medicine, North Carolina State University,
4700 Hillsborough Street, Raleigh, North Carolina 27606 USA

ABSTRACT: This brief review presents major examples of fungi and related organisms that have been reported to be associated with diseases of marine fish and invertebrates. Fungi are generally opportunistic invaders, especially of immunologically compromised hosts. Moreover, the invasion of hosts by fungi commonly represents a "dead end" in the life cycle of these organisms. Current knowledge about fungal pathogens of the categories of marine hosts under consideration is reviewed.

INTRODUCTION

Most descriptions and research on fungal diseases of marine animals have focused on commercially important aquaculture or fishery populations. Thus, most pathogens have been described from either finfishes, bivalve molluscs, or decapod crustaceans (Table 1). While the recorded list of pathogenic fungi affecting some marine groups, such as fishes, is relatively short, this may reflect the smaller degree of attention directed at these pathogens compared to their freshwater counterparts. If so, this discrepancy will probably disappear with increasing interest in marine aquaculture and studies of marine fishery populations. However, less than 500 species of fungi (including nonpathogenic species) have been isolated from oceans and estuaries (Kohlmeyer and Kohlmeyer, 1979). This constitutes less than 1% of all fungal taxa, suggesting that the fungal fauna of marine ecosystems may be relatively depauperate compared to freshwater or terrestrial environments (Kohlmeyer and Kohlmeyer, 1979).

This overview is not intended to be a comprehensive discussion of fungal agents that infect marine animals. The reader should refer to recent reviews such as those of Alderman (1982), Alderman and Polglase (1986), Neish and Hughes (1980), Sindermann and Lightner (1988), Porter (1986), and Sparks (1985) for this information.

Pathology in Marine Science

Table 1. Fungal pathogens isolated from marine and estuarine fishes and invertebrates.

Taxon	Pisces	Bivalvia	Decapoda	Other Invertebrates	References
Myxomycota					
Thraustochytriales					
Schizochytrium sp. (?)	*Salmo gairdneri*				Polglase (1981)
Ulkenia amoeboidea				Cephalopoda: *Eledone cirrhosa*	Polglase (1981)
Unknown				Cephalopoda: *Illex illecebrosus*	Jones and O'Dor (1983)
				Nudibrachia: *Tritonia diomedea*	McLean and Porter (1987)
Eumycota					
Mastigomycotina					
Oomycetes					
Leptolegniella marina (= *Leptolegnia marina*)		*Cardium echinatum*[1] *Barnea candida*[1]	*Pinnotheres pisum*[1] *Callinectes sapidus*[2]		[1]Atkins (1954a); [2]Johnson and Pinschmidt (1963)
Leptolegnia baltica				Copepoda: *Eurytemora hirundoides*	Höhnk and Vallin (1953)
Haliphthoros milfordensis			*Pinnotheres* (E)[1] *Homarus americanus*[2] *Homarus vulgaris*[2] *Penaeus setiferus*[3] *Callinectes sapidus* (E)[3]	Gastropoda: *Urosalpinx cinerea*[1] Branchiopoda: *Artemia salina* (E)[3]	[1]Vishniac (1958) [2]Fisher (1988); [3]Tharp and Bland (1977)
Haliphthoros phillipinensis			*Penaeus monodon*		Hatai et al. (1980)

(E) indicates that this was an experimentally infected host.

(continued)

Table 1 (continued)

Taxon	Pisces	Bivalvia	Decapoda	Other Invertebrates	References
Atkinsiella dubia (= *Plectospira dubia*)			*Pinnotheres pisum* *Gonoplax rhomboides* *Trypton spongicola* (E) *Crangon vulgaris* (E) *Leander serratus* (E) *Portunus depurator* (E) *Macropodia* sp. (E)		Atkins (1954b)
Atkinsiella hamanaensis			*Scylla serrata*		Bian and Egusa (1980)
Lagenidium callinectes			*Callinectes sapidus*[1] *Penaeus* spp. (All?)[3] *Geryon quinquidens*[4] *Palaemon macro-dactylus*[5] *Panopeus herbstii*[6] *Libinia dubia*[6] *Menippe mercenaria* (E)[6] *Neopanope texiana* (E)[7] *Pinnotheres ostreum* (E)[7]	Cirripedia: *Chelonibia patula*[2]	[1]Couch (1942) [2]Johnson and Bonner (1960) [3]Lightner (1988a) [4]Cassels and Krebs (1983) [5]Fisher (1983) [6]Bland and Amerson (1974) [7]Sindermann (1988a)
Lagenidium chthamalophilum				Cirripedia: *Chthamalus fragilis denticulata*	Johnson (1958)
Lagenidium scyllae			*Scylla serrata*		Bian et al. (1977)
Lagenidium sp.			*Homarus americanus*[1] *Homarus vulgaris*[1] *Cancer magister*[2]	Branchiopoda: *Artemia salina*[1]	[1]Fisher (1988) [2]Armstrong and Fisher (1988)

(E) indicates that this was an experimentally infected host.

(continued)

Table 1 (continued)

Taxon	Pisces	Bivalvia	Decapoda	Other Invertebrates	References
Sirolpidium zoophthorum		*Mercenaria mercenaria* *Mercenaria mortoni* *Crassostrea virginica* *Teredo* (E) *Argopecten* (E) *Tapes* (E)			Davis et al. (1954)
Sirolpidium sp.			*Penaeus* spp. (All?)		LeBitoux (pers. comm., quoted in Lightner [1988a])
Pythium thalassium			*Pinnotheres pisum* *Crangon vulgaris* (E) *Leander serratus* (E) *Portunus depurator* (E) *Macropodia* sp. (E) *Palaemon serratus*		Atkins (1955)
Pythium afertile (?)					Delves-Broughton and Poupard (1976)
Saprolegnia sp.	*Brevoortia tyrannus*				Dykstra et al. (1989)
Aphanomyces sp.	*Dorosoma cepedianum* *Brevoortia tyrannus*				Dykstra et al. (1989)
Deuteromycotina					
Hyphomycetes					
Fusarium solani	*Melichthys vidua*[1] *Sphyrna tiburo*[4]		*Penaeus californiensis*[2] *Penaeus japonicus*[2] *Penaeus monodon*[2] *Penaeus stylirostris*[2] *Penaeus vannamei*[2] *Homarus americanus*[3]		[1]Ostland et al. (1987) [2]Hose et al. (1984) [3]Sinderman (1988b) [4]Muhvich et al. (1989)

(E) indicates that this was an experimentally infected host.

(continued)

Table 1 (continued)

Taxon	Pisces	Bivalvia	Decapoda	Other Invertebrates	References
Fusarium oxysporum	*Pagrus* sp.				Hatai et al. (1986)
Fusarium sp.			*Penaeus duorarum*		Lightner (1981)
Cladosporium sphaerospermum				Cephalopoda: *Eledone cirrhosa*	Polglase et al. (1984)
Cladosporium cladosporioides			*Homarus americanus*		Miller and Fleming (1983)
Cladosporium sp.	*Gadus morhua*				Reichenbache-Klinke (1956)
Septocylindricum eriocheir			*Eriocheir sinensis*		Mann and Pipelow (1938)
Ramularia branchialis			*Homarus vulgaris* (= *H. gammarus*)		Sordi (1958)
Didymaria palinuri			*Palinurus vulgaris*		Sordi (1958)
Aureobasidium sp.(?)	*Trygon pstonacea*				Alderman (1982)
Exophiala salmonis	*Salmo salar*				Richards et al. (1978)
Exophiala piscifila	*Plotosus anguillaris* *Mustelus canis*				Gaskins and Cheung (1986)
Exophiala-like	*Stenotomus chrysops* *Gadus morhua* *Hippocompus hudsonius* *Xanthichthys ringens* *Amphiprion sebae* *Tautoglabrus adspersus* (E) *Pseudopleuronectes americanus* (E) *Fundulus heteroclitus* (E)				Blazer and Wolke (1979)
Sarcinomyces crustaceus	*Spondyliosoma cantharus*				Todaro et al. (1983)

(E) indicates that this was an experimentally infected host.

(continued)

Table 1 (continued)

Taxon	Pisces	Bivalvia	Decapoda	Other Invertebrates	References
Blastomycetes					
Cryptococcus gammari				Amphipoda: *Gammarus*	Pixell-Goodrich (1928)
Rhodotorula minuta				Amphipoda: *Podoceras brasiliensis*	Roth et al. (1962)
Ascomycotina					
Hemiascomycetes					
Metschnikowia artemiae				Branchipoda: *Artemia salina*	Codreanu & Codreanu-Balcescu (1981)
Metschnikowia bicuspidata var. *australis*				Branchiopoda: *Artemia salina*	Lachance et al. (1976)
Metschnikowia sp.				Copepoda: *Calanus plumchrus*	Seki and Fulton (1969)
Metschnikowia sp.				Copepoda: *Eurytemora velox*	Fize et al. (1970)
Pyrenomycetes					
Trichomaris invadens			*Chionoecetes bairdi* *Chionoecetes opilio* *Chionoecetes tanneri*		Hibbits et al. (1981)
Uncertain taxonomy					
Ichthyophonus hoferi	*Clupea harengus* *Alosa pseudoharengus* *Scomber scombrus* *Pleuronectes platessa* many others				McVicar (1982)

(E) indicates that this was an experimentally infected host.

(continued)

Table 1 (continued)

Taxon	Pisces	Bivalvia	Decapoda	Other Invertebrates	References
Ostracoblabe implexa		*Ostrea edulis*[1] *Crassostrea cucullata*[2]			[1]Alderman and Jones (1971) [2]Raghukumar and Lande (1988)
Rosette Agent	*Oncorhynchus tshawytscha*				Harrell et al. (1986)
Spongiophaga communis				Porifera: *Ircinia Cacospongia*	Polejaeff (1884, cited in Lauckner (1980))
Spongiophaga (?)				Porifera: *Hippospongia gossypine* *Hippospongia lachne* *Spongia graminae* *Spongia barbara* *Spongia dura* *Spongia obliqua*	Galtsoff (1942)

(E) indicates that this was an experimentally infected host.

In this review, I will summarize the various fungal agents that are currently known to infect marine and estuarine fishes and invertebrates, with emphasis on recent information that has not been covered in previous reviews. Hopefully, this discussion will provide some insight into biological relationships shared by these target groups and how they relate to clinical management of these diseases.

General characteristics of marine fungal infections. Our knowledge of marine fungal diseases is rudimentary, consisting primarily of the identification and pathology of the etiological agents. The mechanisms regulating their invasiveness and pathogenicity are poorly understood. In mammals, fungi are classical opportunists, which usually infect immunocompromised individuals (Emmons et al., 1977). There is increasing evidence that such is also the case for fungal infections of freshwater fishes (Pickering and Duston, 1983). However, while we know little about the importance of immunity on the pathogenicity of marine fungi, there is some empirical evidence that supports a similar opportunistic role.

First, the host may often be immunosuppressed. For example, infections of adult crustaceans often occur in conjunction with either a concurrent infection, some stressful environment (e.g., crowding), or otherwise damage to immunological barriers (e.g., wounds). Immature invertebrates (e.g., eggs or larvae) are also a common target. Little is known about the immunological competence of immature marine invertebrates; we do know that very young fish often have incompletely developed immune systems (Van Muiswinkel et al., 1985).

Also supporting the opportunistic nature of these fungi is the fact that most are probably facultative pathogens. *Ichthyophonus hoferi* and *Trichomaris invadens* are the only marine fungi that are considered to be obligate pathogens (Alderman, 1982). In most cases of fungal pathogenicity in mammals, the spores that initiate infections come not from a diseased animal, but rather from a saprophytic colony; the infection is, to all intents and purposes, a "dead-end" for the fungus and entirely incidental to its normal range of activities as a saprophyte. There is evidence to suggest that this is also the case in some marine host-parasite relationships. For example, in ulcerative mycosis of estuarine fishes, the oomycete fungi in the lesions appear to die near the end of the course of the disease without forming any reproductive stages (Noga et al., 1988, Noga, unpubl. data). Furthermore, many consider dead animals, uneaten food, and other detritus as an important source of inoculum for certain fungal infections (Armstrong and Fisher, 1988).

However, in other cases the relative importance of "live" versus "dead" hosts is unclear. For example, gravid female crustaceans can harbor Oomycetes as an infection of the egg mass that they carry (Armstrong and Fisher, 1988; Bland and Amerson, 1974) (Fig. 1). Also, some feel that juvenile and adult shrimp may carry *Lagenidium* as a subclinical infection and, thus, are a source of the pathogen (Lightner, 1988a). In summary, while some fungal pathogens have been isolated from marine surfaces (Fuller et al., 1964), and we consider defacto that most are facultative parasites, we still know little about the relative importance of various substrates to the ecological distribution of these agents. Nonetheless, the potential importance of saprotrophy on pathogen load has management implications, since isolation and quarantine may not be totally successful in eliminating the pathogen. Good sanitation practices and stress reduction are often critical for successful control.

Once established in an individual host, marine fungal infections are often fatal because there are no safe yet effective methods for treating systemic infections. Being

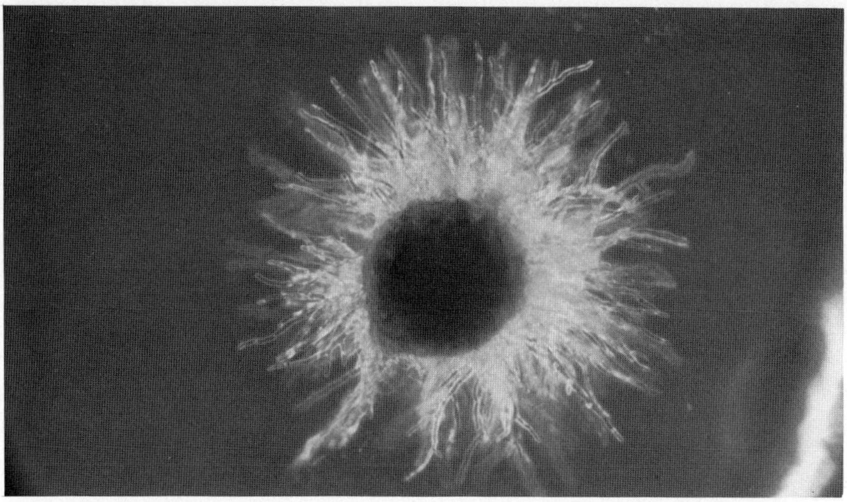

Fig. 1. *Lagenidium callinectes* infection of a blue crab *Callinectes sapidus* egg.

eucaryotes, fungi have a number of metabolic pathways that are similar to the Metazoa, which can make it difficult to target specific modes of therapy. Prophylaxis is often the treatment of choice. Another potential problem is the public health aspects of using some antimycotic agents, such as malachite green oxalate, which may be dangerous to the consumer (i.e., man), if residues are present in edible tissues.

Host ranges of pathogenic fungi. The fungi infecting marine finfish and invertebrates include representatives of three of the five major "groups" of true fungi (Eumycota): Mastigomycotina, Deuteromycotina, and Ascomycotina. The Zygomycotina and Basidiomycotina are not represented. Most important pathogens belong to the Classes Oomycetes (Mastigomycotina) and Hyphomycetes (Deuteromycotina). Some of the Myxomycota, as well as a number of fungus-like agents, have also been reported to be pathogens.

Myxomycota

Thraustochytriales

The thrautochytrids are members of the so-called "wall-less fungi." Their classification as fungi is in doubt since they have many features in common with certain algae (Perkins, 1974). Nonetheless, they have been included in previous reviews of fungal diseases and do deserve mention since they are increasingly being associated with marine animal diseases, especially skin lesions (Polglase, 1981; McLean and Porter, 1987; Jones and O'Dor, 1983). Being ubiquitous saprophytes on marine surfaces, their role in the pathogenesis of these diseases requires careful evaluation.

Eumycota

Mastigomycotina

Oomycetes. Oomycetes are probably the most important fungal pathogens of marine invertebrates. The early life stages (egg, larva, and postlarva) are the most commonly targeted group. Included is *Lagenidium,* responsible for larval mycosis of shrimp and crabs, *Haliphthoros* disease of lobsters and shrimp, and *Sirolpidium*, which causes larval mycosis of oysters and clams.

Many Oomycetes have a broad host range and naturally infect several major taxa. Examples include *Lagenidium callinectes* (Fig. 1) and *Haliphthoros milfordensis* (Table 1). Many Oomycetes can also experimentally infect a wide range of hosts (e.g, *Atkinsiella* [= *Plectospiral*] *dubia* and *Pythium thallasium*), suggesting that their actual host range may be greater than the presently reported outbreaks would indicate.

Interestingly, there are apparently no reports in either wild or aquaculture populations of oomycete infections of fish eggs or larvae, even though the Oomycetes are among the most important pathogens of freshwater fish ova. The majority of commercially important cultured marine fishes (e.g., yellowtail, *Seriola quinqueradiata*; milkfish, *Chanos chanos*; mullet, *Mugil cephalus*; and eel, *Anguilla* sp.) are collected in the wild as fry or fingerlings (Brown, 1983). As artificial spawning of marine fishes becomes more common, will Oomycetes also plague such hatcheries?

While some oomycete genera can infect both fishes and invertebrates (e.g, *Pythium*), species are restricted to either vertebrate or invertebrate hosts. Similarly, while some oomycete genera infect both marine and freshwater animals (e.g., *Atkinsiella, Lagenidium*), particular species are restricted to either a marine or freshwater environment. This is felt to be primarily due to the inability of freshwater species to withstand the osmotic conditions of the marine environment and visa versa (Harrison and Jones, 1975).

Oomycetes are the most common fungal pathogens affecting freshwater fishes, but until recently, they were rarely reported from subadult or adult marine fishes. Within the last several years, massive epidemics of skin ulcers have been reported in estuarine fishes along the Atlantic coast of the Unites States. This disease, called ulcerative mycosis (UM), is associated with infections by the Saprolegniaceous fungi *Aphanomyces* and *Saprolegnia*. Atlantic menhaden, *Brevoortia tyrannus*, appears to be the most commonly affected species, although a number of other estuarine species have been observed with similar lesions (Noga, 1988). Documented occurrences of UM have been reported from three major estuarine systems of the northwestern Atlantic Ocean: Chesapeake Bay, Albemarle-Pamlico Sounds, and the St. Johns River.

Instead of the relatively superficial lesions associated with typical oomycete infections of freshwater fishes, UM produces a deep, penetrating skin lesion that often involves underlying internal organs. Unlike the mild mononuclear inflammatory response typically associated with oomycete infections of freshwater fishes, UM induces an intense chronic inflammation (Noga et al., 1988).

Other epidemics, termed red spot, have been repeatedly observed in estuarine fishes of Australia and New Guinea since 1972 (Callinan, 1985). Red spot causes a large, usually relatively shallow ulcer in mullet (*Mugil cephalus*) and other estuarine fishes (Roberts et al., 1986), which is associated with an intense chronic inflammatory response to broad (12-18 μm), aseptate hyphae. While the fungi in red spot have not been successfully cultured, their morphology strongly suggests that they may be Oomycetes.

The role these fungi play in the development of the skin lesions is uncertain. While fungi have been identified from all stages of ulcerative mycosis (Fig. 2), it appears as if only advanced lesions of red spot have fungi. Whereas Koch's postulates have not been fulfilled for UM or red spot, Miyazaki, Hatai, and co-workers provided evidence that an *Aphanomyces* species was responsible for mycotic granulomatosis, a deep, ulcerative, granulomatous dermatopathy of several freshwater fishes in Japan

Fig. 2. Ulcerative mycosis infection of Atlantic menhaden *Brevoortia tyrannus*.

(Hatai et al., 1984). Thus, it appears that this "atypical" form of oomycete infection may be relatively common in fishes.

Deuteromycotina

Hyphomycetes

In the Hyphomycetes, there is no true distinction between marine and freshwater pathogens. Unlike the Oomycetes, the Hyphomycetes are not well adapted to an aquatic mode of existence. Many of the agents that cause disease are common saprophytes in freshwater and terrestrial environments (e.g., *Fusarium*, *Exophiala*). However, they are still capable of causing serious diseases in marine animals. Unlike the Oomycetes, which mostly infect larval invertebrates, the Hyphomycetes most commonly affect subadult and adult Crustacea. These agents are considered to have a low pathogenicity. Thus, they often cause disease in concert with some presumptively immunosuppressive event. For example, *Fusarium* infection of penaeids is usually associated with REO (reo-like virus) infection, chronic exposure to heavy metals, or overcrowding (Lightner, 1988b). Also, wounds appear to be a common portal of entry and may be necessary for the initiation of an infection. *Cladosporium* infection of *Octopus* (Polglase et al., 1984) and experimental *Exophiala* infection of marine fishes (Blazer and Wolke, 1979) have also been associated with wounds.

The most clinically important hyphomycete is *F. solani*, which is responsible for black gill disease of shrimp (Fig. 3) and black (burn) spot disease of lobsters (Lightner, 1981). As for the oomycete larval mycoses, *Fusarium* infections are difficult to pharmaceutically treat in infected individuals; proper hygiene appears to be the best method of control. Isolated cases of *Fusarium* have also been recently reported from several fishes. Ostland et al. (1987) reported a kidney infection in an aquarium-held triggerfish, *Melichthys vidua*, associated with a granulomatous peritonitis. Muhvich et al. (1989) isolated *F. solani* from bonnethead sharks, *Sphyrna tiburo*. Deep infections causing a chronic myositis and chondritis were associated with skin lesions. Adjacent tissues (gill, orbit, or kidney) were also affected. Hatai et al. (1986) isolated *F. oxysporum* from fingerling red sea bream, *Pagrus* sp. During the epidemic, *Edwardsiella tarda* was also cultured from affected fish, so the relative importance of the fungus was uncertain.

Exophiala includes several species reported as fungal pathogens of marine fishes (Gaskins and Cheung, 1986). A related pathogen, *Sarcinomyces crustaceus*, was recently isolated from the exophthalmic eye of a cultured black sea bream, *Spondyliosoma cantharus* (Todaro et al., 1983). However, the authors did not mention whether this same organism had been observed in clinical material (histology, etc.) from the lesion. Other Hyphomycetes (*Cladosporium* spp., *Septocylindricum*, *Ramularia, Didymaria*, and *Aureobasidium*) have been documented from single cases.

Blastomycetes

The Blastomycetes include yeast-like forms that have infrequently been reported as pathogens of marine animals. However, many blastomycetous, as well as ascomycetous yeasts, have been isolated from marine and estuarine environments. Many are probably transient population derived from terrestrial sources (Kohlmeyer and Kohlmeyer, 1979).

Ascomycotina. A number of ascomycetous yeasts (*Metschnikowia* spp.) have been reported as individual cases (Table 1). However, the most important marine pathogen in this group is *Trichomaris invadens*, a common disease of the tanner crab (*Chinocetes bairdi*), which is responsible for black mat syndrome (Hibbits et al., 1981). The identification of this disease points out the importance of differentiating true fungal pathogens from fungal contaminants. When this disease was first described, it was attributed to a *Phoma* species, based upon the culture of *Phoma* from lesions (van Hyning and Scarborough, 1973). While *Phoma* is a known pathogen of aquatic animals, it is also a ubiquitous saprophyte. Thus, identification of such facultative pathogens as disease agents requires a careful examination of clinical material, and eventually reproduction of the disease with the suspected pathogen.

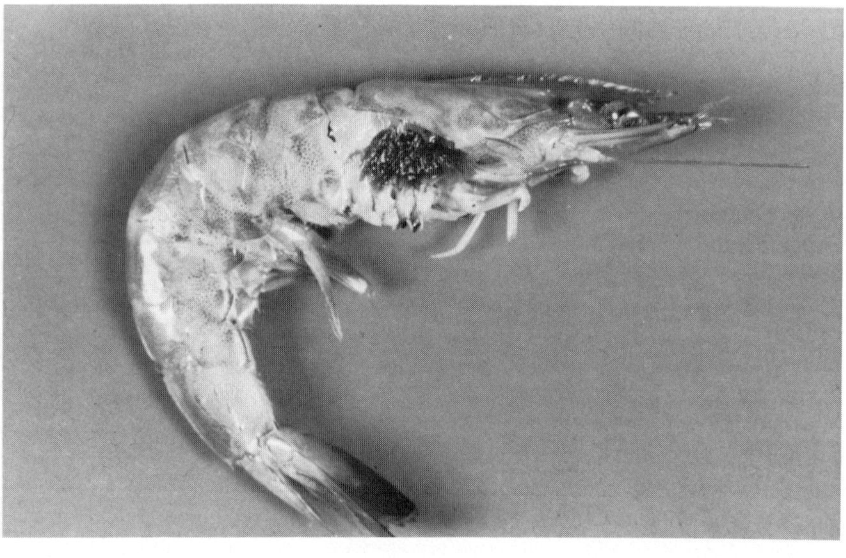

Fig. 3. *Fusarium* infection of shrimp *Penaeus* sp. causing "black gill" disease.

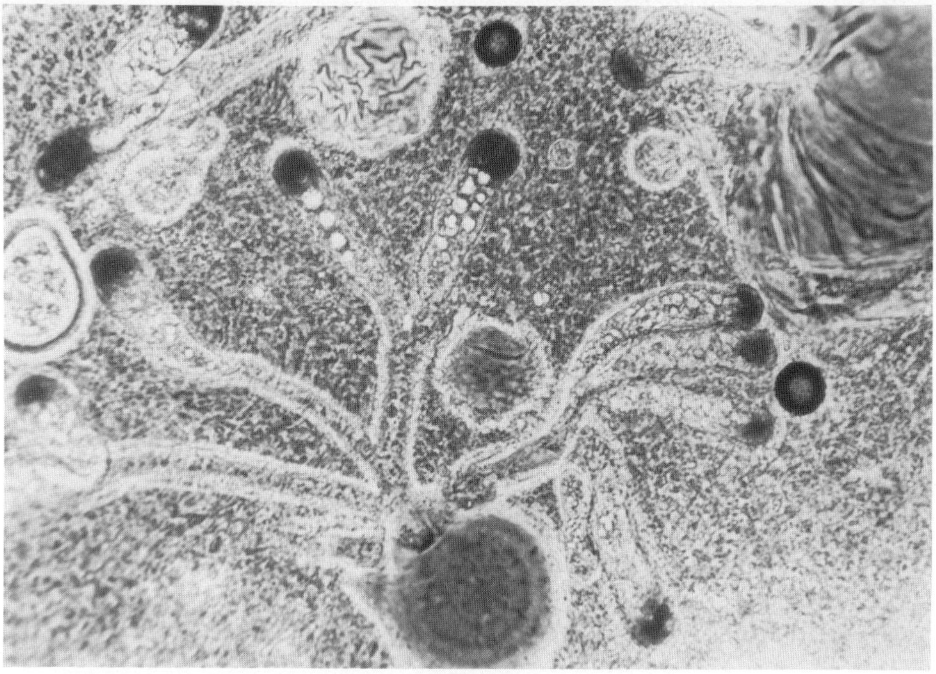

Fig. 4. A germinating spore of *Ichthyophonus hoferi* taken from the tissue of a dead plaice *Pleuronectes platessa* (from McVicar, 1982).

Uncertain taxonomy. This is a heterogenous group of organisms, some of which may not actually be fungi, but have been traditionally considered as such in most reviews. The most important is *Ichthyophonus hoferi* (Fig. 4), which has been responsible for a number of epidemics in wild marine fish populations. Most notable include outbreaks in Atlantic herring, *Clupea harengus*, in the northwestern Atlantic Ocean and in plaice, *Pleuronectes platessa,* in the North Sea (McVicar, 1982). *Ichthyophonus* is the most common fungal pathogen of marine fishes and may have a substantial impact upon some natural populations (McVicar, 1982).

Harrell et al. (1986) have reported the rosette agent, a fungus-like organism responsible for chronic, high mortality of adult chinook salmon raised in net-pens in Puget Sound, Washington, USA. It is the only obligately intracellular eucaryotic pathogen reported from fish and primarily affects the lymphoreticular tissue (kidney and spleen). Organisms accumulate and replicate in fixed macrophages; in severe infections, it can be found in many other organs. It is a 3-7 μm, spherical, Gram-positive, PAS-positive, and GMS-positive agent that produces characteristic clusters or rosettes both *in vitro* and *in vivo*.

Ostracoblabe implexa is the cause of shell disease in oysters. It can grow in a wide range of shell substrates, but does not cause serious disease in most species. The severe proliferative and necrotic host reaction that characterizes shell disease is only seen in the European flat oyster, *Ostrea edulis*, from the North Sea (Alderman and Jones, 1971). An apparently milder form of shell disease, associated with the development of small black (melanized?) flecks or raised white nodules on the inner shell surface, has recently been described in rock oysters, *Crassostrea cucullata*, from the Indian Ocean (Raghukumar and Lande, 1988).

Spongiophaga is the designation used for a fungus-like agent that has been associated with mass mortality of sponges in many tropical and warm-temperate seas. Most notable have been epidemics in the Caribbean Sea, the Florida Keys (USA), and the Gulf of Mexico (Galtsoff, 1942), as well as separate earlier outbreaks in the Indian Ocean (Poltjaeff, 1884, cited in Lauckner, 1980). The organism(s) has (have) never been cultured nor have any reproductive structures ever been identified.

SUMMARY

Fungi are ubiquitous pathogens of both wild and cultured populations of marine fishes and invertebrates. The Oomycetes (Mastigomycotina) and the Hyphomycetes (Deuteromycotina) contain most of the clinically important agents. Several lines of empirical evidence suggest that these agents are opportunistic pathogens that primarily affect hosts with suboptimal immunity. Under the proper conditions, they are highly pathogenic and difficult to treat. This suggests that they will continue to be problematic pathogens under the stressful conditions of aquaculture.

ACKNOWLEDGMENTS

I wish to thank Dr. C.E. Bland for providing photographs of *Fusarium* and *Lagenidium*, Dr. A.H. McVicar for providing the photograph of *Ichthyophonus*, Dr. R. Bennett and colleagues for permission to quote from their unpublished manuscript, and Dr. J. Landsberg for critical comments. Supported in part by the North Carolina Department of Natural Resources and Community Development, Division of Marine Fisheries, Contract #88-0670.

LITERATURE CITED

Alderman, D.J. 1982. Fungal diseases of aquatic animals. In: "Microbial Diseases of Fish" (R.J. Roberts, ed.), pp. 189-242. Society for General Microbiology, Special Publication 9, Academic Press, New York.

Alderman, D.J. and E.B.G. Jones. 1971. Shell disease of oysters. Fishery Invest., London, Ser. II, 26, 1-19.

Alderman, D.J. and J.L. Polglase. 1986. Are fungal diseases significant in the marine environment? In: "The Biology of Marine Fungi" (S.T. Moss, ed.), pp. 189-198. Cambridge Univ. Press, New York.

Armstrong, D.A. and W.S. Fisher. 1988. Fungus (*Lagenidium*) disease of dungeness crabs. In: "Disease Diagnosis and Control in North American Marine Aquaculture" (C.J. Sindermann and D.V. Lightner, eds.), pp. 215- 219. Elsevier, New York.

Atkins, D. 1954a. Further notes on a marine member of the Saprolegniaceae, *Leptolegnia marina* n. sp. infecting certain invertebrates. J. Mar. Biol. Assoc. U.K., 33, 613-625.

Atkins, D. 1954b. A marine fungus *Plectospira* dubia n. sp. (Saprolegniaceae) infecting crustacean eggs and small Crustacea. J. Mar. Biol. Assoc. U.K., 33, 721-732.

Atkins, D. 1955. *Pythium thallasium* n. sp. infecting the egg mass of the pea crab *Pinnotheres pisum*. Trans. Brit. Mycol. Soc., 38, 31-46.

Bian, B.Z. and S. Egusa. 1980. *Atkinsiella hamanaensis* sp. nov. isolated from cultivated ova of the mangrove crab, *Scylla serrata* (Forsskal). J. Fish Dis., 3, 373-386.

Bian, B.Z., K.P.G.L. Hatai and S. Egusa. 1977. Studies on the fungal diseases in crustaceans. I. *Lagenidium scyllae* sp. nov. isolated from cultivated ova and larvae of the mangrove crab (*Scylla serrata*). Trans. Mycol. Soc. Japan, 20, 115-124.

Bland, C.E. and H.V. Amerson. 1974. Occurrence and distribution in North Carolina waters of *Lagenidium callinectes* Couch, a fungal parasite of blue crab ova. Chesapeake Sci., 15, 232-235.

Blazer, V.S. and R.E. Wolke. 1979. An *Exophiala*-like fungus as the cause of a systemic mycosis of marine fish. J. Fish. Dis., 2, 145-152.

Brown, E.E. 1983. "World Fish Farming – Cultivation and Economics." AVI Publications, Westport, Connecticut.

Callinan, R.B. 1985. Diseases of native Australian fishes. In: "Diseases of Australian Fish and Shellfish" (J.D. Humphrey and J.S. Langdon, eds.), pp. 102-117. Australian Fish Health Reference Laboratory, Regional Veterinary Laboratory, Benalla, Victoria.

Cassels, F. and C.T. Krebs. 1983. A comparison of artificial incubation methods using ova of the red crab *Geryon quinquedens* (Decapoda Brachyura). J. Crust. Biol., 3, 565-574.

Codreanu, R. and D. Codreanu-Balcescu. 1981. On two *Metschnikowia* yeast species producing hemocoelic infections in *Daphnia* magna and *Artemia salina* (Crustacea, Phyllopoda) from Romania. J. Invert. Path., 37, 22-27.

Couch, J.N. 1942. A new fungus on crab eggs. J. Elisha Mitchell Sci. Soc., 58, 158-162.

Davis, H.C., V.L. Loosanoff, W.H. Weston and C. Martin. 1954. A fungus disease in clam and oyster larvae. Science, 120, 36-38.

Delves-Broughton, J. and C.W. Poupard. 1976. Disease problems of prawns in recirculation systems in the U.K. Aquaculture, 7, 201-217.

Dykstra, M.J., J.F. Levine, E.J. Noga, J.H. Hawkins, P. Gerdes, W.J. Hargis, Jr., H.J. Grier and D. Te Strake. 1989. Ulcerative mycosis: A serious menhaden disease of the southeastern coastal fisheries of the United States. J. Fish. Dis., 12, 175-178.

Emmons, C.W., C.H. Binford, J.P. Utz, and K.J. Kwon-Chung. 1977. "Medical Mycology," 3rd edition. Lea and Febiger, Philadelphia.

Fisher, W.S. 1983. Eggs of *Palaemon macrodactylus*: III. Infection by the fungus, *Lagenidium callinectes*. Biol. Bull., 164, 214-226.

Fisher, W.S. 1988. Fungus (*Lagenidium*) disease of lobsters. In: "Disease Diagnosis and Control in North American Marine Aquaculture" (C.J. Sindermann and D.V. Lightner, eds.), pp. 247-250. Elsevier, New York.

Fize, A., J.F. Manier and J. Maurand. 1970. Sur un cas d'infestation du Copepode *Eurytemora velox* (Lillj) par une levure du genre *Metschnikowia* (Kamienski). Ann. Parasitol. Hum. Comp., 45, 357-363.

Fuller, M.S., B.E. Fowles and D.J. McLaughlin. 1964. Isolation and pure culture studies of marine phycomycetes. Mycologia, 56, 745-756.

Galtsoff, P.S. 1942. Wasting disease causing mortality of sponges in the West Indies and Gulf of Mexico. Proc. of 8th Amer. Sci. Conf., 3, 411-421.

Gaskins, J.E. and P.J. Cheung. 1986. *Exophiala pisciphila*. A study of its development. Mycopathologica, 93, 173-184.

Harrell, L.W., R.A. Elston, T.M. Scott and M.T. Wilkinson. 1986. A significant new systemic disease of net-pen reared chinook salmon *Oncorhynchus tshawytscha* brood stock. Aquaculture, 55, 249-262.

Harrison, J.L. and E.B.G. Jones. 1975. The effect of salinity on sexual and asexual sporulation in members of the Saprolegniaceae. Trans. Brit. Mycol. Soc., 65, 389-394.

Hatai, K., B.Z. Bian, C.A. Baticados and S. Egusa. 1980. Studies on the fungal diseases in crustaceans. II. *Haliphthoros phillipinensis* sp. nov. isolated from cultivated larvae of the jumbo tiger prawn. Trans. Mycol. Soc. Japan, 21, 47-55.

Hatai, K., S. Takahashi and S. Egusa. 1984. Studies on the pathogenic fungus of mycotic granulomatosis. IV. Changes of blood constituents of any *Plecoglossus altivelis*, experimentally infected and normally infected with *Aphanomyces piscida*. Fish Pathol., 19, 17-23.

Hatai, K., S.S. Kubota, N. Kida and S.-I. Udagawa. 1986. *Fusarium oxysporum* in Red Sea bream *Pargus* sp. J. Wildlife Dis., 22, 570-571.

Hibbits, J., G.C. Hughes and A.K. Sparks. 1981. *Trichomaris invadens* gen. sp. nov. An ascomycete parasite of the tanner crab *Chionocetes bairdi* Rathbun (Crustacea, Brachyura). Can. J. Bot., 59, 2121-2128.

Höhnk, W. and S. Vallin. 1953. Epidemisches absterben von *Eurytemora* in bothnischen meerbusen, verursacht durch *Leptolegnia baltica* nov. spec. Veroff. Inst. Meeresforsch. Bremerh., 2, 215-223.

Hose, J.E., D.V. Lightner, R.M. Redman and D.A. Danald. 1984. Observations on the pathogenesis of the imperfect fungus, *Fusarium solani.* in the Californian brown shrimp, *Penaeus californiensis*. J. Invert. Path., 44, 292-303.

Johnson, T.W., Jr. 1958. A fungus parasite in the ova of the barnacle *Chthalmus fragilis denticulata*. Biol. Bull., 114, 205-214.

Johnson, T.W. and R.R. Bonner, Jr. 1960. *Lagenidium callinectes* Couch in barnacle ova. J. Elisha Mitchell Sci. Soc., 76, 147-149.

Johnson, T.W. and W.C. Pinscmidt. 1963. *Leptolegnia marina* Atkins in blue crab ova. Nova Hedwigia, 5, 413-418.

Jones, G. and R.K. O'Dor. 1983. Ultrastructural observations on a Thraustochytrid fungus parasitic in the gills of squid (*Illex illecerebrosus* LeSueur). J. Parasit., 69, 903-911.

Kohlmeyer, J. and E. Kohlmeyer. 1979. "Marine Mycology. The Higher Fungi." Academic Press, New York.

Lachance, M.A., M. Miranda, M.W. Miller and H.J. Phaff. 1976. Dehiscence and active spore release in pathogenic strains of the yeast *Metschnikowia bicuspidata* var. *australis*: possible predatory implication. Can. J. Microbiol., 22, 1756-1761.

Lauckner, G. 1980. Diseases of Porifera. In: "Diseases of Marine Animals" (O. Kinne, ed.), pp. 139-165. John Wiley and Sons, New York.

Lightner, D.V. 1981. Fungal diseases of marine Crustacea. In: "Pathogenesis of Invertebrate Microbial Diseases" (E.W. Davidson, ed.), pp. 451-484, Allanheld, Osmun Publ., Totowa, New York.

Lightner, D.V. 1988a. Larval mycosis of penaeid shrimp. In: "Disease Diagnosis and Control in North American Marine Aquaculture" (C. Sindermann and D.V. Lightner, eds.), pp. 58-63. Elsevier, New York.

Lightner, D.V. 1988b. Fungus (*Fusarium*) disease of juvenile and adult penaeid shrimp. In: "Disease Diagnosis and Control in North American Marine Aquaculture" (C. Sindermann and D.V. Lightner, eds.), pp. 64-69. Elsevier, New York.

Mann, H. and U. Pieplow. 1938. Die Brandenflekenkrankheit bei Krebsen und ihre Erreger. Zeitschr. für Fischerei, 36, 225-240.

McLean, N. and D. Porter. 1987. Lesions produced by a thraustochytrid in *Tritonia diodomedea* Mollusca, Gastropoda, Nudibranchia. J. Invert. Path., 49, 223-225.

McVicar, A.H. 1982. *Ichthyophonus* infections of fish. In: "Microbial Diseases of Fish" (R.J. Roberts, ed.), pp. 243-269. Society for General Microbiology, Special Publication 9, Academic Press, London.

Miller, J.D. and L.C. Fleming. 1983. Fungi associated with an infestation of *Pseudocarcinonemertes homari* on *Homarus americanus*. Trans. Brit. Mycol. Soc., 80, 9-12.

Muhvich, A.G., R. Reimschuessel, M.M. Lipsky and R.O. Bennett. 1989. *Fusarium solani* isolated from newborn bonnethead sharks *Sphyrna tiburo* (L.). J. Fish Dis., 12, 57-62.

Neish, G.A. and G.C. Hughes. 1980. "Fungal Diseases of Fishes." TFH Publications, Neptune, New Jersey.

Noga, E.J. 1988. Determining the relationship between water quality and infectious disease in fishery populations. Water Res. Bull., 24, 967-973.

Noga, E.J., J.F. Levine, M.J. Dykstra and J.H. Hawkins. 1988. Pathology of ulcerative mycosis in Atlantic menhaden *Brevoortia tyrannus*. Dis. Aquat. Organ., 4, 189-197.

Ostland, V.E., H.W. Ferguson, R.D. Armstrong, A. Asselin and R. Hall. 1987. Case report: Granulomatous peritonitis in fish associated with *Fusarium solani*. Vet. Rec., 121, 595-596.

Perkins, F.O. 1974. Phylogenetic considerations of the problematic thraustochytriaceous-labyrinthulid-*Dermocystidium* complex based on observations of fine structure. Veroff. Inst. Meeresforsch. Bremerh., Suppl 5, 45-63.

Pickering, A.D. and J. Duston. 1983. Administration of cortisol to brown trout, *Salmo trutta* L., and its effects on the susceptibility to *Saprolegnia* infection and furunculosis. J. Fish Biol., 23, 163-175.

Pixell-Goodrich, H. 1928. Reactions of *Gammarus* to injury and disease with notes on some microsporidal and fungoid diseases. Quart. J. Microscop. Soc., 72, 325-353.

Polglase, J.L. 1981. Thraustochytrids as potential pathogens of marine animals. Bull. Brit. Mycol. Soc., 16, (Suppl. 1), 5.

Polglase, J.L., N.J. Dix and A.M. Bullock. 1984. Infection of skin wounds in the lesser octopus *Eledone cirrhosa* by *Cladosporium sphaerospermum*. Trans. Brit. Mycol. Soc., 82, 577-580.

Porter, D. 1986. Mycoses of marine organisms: an overview of pathogenic fungi. In: "The Biology of Marine Fungi" (S.T. Moss, ed.), pp. 141-153. Cambridge Univ. Press, New York.

Raghukumar, C. and V. Lande. 1988. Shell disease of rock oyster *Crassostrea cucullata*. Dis. Aquat. Organ., 4, 77-81.

Reichenbache-Klinke, H.H. (1956). Uber einige bisher ubekannte Hyphomyceten bei Verscheidenen Suswasser und Meeresfischen. Mycopath. Mycol. Appl., 7, 333-368.

Richards, R.H., A. Holliman and S. Helagson. 1978. *Exophiala salmonis* infection in Atlantic salmon, *Salmo salar* L. J. Fish Dis., 1, 357-368.

Roberts, R.J., D.J. Macintosh, K. Tonguthai, S. Boonyaratpalin, N. Tayaputch, M.J. Phillips and S.J. Millar. 1986. Field and laboratory investigations into ulcerative fish diseases in the Asia-Pacific region. FAO, Bangkok, Tech. Rept., FAO Proj. TCP/RAS/4508, 214 p.

Roth, F.J., Jr., D.G. Ahearn, J.W. Fell, S.P. Meyers and S.A. Meyer. 1962. Ecology and taxonomy of yeasts isolated from various marine substrates. Limnol. Oceanogr., 7, 178-185.

Seki, H. and J. Fulton. 1969. Infection of marine copepods by *Metschnikowia* sp. Mycopathol. Mycol. Appl., 38, 61-70.

Sindermann, C.J. 1988a. Fungus (*Lagenidium*) disease of blue crab eggs and larvae. In: "Disease Diagnosis and Control in North American Marine Aquaculture" (C. Sindermann and D.V. Lightner, eds.), pp. 197-199. Elsevier, New York.

Sindermann, C.J. 1988b. Fungus (*Fusarium*) disease of juvenile lobsters. In: "Disease Diagnosis and Control in North American Marine Aquaculture" (C. Sindermann and D.V. Lightner, eds.), pp. 255-257. Elsevier, New York.

Sindermann, C.J. and D.V. Lightner (eds.). 1988. "Disease Diagnosis and Control in North American Marine Aquaculture." Elsevier, New York.

Sordi, M. 1958. Micosi dei crostacei decapodi marini. Riv. di Parassitol., 19, 132-137.

Sparks, A.K. 1985. "Synopsis of Invertebrate Pathology Exclusive of Insects." Elsevier, New York.

Tharp, T.P. and C.E. Bland. 1977. Biology and host range of *Haliphthoros milfordensis*. Can. J. Bot., 55, 2936-2944.

Todaro, F., A. Berdar, A. Cavaliere, G. Criseo and L. Pernice. 1983. Gas ophthalmus in black sea bream *Spondyliosoma cantharus* caused by *Sarcimomyces crustaceus*. Mycopathologia, 81, 95-98.

van Hyning, J.M. and A.M. Scarborough. 1973. Identification of fungal incrustations on the shell of the snow crab (*Chionoxcetes bairdi*). J. Fish. Res. Bd. Can., 30, 1738-1739.

Van Muiswinkel, W.B., D.P. Anderson, C.H.J. Lamers, E. Egberts, J.J.A. Van Loon and J.P. Ijssel. 1985. Fish immunology and fish health. In: "Fish Immunology" (M.J. Manning and M.F. Tatner, eds.), pp. 1-8. Academic Press, New York.

Vishniac, H.S. 1958. A new marine phycomycete. Mycologia, 50, 66-79.

PROTOZOAN DISEASES SECTION

PROTOZOAN DISEASES: INTRODUCTORY REMARKS

Ralph Elston

Battelle Marine Sciences Lab., 439 West Sequim Bay Road,
Sequim, Washington 98382 USA

Those who have enthusiasm and excitement for marine protozoan diseases include some who would probably describe themselves first and foremost as parasitologists, but also include many others whose primary calling may be descriptive or experimental pathology, immunology, molecular biology, or other disciplines.

Most of us have an interface with science applications in resource management or aquaculture. It is through these activities that I have been personally blessed with the gifts of exciting subject matter in the form of new parasitic diseases of marine fish and invertebrates. It is always delightful to meet colleagues to share this excitement. Like many of you, however, I have learned to conceal this glee when interacting with resource managers and aquaculturists. So while assuming a concerned and serious facade, as we dutifully note the details of the latest catastrophe, we find ourselves joyfully looking forward to many hours of exciting research – not to mention the possibility of obtaining additional research funds to further our avocation.

Both invertebrates and fish are represented as hosts of protozoa. Therefore, we play an important role in solving the applied problems in resource management and aquaculture.

I would like to briefly summarize my outlook on some of the important issues this field faces today in terms of both fundamental and applied research. Although I have oriented these comments toward protozoan disease, some apply to marine pathology in general.

First, the importance of protozoan diseases of marine fish and shellfish is staggering on a global economic scale. Impacts of *Haplosporidium, Perkinsus, Bonamia, Marteilia,* and others on oysters (and other species such as abalone) are measured in the hundreds of millions of dollars in lost livelihoods and ways of life. Protozoan diseases of fishes are also extremely important, especially as these pertain to farmed salmon and wild fish resources. These fascinating microorganisms will continue to be a blight on man's ability to utilize and domesticate natural resources and provide food for a hungry world. Thus, there is no shortage of work today in this field and even as many of us struggle to obtain research funds, I am optimistic about the future of the discipline as we observe the worldwide growth of aquaculture and the continued importance of marine resources.

Pathology in Marine Science

On the side of fundamental research, one of the more perplexing problems for me has been the taxonomic position of some of the newly discovered protistans. While many of the newly discovered organisms fit more or less neatly into defined taxa, many appear simply as "little round things" that are morphologically non-descript. Really, because we are on a frontier of marine pathology, we lack a higher taxonomic framework that will readily accommodate many of these organisms. I would encourage those of you with the interest to apply some of your research talents and imagination to this problem, although, to retain comtemporary respectability, you will need to apply tools of the so-called "New Biology" along with more traditional methods. In addition, to help solve problems, we still need invertebrate cell culture to help sort out life cycles of the many intracellular parasites and to characterize aspects such as the type of mitosis, which is a key characteristic. A better view of the relatedness of these "little round things" will give us a better perspective of their significance and mechanism of pathogenicity.

Mentioning the "New Biology" and use of molecular tools brings up a second important point that I would like to touch on. As all of you have observed over the last several years, molecular biology has invaded nearly every facet of biology with substantial advances resulting in many fields. Simultaneously, marine parasitology and marine pathology, in general, have entered a rapid growth phase resulting largely from the worldwide growth of aquaculture. Thus, to advance our field today, we face a double challenge: the first, resulting from an explosion of interest in marine pathology is that we have a backlog of work in areas of basic biology (e.g., life cycles, host specificity), which is absolutely essential, fascinating, and worthwhile in its own right; the second, where appropriate, is that we must adapt and utilize molecular biological approaches in order to advance our science most efficiently, to be contemporary with the other areas of biology, and to be competitive for research funds.

Finally, there's just one point I would like to mention in the application of marine parasitology and pathology. I would like to emphasize the need for numerical and statistically defensible data regarding the impact of diseases on populations of marine animals. Specifically, we need to know what percent die or are moribund by specific criteria. What percent per year is lost from reduced product quality from parasitized tissues? Unfortunately, the temptation to overrate the importance of a disease has been all too often exercised when writing grant proposals or technical manuscripts. The result is that we are regarded with some skepticism when we visit the fish or shellfish farm or write grant proposals. The farmer worries that we will find some alarming disease even though everything seems to be operating smoothly. We simply need to put things in perspective and attempt to rank the various diseases by the order of their economic impact. This is particularly important today since in most parts of the world research funds are limited. Thus, in order to effectively solve problems for the aquaculture industry and support its growth, we must select the most important problems the industry faces for funding and detailed study. In turn, as we solve these problems, we will be the recipients of increased research funding.

This sounds obvious, but in practice the most significant disease problems do not always receive a high priority for funding, although they should, at least when the funding base is aquacultural or agricultural in nature.

COCCIDIA FOUND IN SOME MARINE FINFISH OFF NOVA SCOTIA

C.M. Morrison and V.M. Marryatt

Halifax Fisheries Research Laboratory, P.O. Box 550,
Halifax, Nova Scotia B3J 2S7, Canada

ABSTRACT: Coccidian parasites were found in the livers of *Merluccius bilinearis, Scomber scombrus, Clupea harengus,* and *Alosa pseudoharengus* and in the kidneys and swimbladders of *Gadus morhua* and *Melanogrammus aeglefinus*. A coccidian was also found in the testes of *C. harengus*. All of the host fishes were from off the East Coast of Nova Scotia, Canada.

INTRODUCTION

We have conducted several surveys over the last few years to investigate abnormalities in marine finfish. Such surveys provide baseline information on diseases that not only may cause or contribute to natural fish mortalities but may become more prevalent when fish are cultured (Sindermann, 1987). Although commercial cultivation for food is in the pilot stage, marine fish are reared for experimental purposes such as the study of developmental stages and nutritional requirements; they are also kept in public aquaria. Many of the parasites encountered in our surveys were new, or inadequately described, and there are many new host records (Morrison et al., 1986). In this account we consider the survey information on coccidia in greater detail, and discuss the results of our more recent work.

SURVEY RESULTS AND DISCUSSION

Coccidia were found in the livers of the silver hake, *Merluccius bilinearis*; mackerel, *Scomber scombrus*; herring, *Clupea harengus*; and gaspereau, *Alosa pseudoharengus*. They were also found in the kidneys and swimbladders of the cod, *Gadus morhua*, the haddock, *Melanogrammus aeglefinus*, and in the testes of herring. Coccidia also occurred in the intestinal epithelium of one silver hake and one American plaice, *Hippoglossoides platessoides*, but not in sufficient numbers to permit measurement of spore size or identification of the various stages in the life cycle.

Coccidia in the livers of silver hake, mackerel, and herring were similar in appearance (Figs. 1-3) and similar in size when measured in paraffin sections

(Morrison et al., 1986), but measurements from fresh sporocysts are needed to confirm such observations. Most of the oocysts were sporulated, containing four sporocysts, each with two sporozoites, but occasionally a zygote was seen. Normally there was no host reaction around the oocysts, but often they were either surrounded by macrophages (Fig. 1), which sometimes contained melanin (Fig. 2), or encapsulated with fibroblasts (Fig. 3). Coccidia were found in the livers of four of five mackerel, four of 14 silver hake, and 47 of 79 herring.

The coccidian in herring liver is *Goussia clupearum* (Thélohan, 1890; Labbé, 1896). While it is not easy to see the two valves in sectioned material using light miscroscopy, the wall being ≈240 nm thick, ultrastructural techniques have confirmed the bivalve nature of the sporocyst and the lack of a Stieda body (Morrison and Hawkins, 1984). The coccidian in mackerel liver has also been identified as *G. clupearum* (Dykova and Lom, 1983); coccidia identified in four gaspereau, which died in a public aquarium in Shippagan, New Brunswick, were also probably *G. clupearum*.

The mode of transmission of *G. clupearum* is not known. Herring, mackerel, and young silver hake are pelagic and have a planktonic diet (Vinogradov, 1984), whereas adult silver hake feed on mackerel and herring (Langton and Bowman, 1980). Direct transmission between herring, mackerel, gaspereau, and juvenile silver hake, or to mature silver hake through consumption of infected herring or mackerel, is therefore possible. If an intermediate host is involved, it is likely to be planktonic. Transmission experiments and ultrastructural studies of the coccidians in mackerel and silver hake liver are necessary before the identity of these parasites can be established unequivocally, especially since these fish are not closely related to each other or to herring.

Coccidia were found in the kidney tubules of 42 of 258 (16%) cod and nine of 131 (6.9%) haddock sampled during 1986-1988 (Morrison and Poynton, 1989a,b). The prevalence of the parasite was significantly higher in cod under 30 cm in length than in larger cod. Macrogametes and microgamonts were found at the apical surface of the kidney tubules, whereas zygotes and sporulating stages were free in the lumen of the kidney tubule (Fig. 4). Two sporozoites were present in the sporocyst, which was thin-walled (≈40 nm in thickness). There was no evidence of a suture or a Stieda body with light microscopy, but electron microscopy revealed the presence of a suture establishing this coccidian as a member of the genus *Goussia*.

In a survey of 2,233 haddock from the Nova Scotia banks, the coccidian *G. gadi* was found in the swimbladders of ≈25% of the fish collected in April and 58% of those collected in October (Odense and Logan, 1976). Ninety-two cod, 56 white hake, 34 silver hake, and 33 pollock examined at the time were not infected. We examined swimbladder tissues from one infected cod and two infected haddock. As described by Fiebiger (1913), developing stages including meronts, macrogamonts, microgamonts, sporocysts, and free sporozoites were seen in the layer of loose connective tissue in the swimbladder wall (Fig. 5) and also in the lumen of the organ. The sporocysts were thick-walled (up to 500 nm), and the two valves of the opened sporocysts were evident in light micrographs (Fig. 6). The mode of development of *G. gadi* differs from that of *G. cichlidarum*, which occurs in the swimbladders of cichlid fish (Landsberg and Paperna, 1985). The latter coccidian, like the one we found in cod and haddock kidney, develops in an epicellular position.

Haddock under 34 cm were not infected with *G. gadi* (Odense and Logan, 1976) as young fish are usually more susceptible to disease. It is probable that, as the swimbladder in both cod and haddock is physoclistous, *G. gadi* can only be

Figs. 1-6. Light micrographs of paraffin sections. Haematoxylin and eosin stain.

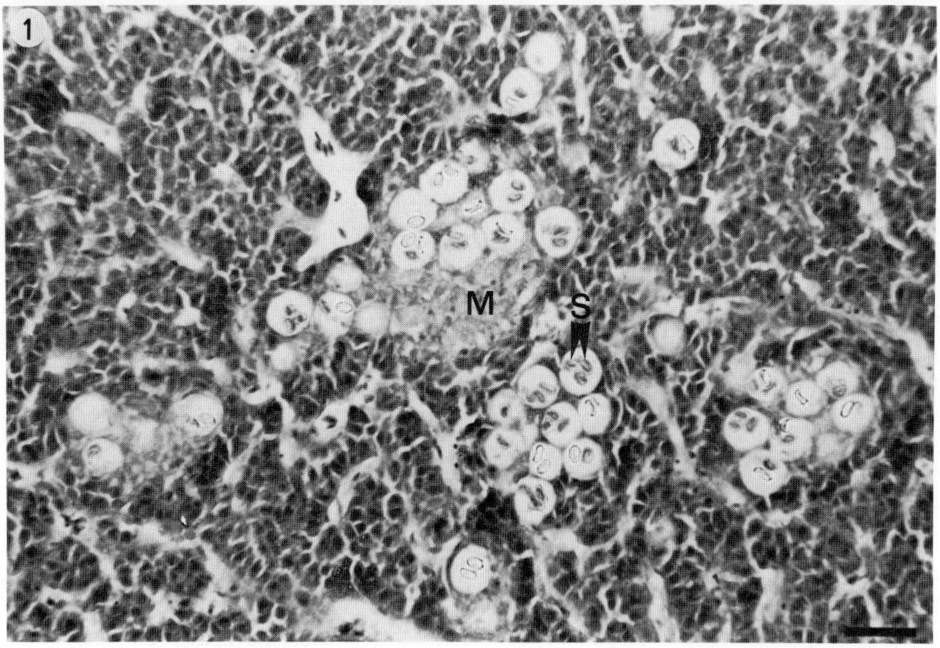

Fig. 1. *Goussia clupearum* in liver of herring. Several oocysts containing sporocysts (S) can be seen. One group is surrounded by macrophages (M). Bar = 50 μm.

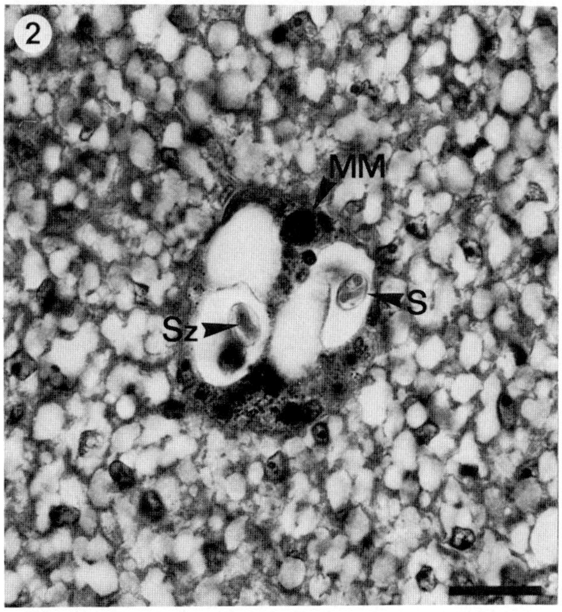

Fig. 2. Coccidian in liver of silver hake. A group of oocysts containing sporocyst (S) with fairly thin walls is surrounded by melano-macrophages (MM). Sporozoites (Sz) can be seen inside the sporocysts. Bar = 20 μm.

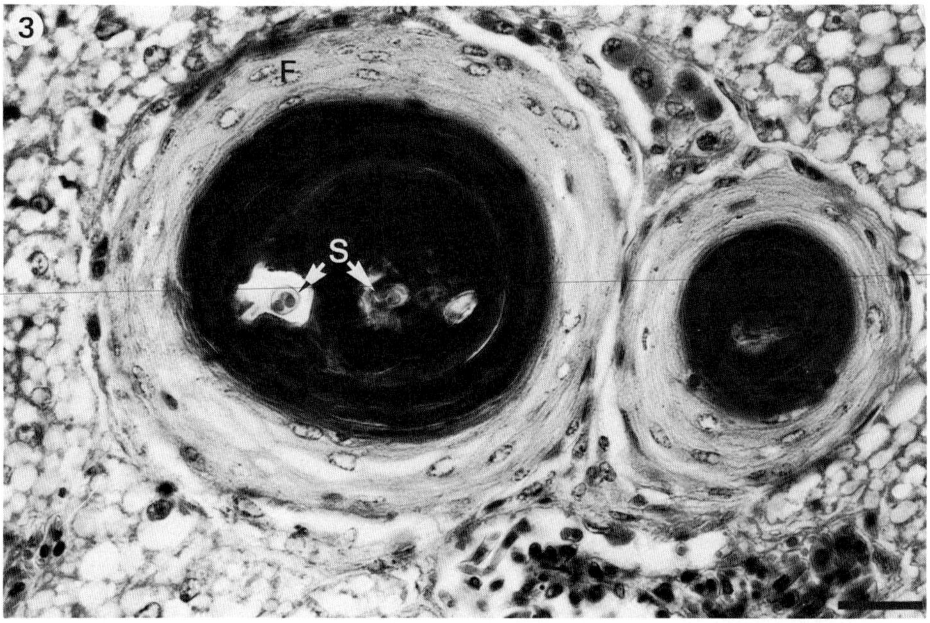

Fig. 3. Coccidian in liver of silver hake. Sporocysts (S) are surrounded by fibroblasts (F). Bar = 20 μm.

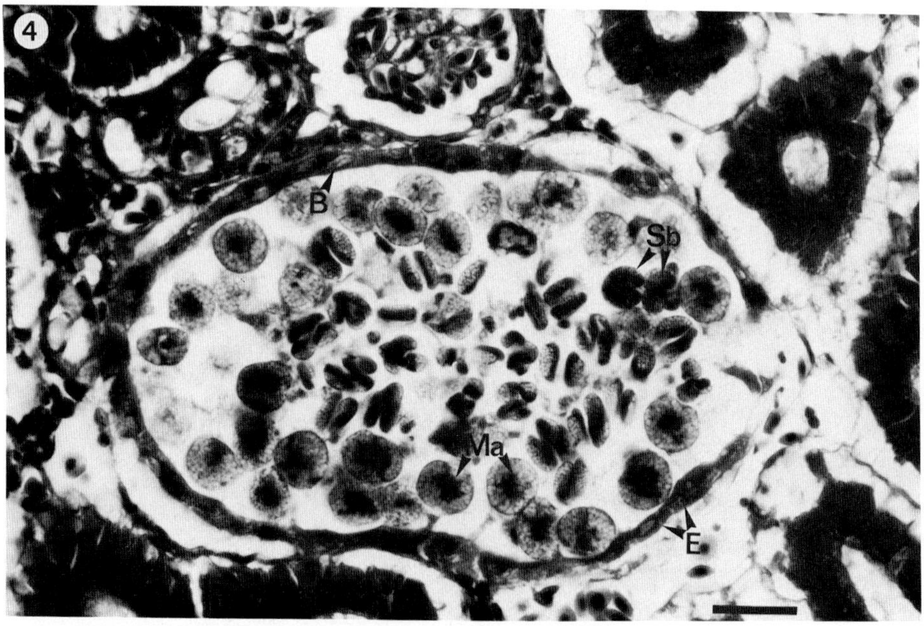

Fig. 4. *Goussia* sp. in the kidney of cod. The macrogametes (Ma) develop attached to the apical surfaces of the epithelial cells (E) of the kidney tubule, which are flattened, and contain membrane-bound bodies (B). The lumen of the tubule contains zygotes and sporoblasts (Sb). Bar = 20 μm.

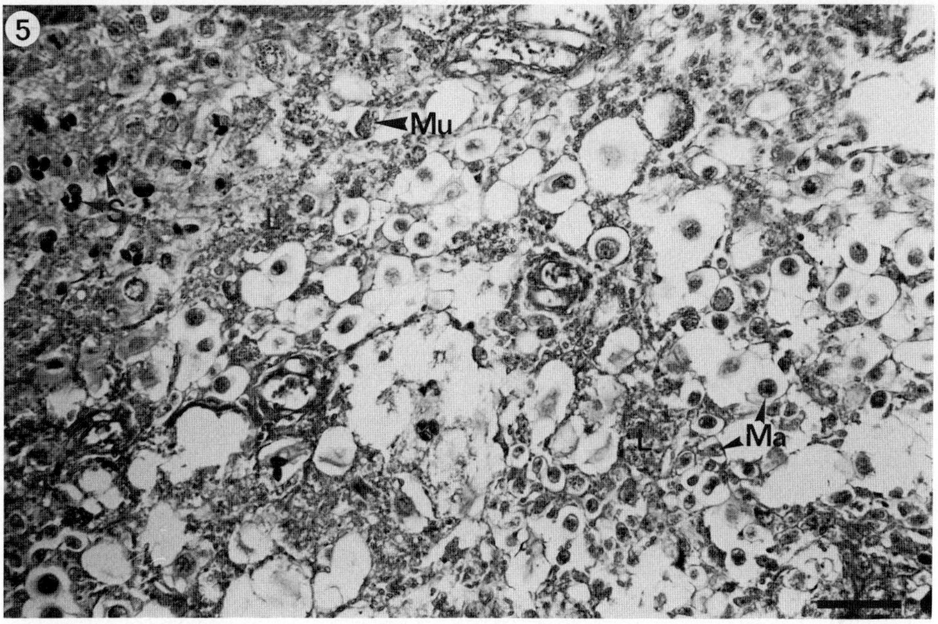

Fig. 5. *Goussia gadi* in the loose connective tissue in the wall of the swimbladder of haddock. Macrogametes (Ma), multi-nucleate stages (Mu), which may be parasites undergoing merogony or microgametogenesis, and sporocysts (S) can be seen. Leukocytes (L) surround many of the oocysts. Bar = 20 μm.

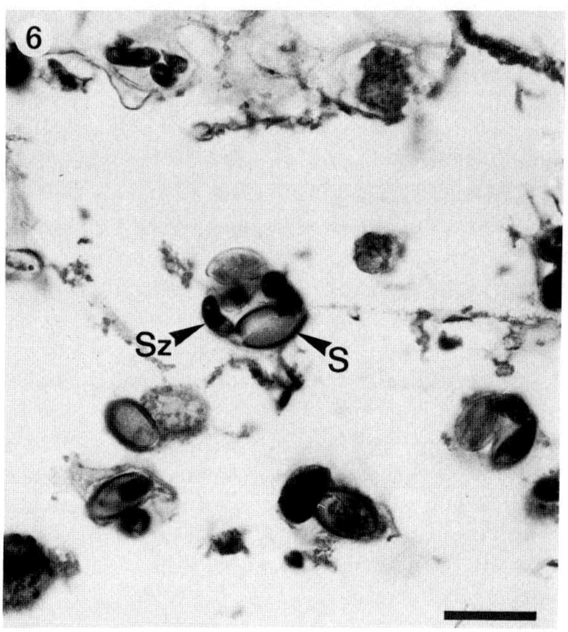

Fig. 6. *Goussia gadi* in swimbladder wall of haddock. One of the bivalved sporocysts (S) has opened to release two sporozoites, one of which can be seen clearly (Sz). The thick wall of the sporocyst is easily seen. The parasites are surrounded by remnants of loose connective tissue. Bar = 20 μm.

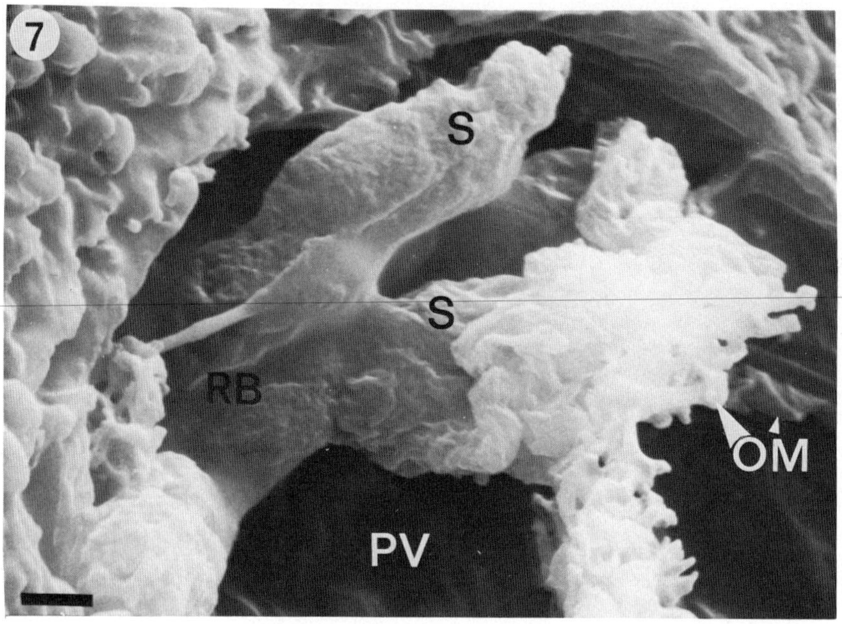

Fig. 7. *Eimeria sardinae* from testis of herring. Scanning electron micrograph of freeze-fractured specimen. Sporocysts (S) are attached together at one end by a residual body (RB), which also attaches them to one side of the parasitophorous vacuole (PV). The oocyst membrane (OM) covers some of the sporocysts. The wall of the sporocyst is thin and collapsed onto the sporozoites inside. No indication of an opening can be seen in the sporocyst wall. Bar = 2 μm.

transmitted to other fish after the death of the host. Possibly an intermediate host that scavenges on the dead fish but is too large for small haddock to eat is involved in the life cycle.

The last coccidian we studied, from the testis of herring, remains an enigma. We retain the name *Eimeria sardinae* for this parasite until more evidence is available on the method by which sporozoites are released from the sporocyst. However, as no suture or Stieda body was visible using light or transmission electron microscopy (Morrison and Hawkins, 1984), this coccidian may belong to a new genus. It was difficult to see the sporocysts using scanning electron microscopy because they were usually covered by the oocyst membrane, but there was no suture on one sporocyst that could be clearly seen (Fig. 7). The sporocyst wall consisted of only two membranes with a total thickness of about 16 nm, so an opening mechanism may not be necessary. Ultrastructural studies of other coccidia with a similar appearance such as *E. myoxocephali* from the "great sculpin" (Fitzgerald, 1975) and *E. brevoortiana* (Hardcastle, 1944) would be of interest in that the sporocysts of these latter species may also lack an opening mechanism.

In a squash preparation of minced herring testis, which had been left in sea water at 4°C for 24 hr, some oocyst walls had opened to release sporocysts, but the sporocysts were intact (Fig. 8) and still attached at one end to a residual body as described by Thélohan (1890). In the same preparation, the sporozoites could be seen moving around in the sporocyst in a head-to-tail fashion (Fig. 9). Only one free

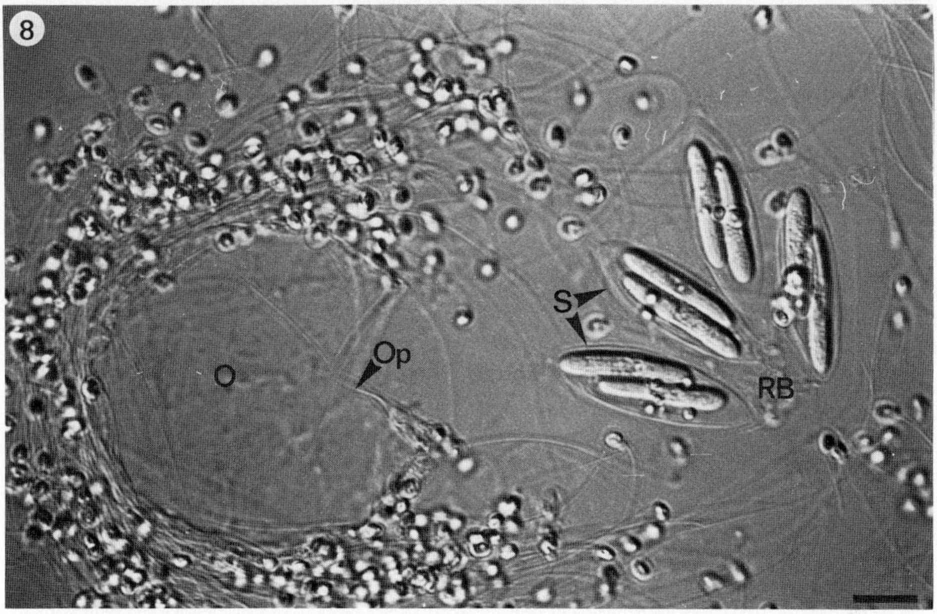

Fig. 8. *Eimeria sardinae*. Squash preparation of minced testis which had been left at 4°C in seawater for 24 hr. Nomarski illumination. The sporocysts (S) have been released from the oocyst (O) through an opening (Op) in one side. They are still attached at one end by the residual body (RB). Two elongate sporozoites and a small sporocyst residuum can be seen in each sporocyst. Many spermatozoa can be seen. Bar = 10 μm.

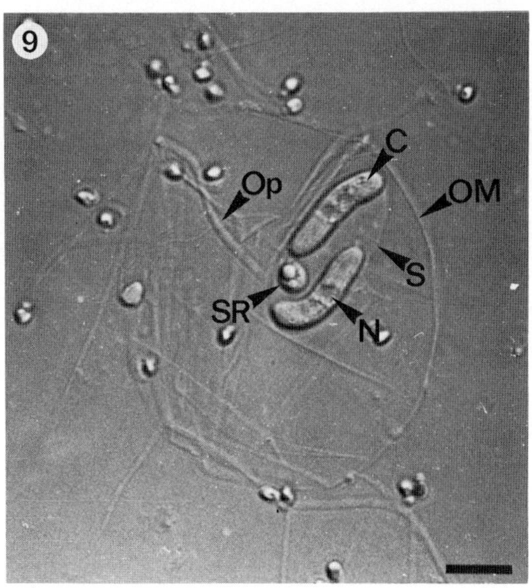

Fig. 9. *Eimeria sardinae* from same preparation as Figure 8. The nucleus (N) and conoid apparatus (C) can be seen in each sporozoite, as well as the sporocyst residuum (SR). This sporocyst (S) is caught in the oocyst membrane (OM), but the others have been released through the opening (Op). Bar = 10 μm.

sporozoite was seen, and we were not able to induce release of the sporozoites from the sporocyst by immersion in distilled water, acid, alkali, or trypsin. Possibly, as in the case of *Calyptospora funduli* (Fournie and Overstreet, 1983), the sporocyst will open only when exposed to chemical stimuli peculiar to the gut of the intermediate or final host.

All coccidia discussed here cause damage by displacement of host tissue. Up to 19% of the testis was replaced by *E. sardinae* (Morrison and Hawkins, 1984), but there was no evidence of castration as reported by Pinto (1956) in *Sardina pilchardus*. A host reaction was often apparent in the livers of herring infected with *G. clupearum*, and in cod and haddock kidneys infected with *Goussia* sp. Obstruction of ducts or blood vessels in the liver or tubules in the kidney by parasites or parasites surrounded by host cells may lead to reduced function in small, heavily infected fish. The fact that the kidney parasite occurs mainly in small cod indicates that most hosts either recover or die from the infection.

The asexual stages of these parasites have not been found. However, the merogony cycle may be very short, as in *Calyptospora funduli* where macrogamont formation takes place about 14 days after infection (Hawkins et al., 1981), and hence difficult to find in wild fish. Merogony may take place in different organs from sporogony, but we have not found meronts in the digestive tract, liver, and kidney of cod or herring examined at different times of the year.

Goussia gadi is unusual in that all stages in the life cycle appear to be present in the wall of the swimbladder, and reinfection may occur, resulting in heavy infections. As the number of heavily infected fish decreased during the winter (Odense and Logan, 1976), it was suggested by these authors that infected fish die. However, we found a host response in some infected haddock (Fig. 5), so some fish may recover.

ACKNOWLEDGEMENTS

We would like to thank the boat and aquarium crews for obtaining and maintaining the fish, Dr. P. Odense for help with the preparation of the herring testis for scanning electron microscopy and for the use of the scanning electron microscope, and Vikki Clayton for her careful typing of the manuscript.

LITERATURE CITED

Dykova, I. and J. Lom. 1983. Fish coccidia: an annotated list of described species. Folia Parasitol., 30, 193-208.

Fiebiger, J. 1913. Studien ueber die schwimmblasen-cocciden der Gadusarten (*Eimeria gadi* n.sp.). Arch. Protistenk, 31, 95-137.

Fitzgerald, P.R. 1975. New coccidia from the spiny dogfish shark (*Squalus acanthias*) and great sculpin (*Myoxocephalus polycanthocephalus*). J. Fish. Res. Bd. Canad., 32, 649-651.

Fournie, J.W. and R.M. Overstreet. 1983. True intermediate hosts for *Eimeria funduli* (Apicomplexa) from estuarine fishes. J. Protozool., 30, 672-675.

Hardcastle, A.B. 1944. *Eimeria brevoortiana*, a new sporozoan parasite from menhaden (*Brevoortia tyrannus*), with observations on its life history. J. Parasitol., 30, 60-68.

Hawkins, W.E., M.A. Solangi and R.M. Overstreet. 1981. Ultrastructural effects of the coccidium, *Eimeria funduli* Duszynski, Solangi and Overstreet, 1979 on the liver of killifishes. J. Fish. Dis., 4, 281-295.

Labbé, A. 1896. Recherches Zoologiques, Cytologiques et Viologiques Sur les Coccidies. Arch. Zool. Exp. Gén., 4, 517-654.

Landsberg, J.H. and I. Paperna. 1985. *Goussia cichlidarum* n.sp. (Barrouxiidae, Apicomplexa), a coccidian parasite in the swimbladder of cichlid fish. Zeit. Parasitenkd., 71, 199-212.

Langton, R.W. and R.E. Bowman. 1980. Food of fifteen Northwest Atlantic gadiform fishes. NOAA Technical Report NMFS SS RF-740.

Morrison, C.M. and W.E. Hawkins. 1984. Coccidians in the liver and testis of the herring *Clupea harengus* L. Canad. J. Zool., 62, 480-493.

Morrison, C.M. and S.L. Poynton. 1989a. A new species of *Goussia* (Apicomplexa, Coccidia) in the kidney tubules of the cod, *Gadus morhua*. J. Fish. Dis., in press.

Morrison, C. and S. Poynton. 1989b. A coccidian in the kidney of the haddock, *Melanogrammus aeglefinus*. J. Fish. Dis., in press.

Morrison, C.M., G. McClelland, J. Cornick and D. Marcogliese. 1986. Parasites and diseases of some finfish off Nova Scotia. Can. Tech. Rep. Fish. Aquat. Sci., 1424, v + 36 pp.

Odense, P.H. and V.H. Logan. 1976. Prevalence and morphology of *Eimeria gadi* (Fiebiger, 1913) in the haddock. J. Protozool., 23, 564-571.

Pinto, J., dos Santos. 1956. Parasitic castration in males of *Sardina pilchardus* (Walb.) due to testicular infestation by the coccidia *Eimeria sardinae* (Thélohan). Rev. Fac. Cienc. Univ. Lisboa, Ser. C, 5, 209-226.

Sindermann, C.J. 1987. Effects of parasites on fish populations: practical considerations. Intl. J. Parasitol., 17, 371-382.

Thélohan, P. 1890. Sur deux coccidies nouvelles parasites de l'épinoche et de la sardine. C.R. Acad. Sci., Paris, 110, 1214-1216.

SOME ULTRASTRUCTURAL ASPECTS OF THE LIFE CYCLE OF *HENNEGUYA* SP. (MYXOZOA), A PARASITE OF AN ESTUARINE FISH OF THE AMAZON RIVER

C. Azevedo[1], L. Corral[1], E. Matos[2], and S. Gusmão[2]

[1]Department of Cell Biology, Institute of Biomedical Sciences, University of Oporto, 4000 Porto, Portugal
[2]Department of Cell Biology, Center of Biological Sciences, Federal University of Pará, 66000 Belém, Brazil

ABSTRACT: Among the various ultrastructural studies on different species of the genus *Henneguya* (Myxozoa), nothing has been published about the ultrastructure of the Brazilian myxozoan species. Only some light microscopy studies have been recorded. Our study describes some particularities of the life cycle stages of an unidentified species of *Henneguya* found in gill filaments of a fish of commercial importance, *Haplosternum littorale*, obtained from the low Amazon River near Belém, Brazil.

Parasitized tissues and free mature spores were prepared by usual techniques. Trophozoites were formed by several plasmodia with numerous vegetative nuclei, generative cells, and sporogony stages. During valvogenesis two valvogenic cells gave rise to two valves of each spore. These cells move to the periphery, when the five-celled pansporoblast is formed. Two valvogenic cells surround each diplokaryotic sporoplasm and two capsulogenic cells give rise to two valves surrounding each spore. The ultrastructure of developmental stages and the morphogenesis of these are described herein.

INTRODUCTION

Among the various ultrastructural studies on different species of the genus *Henneguya* (Myxozoa, Myxobolidae), nothing has been published about the ultrastructure of the Brazilian myxozoan species. Only some light microscopy studies have been recorded (Walliker, 1969; Cordeiro et al., 1984; Kent and Hoffman, 1984). In this paper we report the ultrastructure of some developmental life cycle stages of a species not yet identified and the host tissue reaction.

MATERIALS AND METHODS

Several specimens of a teleost fish of great commercial importance, *Haplosternum littorale* (Siluriformes, Callichthyidae) obtained from the estuary of the Amazon River near Belém, Brazil, were used for this study. For transmission electron microscopy (TEM), small fragments of the parasitized gill were fixed in 3% glutaraldehyde in 0.1 M sodium cacodylate buffer (pH 7.4) for 3 hr at 4°C, washed in the same buffer for 2 hr at 4°C and post-fixed in 2% osmium tetroxide buffered with 0.1 M sodium cacodylate for 2 hr at 4°C. Following ethanol and propylene oxide dehydration, samples were embedded in Epon. Ultrathin sections were cut with a diamond knife and after double staining, observed in a JEOL 100CXII electron microscope operated at 60 kV.

For scanning electron microscopy (SEM), isolated spores prepared by techniques previously reported by Azevedo (1984) were observed in a JEOL 35C SEM at 20 kV.

RESULTS

Numerous white trophozoites, polysporous cysts, round to elipsoidal, were found in the gill filaments. The trophozoites were formed by several plasmodia with numerous vegetative nuclei, generative cells, and sporogony stages.

During valvogenesis, the two valvogenic cells (VC) (Fig. 1) gave rise to two valves of each spore. These cells move to the periphery, when the five-celled pansporoblast is formed. Two VC surround each diplokaryotic sporoplasm and two capsulogenic cells, forming a peripheral microtubular manchet (MM) (Fig. 2), are present. Later, a well-developed longitudinal MM was observed among the electron-dense and granular material that appears within the VC tails (Figs. 3, 4). Gradually, the VC become more elongated, each originating a tail (Fig. 4). The suture line was simultaneously differentiated between two valves. During these developmental stages, the surrounding host cells show some vacuolarization and signs of lysis (Fig. 3).

While the tails grow, the longitudinal MM becomes more evident and the dense granular masses gradually disappear (Fig. 4). At the final phase of maturation, the spore body is very dense and its internal structures are hardly visible (Figs. 5, 6). When the spores attain full maturation, the surrounding host tissue shows evident signs of lysis (Figs. 5, 6). The spore body has an ellipsoidal form ≈ 13.5 µm long and 5.3 µm wide. The two valves are thin, equal, and symmetric, each tapering to a long tail ≈ 45.2 µm long (Figs. 5-7).

Fig. 1. Ultrathin section of a pansporoblast showing two valvogenic cells (VC) and the sporoplasm cell (SC) surrounded by the pericyte.

Fig. 2. Pansporoblast showing the valvogenic cell (VC) with some sections of microtubules (Mt) near the suture line (arrows). The sporoplasm cell (SC) shows the nucleus with a nucleolus and several sporoplasmasomes (Ss).

Fig. 3. Disporoblastic pansporoblast showing the valvogenic cells (VC) and their tail, capsulogenic cells (CC) and a diplokaryotic sporoplasm cell (SC).

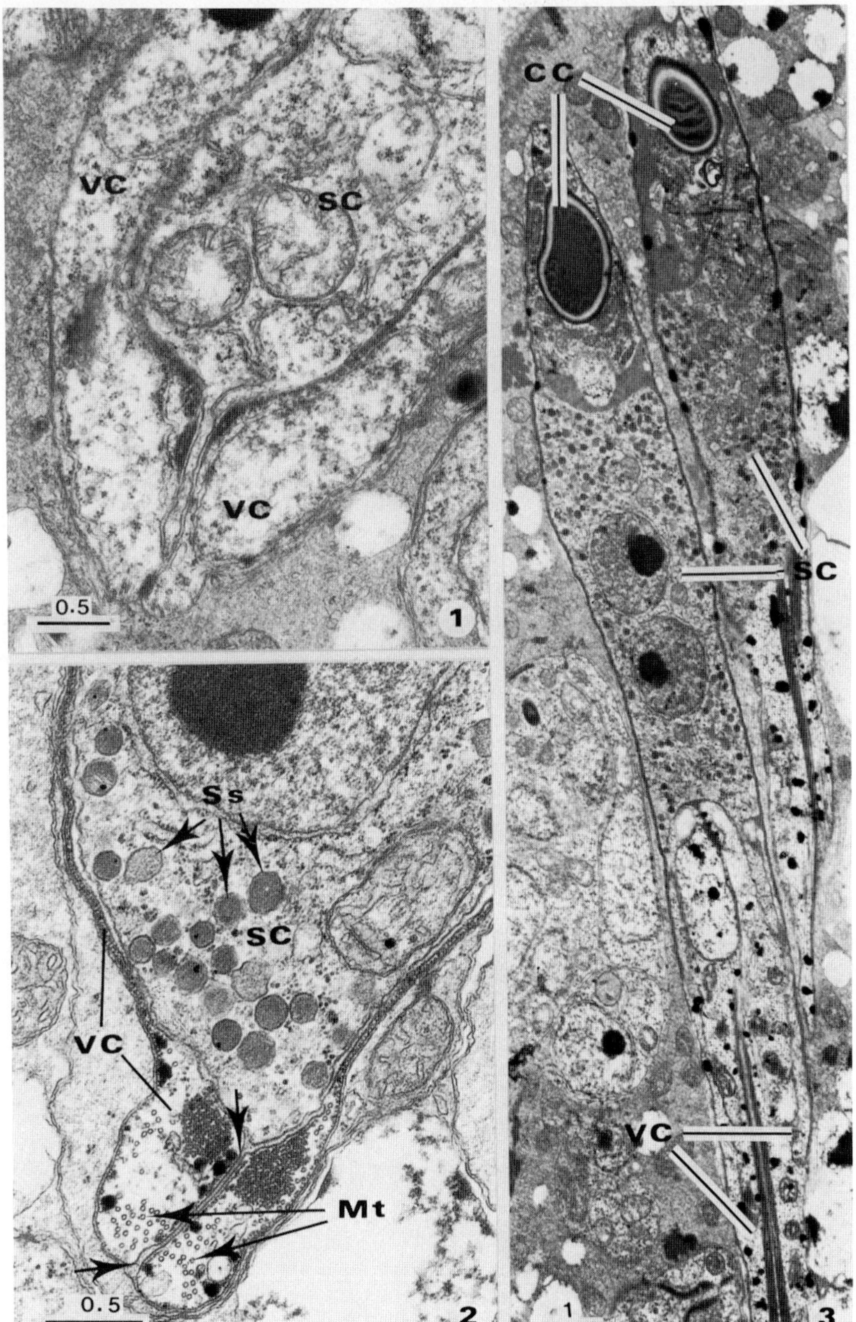
VC
SC
VC
0.5
1
Ss
SC
VC
Mt
0.5
2
CC
SC
VC
1
3

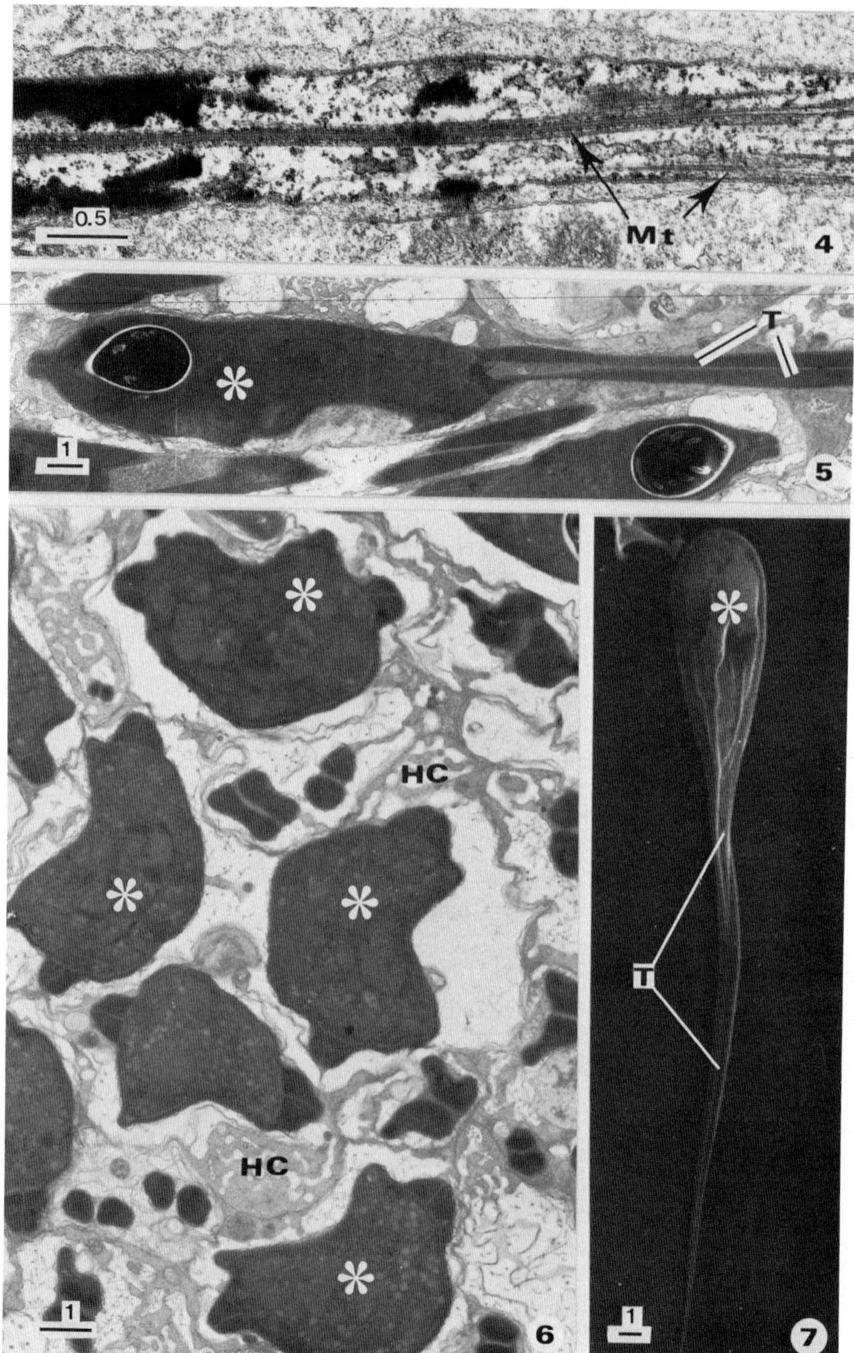
0.5
Mt
4
T
1
5
HC
HC
1
6
T
1
7

DISCUSSION

The ultrastructural organization of the histozoic parasite studied corresponds to that of members of the genus *Henneguya* Thelohan, 1892 (Lom and Noble, 1984; Current, 1979).

One of the most important features is the appearance of the granular dense masses in tail development, which gradually disappeared during the microtubulogenesis and were never seen after the tails had formed. They seem to represent the microtubular organizer centers; however, only an autoradiographic study can confirm this suggestion. A similar morphogenesis of the tails in other myxosporeans has been described (Current, 1979; Lom and de Puytorac, 1965; Lom, 1969; Desportes and Théodoridès, 1982; Desser et al., 1983).

The prevalence and pathogenicity of this parasite in the fish were not determined, but great ultrastructural alterations associated to lysis in surrounding host tissues seem to contribute to fish mortalities.

ACKNOWLEDGEMENTS

This work was partially supported by University of Oporto and the Eng. A. Almeida Foundation.

LITERATURE CITED

Azevedo, C. 1984. Ultrastructure of the spore of *Haplosporidium lusitanicum* sp. n. (Haplosporida, Haplosporidiidae), parasite of a marine mollusc. J. Parasitol., 70, 358-371.

Cordeiro, N.S., P.T. Artigas, I. Gibia and R.S. Lima. 1984. *Henneguya pisciforme* n. sp. microsporideo parasita de branquia do lambari *Hyphessobrycon anisitsi* (Pisci, Characidae). Mem. Inst. Butantan, 48, 61- 69.

Current, W.L. 1979. *Henneguya adiposa* Minchew (Myxosporida) in the channel catfish: ultrastructure of the plasmodium wall and sporogenesis. J. Protozool., 26, 209-217.

Desportes, I. and J. Théodoridès. 1982. Données ultrastructurales sur la sporogenèse de deux myxosporidies rapportées aux genres *Leptotheca* et *Ceratomyxa* parasites de *Merluccius merluccius* (L.) (Téléostéen Merluciidae). Protistologica, 28, 533-557.

Fig. 4. Detail of the developmental stage of the tail where several longitudinal microtubules (Mt) are present.

Fig. 5. Longitudinal section of an immature spore showing the spore body (*) and tail (T).

Fig. 6. Several transversal sections of the spores (*) among the vacuolarizated host cells (HC).

Fig. 7. Mature isolated spore observed in SEM showing the body (*) and tail spore (T).

Desser, S.S., K. Molnar and I. Weller. 1983. Ultrastructure of sporogenesis of *Thelohanellus nikolskii* Akhmerov, 1955 (Myxozoa: Myxosporea) from the common carp, *Cyprinus carpio*. J. Parasitol., 69, 504- 518.

Kent, M.L. and G.L. Hoffman. 1984. Two new species of *Myxobolus inaequus* sp. n. and *Henneguya theca* sp. n. from the brain of South America knife fish, *Eigemannia virescens* (V.). J. Protozool., 31, 91-94.

Lom, J. 1969. Notes on the ultrastructure and sporoblast development in fish parasitizing myxosporidian of the genus *Sphaeromyxa*. Z. Zellforsch., 97, 416-437.

Lom, J. and E.R. Noble. 1984. Revised classification of the class Myxosporea Butschli, 1881. Folia Parasitologica (Praha), 31, 193-205.

Lom, J. and P. de Puytorac. 1965. Studies on the myxosporidian ultrastructure and polar capsule development. Protistol., 1, 53-65.

Walliker, D. 1969. Myxosporidea of some Brazilian freshwater fishes. J. Parasitol., 55, 942-948.

FINE STRUCTURE OF A NEW PARASITE (*PERKINSUS*-LIKE SPECIES) OF *RUDITAPES DECUSSATUS* (BIVALVIA) FROM PORTUGAL

C. Azevedo[1], L. Corral[1], R. Cachola[2], and Frank O. Perkins[3]

[1]Department of Cell Biology, Institute of Biomedical Sciences, University of Oporto, Porto, Portugal
[2]National Institute of Fisheries Research, Faro, Portugal
[3]Virginia Institute of Marine Science, School of Marine Science, College of William and Mary, Gloucester Point, Virginia, USA

ABSTRACT: What have been interpreted to be three phases in the developmental cycle of an unidentified protistan parasite of *Ruditapes decussatus* from Portugal are described as a result of transmission electron microscopy. This parasite shows similarities to *Perkinsus marinus*; however, differences also exist.

INTRODUCTION

Mortalities, occurring during the last 4 years in Portuguese clams, *Ruditapes decussatus*, and caused by the presence of parasites, have been described (Azevedo et al., 1988; Chagot et al., 1987; Comps and Chagot, 1987; Ruano and Cachola, 1986). The most important pathogens found in these clams are *Perkinsus*-like organisms that are similar to those species found in the bivalve molluscs *Crassostrea virginica* and *Macoma balthica* (Perkins 1976, 1988). The genus *Perkinsus* belongs to class Perkinsasida and phylum Apicomplexa (Levine, 1988), which parasitizes at least 34 species of bivalve molluscs worldwide (Perkins, 1988). Although they are serious molluscan pathogens that cause extensive mortalities, many of their life cycles are not known.

The aim of this work is to provide an ultrastructural description of some life cycle stages of the *Perkinsus*-like organism.

MATERIALS AND METHODS

Some parasitized clams, *Ruditapes decussatus* (Mollusca, Bivalvia), obtained from the estuarine region in the south of Portugal (Algarve) from a population with high mortality, were used for light microscopy (LM) and transmission electron microscopy (TEM). For TEM, small fragments of the heavily infected gill tissues were fixed in 3% glutaraldehyde in 0.1M sodium cacodylate buffer (pH 7.6), for 2 hr at 4°C, washed for 2 hr at 4°C and postfixed in 2% osmium tetroxide in the same buffer for 2 hr at 4°C. All pieces were dehydrated by a graded series of ethanol and embedded in Epon. Semithin sections for LM, stained with methylene blue-azur II, were used to locate the different life cycle stages. Ultrathin sections were double stained with uranyl acetate and lead citrate and observed with a JEOL 100CXII electron microscope operated at 60 kV.

RESULTS

The parasitized areas, appearing as white cysts (the largest ≈ 1mm diameter) situated in the connective tissue of the gill filaments, could be observed macroscopically. Observed with LM, each cyst was composed of one or more parasite cells surrounded by several dense host cells (Fig. 1). Based on ultrastructural observations, it was possible to find three life cycle stages (Figs. 2-12). The earliest phase observed was in the form of a 10-15 μm diameter spherical or elliptical cell surrounded by a thin wall that was in direct contact with the host cells. The nucleus was generally central and spherical, with little chromatin, and a large, central nucleolus. The latter typically included fibrillar and granular components. The cytoplasm contained some lipid, numerous ribosomes, vesicles, and tubulovesicular mitochondria (Figs. 2-4). Binary fission was the only form of cytokinesis observed and was only observed in this earliest phase. During this phase, the vesicles fused, giving rise to numerous vacuoles (Fig. 5). Sometimes it was possible to observe host material that showed signs of lysis surrounding the parasite cell (Figs. 4, 5).

The middle phase was characterized by the presence of a denser fibrogranular wall in close contact with host cells. The nucleus was more eccentric and the nucleolus was generally larger (Fig. 7). A fibrillar mass situated between the nucleolus and the nuclear envelope was frequently observed. In this stage the nuclear

---

Fig. 1. Semithin section of a cyst containing several parasite cells. Host cells (*). Magnification scale is in microns on this and subsequent micrographs.

Fig. 2. Spherical parasite cell with a spherical nucleus and a central nucleolus. The cytoplasm has some lipid droplets and several mitochondria. The host cellular material, which has signs of lysis, surrounds the parasite (*).

Fig. 3. Two parasite cells, one of which has several lipid droplets. The surrounding host cells exhibit lysis (*). Wall of parasite (W).

Fig. 4. Thin wall (W) in direct contact with the lysed host cell.

Fig. 5. Cytoplasm of parasite with numerous vacuoles (prior to cell division). Cell wall (W); host cell (*).

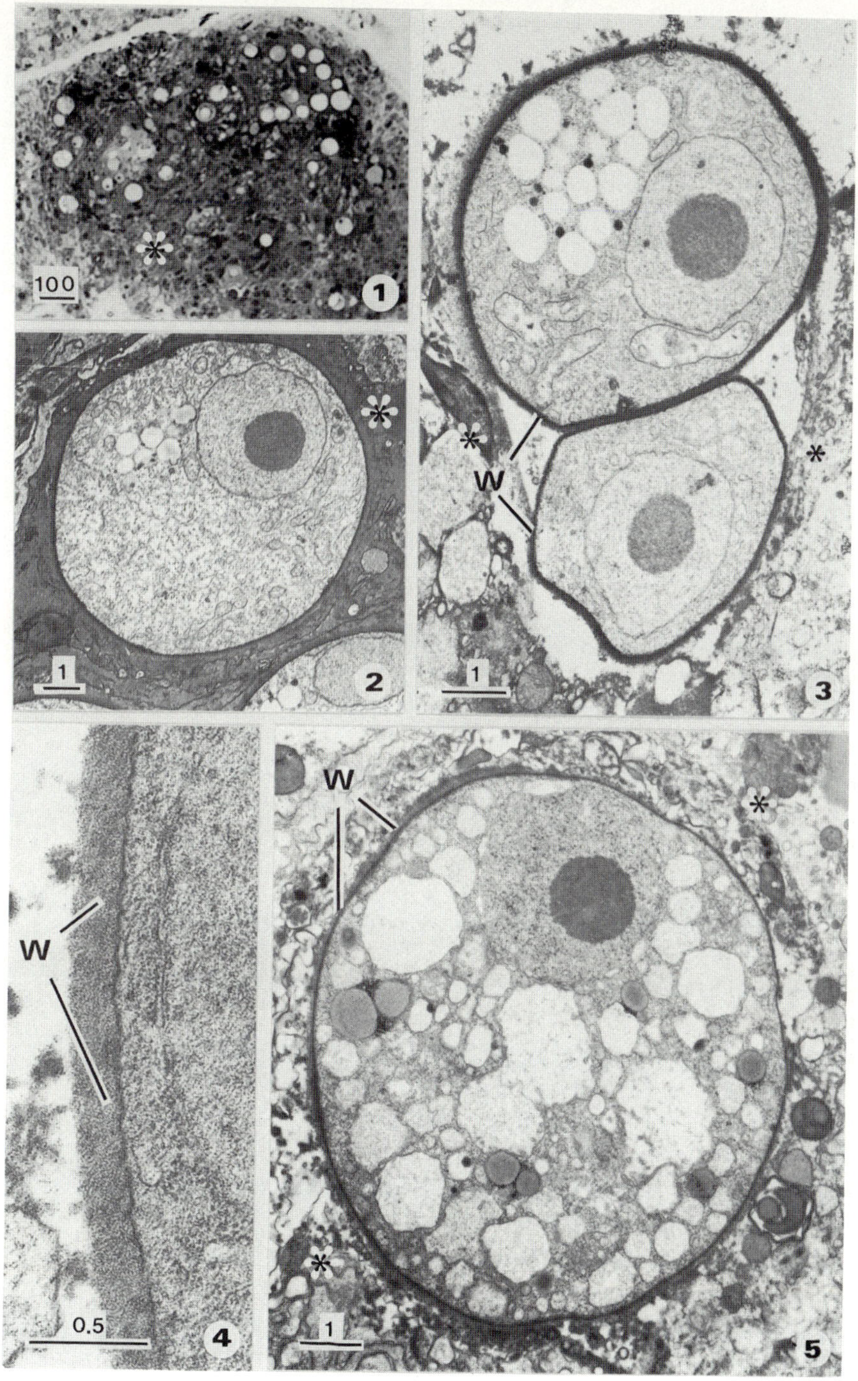
100
1
2
3
4
5
W
0.5

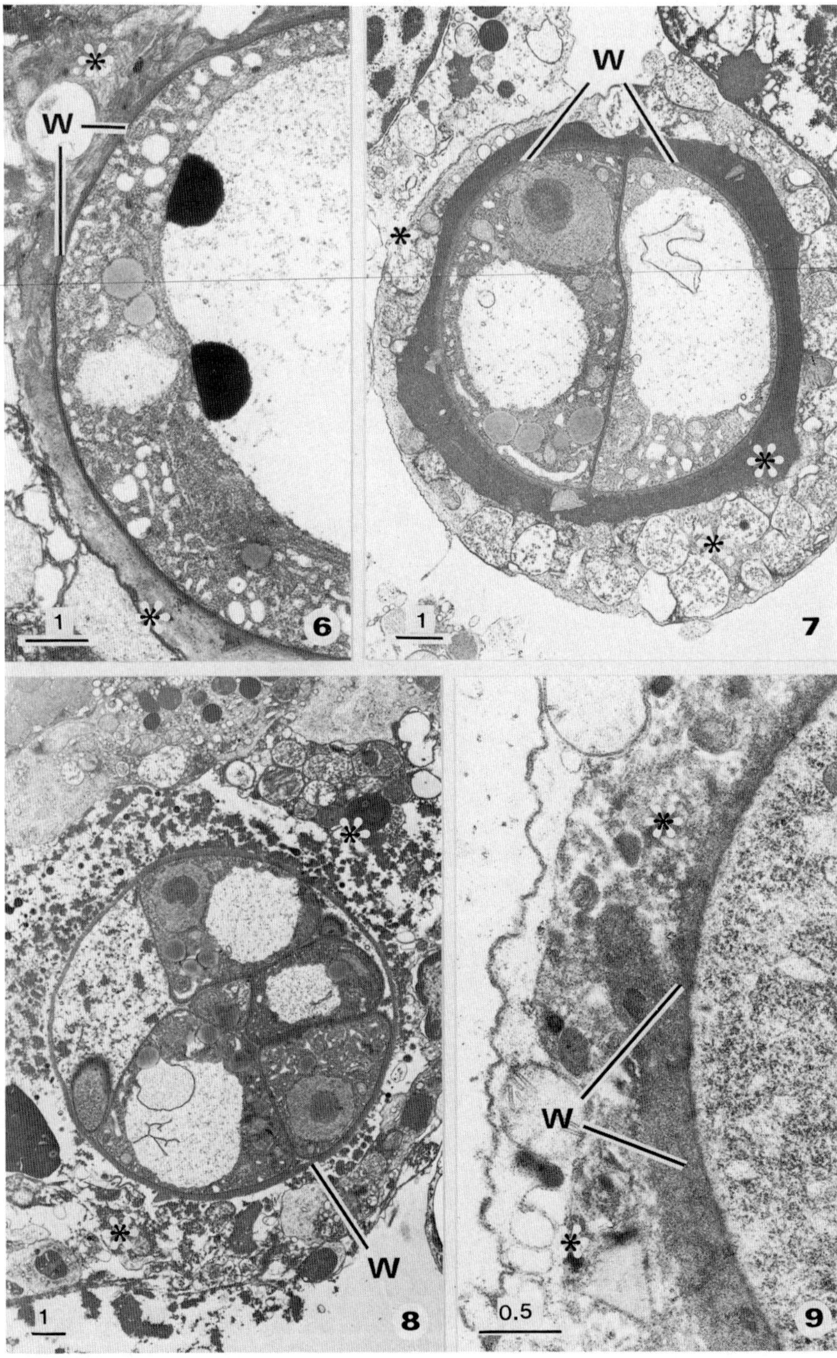
W
*
1
6
W
*
1
7
*
1
W
8
*
W
0.5
9

pores were more evident. The cytoplasm acquired a large, eccentric vacuole, 10-20 μm in diameter, that appeared to result from vacuolar fusion (Figs. 6, 7). In the granular matrix of the vacuole, some dense, volutin-like droplets occurred in close contact with the vacuole membrane (Fig. 6). These coalesced to form large vacuoplasts. During this phase, the wall became thicker and the host cells around the parasite showed increased signs of lysis (Figs. 8, 9). In sections, groups of 8-10 parasite cells may be observed within the same wall (Fig. 8).

At the final phase, each parasite cell was delimited by a denser and thicker wall composed of an internal fibrogranular matrix in contact with the plasmalemma and an external zone composed of a denser heterogeneous material that appeared to be derived from the host cells (Figs. 10-12). Around the dense material, representing the host reaction, some presumptive granulocytes were found. During this phase, the parasite's cytoplasm became a thin layer pressed against the cell wall with most of the cell volume being represented by the vacuole (Fig. 12). When the parasite reached this stage, generally the parasitized clams showed obvious signs of the disease in that their shell valves were gaping.

At this time, zoosporulation has not been observed in the cells released in sea water after incubation in fluid thioglycollate medium.

DISCUSSION

The unidentified, presumptive *Perkinsus*-like organism parasitizing the Portuguese clam, *R. decussatus*, seems to be responsible for high mortalities that have occurred in recent years. There have been few studies on the life cycle of this organism, which ultrastructurally resembles *Perkinsus marinus*, a parasite of the American oyster, *Crassostrea virginica*, and the clam, *Macoma balthica* (Perkins, 1988). However, the dense host reaction to one developmental stage of the *Perkinsus*-like organism is different from oysters infected with *P. marinus* where no signs of such a host reaction occur (Perkins, 1976).

Neither natural nor laboratory-induced prezoosporulation and zoosporulation have yet been observed, but the three different life cycle stages seem to be similar to those described for *P. marinus*. Furthermore, pathogenicity resulting from the parasite increased with summer temperature as it does with *P. marinus*-infected molluscs.

Fig. 6. Periphery of a parasitic cell with two dense volutin-like droplets adhering to vacuole membrane. Wall (W); host cell (*).

Fig. 7. Two parasitic cells, one of which has an eccentric nucleus, completely surrounded by a dense fibrogranular wall (W) in close contact with host cell (*).

Fig. 8. Group of 8-10 parasite cells surrounded by a thick wall (W). A thin wall is present around each cell.

Fig. 9. Detail of a parasite cell wall (W) in close contact of the lysed host cell (*).

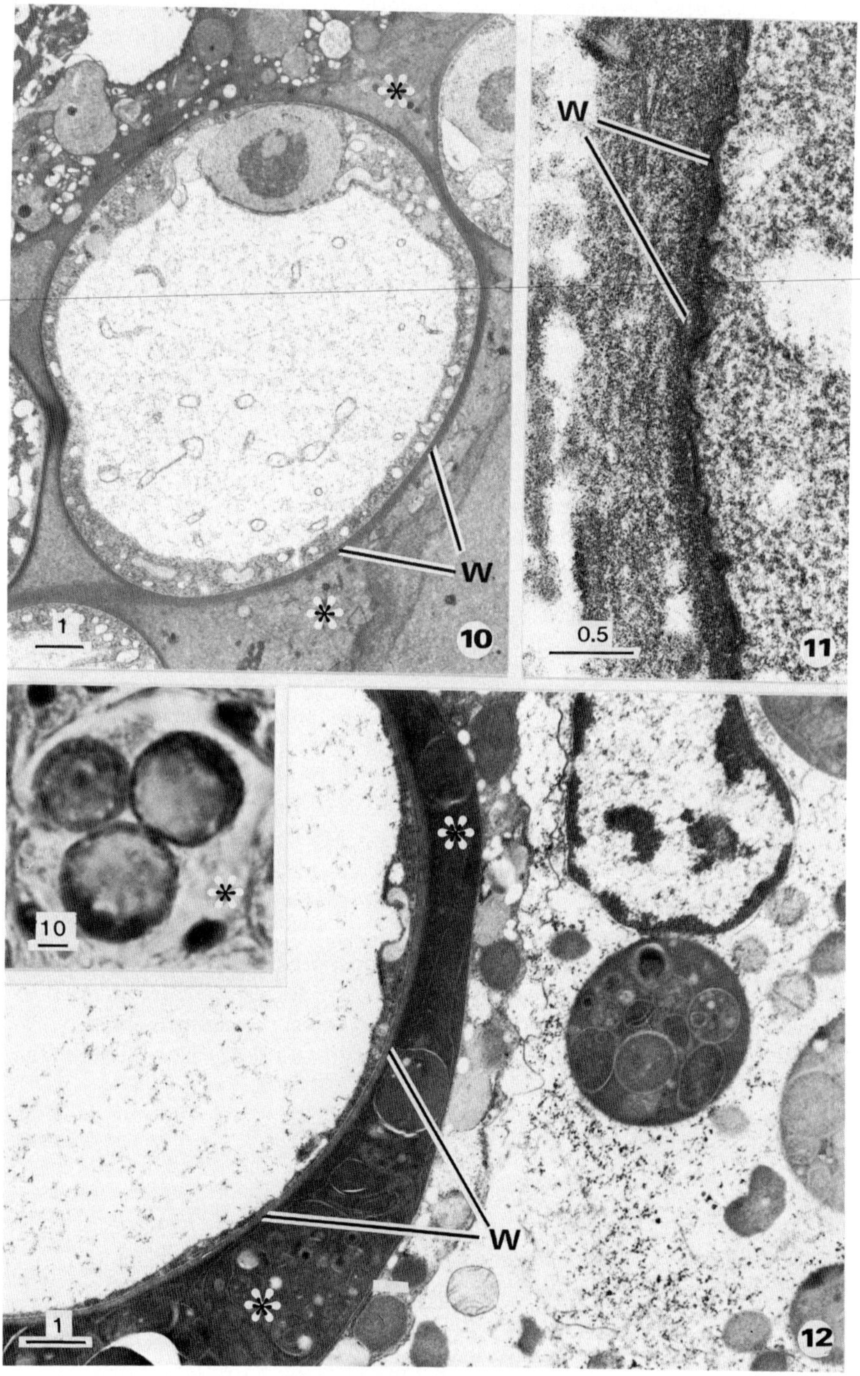
W
W
1
10
0.5
11
10
W
1
12

No other bivalve living in the region, including *Crassostrea angulata*, was found to be infected with *Perkinsus*-like organisms. Further and more detailed studies are needed to understand the complete life cycle of this protist that causes serious mortalities of *R. decussatus.*

ACKNOWLEDGEMENT

This work was partially supported by JNICT and Fundação Luso-Americana para o Desenvolvimento.

LITERATURE CITED

Azevedo, C., L. Corral and R. Cachola. 1988. Estudo ultrastructural de protozooses da fauna da Ria Formosa. Proc. 5th Congress Algarve, Albufeira, 20-23 January, pp. 531-536.

Chagot, D., M. Comps, V. Boulo, F. Ruano and H. Grizel. 1987. Histological study of a cellular reaction in *Ruditapes decussatus* infected by a protozoan. Aquaculture, 67, 260-261.

Comps, M. and D. Chagot. 1987. Une parasitose nouvelle chez la Palourde *Ruditapes decussatus* L. C.R. Acad. Sci., Paris, 304, 41-44.

Levine, N.D. 1988. The protozoan phylum Apicomplexa. CRC Press, 1, 9.

Perkins, F.O. 1976. *Dermocystidium marinum* infection in oysters. Mar. Fish. Rev., 38, 19-21.

Perkins, F.O. 1988. Structure of protistan parasites found in bivalve molluscs. In: "Disease Processes in Marine Bivalve Molluscs" (W.S. Fisher, ed.), American Fisheries Society, Special Publication 18, pp. 93-111. Bethesda, Maryland.

Ruano, F. and R. Cachola. 1986. Outbreak of a severe epizootic of *Perkinsus marinus* (Levin - 78) at Ria de Faro clam's culture beds. Proc. PAMAQ II, pp. 41-42, Porto-Portugal.

Fig. 10. Parasite cell showing a thin layer of the cytoplasm pressed between the cell wall (W) and vacuole. Dense host material (*), showing signs of lysis, is present around the cell.

Fig. 11. Detail of the wall (W) of a cell in the same stage of development as in Fig. 10.

Fig. 12. Cell surrounded by a dense host material (*). Wall (W). Inset is a light micrograph of a cell grouping in the same stage of development.

PERKINSUS IN AUSTRALIA

R.J.G. Lester, C.L. Goggin, and K.B. Sewell

Department of Parasitology, University of Queensland
Brisbane, Australia 4067

ABSTRACT: *Perkinsus* spp. are widespread in Australian molluscs. They have been associated with deaths of abalone, *Haliotis laevigata,* in South Australia and of clams, *Tridacna gigas,* in Queensland. Infection in *H. laevigata* is linked to high environmental temperature. The number of *Perkinsus* species present in Australia is not known although several morphological types of trophozoites occur in different hosts. Experiments on transmission indicate that some *Perkinsus* species can infect many species of mollusc.

INTRODUCTION

The apicomplexan protist *Perkinsus marinus* is recognized as an important pathogen of the American oyster, *Crassostrea virginica*, along the Gulf and Atlantic coasts of the United States (Perkins, 1987). Related parasites are considered a threat to cultured bivalves in the Mediterranean (da Ros and Canzonier, 1985), on the Atlantic coasts of Spain (Villalba and Navas, 1988) and Portugal (Comps and Chagot, 1987), and in Hawaii (Kern et al., 1973).

In Australia, *Perkinsus olseni* causes abscesses up to 10 mm across in the muscle of the blacklip abalone, *Haliotis rubra*, and this reduces their market value (Lester and Davis, 1981). Semi-isolated pockets of infection have been known in South Australia for 15 years (Fig. 1) though no field mortalities have been associated with the infections.

Haliotis rubra is one of two abalone species that support the Australian abalone fishery, worth A$84 million (US$70 million) in 1986-87. A second species, the greenlip abalone, *H. laevigata*, is more valuable than *H. rubra* but less widespread.

Between 1982 and 1984, *H. laevigata* largely disappeared from the west coast of the Gulf of St Vincent, South Australia. Shells of freshly dead animals, some still containing tissue, were common. Tests for heavy metal and pesticide poisoning were negative (Shepherd, 1985). In 1986, *Perkinsus* parasites, similar to *P. olseni*, were found in 70% of *H. laevigata* at the edge of the die-back area (Lester, 1986).

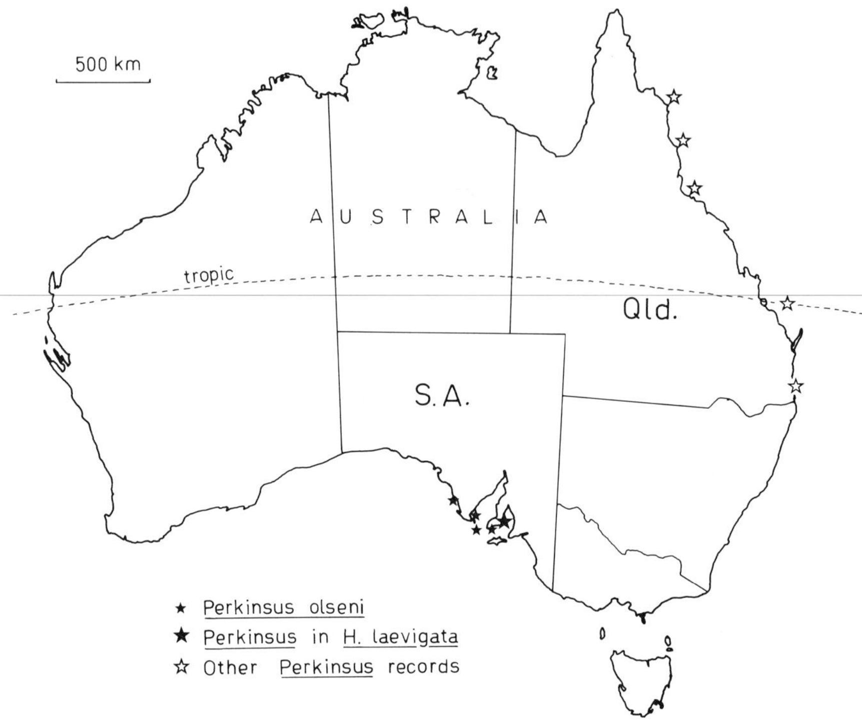

Fig. 1. Known distribution of *Perkinsus* spp. in Australia. Small black stars: *P. olseni* in *Haliotis rubra*; large black star: *P. olseni* in *H. laevigata*; other stars: *Perkinsus* spp. in reef and subtropical bivalves.

On the Great Barrier Reef, *Perkinsus* parasites were first found by Perkins (1985) who reported them in the giant clam, *Tridacna maxima*, and the oyster, *Saccostrea cucculata*. Dying giant clams of a second species, *T. gigas*, were later found infected and a subsequent survey showed that *Perkinsus* spp. were widespread in bivalves on the Reef (Goggin and Lester, 1987; Fig. 1).

In this paper we present recent observations on infections in abalone and relate them to infections in other molluscs.

OBSERVATIONS

Die-back in *H. laevigata*. The die-off of *H. laevigata* occurred along the coast of the Yorke Peninsula, South Australia (Fig. 2). Samples for thioglycollate culture (Ray, 1966) were taken from various sites on five occasions (Fig. 2, Table 1). Greenlip abalone with *Perkinsus* infections were found only in the die-back area. A histological survey (Lewis et al., 1987) of 125 *H. laevigata* from seven areas in South Australia found infection in only the die-back area. Thus, there is a strong correlation between infection and die-back.

In July (winter), 1987, 140 uninfected adult *H. laevigata*, 130 to 160 mm shell length, were introduced into the sea off Stansbury (Fig. 2) in the center of the depleted area to aid recolonization. The following spring (October, 1987; Table 1), thioglycollate culture of tissues from 20 animals indicated they were still free of infection, but by late summer 15 of 20 sampled (75%) were carrying *Perkinsus* and many had died. By August, after removing a further 20 animals for tissue culture, only six of the 80 that should have remained could be found alive.

The translocated abalone had apparently become infected only in the summer. Our winter field collections totalled 157 abalone (of four species) from the die-back area of which nine were infected (6%); in summer we examined 104 and found 43 infected (41%), providing further evidence that there was a relationship between *Perkinsus* prevalence and season. Lester and Davis (1981) observed that *H. rubra* infected with *P. olseni* and kept at 20°C died with heavy infections whereas those kept at 15°C encapsulated the parasites and survived. Mortalities of *C. virginica* due to *P. marinus* are prevalent at high temperatures (Beckert et al., 1972).

Though there are no long-term records of water temperatures for the Yorke Peninsula, air temperatures at the local meteorological station (Warooka) indicate that winters have been shorter and summers longer for the last decade compared with the previous 25 years (Fig. 3). This may have influenced the water maximum temperatures and thus made *H. laevigata* more susceptible to the disease.

In the die-back area, four species of abalone carried *Perkinsus* infections (Table 1). The morphology of the trophozoites in *H. laevigata* was identical to that described for *P. olseni*. No significant differences in development were noticed between parasites from all four species of abalone and we believe the same parasite infects them all.

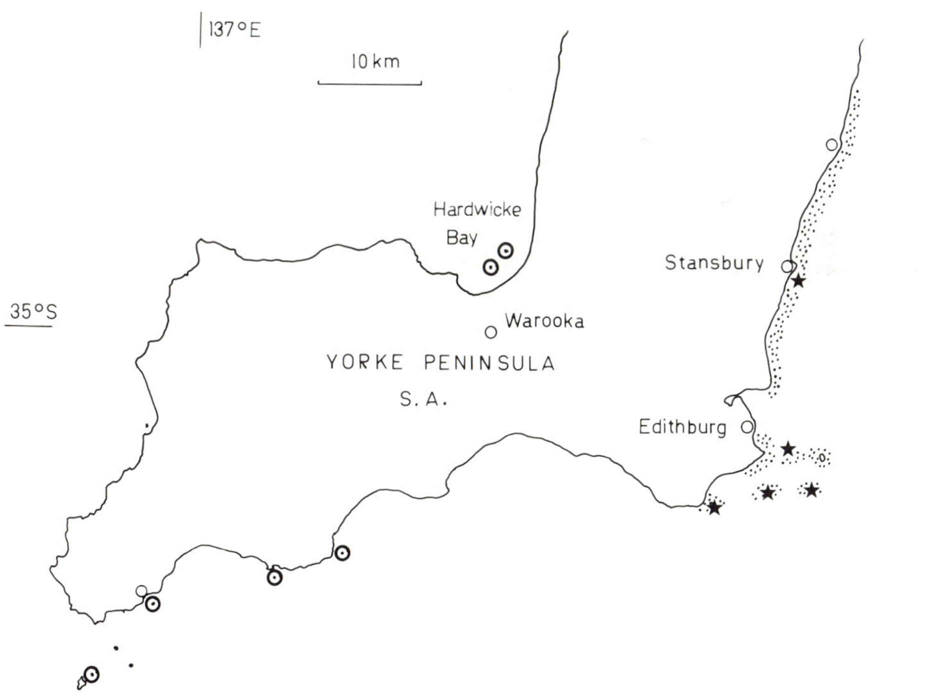

Fig. 2. Tip of Yorke Peninsula, South Australia, showing extent of die-back area (stippled), sites where *Haliotis laevigata* have been found infected with *Perkinsus olseni* (stars) and sites where *H. laevigata* were not infected (centred circles).

Table 1. Prevalence of *Perkinsus* infection in *Haliotis* species from the Yorke Peninsula, South Australia. "D" = die-back area; "O" = other areas; "–" = no sample.

		March 1986	July 1987	October 1987	March 1988	August 1988
H. laevigata	D	22/32	1/16	0/17	1/10	3/46
translocated*		–	–	0/20	15/20	2/20
	O	0/16	–	–	–	0/109
H. rubra	D	–	–	–	1/1	0/1
	O	–	–	–	–	2/32
H. scalaris	D	–	0/5	0/3	0/2	1/10
	O	–	–	–	–	0/28
H. cyclobates	D	4/15	0/10	0/30	2/23	2/49
	O	–	–	–	–	–

*Moved from unaffected area (Hardwicke Bay) to center of die-back area (Stansbury) in July, 1987.

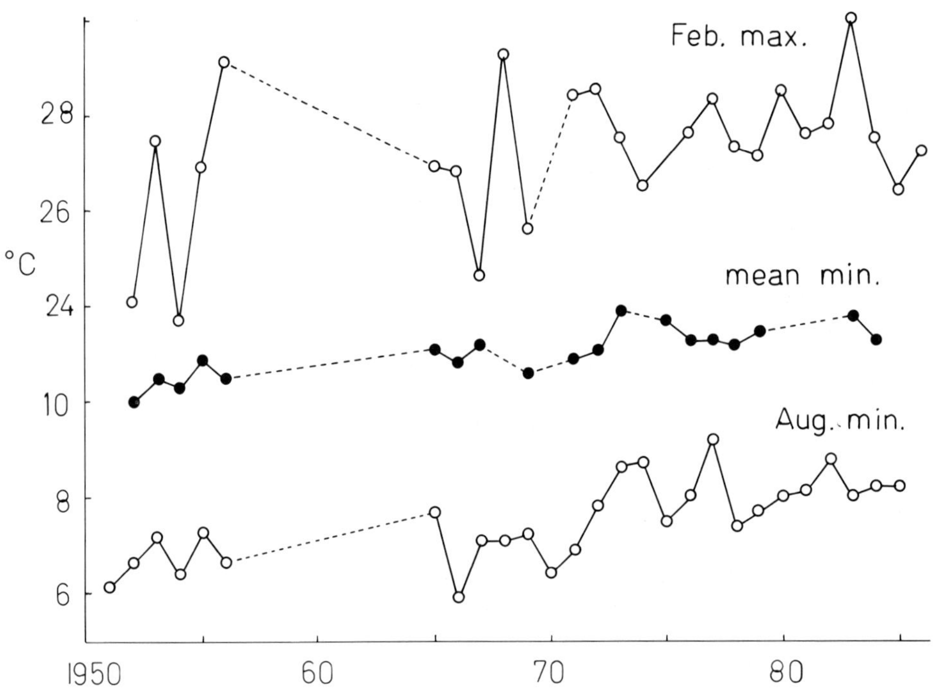

Fig. 3. Air temperature records for Warooka, South Australia, for 1952 to 1985. Top: mean daily maximum temperature in February; bottom: mean daily minimum temperature in August; center: mean daily minimum temperature for the year. In all three cases temperature and year are positively correlated at the 98% level of confidence.

Though individuals of *H. rubra* are infected in several parts of South Australia (Fig. 1) and two infections were found near but outside the die-back area (Table 1), *H. laevigata* has been found infected only in the depleted area and not elsewhere in the state, suggesting there is little cross infection between *H. rubra* and *H. laevigata*.

Haliotis rubra is an abalone of exposed steep rocky shores and usually occurs at 2 to 5 meters; *H. laevigata* prefers a flatter bottom and in exposed sites is in water over 4 meters deep (Shepherd, 1973). The die-back area is sheltered and is composed of relatively shallow sea grass beds 3 to 6 meters deep with numerous specimens of *Pinna bicolor* to which are attached a small noncommercial abalone, *H. cyclobates*. The southern limit of the die-back more or less corresponds to the limit of *H. cyclobates*. *Haliotis cyclobates* could be a reservoir for the infection in *H. laevigata*. It is likely that the shallow sea grass beds reach a higher temperature than the deeper water off the rocky headlands to the west and this predisposed *H. laevigata* to infection. Whether the parasite is pathogenic to *H. cyclobates* is not known. Specimens are still common in the die-back area.

Molluscs in the depleted area other than abalone were examined for *Perkinsus* spp. in August, 1988. Infections were found in two of five ark shells, *Barbatia pistachia*, and three of 20 cockles, *Katelysia rhytiphora*. Whether these are *P. olseni* has yet to be determined.

The species problem. Parasites similar to *P. marinus* have been recorded from 24 species of bivalves in North America (Ray, 1954; Andrews, 1955), six in the Mediterranean (da Ros and Canzonier, 1985), and 30 in Australia (Goggin and Lester, 1987); how many parasite species are involved is not known. Only two species have been named, *Perkinsus marinus* and *P. olseni*. The latter differs from the former in several ways but particularly in the large size of its trophozoite. Trophozoites of *P. marinus* vary in size and are commonly 5 to 7 μm in diameter (Mackin et al., 1950). They are frequently seen to be within host cells, particularly amoebocytes, and are spread throughout the oyster's connective tissues. Trophozoites of *P. olseni* (Fig. 4) are 13 to 16 μm in diameter. They appear to be extracellular and are often free in the hemolymph. When described, *Perkinsus olseni* was the only member of the genus from a gastropod.

Trophozoites of *Perkinsus* sp. that we have found in *Tridacna maxima* are only 2 to 6 μm in diameter (Goggin and Lester, 1988) and are usually surrounded by a halo of digitate processes (Fig. 5) that appear to be the remains of a host cell. Trophozoites are primarily restricted to the digestive gland.

Trophozoites in *Chama pacificus* are also small, 2 to 4 μm in diameter (Goggin and Lester, 1988), and are embedded in the digestive gland in a PAS-positive matrix, which is apparently a reaction by the host to the parasite (Fig. 6). At Heron Island, Great Barrier Reef, *C. pacificus* frequently have very heavy infections but parasite associated mortalities have not been observed.

Perkins (1989) observed that trophozoites of *Perkinsus* sp. from *Macoma balthica* were larger than those of *P. marinus*, were pyriform or egg shaped rather than coccoid, and were confined within pseudocapsules in the host tissue. He concluded they probably belonged to a different species.

P. marinus is widely believed to show strong host specificity (Lauckner, 1983). This originates from the work of Ray (1954) and Andrews and Hewatt (1957) who, after demonstrating the ease with which *P. marinus* could be transferred by injection from oyster to oyster, failed to produce infection in other bivalves by the same method. Perkins and Menzel (1966) later showed that *P. marinus* produces motile

Figs. 4 to 7. *Perkinsus* species in molluscs. All at the same magnification.

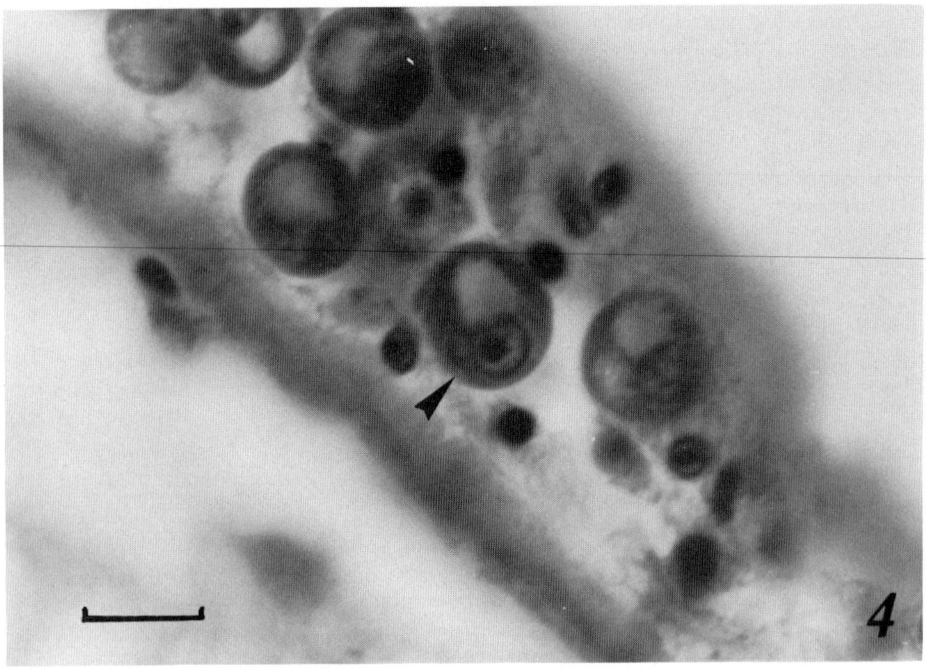

Fig. 4. Trophozoite of *Perkinsus olseni* in the gill of *Haliotis laevigata* from South Australia. Bar = 10 μm.

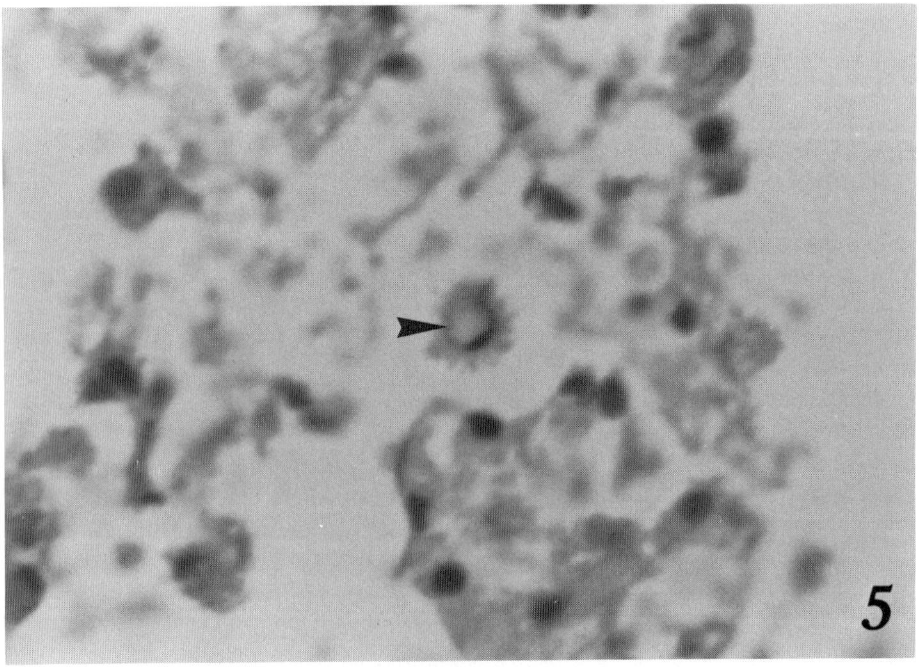

Fig. 5. Trophozoite of *Perkinsus* sp. in the digestive gland of *Tridacna maxima* from Lizard Island, Great Barrier Reef.

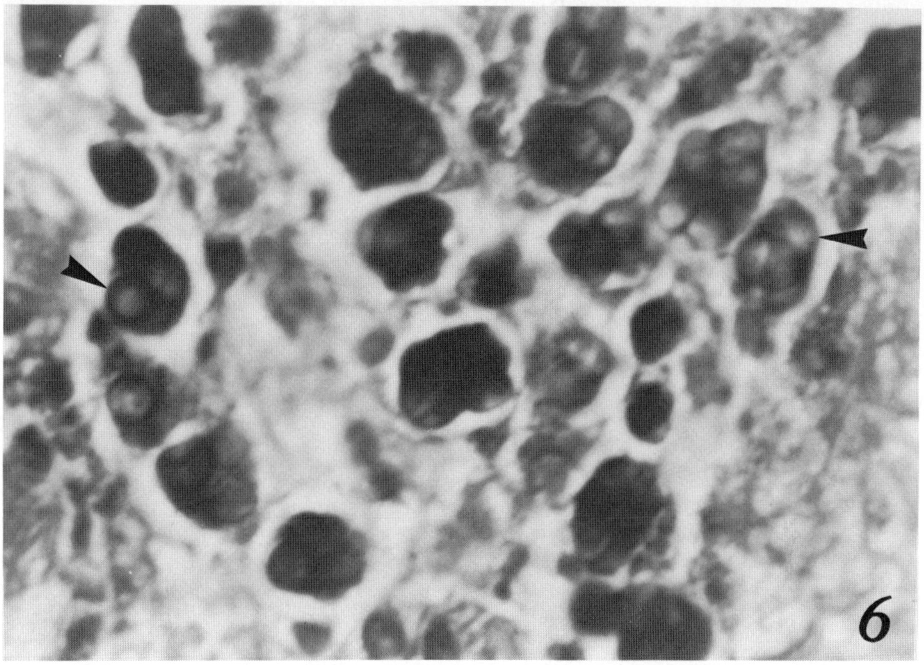

Fig. 6. Trophozoites of *Perkinsus* sp. in the digestive gland of *Chama pacificus* from Heron Island, Great Barrier Reef.

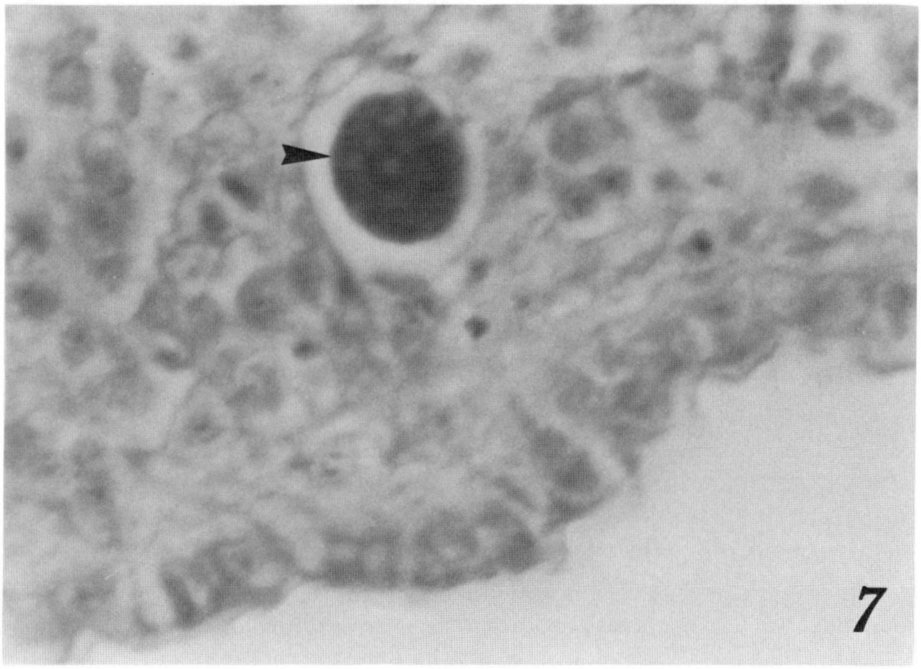

Fig. 7. Schizont of *Perkinsus* sp. in the mantle of *Anadara trapezia*; 3 week old experimental infection.

zoospores, which are now known to spread infection between oysters. As far as we can determine, no one has exposed molluscs other than *C. virginica* to zoospores of *P. marinus* and looked for infection. Recently, White et al. (1987) showed that pyramidellid gastropods that fed on infected oysters became infected themselves, demonstrating that *P. marinus* has a much wider host range than was hitherto believed.

Our cross-infection experiments using zoospores of both *P. olseni* and a *Perkinsus* sp. found in *Anadara trapezia* in Moreton Bay showed that both species have low host specificity (Table 2). Infected tissues from the original hosts were cultured in thioglycollate and sea water at 20 and 27°C. After development of sporangia, the Petri dishes to which they were attached were placed in aquaria containing the target host and left for 2 to 3 days until all sporangia were discharged. For the following weeks, recipient animals and controls under identical conditions were maintained at either 20°C (abalone and some *A. trapezia*) or 27°C (other molluscs). They were then examined for infection using the thioglycollate method (Ray, 1966). The recipient abalone used were 37 to 108 mm in shell length. None of the controls became infected. We found that we were able to transmit *P. olseni* to other abalone and even to bivalves (Table 2). A *Perkinsus* sp. from *A. trapezia* developed in the abalone *H. scalaris*, was reisolated and used to produce infections in two species of oysters. Parasites from *A. trapezia* also produced infection in the gastropod *Pyrazus ebeninus*.

Table 2. Successful cross-infection experiments with *Perkinsus* species. In parentheses, number infected/number exposed.

SOURCE		RECIPIENTS	
Haliotis laevigata		*H. laevigata* (5/6)	
		H. scalaris (3/3)	
		Pinctada sugillata (19/22)	
		Anadara trapezia (6/7)	
Haliotis cyclobates		*H. rubra* (3/3)	
		P. sugillata (9/9)	
		A. trapezia (2/2)	
Anadara trapezia	*per*	*H. scalaris* (3/3) to	*Crassostrea commercialis* (7/10)
			Pinctada sugillata (7/7)
Anadara trapezia		*A. trapezia* (3/3)	
		C. commercialis (30/30)	
		Pyrazus ebeninus (12/13)	

We have not yet found trophozoites in histological sections of recipient molluscs to determine whether or not they retain their distinctive morphology. However, preliminary observations suggest that the time for development from prezoosporangium to zoospore discharge in *P. olseni* takes 1 day longer than in *Perkinsus* sp. from *A. trapezia* when both are cultured at 27°C.

Pathology. *P. marinus* kills oysters by extensive invasion and lysis of host tissues (Mackin, 1951). Abalone dying with *P. olseni* infections similarly have many parasites free in the haemal spaces. In recovering abalone, dead parasites are found in abscesses in the muscle (Lester and Davis, 1981).

We compared the mortality rate of heavily infected *A. trapezia* with that of controls. Twenty *A. trapezia* were infected using zoospores from naturally infected *A. trapezia* so heavily that after the Ray Test, the surface of the foot was almost black with prezoosporangia. Infected and control cockles were kept for 4 weeks at 30°C to stress them and their mortality rates monitored. There was no significant difference. Histological sections after three weeks showed that numerous schizonts had developed in cavities just beneath the surface of the epithelium of the foot, mantle, and gill (Fig. 7). However, no parasites were found in the digestive gland or elsewhere in the internal organs, nor were any in the hemolymph. The lack of pathogenicity shown in this experiment is in accord with the observation of Mackin (1951) that pathogenesis of *P. marinus* is associated with the widespread dissemination of parasites through the host tissue. After three weeks, all the parasites found were mature schizonts. We are currently investigating whether these eventually release a wave of merozoites that spread to other organs and produce more extensive tissue damage.

The subepidermal location of the schizonts in the experimentally infected *A. trapezia* indicates that it was unlikely the cockles were infected *per os*. Free-swimming zoospores apparently penetrated a short distance into whatever tissue they contacted first and there they remained until schizogony was complete. This differs from the entry of zoospores of *P. marinus* into *C. virginica*, where the first schizonts are normally found in the walls of the digestive tract (Mackin, 1951).

CONCLUSIONS

Differences in morphology of trophozoites and in development times suggests that several species of *Perkinsus* occur in Australia. They may have different degrees of pathogenicity in different hosts.

The discrimination of species remains a major challenge. Though DNA studies may help, it is likely that cross-infection experiments such as those we have described followed by morphological observations on the resulting stages will give a much better concept of the type species, *P. marinus*, than is available now. This, in turn, will lead to a clarification of the status of other species in this enigmatic but widespread genus.

ACKNOWLEDGEMENTS

For assistance with our field work in South Australia, we thank Mr. A. Geering and Mr. C. Johnson, Dr. S. Shepherd and other officers of the South Australian Department of Fisheries, and Ms. C. Gannon and Dr. A. Butler of the University of Adelaide. The project was funded by grants from the Fishing Industry Research Trust Account (FIRTA 87/9) and the Marine Sciences and Technologies Grants Scheme (MST A08600937).

LITERATURE CITED

Andrews, J.D. 1955. Notes on fungus parasites of bivalve mollusks in Chesapeake Bay. Proc. Natl. Shellfisher. Assoc., 45, 157-163.

Andrews, J.D. and W.G. Hewatt. 1957. Oyster mortality studies in Virginia. II. The fungus disease caused by *Dermocystidium marinum* in oysters of Chesapeake Bay. Ecol. Monogr., 27, 1-25.

Beckert, H., D.G. Bland and E.B. May. 1972. The incidence of *Labyrinthomyxa marina* in Alabama. Alabama Mar. Res. Bull., 8, 18-24.

Comps, M. and D. Chagot. 1987. Une parasitose nouvelle chez la Palourde *Ruditapes decussatus* L. C.R. Acad. Sci., Paris, 304, Serie III, No. 1, 41-44.

da Ros, L. and W.J. Canzonier. 1985. *Perkinsus*, a protistan threat to bivalve culture in the Mediterranean basin. Bull. Europ. Assoc. Fish Pathol., 5, 23-25.

Goggin, C.L. and R.J.G. Lester. 1987. Occurrence of *Perkinsus* species (Protozoa, Apicomplexa) in bivalves from the Great Barrier Reef. Dis. Aquat. Organ., 3, 113-117.

Goggin, C.L and R.J.G. Lester. 1988. Parasites of the genus *Perkinsus* from reef bivalves. Proc. 6th Intl. Coral Reef Symposium, p. 36 (Abstract).

Kern, F.G., L.C. Sullivan and M. Takata. 1973. *Labyrinthomyxa*-like organism associated with mass mortalities of oysters, *Crassostrea virginica*, from Hawaii. Proc. Natl. Shellfisher. Assoc., 63, 43-46.

Lauckner, G. 1983. Diseases of Mollusca: Bivalvia. In: "Diseases of Marine Animals" (O. Kinne, ed.), pp. 477-961. Biologische Anstalt Helgoland, Hamburg, Germany.

Lester, R.J.G. 1986. Abalone die-back caused by protozoan infection? Austral. Fisher., 45(6), 26-27.

Lester, R.J.G. and G.H.G. Davis. 1981. A new *Perkinsus* species (Apicomplexa, Perkinsea) from the abalone *Haliotis ruber*. J. Invert. Pathol., 37, 181-187.

Lewis, R.K., S.A. Shepherd, P. O'Donoghue and P.H. Phillips. 1987. Protozoan parasite (*Perkinsus*) infection in abalone: a progress report. Safish, 11(1), 7-8.

Mackin, J.G. 1951. Histopathology of infection of *Crassostrea virginica* (Gmelin) by *Dermocystidium marinum* Mackin, Owen and Collier. Bull. Mar. Sci. Gulf Caribb., 1, 72-87.

Mackin, J.G., H.M. Owen and A. Collier. 1950. Preliminary note on the occurrence of a new protistan parasite, *Dermocystidium marinum* n.sp. in *Crassostrea virginica* (Gmelin). Science, 111, 328-329.

Perkins, F.O. 1985. Range and host extensions for the molluscan bivalve pathogens, *Perkinsus* spp. VII Intl. Congr. Protozoology, p. 81 (Abstract).

Perkins, F.O. 1987. Protistan parasites of commercially significant marine bivalve molluscs - life cycles, ultrastructure, and phylogeny. Aquaculture, 67, 240-243.

Perkins, F.O. 1989. Structure of protistan parasites found in bivalve molluscs. In: "Disease Processes in Marine Bivalve Molluscs" (W. Fisher, ed.). American Fisheries Society, Washington, D.C., special publication, in press.

Perkins, F. O. and R.W. Menzel. 1966. Morphological and cultural studies of a motile stage in the life cycle of *Dermocystidium marinum*. Proc. Natl. Shellfisher. Assoc., 56, 23-30.

Ray, S.M. 1954. Biological studies of *Dermocystidium marinum*, a fungus parasite of oysters. Rice Inst. Pamph., No. 41, Special Issue.

Ray, S.M. 1966. A review of the culture method for detecting *Dermocystidium marinum* with suggested modifications and precautions. Proc. Natl. Shellfisher. Assoc., 54, 55-69.

Shepherd, S.A. 1973. Studies on southern Australian abalone (genus *Haliotis*) I. Ecology of five sympatric species. Austral. J. Mar. Freshwater Res., 24, 217-257.

Shepherd, S.A. 1985. No evidence of contamination of abalone stocks in western St Vincent Gulf. Safish, 9(6), 5.

Villalba, A. and J.I. Navas. 1988. Occurrence of *Minchinia tapetis* and a *Perkinsus*-like parasite in cultured clams, *Ruditapes decussatus* and *R. philippinarum*, from south Atlantic coast of Spain, preliminary results. 3rd Intl. Colloq. Pathol. Mar. Aquacult., pp. 57-58 (Abstract).

White, M.E., E.N. Powell, S.M. Ray and E.A. Wilson. 1987. Host to host transmission of *Perkinsus marinus* in oyster (*Crassostrea virginica*) populations by the ectoparasitic snail *Boonea impressa* (Pyramidellidae). J. Shellfish Res., 6, 1-5.

PRELIMINARY REPORT ON THE DISEASES AND PARASITES OF JUVENILE WALLEYE POLLOCK, *THERAGRA CHALCOGRAMMA*, FROM THE GULF OF ALASKA

J. Frank Morado and Albert K. Sparks

National Marine Fisheries Service, Northwest & Alaska Fisheries Center,
Division of Resource Assessment & Conservation Engineering,
7600 Sand Point Way N.E., Seattle, Washington 98115, USA

ABSTRACT: In 1986, a baseline survey was begun to investigate the possible role of parasites and diseases as regulators of wild, commercially important fish populations. Juvenile walleye pollock, *Theragra chalcogramma*, were chosen because much is known about their growth rate, life history, and distribution. A total of 142 fish were collected in 1986, and 726 in 1987.

A microsporidian, *Pleistophora* sp., that infects body muscle may have the greatest impact on the viability of infected juveniles and could be the cause of significant mortalities. Highest prevalence (25%) occurred in the 50 mm size group and then declined. A myxozoan, *Sphaerospora* sp., was observed in the kidney tubules of juvenile pollock. Prevalences of this parasite increased with increasing host size after being absent or nearly absent in the 40 mm to 60 mm size groups. Highest prevalence (35%) was attained in the 100 mm size group. Numerous metazoans were observed in juvenile pollock and included several species of cestodes, digenetic trematodes, and nematodes. The probable importance of these parasites as mortality factors is discussed.

INTRODUCTION

Historical landing records and abundance projections of exploited fish indicate that large population fluctuations occur, the cause of which are frequently unknown. Therefore, substantial efforts are being exerted to understand more fully the life history and biology of commercially valuable species, and the physical and chemical aspects of their environment. Because of such efforts, disease as a possible regulator of fish population fluctuations is only recently receiving wide attention. For example, a symposium was recently held solely to address the role of disease on fish and shellfish populations (Trans. 15th North American Wildlife Natural Resources Conference, 1985). In addition, Anderson and Gordon (1982), May (1983), and

Munro et al. (1983) have presented excellent accounts on the role of parasites or infectious diseases as regulators of animal populations.

We are in the third year of a broad based study in which the diseases of several juvenile fish species in the Bering Sea and Gulf of Alaska and are being described and enumerated. The walleye pollock, *Theragra chalcogramma*, population that spawns in late March in Shelikof Strait, Gulf of Alaska, is the target of a large domestic fishery and is one species under investigation. The possible impact of disease on this population is one part of a comprehensive cooperative study designed to obtain a better understanding of the factors that influence its abundance. We have focused on juvenile walleye pollock because it is in the early life history stages of fish that mortalities resulting from disease can have their greatest impact (Munro et al., 1983). This report presents the results of the first 2 years of our study.

METHODS

Live juvenile walleye pollock were collected by mid-water trawl from seven stations in the Gulf of Alaska in 1986 and from 33 stations in 1987 (Fig. 1). Only those stations where more than 20 live fish could be collected for our study were sampled. Each fish was killed by insertion of a sharp object (probe) into the brain, followed by an incision on each side of the body cavity before placing the specimen in Dietrich's or Bouin's fixative. Incisions were also made in the intestines, cranium, and body muscle of fish greater than 90 mm to enhance fixation. The specimens were then transported to the laboratory.

At the laboratory, individual fish were measured (standard length) and cut into five cross sections. The four cuts were made through the head through the posterior region of the eye, the pectoral fin area, mid-body, and at the anus. This method allowed us to routinely examine all major organs and tissues. The cross sections were

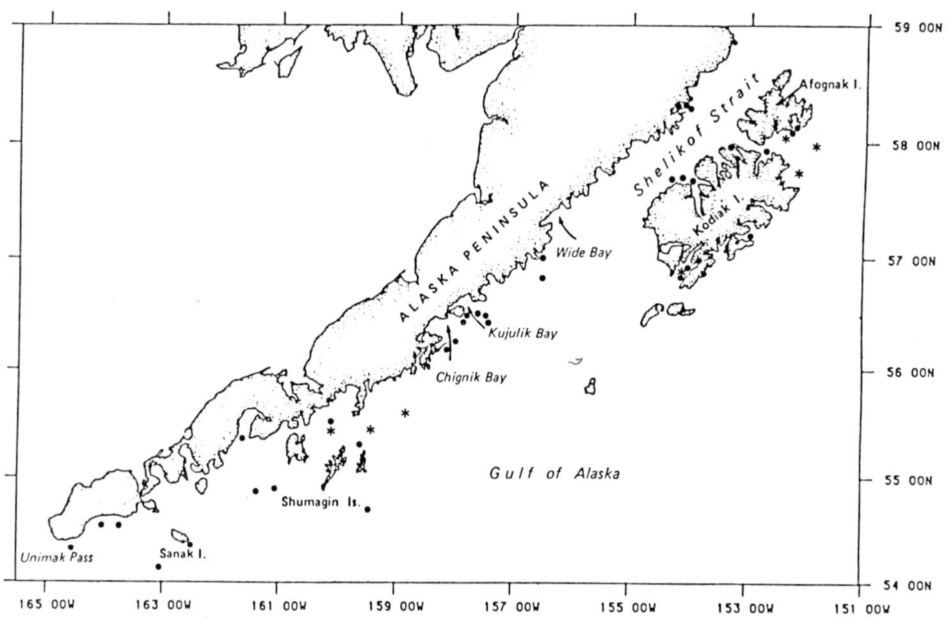

Fig. 1. The location of 1986 (*) and 1987 (●) mid-water trawls during which samples were taken for disease studies.

then processed via standard paraffin histological technique, sectioned at 4-6 μm and stained with hematoxylin and eosin. Special stains (PAS, Geimsa's, Mallory's triple stain) were employed as needed.

RESULTS

A total of 790 specimens were readable (142 in 1986; 648 in 1987). Each year's collections were grouped into seven, 10 mm size groups ranging from 40 mm to 110 mm. The 1986 collections were exploratory, resulting in relatively small sample sizes for each size group, particularly the 40 mm and 110 mm size groups. Most of the 1987 size categories were well represented (N > 30) except for the 40 mm and 110 mm size groups. We will, therefore, limit our comments to the 50 mm to 100 mm size groups.

Microsporidia, myxozoa, digenetic trematodes, cestodes, and nematodes comprised the major parasite groups observed in the organs and tissues of juvenile walleye pollock. The prevalence trends for 1986 and 1987 are presented in Table 1 and Figures 2-4. Although multiple infections (different parasites present in a single specimen) were common, the data are being reported as if lone infections occurred. The figures reveal a general increase for all parasite prevalences with increase in host size, except for microsporidian infections.

Digenetic trematodes were the most prevalent parasites (Table 1) and were typically found in the lumen of the digestive tract. They were occasionally found in the lumen of the pyloric caeca, and rarely in the gall bladder. The only tissue change evident was compression of epithelial cells that were grasped by the suckers. Necrosis was only rarely observed. Only two trematode metacercariae were observed in skeletal muscle resulting in disruption of normal, skeletal muscle architecture. No host response was elicited to either life history stage.

Figure 2a indicates that juvenile walleye pollock become infected by digenetic trematodes early in their development (26% at 50 mm in 1987) and that prevalences increase slightly over time. The highest prevalence (45%) was observed in the 100 mm size group in 1986.

At least two species of digenetic trematodes, *Prodocotyle sinusacca* and *Lecithaster gibbosus*, were identified in juvenile walleye pollock. Most of the trematodes possessed eggs, indicating that they were sexually mature. Body burdens (determined from tissue sections) ranged from 1-9 parasites per fish, with heaviest burdens occurring in larger juveniles.

Nematodes were the second most prevalent parasites (Table 1) and were typically found free in the lumen of the digestive tract and pyloric caeca or encysted in the mesenteries of several organs (liver, gall bladder, pancreas). They were occasionally found in the muscle tissue of ingested crustacean prey items, the liver parenchyma, or free in the coelom. Disruption of normal tissue architecture and parasite encapsulation were typically evident wherever nematodes were embedded.

A general increase in nematode prevalence was observed with increase in size (Fig. 2b). The highest infection rate (38%) was observed in the 80 mm size group in 1986.

A myxozoan of the genus *Sphaerospora* was discovered in the proximal kidney tubules of juvenile walleye pollock (Fig. 6). No infections were diagnosed in the 40 mm and 50 mm size groups but shortly after, prevalences increased rapidly with increased size of the host (Fig. 3a) attaining a maximum of 35% in the 100 mm size group in 1986. Stages of the myxozoan were typically observed free in the lumen of the proximal tubule and only occasionally in the cuboidal epithelium. Distal tubule

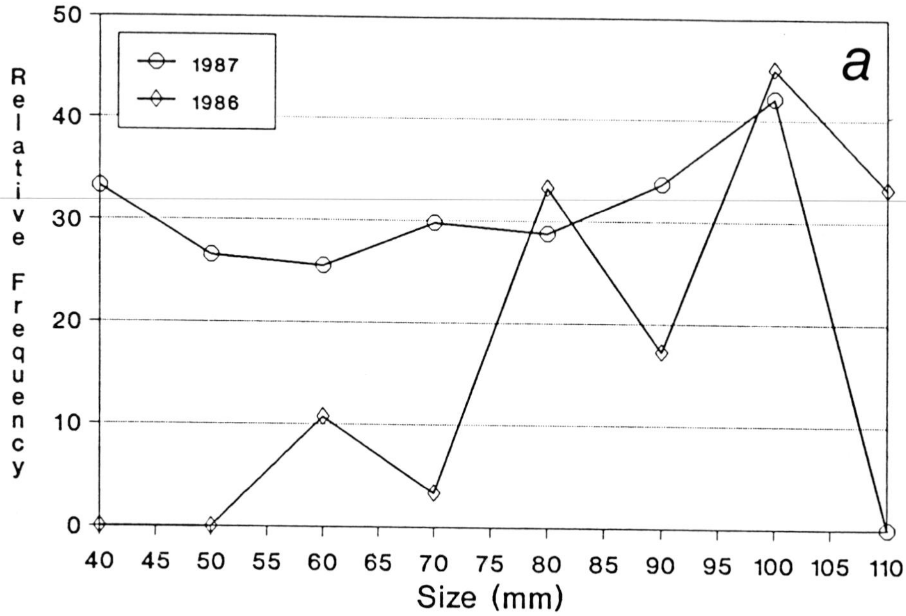

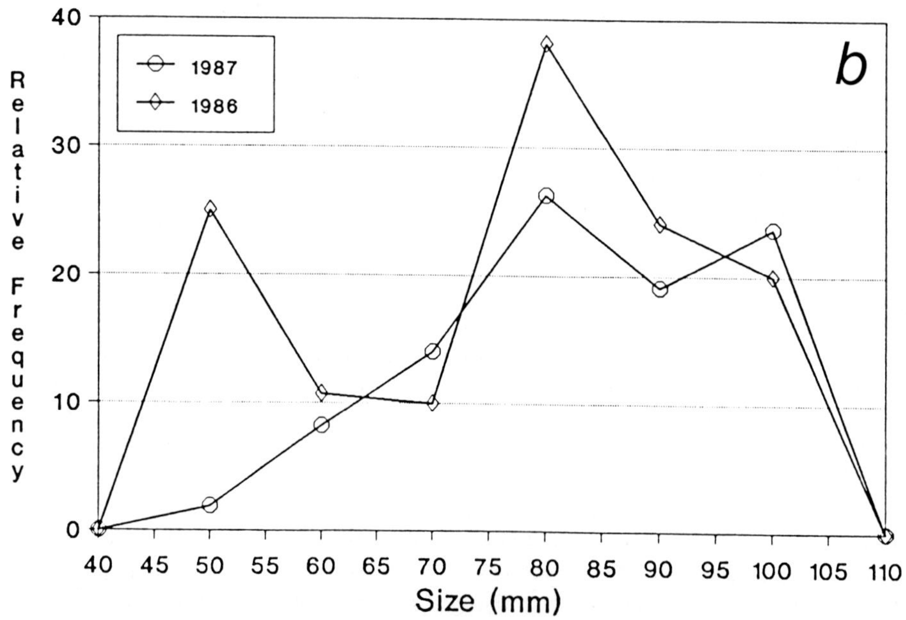

Fig. 2. 1986 and 1987 prevalences of digenetic trematodes (2a) and nematodes (2b) in juvenile walleye pollock.

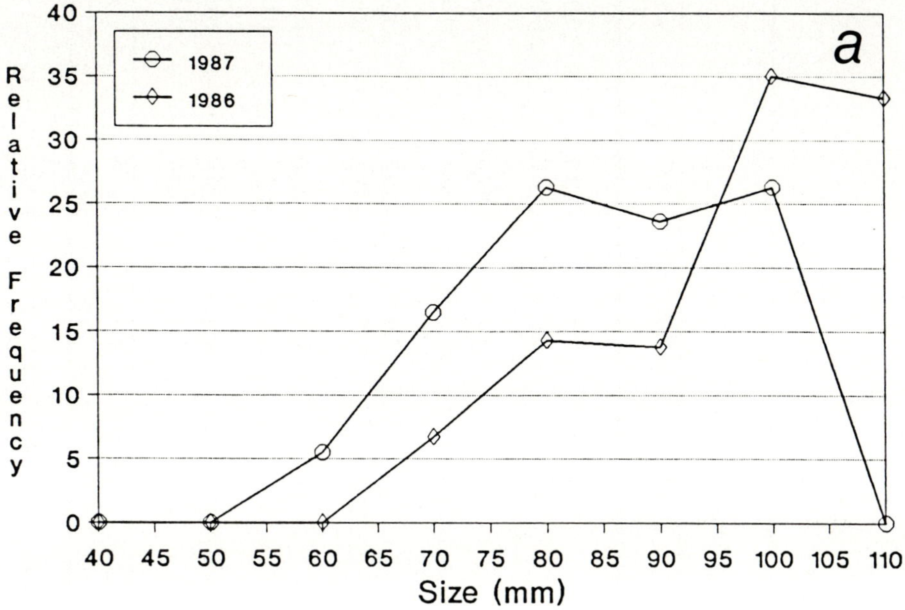

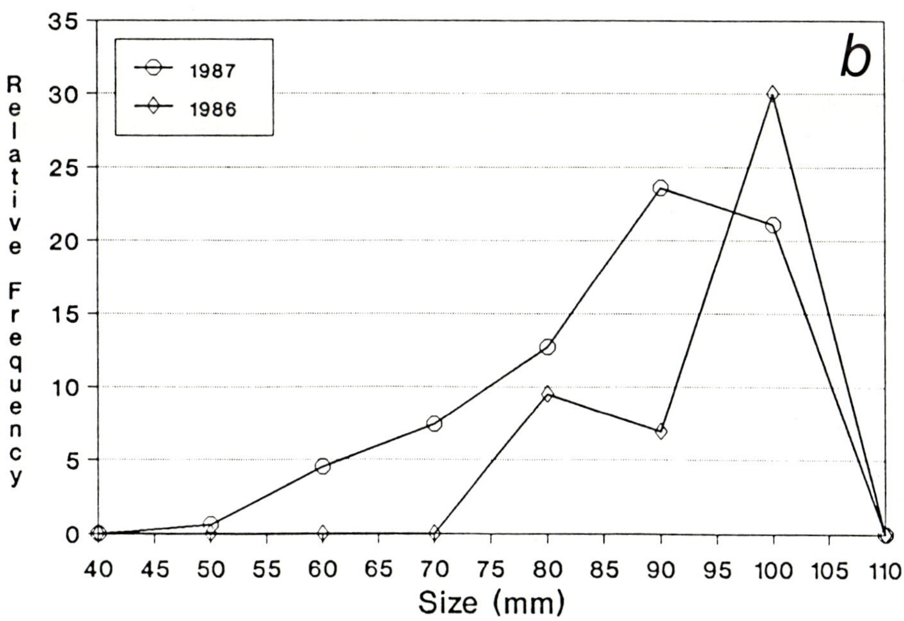

Fig. 3. 1986 and 1987 prevalences of a myxozoan, *Sphaerospora* sp., (3a) and cestodes (3b) in juvenile walleye pollock.

Table 1. Parasite prevalences in 1986, 1987 juvenile walleye pollock, *Theragra chalcogramma*.

	SIZE (mm)							
PARASITE	40	50	60	70	80	90	100	110
Microsporida 86	0.0	25.0	3.6	13.3	4.8	6.9	10.0	0.0
Microsporida 87	0.0	8.6	13.6	3.3	6.8	0.0	0.0	0.0
Myxozoa 86	0.0	0.0	0.0	6.7	14.3	13.8	35.0	33.3
Myxozoa 87	0.0	0.0	5.5	16.5	26.3	23.6	26.3	0.0
Trematode 86	0.0	0.0	10.7	3.3	33.3	17.2	45.0	33.3
Trematode 87	33.3	26.5	25.5	29.8	28.8	33.7	42.1	0.0
Cestode 86	0.0	0.0	0.0	0.0	9.5	6.9	30.0	0.0
Cestode 87	0.0	0.6	4.5	7.4	12.7	23.6	21.1	0.0
Nematode 86	0.0	25.0	10.7	10.0	38.1	24.1	20.0	0.0
Nematode 87	0.0	1.9	8.2	14.0	26.3	19.1	23.7	0.0
SAMPLE SIZE	9	170	138	151	139	118	58	7

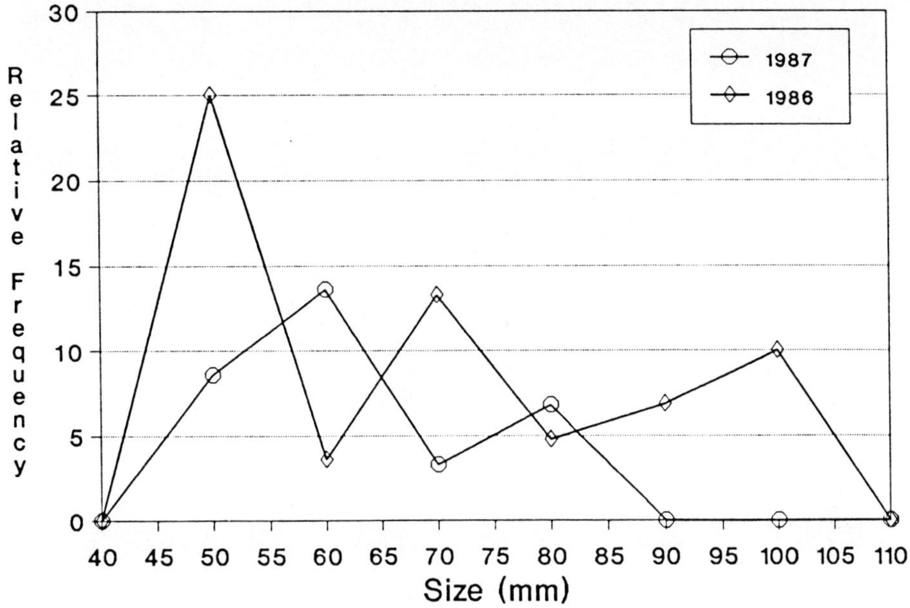

Fig. 4. 1986 and 1987 prevalences of a microsporidian, *Pleistophora* sp., in juvenile walleye pollock.

epithelium was only rarely infected and spores were not apparent. Epithelial necrosis and exfoliation were evident in heavy infections but a host response was not elicited.

Larval cestodes were frequently encountered in the intestine and pyloric ceaca of juvenile walleye pollock with prevalences increasing rapidly with increasing size of the host and eventually exceeded 20% in juveniles greater than 90mm (Fig. 3b). The highest prevalence (30%) was observed in the 100 mm size group in 1986. One apical and four lateral suckers were observed in some cestodes indicating that at least one was a member of the order Proteocephalidea, probably *Bothriocephalus scorpei*.

A microsporidian of the genus *Pleistophora* was observed in the skeletal muscle of juvenile pollock although heart muscle was infected on two occasions. None of the infections were grossly visible. All infections were multi-focal and accompanied by a host response that consisted of histiocyte infiltration and encapsulation when the sarcolemma was breached (Fig. 5). A pansporoblast and numerous spores were frequently visible within infected muscle. Infections were typically light to moderate; however, several cases were severe in which an entire body side was infected. The highest prevalence occurred in the 70 mm size group in 1986 (Fig. 4). Unlike previous parasitic infections, *Pleistophora* infections were more prevalent in smaller sized fish and decreased with increasing host size.

An unidentified protozoan was discovered at low prevalences in the kidney of juvenile walleye pollock. The parasite infected the cuboidal epithelium of the proximal tubule and caused severe hypertrophy (Fig. 7). Greatly enlarged host cells

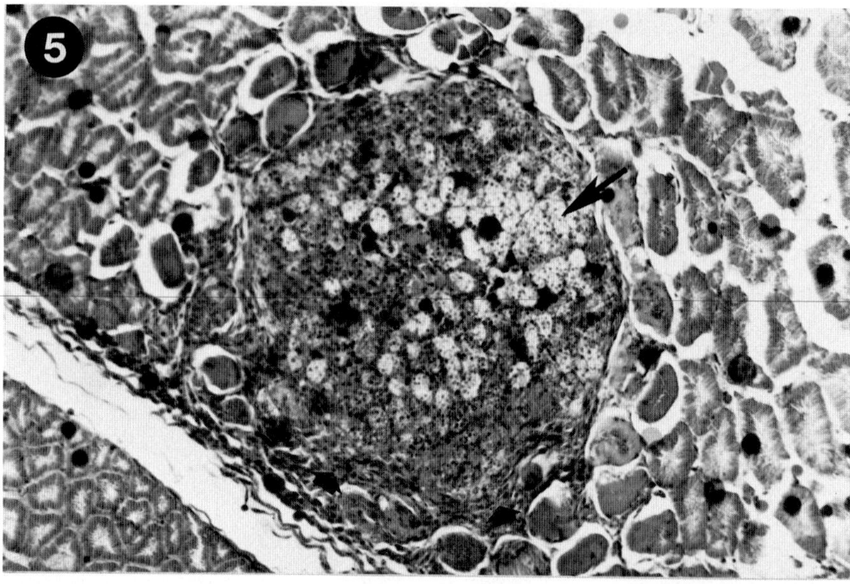

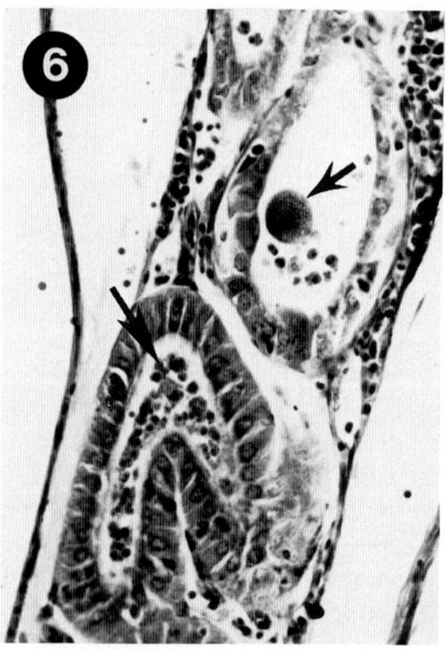

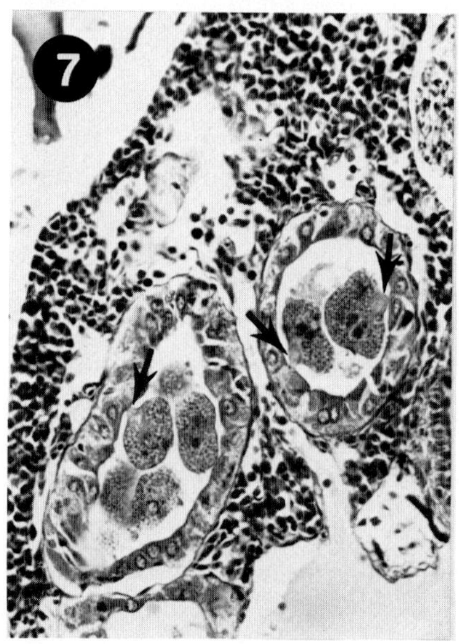

Fig. 5. Heavily infected skeletal muscle of juvenile walleye pollock by a microsporidian, *Pleistophora* sp. Note the spores of the microsporidian (arrow) and the histiocytic infiltrate and encapsulation attempt by the host to wall-off the parasite (arrow heads).

Fig. 6. Heavily infected proximal tubule of juvenile walleye pollock kidney by a myxozoan, *Sphaerospora* sp. Note the presence of the myxozoan in the lumen (long arrow) and the lack of host response. Also note the putative microsporidian in an adjacent proximal tubule (short arrow).

Fig. 7. Unidentified protozoan infecting the proximal tubule epithelium of juvenile walleye pollock. Note the apparent stalk (arrows) and size of the infected host cell.

extended into the lumen of the proximal tubule at which time a stem was observed leading from the cell to the basement membrane. Host cell nuclei were occasionally seen. Infections ranged from light to moderate in intensity but a host response was not elicited. No spores were observed although putative pansporoblastic stages were evident, suggesting a small *Pleistophora* sp. Multiple infections of the putative microsporidian and *Sphaerospora* sp. were common within an infected kidney and stages of both were frequently observed within the same infected tubule.*

DISCUSSION

Sufficient evidence exists demonstrating that mortalities resulting from disease occur in fish and other organisms; however, the significance of disease related mortalities as regulators of animal populations is not always evident, especially for wild marine fish. The effects of mortalities on fish populations are dependent upon the mortality rate, the number of year classes affected, and the species involved. All observed mortalities stimulate concern, but mortality rates need to be determined because greater mortality rates produce more significant reductions and probably more long lasting effects. However, small increases in mortality rates of larvae and juveniles can have potentially devastating effects because of their sheer numbers and vulnerability (Munro et al., 1983). Significant mortalities in several year classes have the effect of depressing pre-recruit and spawning numbers, but it is conceivable that species with short life histories may recover more quickly from disease and other induced mortalities. The difficulty arises in attempting to determine what constitutes a significant mortality rate and its subsequent effects on an affected population. As an example, Hargis (1985) and Munro et al. (1983) present slightly different views on the ensuing effects of *Ichthyophonus*-induced mortalities in herring, *Clupea harengus*.

Providing definitive proof that disease is the cause of mortalities in highly mobile populations is arduous because accurate and precise population enumerations and epizootiological studies are difficult (Hargis, 1985). These difficulties might be circumvented with long term investigations to identify diseases, their causes, predisposing factors, and long term trends. However, van Banning (1987), after completing the first half of a 10-year study, noted that disease prevalences in wild marine fish stocks can vary tremendously within each species, area, and season.

Many reports describe diseases of wild adult fish and their probable causative agents, most are environmental or pollution oriented while fewer are non-pollution based. Of these, the more pertinent include the studies by van Banning (1987), Cali et al. (1986), Takvorian and Cali (1981, 1984), and Laudan et al. (1986a,b, 1987). The works by Takvorian, Laudan, and associates are particularly interesting because theirs represent a multi-disciplinary approach (immunological and surveillance) to the study of the effects of a microsporidian, *Glugea stephani*, in the winter flounder, *Pseudopleuronectes americanus*. In the Bering Sea, several reports describe chronic disorders and report their distribution and prevalences from large but time-limited

---

*Dykova and Lom (1988, Europ. J. Protistol., 23, 258-261) recently reported a myxosporean, *Chloromyxum reticulatum*, from the liver of burbot, *Lota lota*, that possesses a life history stage (plasmodium) that is remarkably similar to the plasmodium of our unidentified protozoan. It therefore appears that the present unidentified protozoan is a myxosporean.

field survey collections (McCain et al., 1978, 1979; McCain and Gronlund, 1981; Aplers et al., 1977a,b; Wellings et al., 1977). To our knowledge, there are no reports on the diseases of wild larval and juvenile fish.

The walleye pollock population of Shelikof Strait is well suited for disease studies and appears to have a density-dependent relationship between adult population size and the number of young recruited to the population. This population spawns at a restricted site within Shelikof Strait in late winter to produce a dense patch of pelagic eggs. Subsequently, eggs, larvae, and juveniles are carried in a general southwest direction by ocean currents (Kendall et al., 1987). Although the distribution of adult Gulf of Alaska walleye pollock is unknown, data are available on the age, growth, feeding, and distribution of larvae (Kendall et al., 1987). These data are important in attempting to determine the impact of disease on a population, especially early life history stages (zygotes, embryos, larvae, and juveniles) where large mortalities are known to occur (May, 1983).

Our data are preliminary because only 2 years of study is presented and it is also unlikely that all disease causing agents of juvenile walleye pollock have been identified. Viruses were not suspected and bacterial infections were not detected. Fungal infections (*Ichthyophonus* sp.) were identified but at very low frequencies. In adopting the sample size recommendations of Amend (1970), we were limiting out ability to detect juvenile walleye pollock diseases to prevalences of those diseases that attained or exceeded 10% (N=30). For the size groups in which we were able to collect more than 60 specimens, detection of disease prevalences from 2-5% was possible.

Although metazoan parasites were the most numerous, they are not the most important with respect to causing significant mortalities in fish. A general reduction in fitness of the host does occur in heavily parasitized fish but mortality threshold levels have not been determined. Their greatest impact is probably in reduced growth rates and fecundity of infested hosts.

Nematodes appear to be the most important of the metazoans and are found throughout the body of infected fish, but adults typically reside in the digestive tract of their final host. Juvenile nematodes are generally histozoic and are typically found encapsulated in body muscle, in various internal organs or on mesenteries (Margolis, 1970). The condition of the host or infected organs is directly proportional to the number of nematodes present in the tissues and nematode induced mortalities in fish have been reported (Margolis, 1970). Immature nematodes found free in the stomach and intestine of juvenile pollock had, almost surely, just left the crustacean intermediate host and were beginning their journey toward their preferred organ for encystment. In their migration, the nematodes have to breach the gut wall, thus causing some tissue damage. The extent of trauma would be directly proportional to the number and size of active nematodes. Perforation through the intestinal mucosa may also provide a portal of entry for bacteria into the coelom. None of the nematodes in juvenile walleye pollock were identified but Margolis and Arthur (1979) have reported several species of nematodes in adult pollock.

Cestodes and digenetic trematodes are also typical parasites of fish. Margolis and Arthur (1979) reported four species of digenetic trematodes and three species of cestodes, one of them being *B. scorpei*. As previously mentioned, it is not believed that cestodes and trematodes are important because adults typically reside in the intestine.

Protozoans are more likely to cause significant mortalities because they typically invade vital organs, have a short developmental period, and spread quickly through infected individuals and an aggregating population such as Shelikof Strait walleye pollock. They are the first to infect larvae and juveniles unless they live in schools separate from adults (Lom, 1970). Microsporidia are capable of producing rapid and massive mortalities in captive and wild fish populations (Canning and Lom, 1986) and are believed to cause significant mortalities in wild populations of the American winter flounder, *Pseudopleuronectes americanus* (Takvorian and Cali, 1981).

The microsporidian in skeletal and cardiac muscle possessed numerous spores within a pansporoblast and was identified as *Pleistophora* sp. It is probably the same species reported by Margolis and Arthur (1979) in adult walleye pollock. Members of the genus are common skeletal muscle parasites of fish and have been reported in numerous marine fish (Canning and Lom, 1986). Pathogenesis appears similar for all *Pleistophora* sp.

The importance of this parasite as a probable and significant regulator of juvenile walleye pollock populations cannot be ignored, particularly with the prevalence levels observed and the host stage infected. To our knowledge, infected fish and other organisms do not recover from microsporidian infections. Our data indicate that juveniles are infected early and because prevalences drop as juveniles increase in size, it is believed that this reduction indicates death of the host. However, further studies are needed to confirm this hypothesis.

The putative microsporidian in kidney proximal tubule epithelium is enigmatic. Sufficient diagnostic features were not present to identify this protozoan with certainty. We are also not aware of any reports on microsporidian infections in fish kidney.

The myxozoan observed in walleye pollock kidney is most likely a member of the genus *Sphaerospora* reported from the kidney of numerous fish species, all apparently possessing similar pathogenesis. *Sphaerospora* infections in adult and, particularly, juvenile fish are a health concern, especially in aquaculture (Bennett and Wolke, 1986; Dykova and Lom, 1982; Grupcheva et al., 1985; Hermanns and Korting, 1985; Wierzbicka, 1986) often causing significant mortalities. Similar effects can be anticipated in wild populations, especially if the host is further stressed by the presence of another parasite, disease, or adverse environmental conditions.

Larval and juvenile walleye pollock grow rapidly, averaging 0.22 mm per day (Kendall et al., 1987). Prolonged reduction of this growth rate resulting from disease or other causes likely reduces the organisms ability to survive. Studies have shown that if mortalities are not produced outright, adverse stress can affect the ability of an affected organism to survive, grow, and eventually, reproduce. Thus, while the mortalities attract the greatest attention, secondary effects and residual mortalities further reduce viability or host numbers. At the present time, we are only attempting to identify and quantify the causes of mortalities. The secondary effects of disease will be addressed as our database expands.

ACKNOWLEDGEMENTS

We would like to thank Mr. Jim Long for the 1986 collections and Ms. Therese Armetta and Ms. Linda Cherepow for assisting in the collections and for technical assistance. We would also like to thank Mr. Mark Myers, Mr. Herb Shippen, and Dr. Art Kendall for their critical reviews of the manuscript.

LITERATURE CITED

Alpers, C.E., B.B. McCain, M.S. Myers and S.R. Wellings. 1977a. Lymphocystis disease in yellowfin sole (*Limanda aspera*) in the Bering Sea. J. Fish. Res. Bd. Can., 34, 611-616.

Alpers, C.E., B.B. McCain, M.S. Myers, S.R. Wellings, M. Poore, J. Bagshaw and C.J. Dawe. 1977b. Pathological anatomy of pseudobranch tumors in Pacific cod, *Gadus macrocephalus*. J. Natl. Canc. Inst., 54, 377-398.

Amend, D.F. 1970. Approved procedure for determining absence of infectious hematopoietic necrosis (IHN) in salmonid fish. U.S. Dept. of Interior, Fish Disease Bull., 31, 1-4.

Anderson, R.M. and D.M. Gordon. 1982. Processes influencing the distribution of parasitic numbers within host populations with special emphasis on parasite-induced mortalities. Parasitology, 85, 373-398.

Bennett, R.O. and R.E. Wolke. 1986. A *Sphaerospora*-like myxosporidean from the kidneys of Atlantic salmon (*Salmo solar*). J. Wildl. Dis., 22, 268-271.

Cali, A., P.M. Takvorian, J.J. Ziskowski and T.K. Sawyer. 1986. Experimental infection of American winter flounder (*Pseudopleuronectes americanus*) with *Glugea stephani* (Microsporida). J. Fish Biol., 28, 199- 206.

Canning, E.U. and J. Lom. 1986. "The Microsporidia of Vertebrates." Academic Press, New York.

Dykova, I. and J. Lom. 1982. *Sphaerospora renicola* n.sp., a myxosporean from carp kidney, and its pathogenicity. Z. Parasitenkd., 68, 259-268.

Grupcheva, G., I. Dykova and J. Lom. 1985. Seasonal fluctuations in the prevalence of *Sphaerospora renicola* and myxosporean bloodstream stages in carp fingerlings in Bulgaria. Fol. Parasitol., 32, 193-203.

Hargis, W.J. Jr. 1985. Quantitative effects of marine diseases on fish and shellfish populations. Trans. N. Am. Wildl. Nat. Res. Conf., 50, 608-640.

Hermanns, W. and W. Korting. 1985. *Sphaerospora tunicae* Plehn, 1925 in tench, *Tinca tinca* L., fry. J. Fish Dis., 8, 281-288.

Kendall, A.W. Jr., M.E. Clarke, M.M. Yoklavich and G.W. Boehlert. 1987. Distribution, feeding, and growth of larval walleye pollock, *Theragra chalcogramma*, from Shelikof Strait, Gulf of Alaska. Fish. Bull., 85, 499- 521.

Laudan, R., J.S. Stolen and A. Cali. 1986a. The immune response of a marine teleost, *Pseudopleuronectes americanus*, (winter flounder) to the protozoan parasite *Glugea stephani*. Vet. Immunol. Immunopathol., 12, 403- 412.

Laudan, R., J.S. Stolen and A. Cali. 1986b. Immunoglobulin levels of the winter flounder (*Pseudopleuronectes americanus*) and the summer flounder (*Paralichthys dentatus*) injected with the microsporidan parasite *Glugea stephani*. Dev. Comp. Immunol., 10, 331-340.

Laudan, R., J.S. Stolen and A. Cali. 1987. The immunomodulating effect of the microsporidan *Glugea stephani* on humoral response and immunoglobulin levels in winter flounder, *Pseudopleuronectes americanus*. J. Fish Biol., 31 (Supl. A), 155-160.

Lom, J. 1970. Protozoa causing diseases in marine fishes. In: "A Symposium on Diseases of Fishes and Shellfishes" (S.F. Snieszko, ed.), pp. 101-123. Am. Fish. Soc. Spec. Publ. 5, American Fisheries Society, Washington, D.C.

Margolis, L. 1970. Nematode disease of marine fishes. In: "A Symposium on Diseases of Fishes and Shellfishes" (S.F. Snieszko, ed.), pp.190-208. Am. Fish. Soc. Spec. Publ. 5, American Fisheries Society, Washington,D.C.

Margolis, L. and J.R. Arthur. 1979. Synopsis of the parasites of fishes of Canada. Fish. Res. Bd. Canada Publ. 199, 269 pp.

May, R.M. 1983. Parasitic infections as regulators of animal populations. Am. Scient., 71, 36-45.

McCain, B.B. and W.D. Gronlund. 1981. Fish diseases in the Bering Sea. In: "The Eastern Bering Sea Shelf: Oceanography and Resources." Office of Marine Pollution Assessment. NOAA. University of Washington Press, Seattle.

McCain, B.B., M.S. Myers, W.D. Gronlund, S.R. Wellings and C.E. Alpers. 1978. The frequency, distribution, and pathology of three diseases of demersal fishes in the Bering Sea. J. Fish Biol., 12, 267-276.

McCain, B.B., W.D. Gronlund, M.S. Myers and S.R. Wellings. 1979. Tumours and microbial diseases of marine fishes in Alaskan waters. J. Fish Dis., 2, 111-130.

Munro, A.L.S., A.H. McVivar and R. Jones. 1983. The epidemiology of infectious disease in commercially important wild fish. Rapp. P.-v. Reun. Cons. Int. Explor. Mer., 182, 21-32.

Takvorian, P.M. and A. Cali. 1981. The occurrence of *Glugea stephani* (Hagenmuller, 1899) in American winter flounder, *Pseudopleuronectes americanus* (Walbaum), from the New York-New Jersey lower bay complex. J. Fish Biol., 18, 491-501.

Takvorian, P.M. and A. Cali. 1984. Seasonal prevalence of the microsporidan *Glugea stephani* (Hagenmuller) in winter flounder, *Pseudopleuronectes americanus* (Walbaum), from the New York-New Jersey lower bay complex. J. Fish Biol., 24, 655-663.

van Banning, P. 1987. Long term recording of some fish diseases using general fishery research surveys in the south-east part of the North Sea. Dis. Aquat. Organ., 3, 1-11.

Wellings, S.R., C.E. Alpers, B.B. McCain and M.S. Myers. 1977. Fish disease in the Bering Sea. Ann. N.Y. Acad. Sci., 298, 290-304.

Wierzbicka, J. 1986. *Sphaerospora anguillae* sp.n. (Myxospora, Bivalvulida), a parasite of eel, *Anguilla anguilla* (L.) Acta Protozool., 25, 119-122.

PARAMOEBIC GILL INFECTION AND ASSOCIATED PATHOLOGY OF ATLANTIC SALMON, *SALMO SALAR*, AND RAINBOW TROUT, *SALMO GAIRDNERI*, IN TASMANIA

B.L. Munday[1], C.K. Foster[2], F.R. Roubal[3], and R.J.G. Lester[3]

[1]Tasmanian State Institute of Technology, Newnham, Tasmania 7250, Australia
[2]Salmon Enterprises of Tasmania Pty. Ltd., Dover, Tasmania 7117, Australia
[3]Department of Parasitology, University of Queensland, Brisbane, Queensland 4067, Australia

ABSTRACT: Severe branchitis associated with *Paramoeba* infection has emerged as the major infectious disease of sea-caged salmonids in Australia. Although amoebae can be found on the gills of fish when water temperatures are in the vicinity of 10°C (winter and spring), clinical disease is associated with water temperatures of 15 to 20°C (spring, summer, and autumn) and salinities of 35 ‰. Affected fish suffer from anorexia and respiratory distress. The histopathology of the disease as well as the morphology of the causative agent are being reported. Also included are notes on the epidemiology, clinical manifestations, treatment, and control of this serious disease.

INTRODUCTION

Extensive salmonid mariculture has been established in Tasmanian waters only since 1984. Prior to this, exploratory, small-scale populations were raised at a seawater site in the southeast of the state (Hortle, 1981) and in the estuary of the Tamar River in the north. It was only with the establishment of salmonid culture in full-strength sea water, and maintenance of the fish during the warmer months of the year, that paramoebic gill disease emerged as a major problem (Munday, 1985).

The only other reported occurrence of this disease is in Coho salmon (*Oncorhynchus kisutch*) in Washington State and California, U.S.A. (Kent et al., 1989). Interestingly, the condition has not been reported in New Zealand where salmonids are kept under very similar environmental conditions to those experienced in Tasmania (Searle, pers. comm.).

Since early 1985, the epidemiology of this disease has gradually been established and related to observed characteristics of the parasitic *Paramoeba* sp. and the response of salmonids to infection of their gills by this organism.

In order to more fully document present knowledge on this subject, the information will be discussed under the headings of characteristics of the parasite, epidemiology, clinical findings, pathology, treatment, and control.

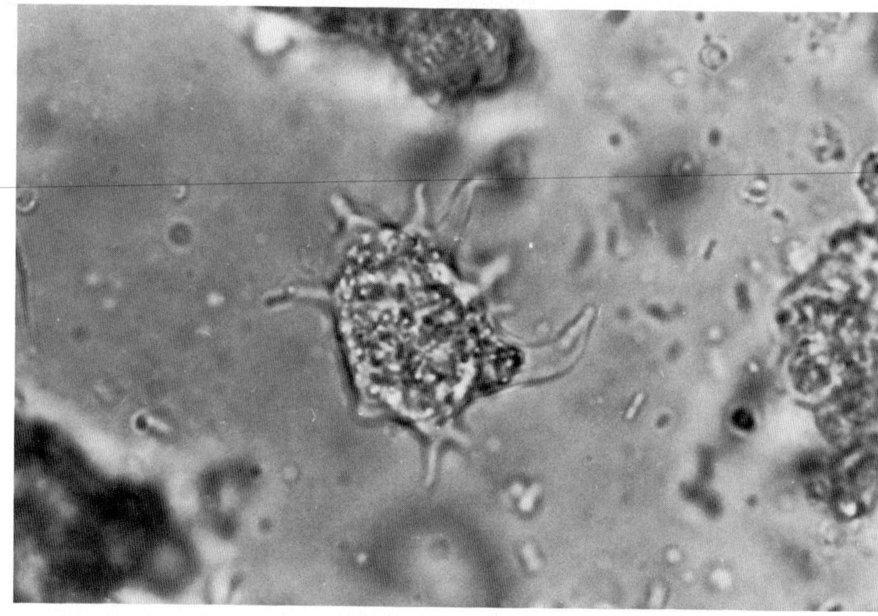

Fig. 1. Transitional form of *Paramoeba* sp. with digitate pseudopodia; 1,000X.

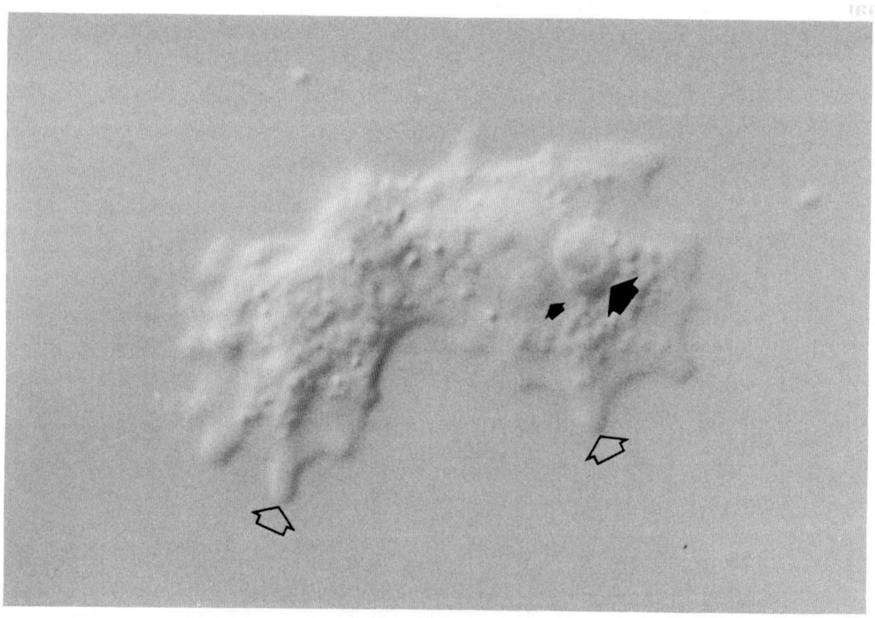

Fig. 2. Differential interference light micrograph of *Paramoeba* sp. showing parasome (large arrow) beside nucleus with central nucleolus (arrow). Pseudopodia (open arrows) free of granules evident in endoplasm; 3,500X.

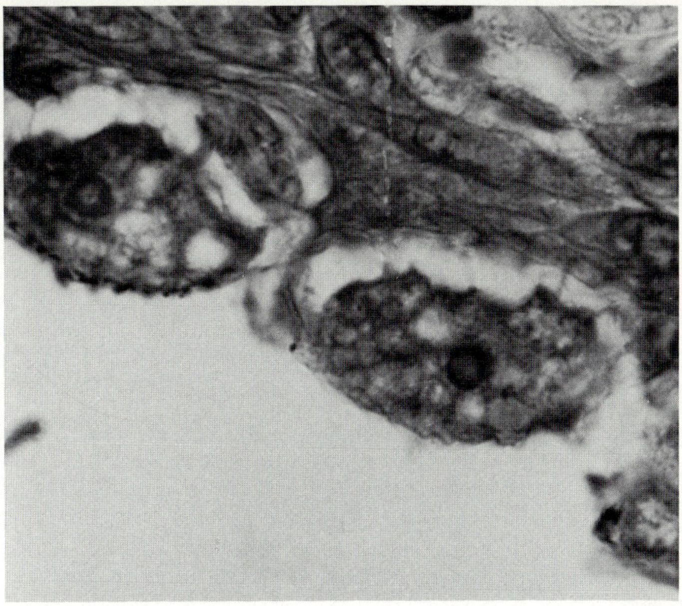

Fig. 3. *Paramoeba* sp. situated within eroded epithelium. Apparent gap between parasite and cell artefactual; H & E, 1,500X.

OBSERVATIONS AND COMMENTS

Characteristics of the parasite. Amoebae freshly removed from infected gills and suspended in sea water are present as subspherical, transitional forms with up to 50 digitate pseudopodia (Fig. 1). These forms are 20 to 30 μm in diameter.

Locomotive forms exhibit typical amoeboid motion, have blunt pseudopodia (Fig. 2), and measure approximately 20 x 15 μm. They contain a round nucleus (≈ 5 μm) with one or two satellite parasomes (≈ 4 μm).

Light microscopical examination of paramoebae on infected gills shows the organisms, which are frequently vacuolated, are closely applied to the gill epithelium (Figs. 3, 4). Depending upon the plane of section, a nucleus and one or two parasomes are present in each organism. Sometimes the parasome has condensations of chromatin at the poles (Fig. 4). The nuclear and parasomal features are more clearly defined by transmission electron micrography (Fig. 5).

Initially, *Paramoeba* sp. were found only on salmonids, but subsequently they were detected at an unpredictable frequency on a range of native marine fish.

Numbers of paramoebae in the water column are never great. Organisms are frequently found on clean nets, but not on nets that have been in the water for some time.

The *Paramoeba* sp. grows on malt-yeast-seawater agar seeded with *Pseudomonas, Klebsiella, Escherichia,* or *Vibrio* spp., but has not been adapted to axenic media.

Epidemiology. Although small numbers of amoebae can be found on the gills of salmonids during winter and spring (water temperatures about 10°C), clinical disease has been seen only when water temperatures are in the range of 12 to 20°C.

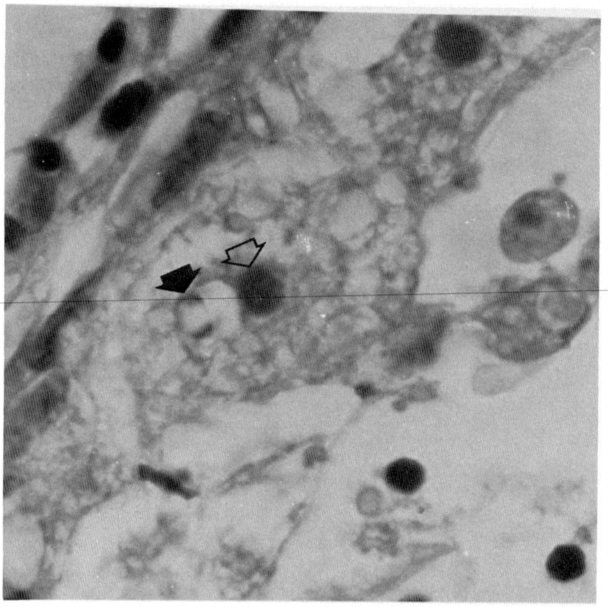

Fig. 4. *Paramoeba* sp. showing nucleus (open arrow) and parasome with polar condensations of chromatin (arrow); H & E, 1,000X.

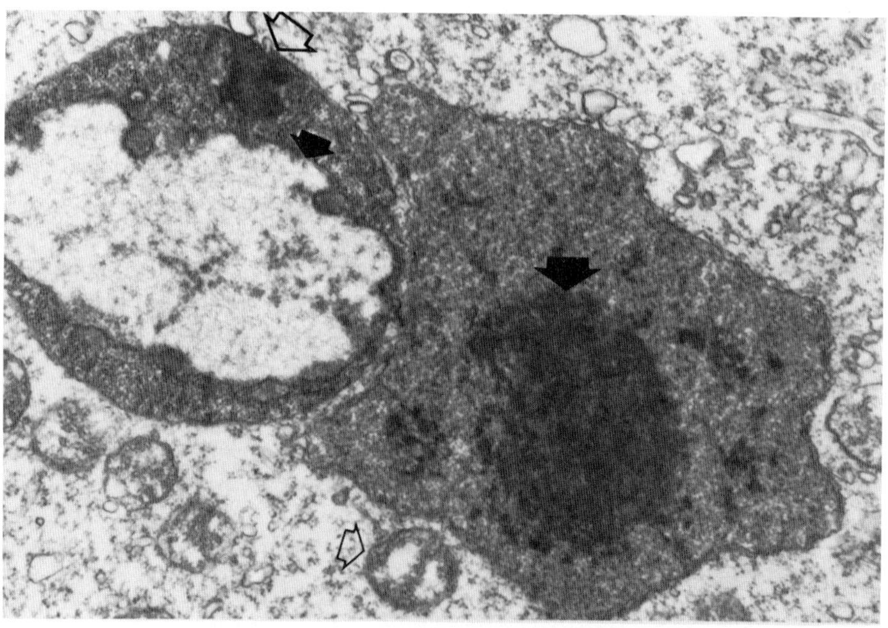

Fig. 5. Electron micrograph showing parasome and nucleus of *Paramoeba* sp. Nucleus, with central nucleolus (large arrow) closely apposed to the parasome containing fibrous central region (small arrow) and nucleus (large open arrow) at one pole. Several mitochondria (small open arrow) evident in cytoplasm of amoeba; 23,000X.

Even more marked is the effect of salinity. Disease is only seen when salinities approach 35 ‰. This was dramatically shown in 1987-88 (southern spring, summer, autumn) when fresh water flows failed at "brackish water" sites. Paramoebic gill disease was observed at these sites some 2 to 3 months after outbreaks occurred at full-seawater sites.

All investigations to date have indicated that under appropriate environmental conditions (water temperature ≥ 12°C, high salinity, poor water exchange) the *Paramoeba* sp. acts as a primary pathogen.

When infection occurs, all or most of the fish in a cage become infected over a period of several weeks. Fish that are subclinically infected or survive clinical disease recover, even though environmental conditions may remain suitable for the occurrence of paramoebic gill disease. Recurrence of disease is unusual, but has been reported in Atlantic salmon 2 to 3 months after an initial outbreak.

Clinical findings. Mortality is variable, reaching 2% per day and up to 50% total in untreated cages. Usually, young fish in their first season in the sea are affected. At this stage, the fish have not had any opportunity to develop immunity to the disease and their osmoregulatory capacity is less than that of larger fish.

The most outstanding signs are related to respiration. The fish are sluggish and swim with open opercula. While there is some depression of appetite, this is variable and dead fish are frequently in good condition with food in their stomachs.

Clinically affected fish have markedly elevated blood sodium levels and subclinically affected animals have lower, but still abnormal levels.

Pathology. Significant pathology is restricted to the gills. Macroscopically, the outstanding feature is the presence of excessive mucus on the gills. In rainbow trout, this mucoid branchitis is relatively diffuse, but with Atlantic salmon, the lesions tend to be patchy.

Histologically, the most prominent lesion is variable epithelial hyperplasia involving the secondary lamellae (Fig. 6). It is notable that *Paramoeba* sp. are only associated with areas of hyperplastic gill epithelium. This hyperplasia leads to clubbing and fusion of the secondary lamellae and even the formation of large vesicles or cysts (Fig. 7). Unlike nodular gill disease described by Daoust and Ferguson (1985), the lesions of paramoebic gill disease are found anywhere along the length of the primary lamellae. Actively secreting mucous cells are numerous in affected areas of the gills, both at the surface and deep within the hyperplastic epithelium. In contrast, chloride cells are reduced in number in the affected areas.

By light microscopy, it is possible to visualize eroded areas of epithelium where the organisms are attached (Fig. 3). Ultramicroscopically, this association is found to be even more intimate with pseudopodia penetrating cavities at the surface of the degenerating epithelial cells or even invading cell junctions (Roubal et al., 1989).

Early in the disease, neutrophils are found infiltrating the hyperplastic epithelium and the filamentar connective tissues. Later the infiltrating cells are mainly of a mononuclear character.

A feature of recovering cases is the presence of mononuclear (possibly lymphoid) nodules along the primary lamellae and especially in the basal, interlamellar tissues (Fig. 8). These reactive nodules are much more obvious in rainbow trout than in Atlantic salmon.

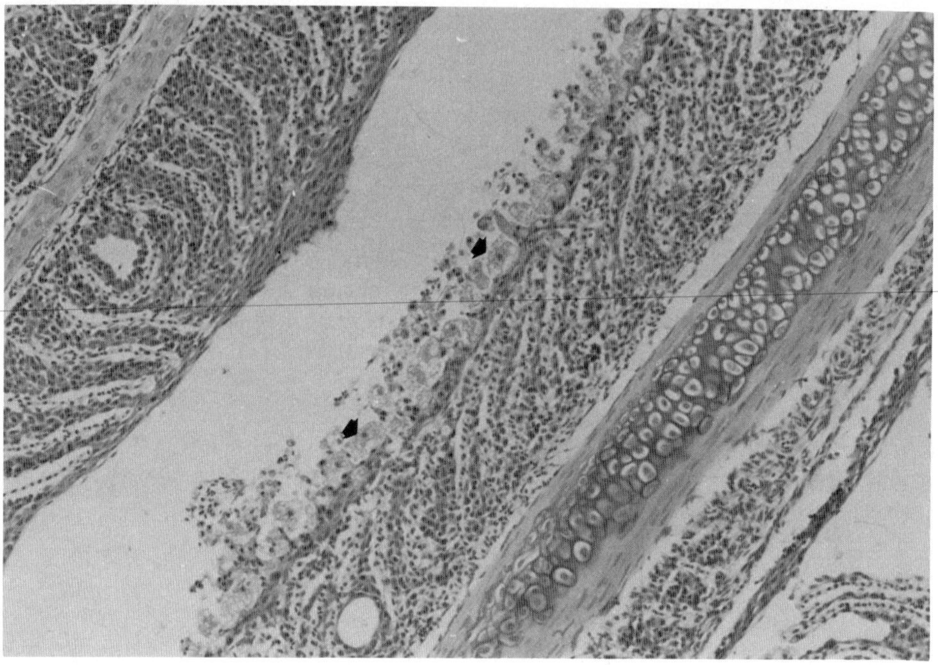

Fig. 6. Hyperplasia of gill epithelium and fusion of lamellae; note adherent amoebae (arrows); H & E, 150X.

Treatment. The only effective treatment to date has been the use of freshwater baths. In some instances cages have been towed into water of relatively low salinity. However, this is not often a practical option and the alternatives are to transfer the fish to land-based tanks or to floating "envelopes" filled with fresh water. It is essential to achieve a substantial reduction in salinity as rapidly as possible. The fish are held in the freshwater bath for 1 to 8 hr. If properly conducted, treatment can lead to a dramatic recovery.

It is believed that the treatment acts in three ways, namely: (1) the numbers of amoebae on the gills are greatly reduced and do not return to pre-treatment levels; (2) the mucus cover on the gills, which is stable in sea water, is removed by freshwater; and (3) the hypernatraemia is reduced, at least temporarily.

Control. Control at this stage is restricted to evasion of the parasite and controlled exposure.

Raising fish in water of low salinity should permit them to grow to harvest without experiencing amoebic gill disease. This appears to be the case in Macquarie Habour where there is usually a surface wedge of low salinity water in areas close to the mouths of rivers emptying into the harbour.

Some fish farmers intend to keep young fish at brackish water sites until they are large enough to withstand the deleterious osmoregulatory effects of the disease or until seasonal conditions are unsuitable for rapid proliferation of the *Paramoeba* sp. This should then permit transfer to full seawater sites without occurrence of the disease.

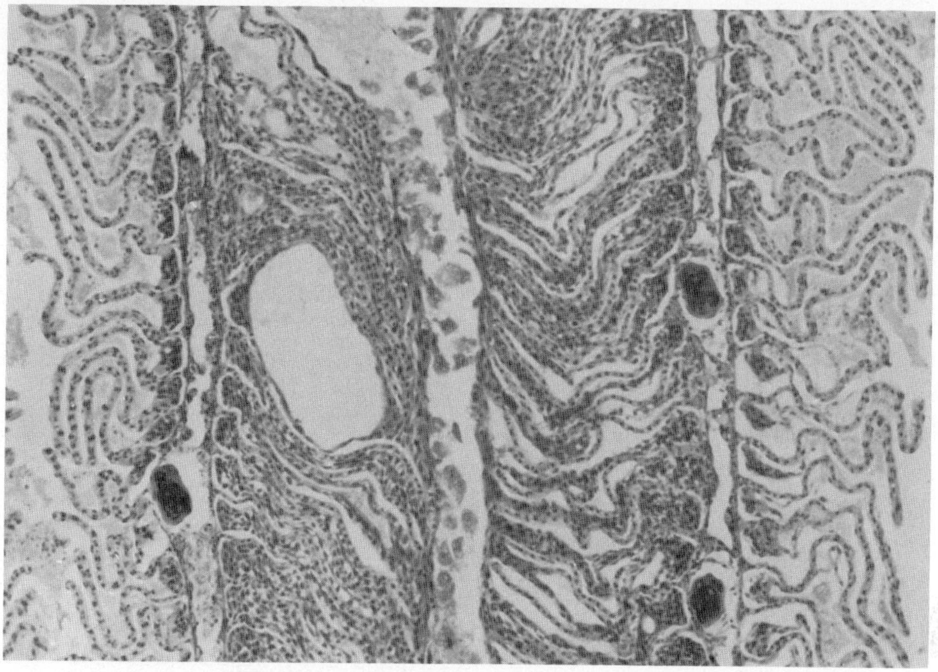

Fig. 7. Hyperplasia of gill epithelium with cyst formation; H & E, 400X.

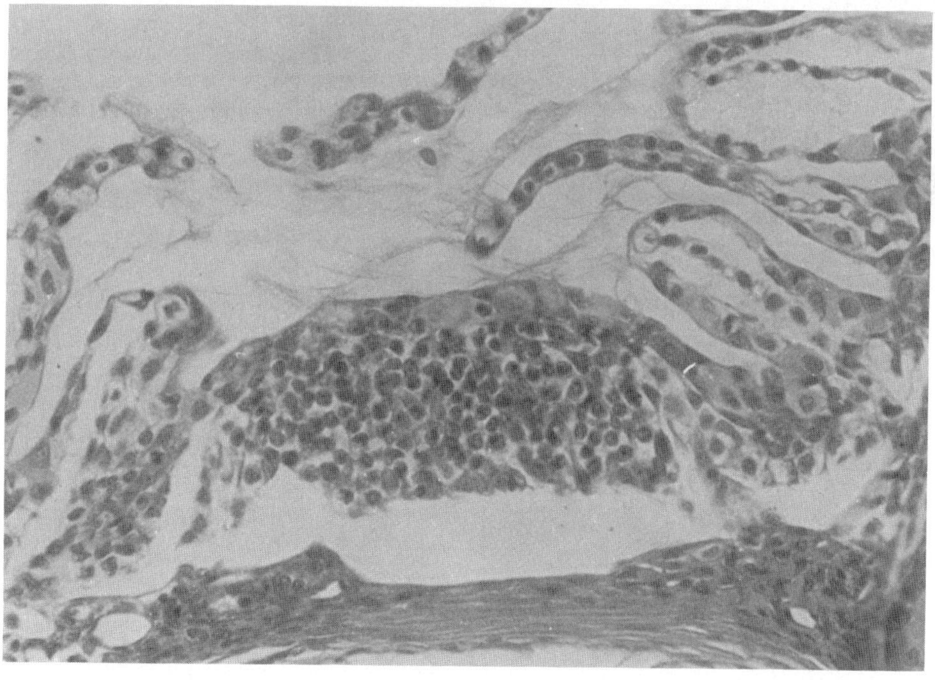

Fig. 8. Mononuclear cell nodule in the interlamellar region; H & E, 400X.

As both the epidemiological and histological features of the disease suggest that there is development of immunity, judicious use of freshwater baths can be employed to produce a degree of "natural" vaccination.

Although this disease has caused an unexpected impediment to the development of salmonid mariculture in Tasmania, active cooperation between scientists has elucidated the cause and provided practical procedures for its control within a period of four years. Future progress will depend upon continuation of this cooperative effort.

ACKNOWLEDGEMENTS

A number of people have provided valuable data used in the preparation of this paper. In particular, the authors would like to thank Dr. J. Handlinger, Mr. M. Hortle, Dr. M. Kent, Mr. T.C. Jones, Dr. K. Lange, and Mr. L. Searle.

LITERATURE CITED

Daoust, P.T. and H.W. Ferguson. 1985. Nodular gill disease: a unique form of proliferative gill disease in rainbow trout, *Salmo gairdneri* Richardson. J. Fish Dis., 8, 511-522.

Hortle, M. 1981. Seawater culture of rainbow trout. Technical Report No. 21. Tasmanian Department of Sea Fisheries, Hobart, Australia.

Kent, M.L., T.K. Sawyer and R.P. Hedrick. 1989. *Paramoeba pemaquidensis* (Sarcomastigophora: Paramoebidae) infestation of the gills of coho salmon *Oncorhynchus kisutch* reared in sea water. Dis. Aquat. Organ., 5, 163-169.

Munday, B.L. 1985. In: "Diseases of Australian Fish and Shellfish" (J.D. Humphrey and J.S. Langdon, eds.). Australian Fish Health Laboratory, Benalla, Australia.

Roubal, F.R., R.J.G. Lester and C.K. Foster. 1989. Studies on cultured and gill-attached *Paramoeba* sp. (Gymnamoebae: Paramoebidae) and the cytopathology of amoebic gill disease in Atlantic salmon, *Salmo salar* L., from Tasmania. J. Fish. Dis., in press.

DEVELOPMENT OF *BONAMIA OSTREAE* PARASITOSIS OF FLAT OYSTER, *OSTREA EDULIS*, FROM GALICIA, NORTHWEST SPAIN

J. Montes

Centro Experimental de Vilaxoan, Conselleria de Pesca,
Apdo. 208, Vilagarcia de Arousa, Spain

ABSTRACT: This study was conducted with a wild oyster (*Ostrea edulis*) population obtained from natural settlement of the "Ría de Ortigueira." Oysters were distributed in three experimental stocks, two of them being kept in rafts and the third on an intertidal bed. Parasitological analyses were performed from May, 1986, to December, 1987. *Bonamia ostreae* was detected throughout the study, with the highest infection rate in December, 1987 (60.6% in stock A) and October, 1987 (57.1% in stock B and 64.3% in stock C). Cumulative mortality reached 94.6%, 89.5%, and 96.0% in stocks A, B, and C respectively. Differences among values of these parameters, infection rates, and cumulative mortality, however, were not statistically significant.

INTRODUCTION

Oyster culture has constituted a traditional activity along the shores of estuaries of Galicia, Spain, being developed on intertidal beds as well as on rafts. Oyster production was reduced in the early 1970s after mortalities caused by *Marteilia refringens* (Massó, 1978; Alderman, 1979). Mortality problems have recently been increased in association with *Bonamia ostreae* parasitosis (Alderman, 1981; Polanco et al., 1984; Montes and Figueras, 1987; Montes and Melendez, 1987; Montes et al., 1989).

The aim of this study, a continuation of that of Montes and Figueras (1987), is to compare *B. ostreae* infection rates and mortality indices of three different experimental stocks. These oyster stocks were obtained from the same initial population and kept in rafts at two culture areas ("Ría de Noya-Muros" and "Ría de Vigo") and on trays in an intertidal bed at a third culture area ("Ría de Ortigueira").

MATERIALS AND METHODS

Oysters employed in this study were collected from a natural settlement during summer, 1985, at Ortigueira estuary. Initial transplantations were performed in May, 1986, and 12,087 specimens were distributed among the following locations (Fig. 1): Ortigueira estuary (stock A), where the oysters were held in trays in an intertidal bed; Muros-Noia estuary (stock B); and Vigo estuary (stock C); at the two latter sites the oysters were cultured on rafts. Since June, 1986, a total of 27 samples have been taken, each included 100 oysters.

The histopathological data being reported are based on the examination of digestive gland and gill pieces of every specimen. Davidson's and Bouin's solutions were employed as fixatives. Prenant's trichrome (Gabe's procedure) and Heidenhein's azan were the staining methods. Smears of the same organs were also examined, using methanol and "hemacolor" for fixation and staining, respectively. The Mann-Whitney test was used for statistical analysis.

RESULTS

Results are compiled in Table 1. These indicate that the lowest infection rate occurred at the beginning of the study and the highest values occurred in December, 1987, with 60.6% in stock A; in October, 1987, 57.1% in stock B; and also in October, 1987, with 64.3% in stock C.

Cumulative mortality reached 94.6%, 89.5%, and 96.0% in stocks A, B, and C, respectively.

Statistical analysis demonstrate that differences among the three stocks are not significant for both indices.

Table 1. Results of the survey.

		INFECTION RATE (%)			CUMULATIVE MORTALITY (%)		
SAMPLE	STOCK	A	B	C	A	B	C
May	1986	0	0	0	–	–	–
June	1986	1.7	0	0	2.0	1.8	1.7
August	1986	4.0	2.0	3.7	6.1	3.9	6.7
October	1986	10.5	3.3	8.6	16.3	8.1	15.4
December	1986	8.6	4.0	–	24.6	12.4	–
February	1987	13.3	5.0	11.4	34.0	17.4	30.4
April	1987	18.5	13.0	24.1	49.2	26.2	55.4
June	1987	37.5	30.0	–	64.7	37.9	–
August	1987	34.6	43.3	–	77.9	53.9	–
October	1987	50.0	57.1	64.3	89.1	71.1	75.7
December	1987	60.6	48.5	53.1	94.6	89.5	96.0

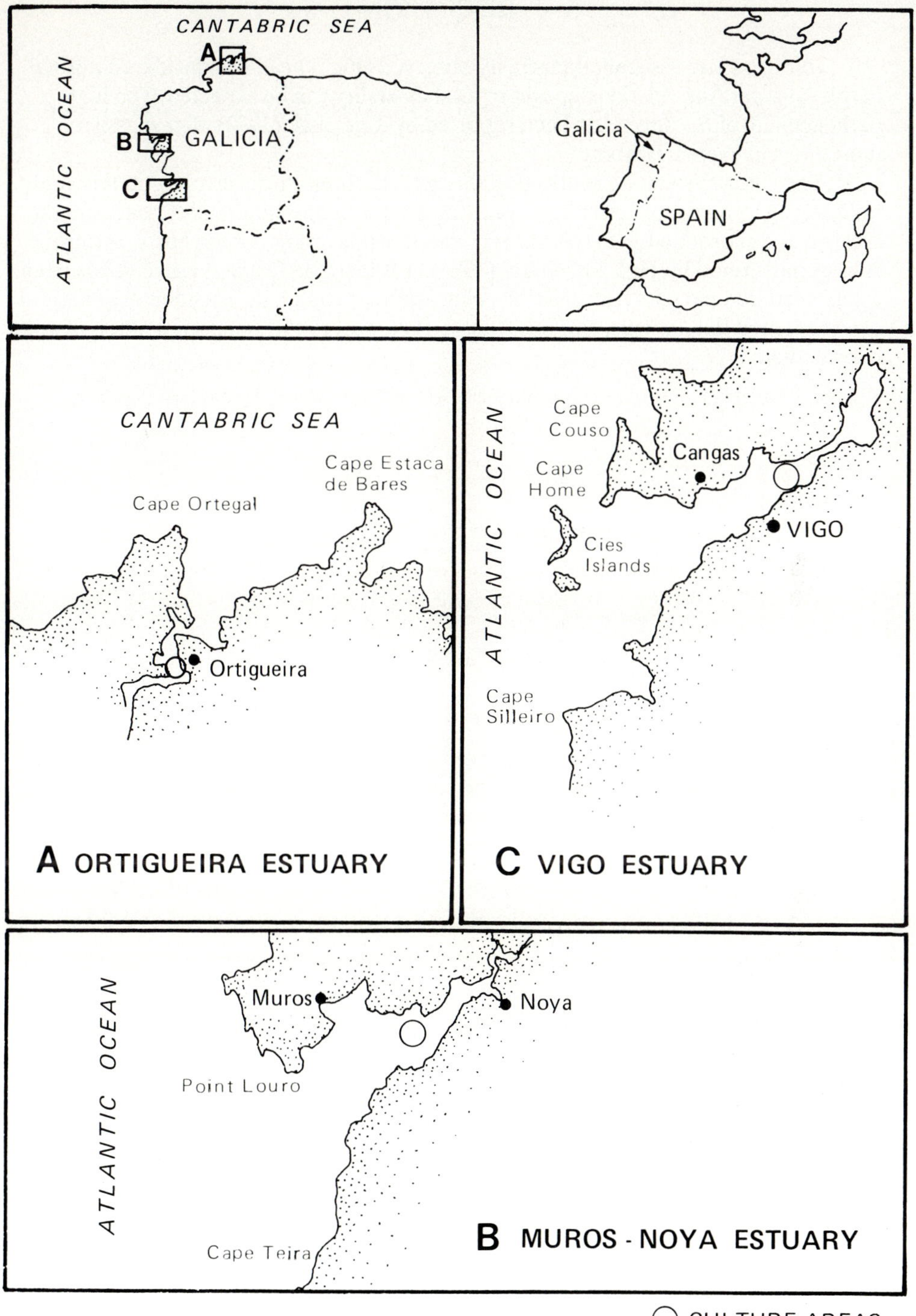

Fig. 1. Location of the estuaries and culture areas.

DISCUSSION

Infection and associated mortality observed (Fig. 2) can be considered normal in this host-parasite relationship where total mortality can be expected. The high pathogenicity of *B. ostreae* has been reported by Tigé et al. (1982) after massive mortalities in French Brittany.

The similarity among results pertaining to all three stocks involved in this study indicates that the culture system is not associated with measurable differences in the survival and reproduction of the parasite and mortality rates among hosts. After the studies performed by Tigé and Grizel (1984) at Rivière de Crach, France, it has been established that parasite transmission occurs all year round, independently of natural environmental fluctuations, and this transmission occurs without participation of any intermediate host.

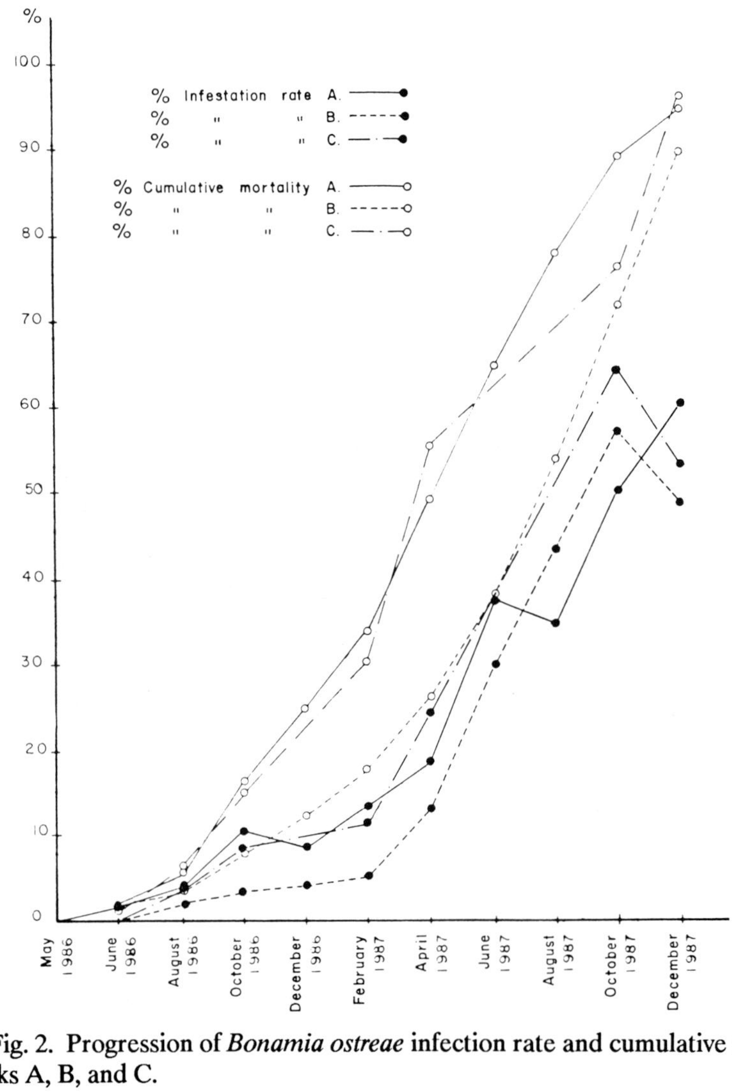

Fig. 2. Progression of *Bonamia ostreae* infection rate and cumulative mortality in stocks A, B, and C.

LITERATURE CITED

Alderman, D.J. 1979. Epizootiology of *Marteilia refringens* in Europe. Mar. Fish. Rev., 41, 67-69.

Alderman, D.J. 1981. Parasite "X", new disease threatens European beds. Fish Farm., 4, 31.

Massó, J.M. 1978. La enfermedad de la glándula digestiva de la ostra plana (*Ostrea edulis* L.) en las rías bajas. Bol. Inst. Esp. Ocean., 251, 125-140.

Montes, J. and A. Figueras. 1987. *Bonamia ostreae* en una población de ostra plana (*Ostrea edulis* L.) cultivada en tres diferentes áreas de la costa gallega. Cuad. Marisq. Publ. Téc., 12, 683-688.

Montes, J. and M.I. Meléndez. 1987. Données sur la parasitose de *Bonamia ostreae* chez l'huitre plate de Galice, cote nord-ouest de l'Espagne. Aquaculture, 67, 195-198.

Montes, J., C.P. Acosta and A. Guerra. 1989. Oyster mortality in Galice (NW Spain). Europ. Aquacult. Soc., in press.

Polanco, E., J. Montes, M.J. Outon and M.I. Meléndez. 1984. Situation pathologique du stock d'huîtres plates en Galice (Espagne) en relation avec *Bonamia ostreae*. Haliotis, 14, 91-95.

Tigé, G. and H. Grizel. 1984. Essai de contamination d'*Ostrea edulis* L. par *Bonamia ostreae* (Pichot et al., 1979) en rivière de Crach (Morbihan). Rev. Trav. Inst. Pêches Marit., 46, 307-314.

Tigé, G., H. Grizel, M.A. Rabouin, N. Cochonnec, G. Audic and A. Langlade. 1982. *Bonamia ostreae* – Evolution de la situation epizootiologique en Bretagne au cours de l'année 1981. Sci. Pêches, 328, 13-13.

AN ELECTRON MICROSCOPE STUDY OF DISEASE PROGRESSION IN THE OYSTER, *CRASSOSTREA VIRGINICA*, INFECTED WITH THE PROTOZOAN PARASITE, *HAPLOSPORIDIUM NELSONI* (MSX)

Robert A. Scro[1] and Susan E. Ford[2]

[1]Division of Science and Research, New Jersey Department of Environmental Protection, CN 409, Trenton, New Jersey 08625 USA;
[2]Rutgers University, Shellfish Research Laboratory, Port Norris, New Jersey 08349 USA

ABSTRACT: Contributed herein is an ultrastructural description of *Haplosporidium nelsoni* (MSX), a highly pathogenic parasite of the American oyster, *Crassostrea virginica*. In addition, the results of an electron microscope study of host cells, particularly hemocytes, which interact with *H. nelsoni,* are being reported. A hypothesis is being advanced as to the pathogenicity of the MSX disease once it penetrates the basal lamina and reaches the systemic stage.

INTRODUCTION

The identification in the American oyster, *Crassostrea virginica*, of the extracellular protozoan parasite, *Haplosporidium* (formerly *Minchinia*) *nelsoni* (MSX) (Haplosporida, Haplosporidiidae) by Haskin et al. (1966) has fostered a multidisciplinary research effort to understand the biology, chemistry, and environmental factors involved in this host-parasite association. Nearly all of the pathological studies of MSX disease have been based on histological information, whereas ultrastructural and cytopathological manifestations of the host-parasite interaction have received very little attention.

Muller (1967) discussed the ultrastructural characteristics of MSX plasmodia and Perkins (1968) described the fine structure of both MSX plasmodia and spores in oysters with heavy infections. Rosenfield et al. (1969) compared the fine structure of spores of three species of *Haplosporidium*, including *H. nelsoni*. These studies were confined primarily to descriptions of parasite morphology rather than to host-parasite interactions.

Histological studies have described the usual course of infection for MSX. Initially, water-borne parasites invade the gill epithelium, being restricted to this surface layer by the basal lamina. Plasmodial stages subsequently proliferate from localized gill epithelial lesions into systemic infections by penetrating this "barrier" and invading the underlying connective tissue and blood spaces (Farley, 1968; Myhre and Haskin, 1968). Infections are rarely lethal until they develop into systemic infections (Ford and Haskin, 1982) and resistance to MSX mortality is associated with restriction of parasites to localized, nonlethal lesions in the gill (Myhre, 1973; Ford and Haskin, 1982, 1987). Therefore, host-parasite interactions in the gill epithelium may provide important information in understanding defense mechanisms against the pathogen.

The present investigation was conducted to clarify cell structural changes in both host and pathogen as infections develop from localized to systemic lesions. It provides a general cytopathological description of the host- parasite interaction that can serve as a baseline for ultrastructural comparisons between genetically resistant and susceptible oysters. The specific objectives of the study were to: (1) identify cellular components of the gill that show ultrastructural change in local or systemic MSX infections; (2) determine whether ultrastructural changes take place in hemocytes located in the gill or hemolymph, which might define their role in MSX disease; and (3) determine if the ultrastructure of the parasite is different in localized compared with systemic infections.

METHODS AND MATERIALS

Determination of MSX infection intensity. Oysters exposed to *H. nelsoni* in Delaware Bay were screened for the presence of systemic parasites using hemolymph diagnosis (Ford and Kanaley, 1988). Representatives with and without systemic infections were selected for ultrastructural examination. Among the latter, oysters with localized infections were identified by examining gill tissue with both light and electron microscopy. Oysters were also collected from Hope Creek Bed in upper Delaware Bay, where low salinity inhibits MSX infections. None of these oysters was diagnosed as having *H. nelsoni* and they were considered as uninfected controls.

Preparation of tissue for transmission electron microscopy. Small (less than 2 mm) pieces of gill were fixed for 18 hr in cold 2.5% glutaraldehyde buffered in 0.1 M sodium cacodylate with 4% sucrose (pH 7.4). The tissue was post-fixed in 1% osmium tetroxide in buffer/sucrose solution for 1 hr, dehydrated, infiltrated, and embedded in Epon. Hemolymph was collected from the adductor muscle sinus with a 1-ml tuberculin syringe and placed in a test tube containing cold buffered 2.5% glutaraldehyde for 1 hr prior to centrifugation at 1000 rpm for 10 min at 4°C. The resultant pellet was then processed in the same manner as that of the tissue, starting with the glutaraldehyde fixation. Thin sections (silver/gold interference colors) were collected on uncoated copper grids, stained with aqueous uranyl acetate and lead citrate (Koulish, 1976), and examined with a Hitachi HU 11F transmission electron microscope at an accelerating voltage of 75 kV.

Digital image processing. Digital image processing was performed on selected electron micrographs to enhance ultrastructural details of certain cellular organelles. The original micrograph was scanned with an Eikonix optical scanner and processed with the Earth Resources Data Analysis System (ERDAS). This was conducted by the Department of Remote Sensing (Rutgers University) and the Geographic and

Statistical Analysis Unit (New Jersey Department of Environmental Protection, Division of Science and Research).

RESULTS

Ultrastructure of gill tissue. The oyster gill filament is comprised of ciliated and nonciliated epithelial cells situated on a basal lamina (Fig. 1). The apical surface of these cells is arranged into microvilli. Adjacent cells are held together by both adhering and septate junctions. Blood spaces and connective tissue lie beneath the basal lamina and the entire structure is supported by a lattice of collagenous rods.

Ultrastructure of *H. nelsoni* (MSX). In early infections, multinucleated plasmodial stages of the MSX parasite were found against the basal lamina between epithelial cells of the gill (Fig. 2). The fine structure of several hundred MSX plasmodia examined in the gill was extremely diverse. The cytoplasm ranged from light to dense, depending on the number of free ribosomes present. In plasmodia with "electron light" cytoplasm, large distended mitochondria with vesicular cristae were common. In those with "dense" cytoplasm, nuclei were usually circumscribed by normally structured mitochondria with vesicular cristae (Fig. 3). Occasionally MSX plasmodia appeared to be dividing, as evidenced by the presence of mitotic figures in the nucleus.

The cytoplasm of all MSX plasmodia contained round, dense membrane-limited bodies, or "haplosporosomes". These structures, found in all species of *Haplosporidium*, had an average diameter of 0.18 ($\pm$0.02 SD, N=30) μm. They were also identified by an inner electron-lucent configuration usually having the

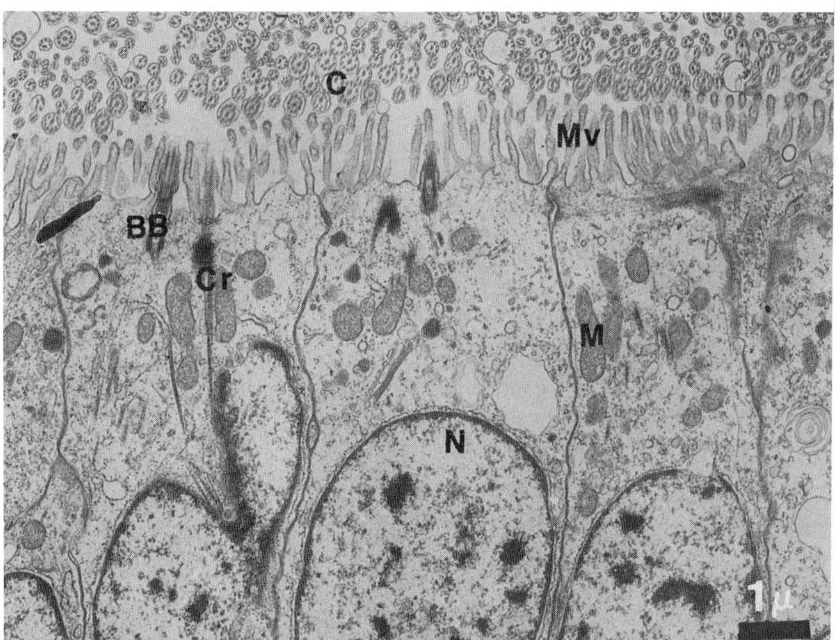

Fig. 1. Epithelial cells of the gill filament in uninfected control oyster. BB, basal body; C, cilia; Cr, ciliary rootlet; M, mitochondria; Mv, microvilli; N, nucleus. 9,075X.

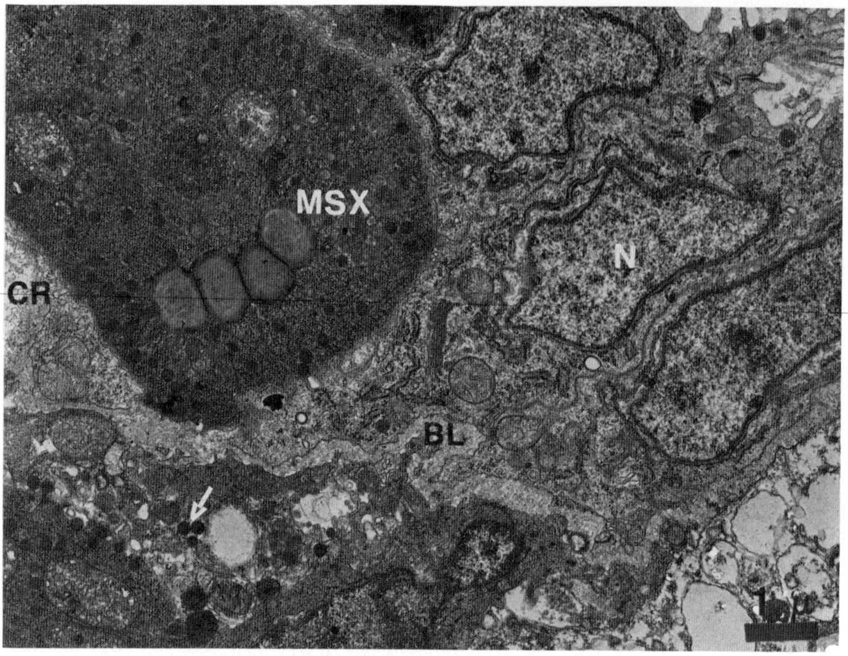

Fig. 2. Cytopathology of nuclei (N) in gill surface epithelial cells of oyster with systemic MSX infection. Note haplosporosome (arrow) outside of MSX plasmodium. BL, basal lamina; CR, collagenous rods. Mv, microvilli. 9,075X.

appearence of a "squat vase" (Perkins, 1968). The area within this configuration was usually not as electron-dense as the area to its exterior. The number of haplosporosomes within any given plasmodium varied greatly. Plasmodia with "electron-light" cytoplasm (few ribosomes) usually had few haplosporosomes. Numerous membranous "spherules" were physically associated with haplosporosomes, which appeared to bud off from the spherules. Lipid-like droplets and digestive lamellae were frequently, but not always, present in the parasite cytoplasm (Figs. 3, 4).

The interface between MSX plasmodia and host cells was one of tight apposition of the respective plasma membranes. Haplosporosomes usually displayed a structural polarity when located close (65 nm or less) to the MSX plasma membrane. This orientation was clearly seen as an electron-lucent configuration shaped like a light-bulb, with the narrow end facing the plasma membrane (Figs. 5, 6).

An additional structure not previously described in the Haplosporida, and herein termed the "concentric body," was consistently found in the MSX cytoplasm. This structure was round, approximately 0.3 μm, membrane-limited, and had a dense core. The core was surrounded by two electron-dense concentric rings. The ring closest to the core appeared to be connected to the core by fine radiating fibers arranged in a spoke-like fashion. As many as four concentric bodies were observed in one thin section of a single MSX parasite (Figs. 4, 7). The identity and function of the concentric bodies is unclear, but because they were consistently observed within MSX plasmodia they undoubtly play a role in the parasite's normal activities.

There were no consistent ultrastructural differences between parasites localized in the gill epithelium and those that had become systemic and were found in the

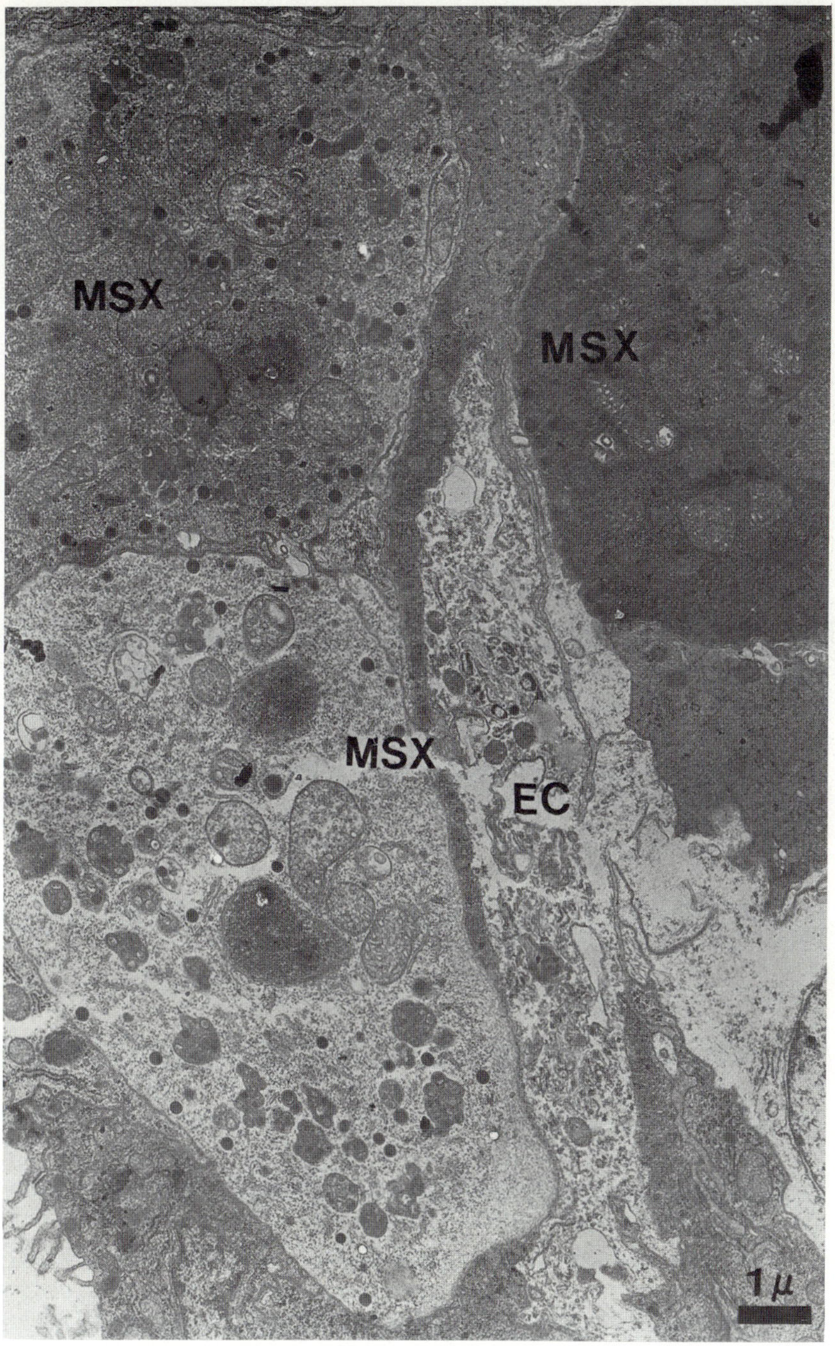

Fig. 3. Three MSX plasmodia in the gill of a systemically infected oyster reveal distinct ultrastructural differences. EC, epithelial cell. 9,075X.

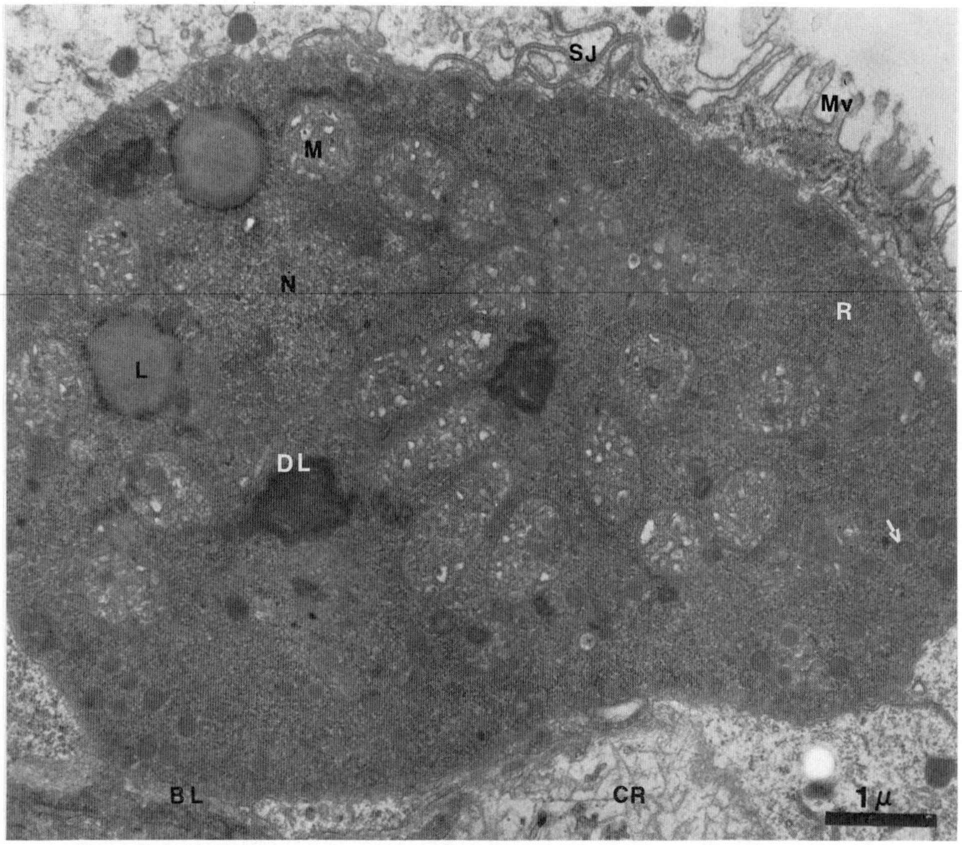

Fig. 4. MSX plasmodium in the gill epithelium of systemically infected oyster. Note dense cytoplasm packed with ribosomes (R), mitochondria (M) circumscribing nucleus (N), and lack of spherule structures. BL, basal lamina; CR, collagenous rod; DL, digestive lamellae; L, lipid; Mv, microvilli; SJ, septate junction; haplosporosomes (arrow). 11,101X.

hemolymph. In both locations, degenerating parasites were occasionally found, with pycnotic nuclei, leached cytoplasmic contents, swollen mitochondria, and deterioration of the plasma membrane (Fig. 8). This condition was not necessarily associated with the presence of hemocytes.

Ultrastructure of host cells in localized infections.

Epithelial cells. Parasite membranes were in close apposition to host cell membranes and there was evidence of epithelial cell deterioration, although junctions between epithelial cells appeared intact (Fig. 2). Cytopathological changes in the gill epithelial cells during local MSX infections were not frequently observed; however, when encountered, these changes involved leaching of the cytoplasmic contents and disorganization of the mitochondria. Mitochondria were also found in the "condensed" form, which may indicate high cellular ADP in an energically depleted cell (Fig. 9). Some epithelial cells in contact with parasites contained digestive lamellae (Fig. 10).

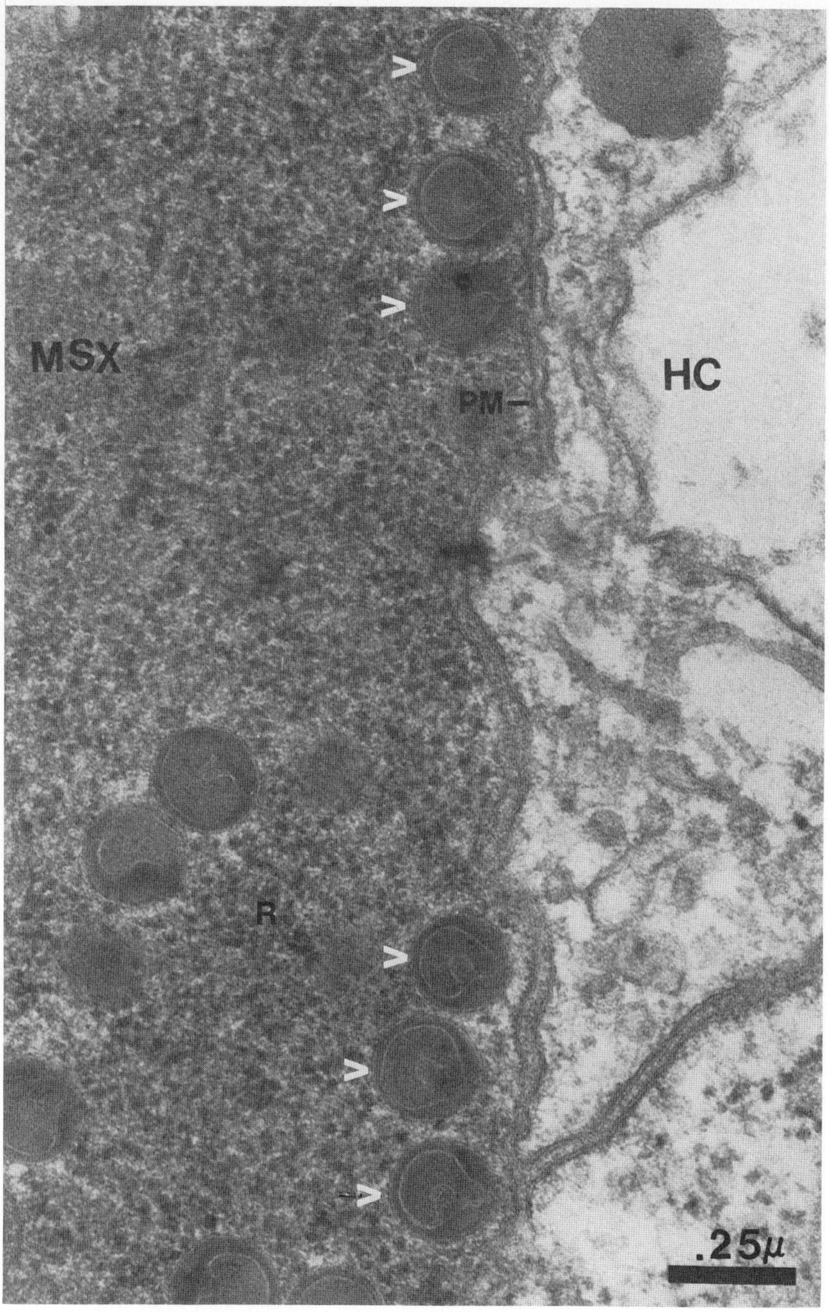

Fig. 5. High magnification of MSX-host cell (HC) interface in gill of systemically infected oyster. Note the polar orientation of haplosporosomes (arrow heads) in close proximity to the plasma membrane (PM). R, ribosomes. 71,500X.

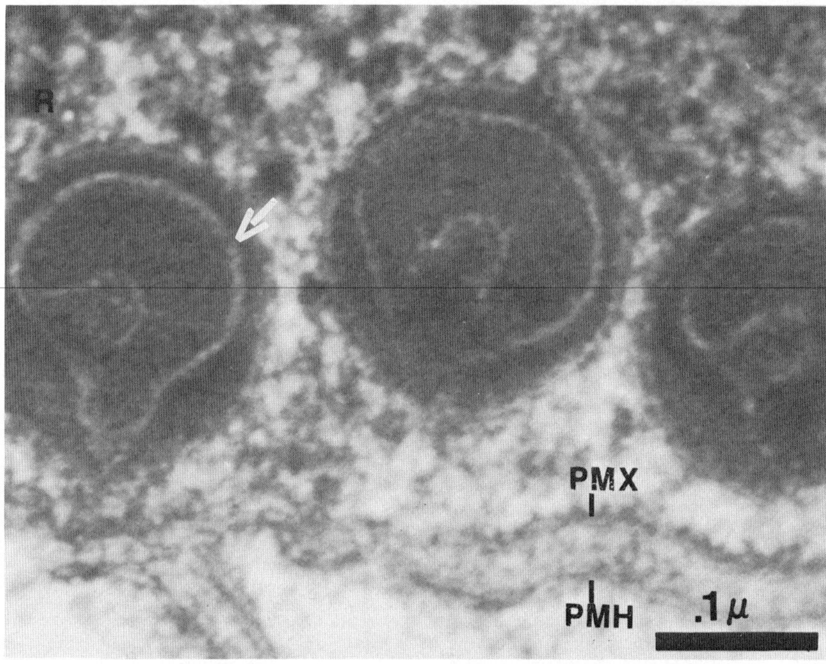
R
PMX
PMH
.1 μ

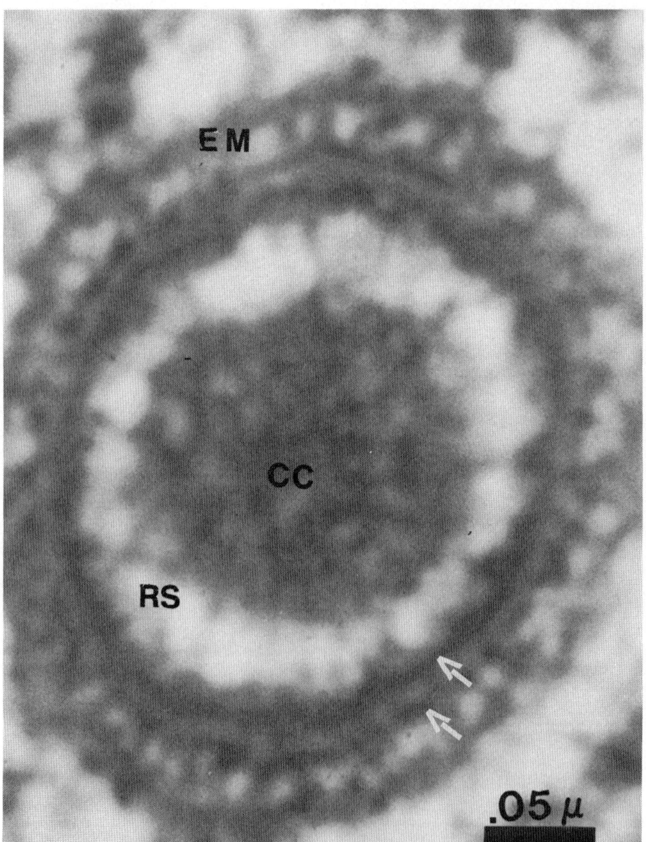
EM
CC
RS
.05 μ

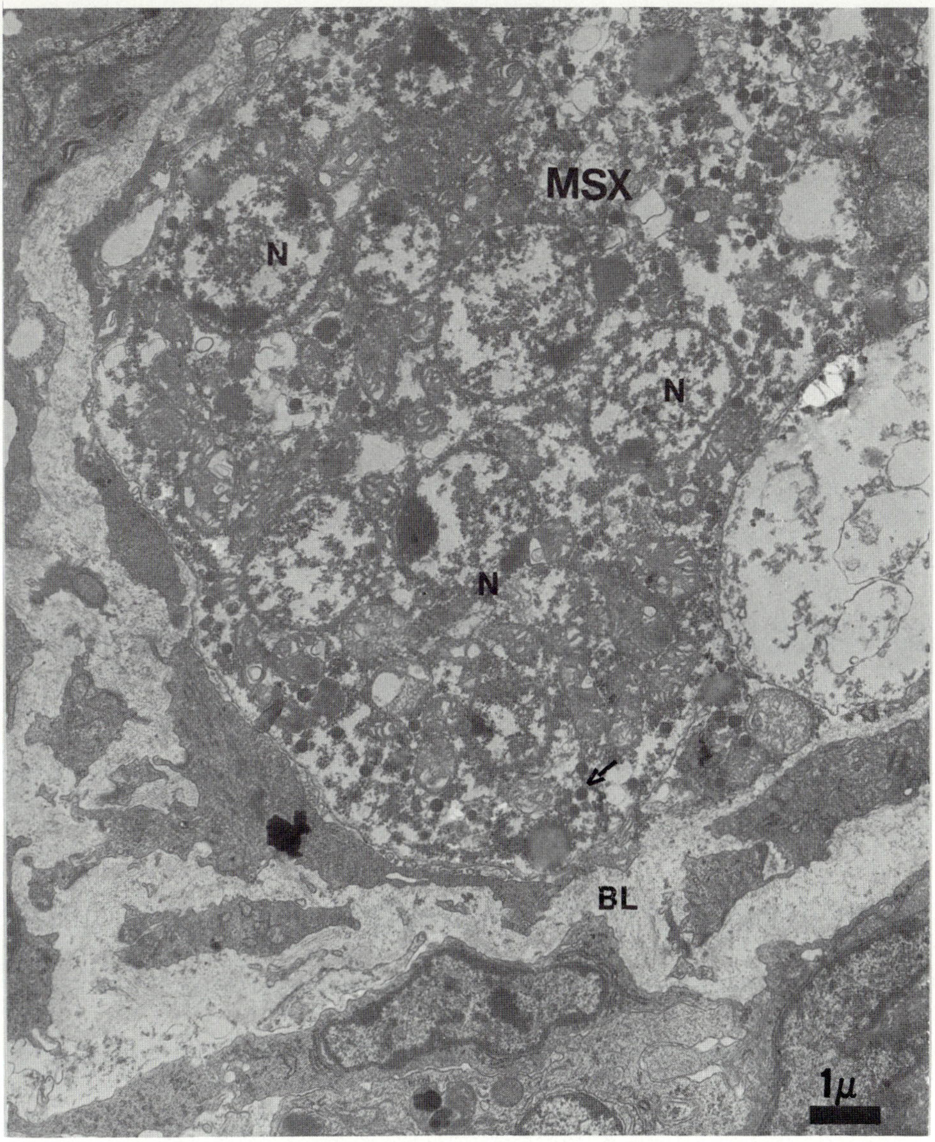

Fig. 8. Degenerating MSX parasite in the gill connective tissue during systemic infection. BL, basal lamina; N, MSX nuclei; haplosporosomes (arrow). 9,075X.

Fig. 6. High magnification digital image of haplosporosomes shown in Fig. 5. Note internal electron-lucent configuration of polar oriented haplosporosomes (arrow). PMH, host cell plasma membrane; PMX, MSX plasma membrane; R, ribosomes. 214,500X.

Fig. 7. Digital image of concentric body in the MSX cytoplasm. CC, central core; EM, external membrane; RS, radiating spokes. 300,250X.

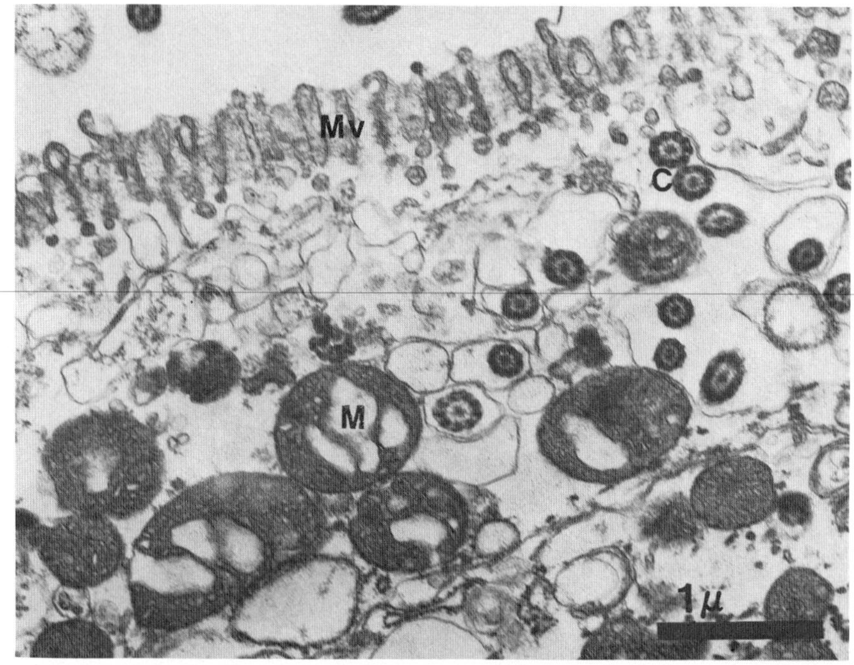

Fig. 9. Cytopathology of gill epithelial cells during local MSX infection. Note extensive crystolysis of mitochondria (M) and leaching of the cytoplasmic matrix. C, cilia; Mv, microvilli. 21,725X.

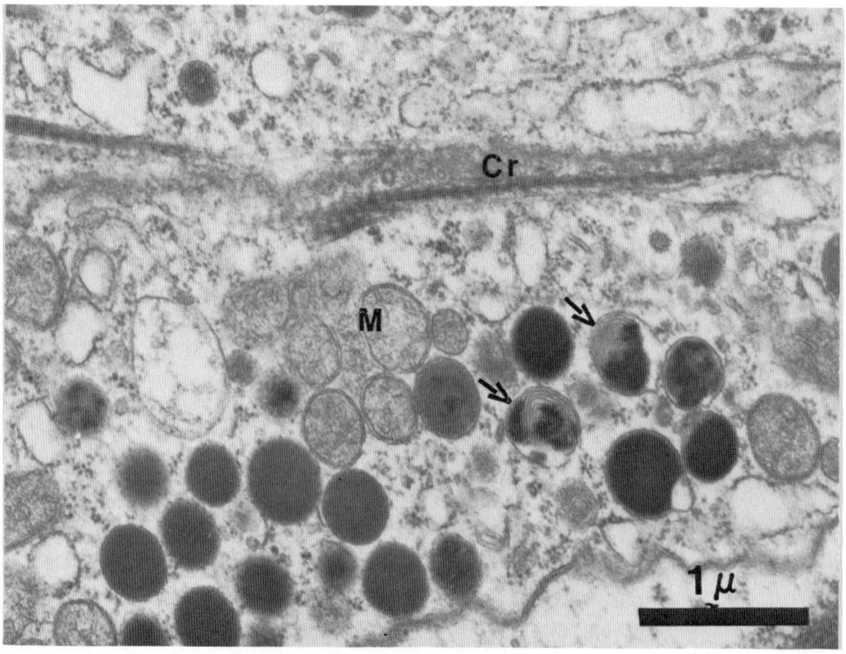

Fig. 10. Ciliated epithelial cells of the gill during local MSX infection showing digestive lamellae (arrows) in the cytoplasm. Cr, ciliary rootlet; M, mitochondria. 21,725X.

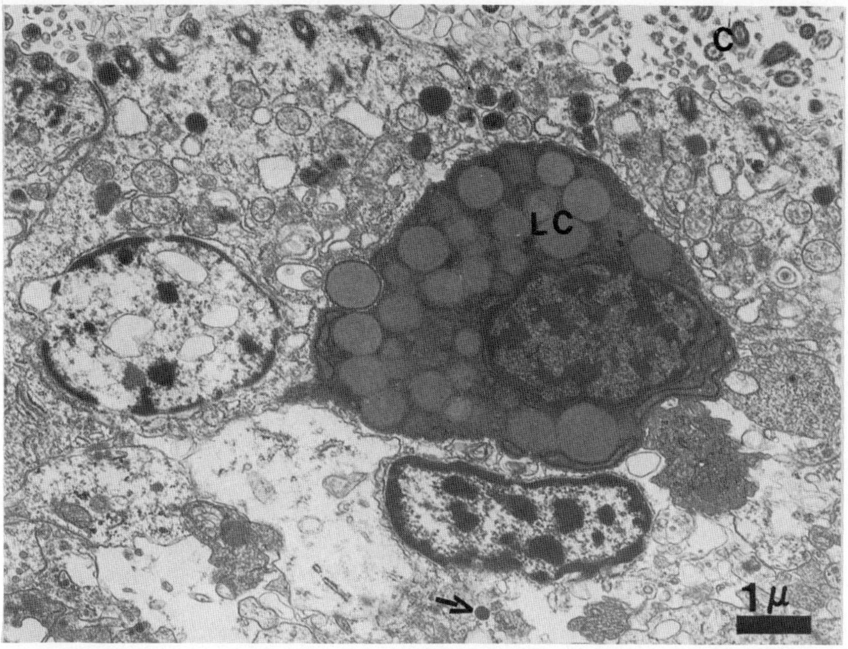

Fig. 11. Cell containing lipid-like globules (LC) in MSX-infected oyster. Note deterioration of nearby epithelial cells, and presence of haplosporosome outside of MSX (arrow). C, cilia. 9,075X.

Another characteristic of infected tissue was the presence of electron dense cells packed with membrane-bound lipid-like globules. These were probably the "brown" or serous cells originally described by Mackin (1951) from oysters infected by *Perkinsus marinus.* They were found in both locally and systemically infected oysters. The cells contained a substantial amount of free ribosomes and rough endoplasmic reticulum. The globules themselves were occasionally found adjacent to, and outside of, the cell (Fig. 11).

Hemocytes. In oysters with localized MSX infections, hemocytosis was clearly evident in the gill epithelium and underlying connective tissue (Fig. 12). The lesions contained several morphological types of granular and agranular hemocytes, but the most prominent were granular cells (granulocytes). The majority of the granulocytes contained round or oval cytoplasmic granules similar to type B granules of Feng et al. (1971), with an average diameter of 0.65 ($\pm$ 0.003 SD, N=30) μm across the long axis. These granules were membrane-limited, with an electron-dense peripheral zone separated from the membrane by an electron-lucent space (Fig. 13). The core of the granules, when found intact, was finely granular and homogenous in appearence. The nucleus of these granulocytes was almost always round, with densely staining heterochromatin lying against the internal nuclear envelope and in several clumps within the nucleoplasm. Euchromatin was evenly dispersed throughout the remainder of the nucleoplasm. The nucleolus was usually oriented peripherally. The cytoplasm of the granulocytes was either rich or sparse in ribosomes, thereby giving these cells an electron light or dense appearance, respectively. Mitochondria were only occasionally seen in these granulocytes. Other granulocytes frequently contained nonmembranous crystaloid granules characteristic of mammalian eosinophilic leucocytes (Fig. 12). Phagocytosis of MSX parasites by hemocytes was never observed in the gill.

Ultrastructure of host cells in systemic infections.

Epithelial cells. Epithelial cells in contact with, or in close proximity to MSX plasmodia often displayed signs of severe nuclear and cytoplasmic degeneration, although pathology was not limited to cells near parasites (Fig. 14). On several occasions free haplosporosomes were found adjacent to degenerating host cells (Fig. 15).

Degeneration of the basal lamina, characterized by disruption and extensive fine granulation, was associated with the presence of MSX plasmodia and free haplosporosomes (Fig. 15). Degeneration of the collagenous rods, which are structurally continuous with the basal lamina, was occasionally observed. Breakdown of this structure was associated with exocytosis of large amounts of membrane-bound MSX material (parasite cytoplasm plus haplosporosomes) into the rods (Fig. 16).

In oysters with systemic MSX infections, the epithelial cells of the gill frequently revealed the presence of double-membrane stacked annulate lamellae in the cytoplasm adjacent to the nucleus (Fig. 17). These structures were only observed on two occasions in non-infected and locally infected oysters.

Hemocytes. In oysters with systemic MSX infections, there was frequent evidence of severe cytopathological changes in the hemocytes, as well as to the epithelial cells, present in the gill. These included extensive vacuolization, leaching of the contents of the cytoplasm, and nuclear inclusions (Fig. 18). Granulocytes,

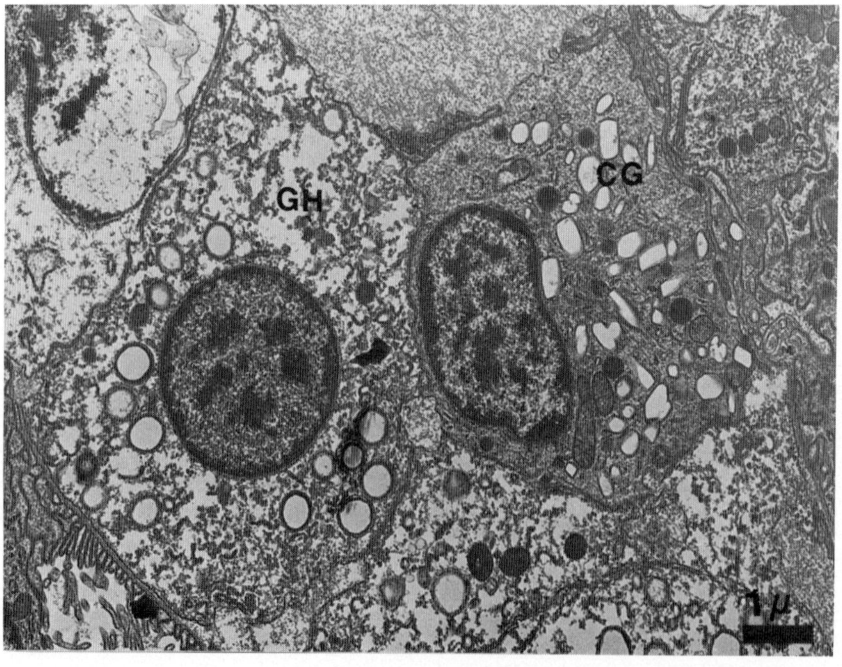

Fig. 12. Granular hemocytes (GH) in the gill epithelium during local MSX infection. Note granulocyte with crystalloid granules (CG). 9,075X.

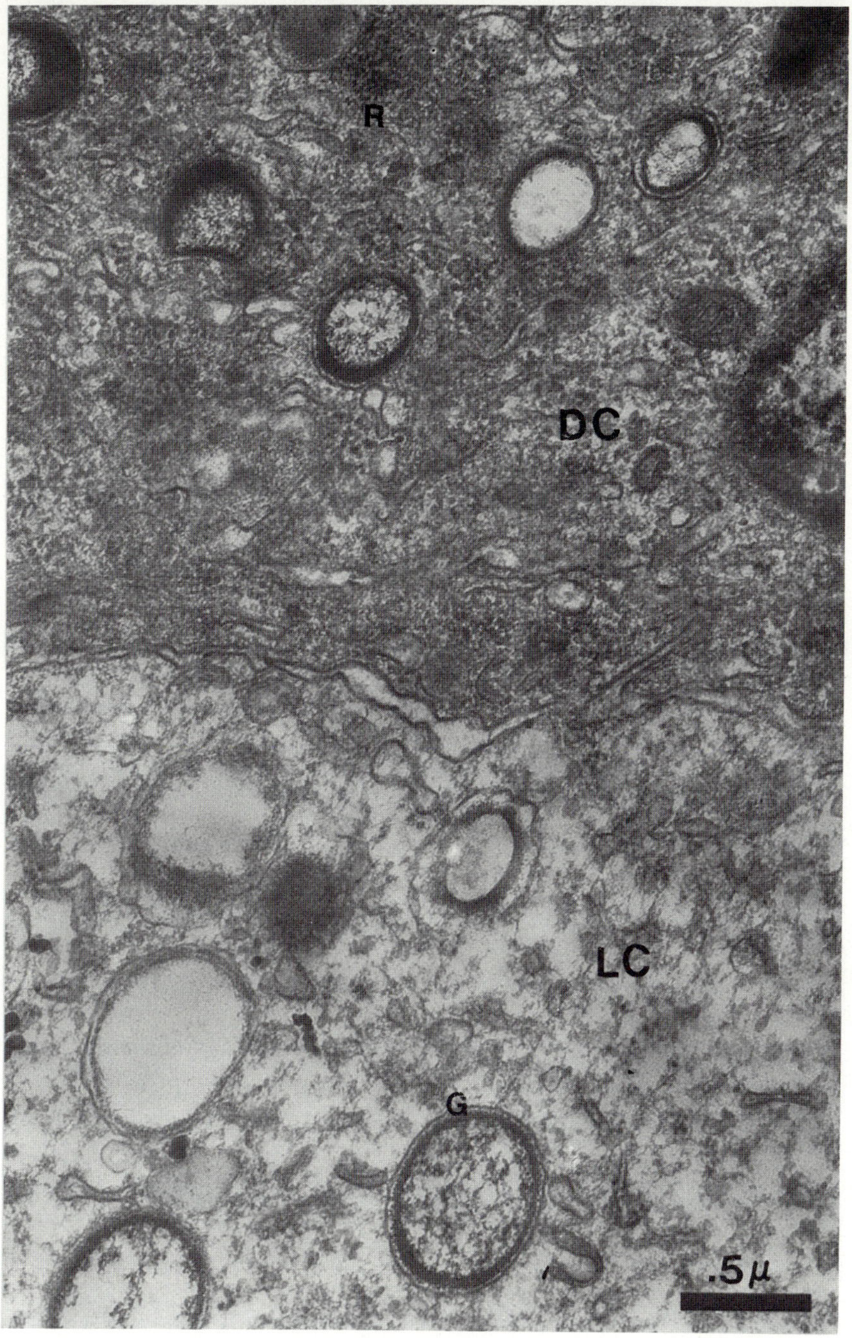

Fig. 13. Adjacent granular hemocytes showing "dense" cytoplasm (DC) and "light" cytoplasm (LC) in locally infected oyster. G, granule; R, ribosome. 35,700X.

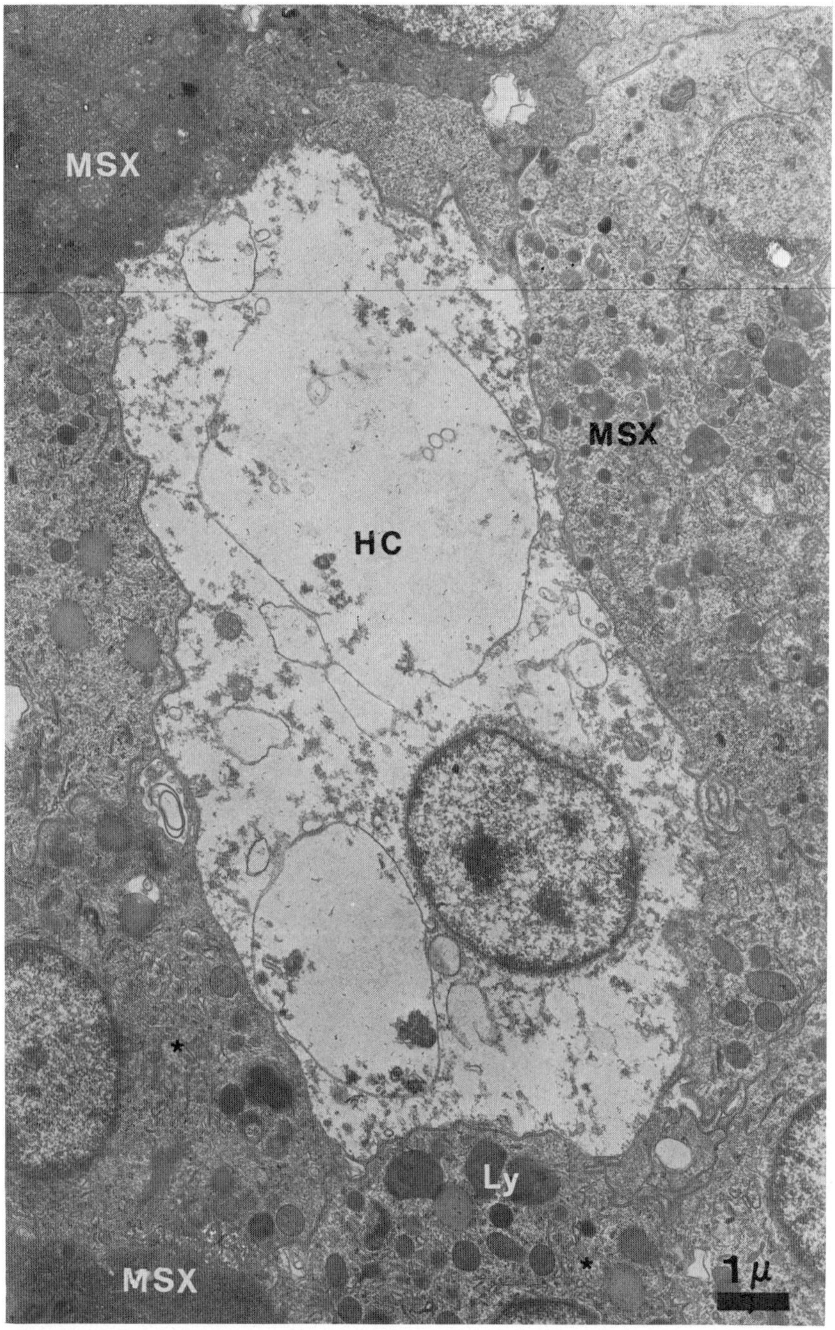

Fig. 14. Cytopathology of host cell (HC) in contact with parasites in oyster gill during systemic MSX infection. Note adjacent hemocytes with high nucleus-to-cytoplasm ratio and dense lysosome-like bodies (Ly). 9,075X.

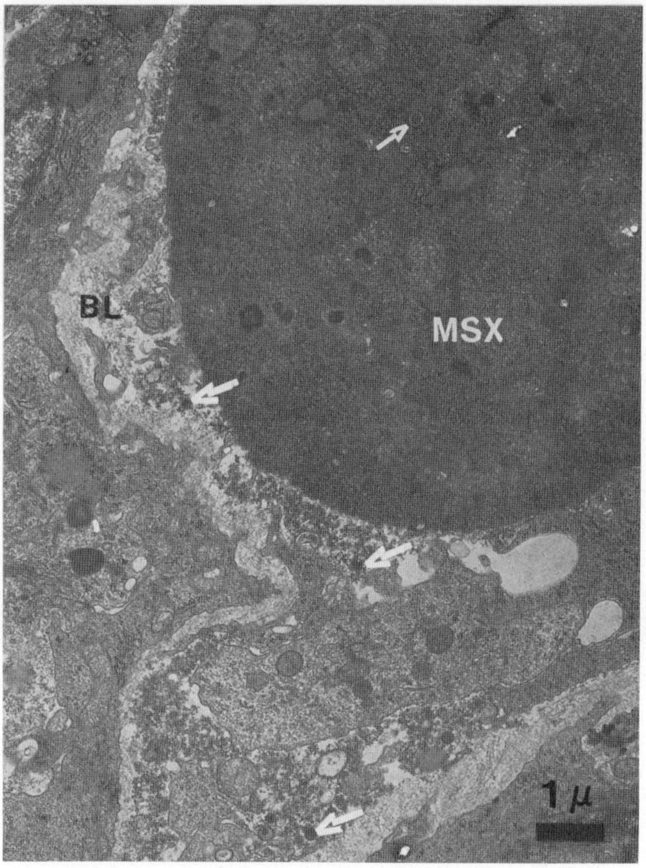
BL
MSX
1 μ

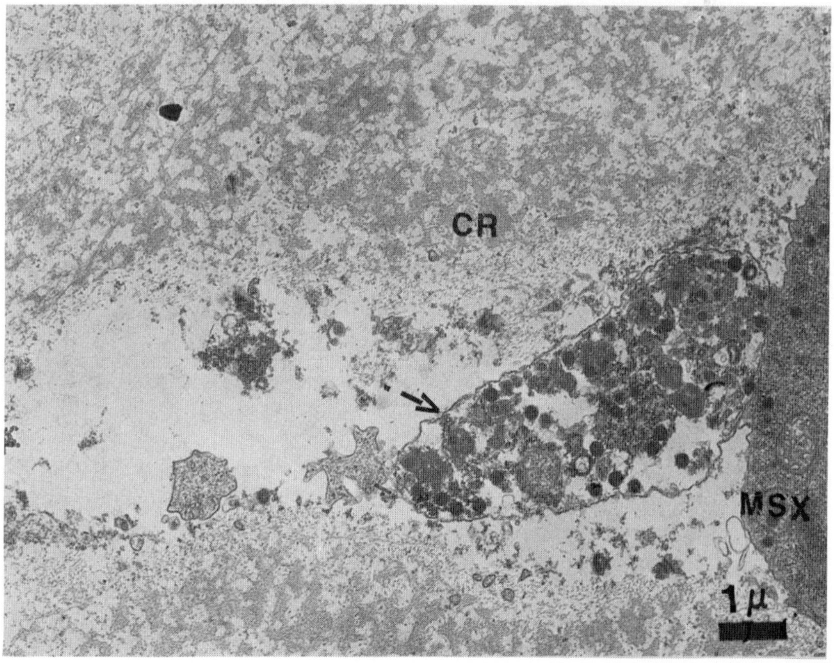
CR
MSX
1μ

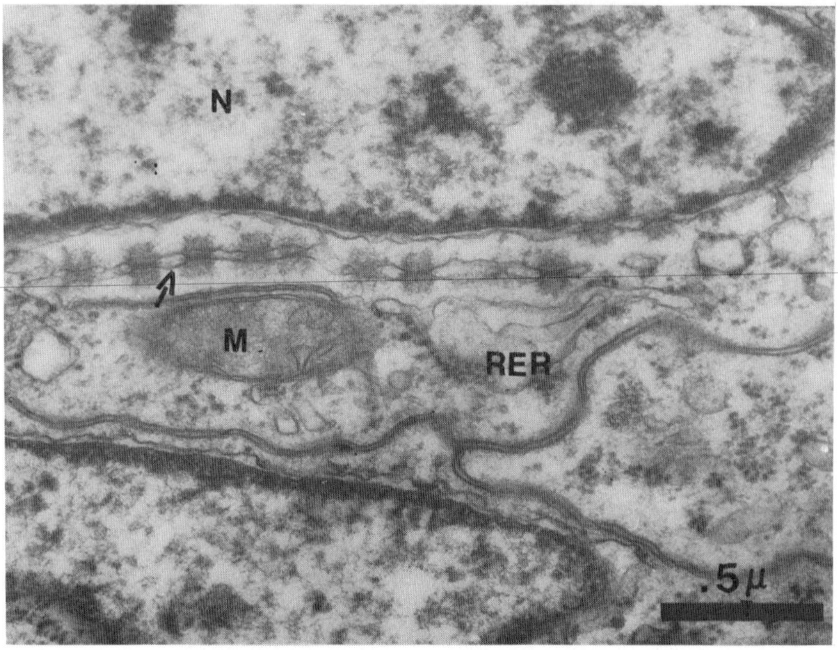

Fig. 17. Annulate lamellae (arrow) in gill connective tissue cell during systemic MSX infection. M, mitochondria; N, nucleus; RER, rough endoplasmic reticulum. 35,750X.

typically found in locally infected oysters, were rare when infections became systemic. Many of the degenerating hemocytes also possessed digestive lamellae, suggesting catabolism of damaged host tissue, parasite waste products, or both.

Agranular hemocytes, not commonly observed in the gill of MSX-infected oysters, were prevalent in the hemolymph. They were characterized by an absence of cytoplasmic granules, an abundance of smooth endoplasmic reticulum, moderate amounts of rough endoplasmic reticulum, large phagosomes containing ingested material, and several digestive lamellae. Glycogen deposits were not observed. Although lacking granules, these cells were not the lymphocyte-like agranular cells, usually called hyalinocytes, which have a relatively large nucleus compared to granulocytes, and a high nucleus-to-cytoplasm ratio (Feng et al. 1971; Cheng, 1981). The size and appearance of nuclei in the agranular hemocytes described in this study

---

(previous page)

Fig. 15 (Top). Cytopathology of basal lamina (BL) in gill during systemic MSX infection. Note haplosporosomes (large arrows) outside of parasite. Concentric body (small arrow). 9,075X.

Fig. 16 (Bottom). Degeneration of collagenous rods (CR) in gill of systemically infected oyster. Note exocytosis of MSX material associated with degenerative area (arrow). 9,075X.

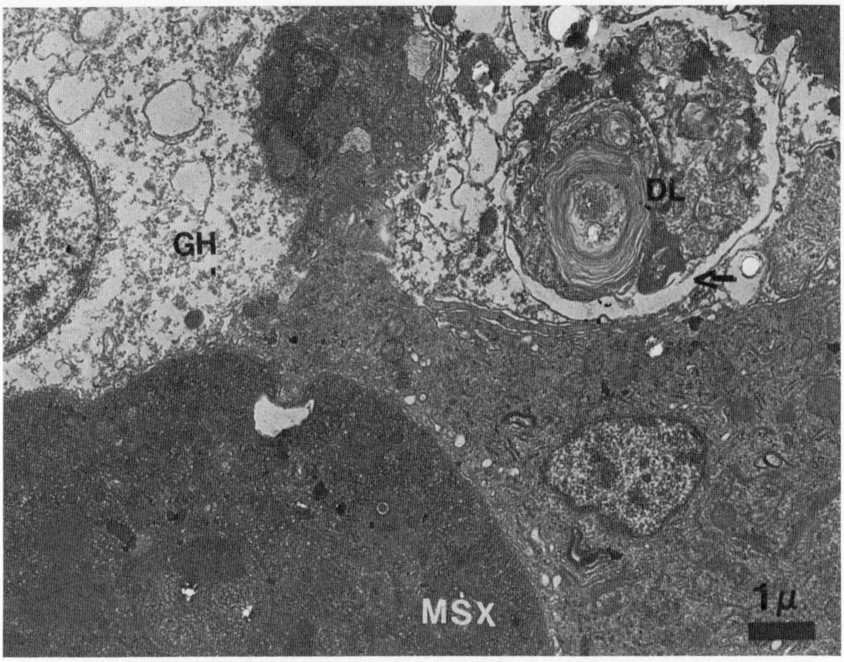

Fig. 18. Cytopathology of granulocyte (GH) in the gill connective tissue of an oyster with systemic MSX infection. Note digestive lamellae (DL) in large phagosome (arrow) within granulocyte cytoplasm. 9,075X.

very much resembled that of granular hemocytes. Agranular cells fitting the description of lymphocyte-like hemocytes, which we called hyalinocytes, were seen but were not common.

Non-hyalonocyte type agranular hemocytes containing ingested MSX plasmodia or fragments of plasmodia were present in the hemolymph of some oysters (Fig. 19). In the same oysters, agranular hemocytes frequently contained one or more hemocytes (Fig. 20), and some were found with both parasites and hemocytes. Ingested parasites or hemocytes were never found in granulocytes or hyalinocytes. In a few instances both phagocytic and phagocytosed agranular blood cells possessed annulate lamellae (Fig. 21).

In several cases, hemocytes were found in the process of being phagocytosed by agranular cells. Microtubules in the cytoplasm of the phagocytosed cells were clearly seen longitudinally arranged between the ingested portion of the cell and the portion not yet ingested (Fig. 22). In one instance the ingested portion of the cell appeared to be encapsulated by several layers or invaginations of the phagocytic cell (Fig. 21).

DISCUSSION AND CONCLUSIONS

Infections of oysters, *C. virginica*, by the parasite *H. nelsoni* result in clear cytopathological alterations in the gill, the site of initial invasion. Cytopathological changes, defined as obvious alterations in normal ultrastructure, became more pronounced as infections progressed from localized areas in the gill into systemic infections. Evidence of metabolic disturbance (condensed mitochondria and annulate

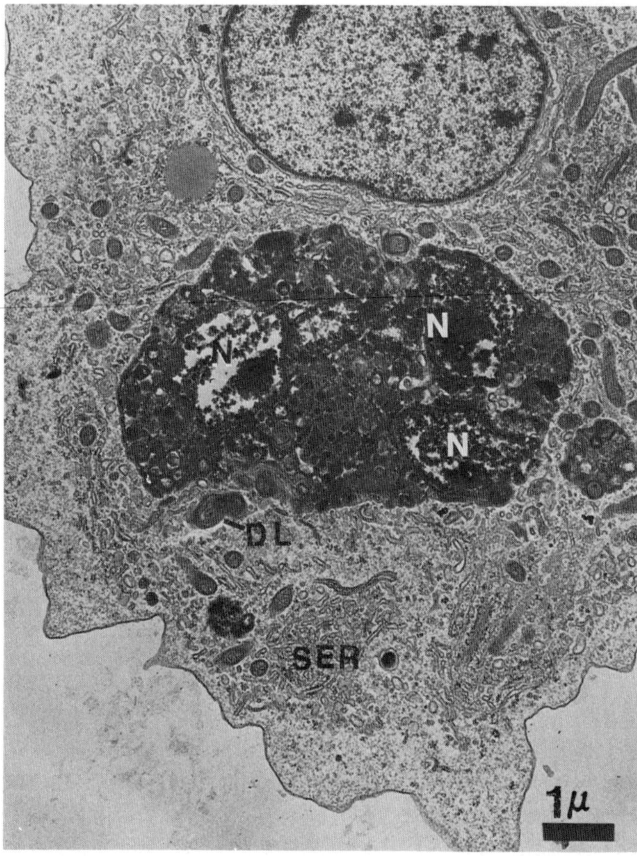

Fig. 19. Phagocytosis of MSX parasite by agranular hemocyte in the blood. Note abundance of smooth endoplasmic reticulum (SER) in the endoplasm. DL, digestive lamellae; N, MSX nuclei. 9,075X.

lamellae) preceeded physical disruption of cells. The morphological appearance of hemocytes, the primary host cellular defense agents, also changed as infections became more advanced.

The foremost difference between infected oysters and uninfected controls was the structural aberration of gill cells and supporting tissue in contact with, or in close proximity to MSX plasmodia. These findings expand the initial electron microscope studies of MSX parasitism in *C. virginica* by Muller (1967) and Perkins (1968). Though limited in his treatment of cytopathology, Perkins (1968) reported cellular deterioration in some cells of the gill epithelium and connective tissue adjacent to the pathogen. Specifically, he found signs of nuclear disruption, loss of cytoplasmic contents, and presence of membrane whorls (digestive lamellae) indicative of intracellular digestion (Gezilius, 1961; Novikoff, 1963; Cheng and Cali, 1974; Ratcliffe and Rowley, 1979). Based on this last observation, Perkins (1968) suggested that the gill epithelial cells may be capable of phagocytosis. Conversely, Yonge (1923) concluded from his original observations on the role of the ciliated epithelial cells in the digestive process of bivalves that these cells were not capable of phagocytosis.

The presence of digestive lamellae in the gill epithelial cells of oysters with localized MSX infections lends support to the view of Perkins that the gill epithelial cells may be involved in phagocytosis, although we never found parasites inside the cells. Epithelial cells of oyster gill filaments contain hydrolytic enzymes (Eble, 1966) and the presence of microvilli on the gill surface indicate an absorptive function and probable need for digestive organelles in normal cell processes. These findings indicate that, in addition to its obvious function as a physical barrier, gill epithelium has the capacity to participate in defense by absorbing and digesting parasite waste products, possibly in conjunction with the hemocytes that infiltrate infection sites (Farley, 1968; Myhre and Haskin, 1968; Ford, 1988).

Hemocytosis preceeds phagocytosis in many protozoan and metazoan parasite infections (Farley, 1968; Cheng and Rifkin, 1970; Cheng, 1981). Nevertheless, we did not observe phagocytosis of *H. nelsoni* by hemocytes in the gill epithelium, although hemocytes were present in large numbers and frequently had phagosomes containing unidentified ingested material. In contrast to the absence of phagocytosis of MSX in the gill, we did find phagocytosis of MSX plasmodia in hemolymph collected from some systemically infected oysters. In the same oysters, we also noted phagocytosis of other hemocytes, which were probably dead or damaged and no longer recognized as "self." Burreson et al. (1988) also reported finding frequent cases of phagocytosed

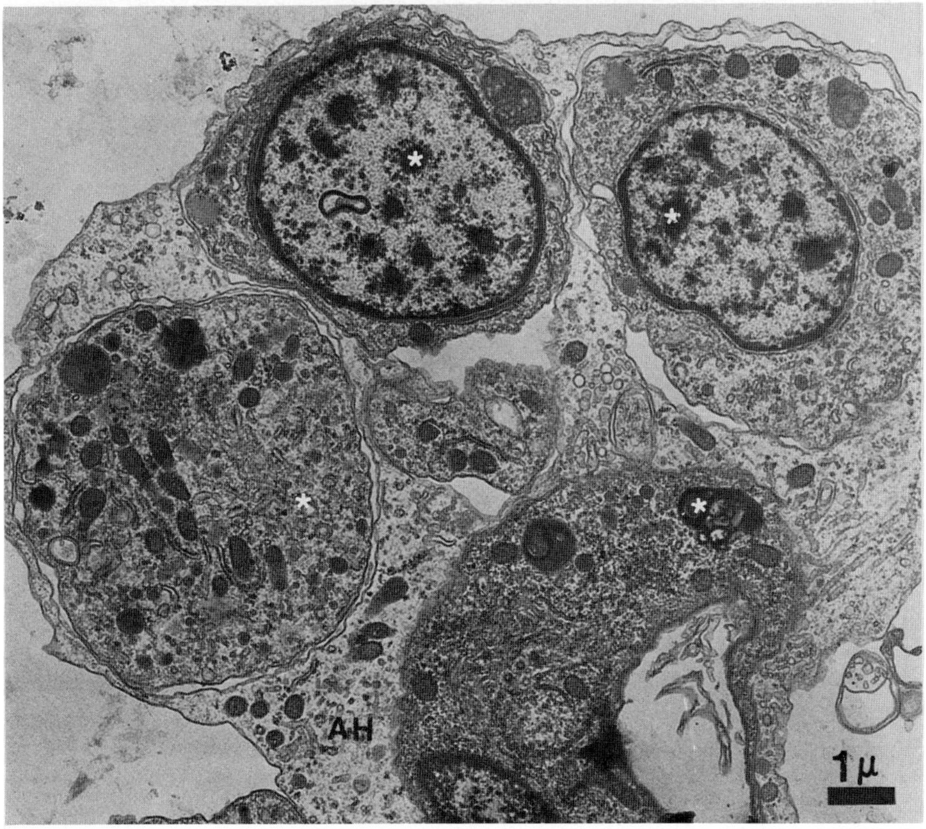

Fig. 20. Phagocytosis of several hemocytes (*) by an agranular hemocyte (AH) in the blood. 9,075X.

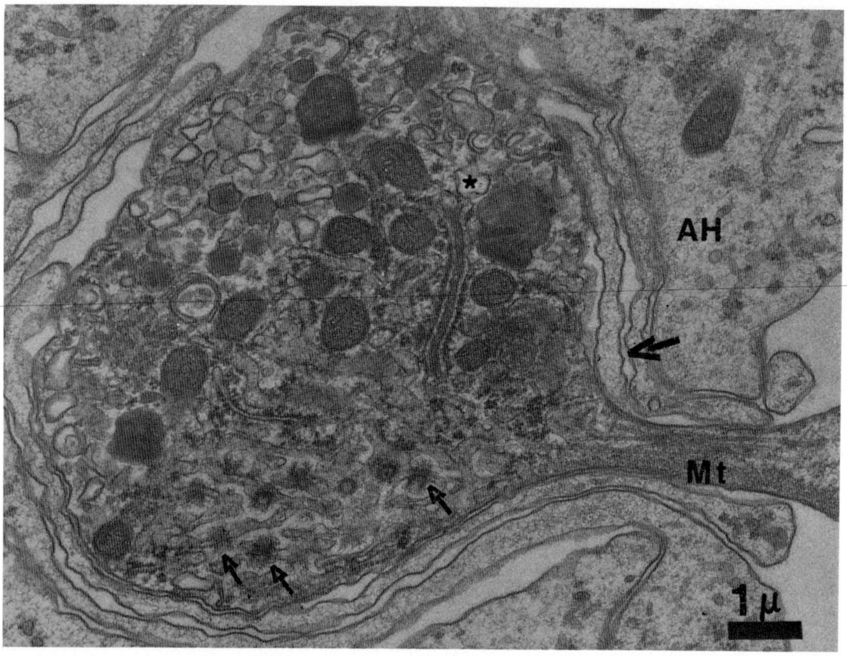

Fig. 21. Agranular hemocyte (AH) in the process of phagocytosing another hemocyte (*) in the blood. Note encapsulation (large arrow) around the ingested portion of the hemocyte, and longitudinal arrangement of microtubules (Mt) in the uningested portion. Annulate lamellae (small arrows). 21,725X.

MSX in hemolymph samples, but Ford and Kanaley (1988) found very few such cases. The irregularity of phagocytosis of *H. nelsoni* by oyster hemocytes may indicate that the former can successfully evade the host's recognition system (Smyth, 1973; Yoshino and Cheng, 1978; Lackie, 1980) until they, too, are damaged and then recognized as foreign.

Of interest in the present study was the finding that phagocytosed *H. nelsoni* and damaged host cells were found almost exclusively in agranular, rather than granular, hemocytes. In systemically infected oysters, most hemocytes were agranular, but circulating granular hemocytes, usually devoid of ingested material, were frequently encountered. The cytoplasm of phagocytic agranulocytes also contained phagosomes and digestive lamellae, indicative of intracellular digestion. The latter were morphologically similar to those found in macrophages, the vertebrate phagocytic blood cells (Novikoff, 1963).

Foley and Cheng (1975) reported that granulocytes are the most actively phagocytic blood cells in *C. virginica*. Certainly, these cells have the required cellular components to digest phagocytosed material since the granules are lysosomes that contain hydrolytic enzymes (Eble and Tripp, 1968; Feng et al., 1971; Cheng and Cali, 1974; Cheng et al., 1974; Cheng, 1975; Feng et al., 1977). Cheng and Cali (1974) and Cheng (1975) described the sequence of events following phagocytosis of bacteria by granular hemocytes of oysters. Bacteria were partially digested within primary phagosomes. The breakdown products were then taken into secondary phagosomes (the cytoplasmic "granules" seen in light and electron microscopy) and further degraded. Digestive lamellae formed around the ingested microorganisms as they were enzymatically degraded and glycogen granules appeared in the cytoplasm. From these and other observations, it has been proposed that granular and agranular hemocytes actually reflect ontogenesis or different stages during the physiologic cycle of an actively phagocytic cell. During the process of intracellular digestion or excytosis of lysozymes, granular hemocytes undergo "degranulation," hence the agranular hemocytes are degranulated granulocytes (Cheng and Foley, 1975; Foley and Cheng, 1977; Mohandus and Cheng, 1985).

Ruddell (1971) found that agranular amoebocytes in *Crassostrea gigas* were associated with intracellular digestion of endocytosed material and the presence of digestive lamellae. Similarly, light microscopical observations on the reduction in the number of cytoplasmic granules within active *M. mercenaria* granulocytes that had phagocytosed *B. megaterium* have been reported (Foley and Cheng, 1977). In fact, a close examination of the data of Foley and Cheng (1975) reveals that fibrocytes (apparently degranulated granulocytes) were associated with bacteria more frequently (56- 57%) than were fully granulated hemocytes (27-31%).

The fact that MSX parasites and damaged host cells were found exclusively in agranular hemocytes may be accounted for by rapid degranulation after phagocytosis by granulocytes; however, we did find several instances of agranular hemocytes in the act of phagocytosis (Figs. 21, 22). The fact that recognizable parasites and hemocytes were common in cells that apparently had little lysosomal enzyme content raises the possibility that the cells could not digest the material. We also failed to find glycogen granules, which would suggest recycling of ingested material.

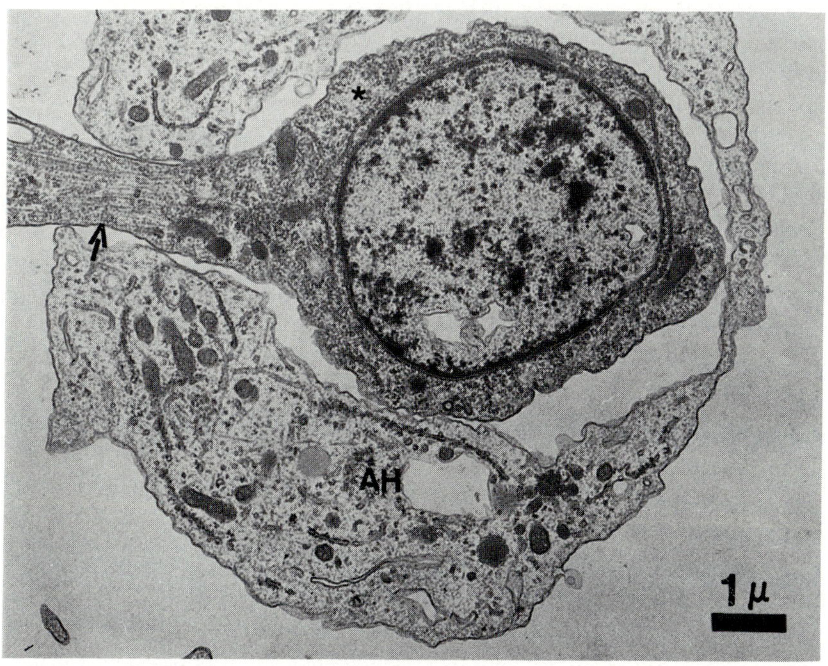

Fig. 22. Agranular hemocyte (AH) in the process of phagocytosing another hemocyte (*). Note longitudinal arrangement of microtubules in portion of hemocyte not yet ingested (arrow). 9,075X.

Douglass and Haskin (1974) and Douglass (1977), studying host-parasite interactions in MSX resistant and susceptible oysters, noted a lack of significant differences in hemocyte responses to MSX in the gill between the two populations. They suggested that hemocytes probably play a secondary role (removal of dead and damaged parasites and host cells) in the resistance mechanism, and that unknown humoral responses may control MSX lesion development in resistant oysters. The findings of the present study appear to support their conclusion, at least in the gill, even though this is the location to which parasites are restricted by mortality-resistant oysters. A humoral defense response to early MSX infections in the gill, closely tied to the hemocytic infiltration, might include cytotoxic secretions (Smyth, 1973; Bayne et al., 1980). We did not, however, find evidence of MSX degradation when they were in contact with hemocytes. Other protective functions for the hemocytes, not specifically directed against the parasite, would include removal of wastes and cellular repair (Cheng, 1981).

The presence of annulate lamellae, identical to nuclear pore complexes, in the cytoplasm of epithelial and blood cells of systemically infected oysters is probably a cellular alteration induced by MSX parasitism. Since these structures were rarely observed in noninfected or locally infected gill tissue, we presume that they are formed during the latter stages of MSX infection when parasite stress is severe. Annulate lamellae are usually found in cells undergoing high protein synthetic activity, such as oocytes and several types of tumor cells (Kessel, 1968). It is thought that these structures play a role in RNA or ribosome processing between the nucleus and cytoplasm. Increase protein synthesis in cells of *H. nelsoni*-infected oysters might produce enzymes for defense or replace substances lost through tissue damage or absorption by the parasite.

The diversity in the ultrastructure of MSX plasmodia extends previous morphological and histochemical observations made on this parasite. Muller (1967) was the first to describe numerous round osmiophilic particles, approximately 0.1-0.15 μm in diameter, in the cytoplasm of *H. nelsoni.* He thought that they resembled viruses. Perkins (1968) presented a more comprehensive and detailed account of their ultrastructure, including their origin in the membranous "spherules," and named them "haplosporosomes" (Perkins, 1971).

The function of haplosporosomes remains a mystery. Several investigators have reported that they disappear during sporulation in *H. nelsoni* (MSX) (Perkins, 1968), *H. coastale* (Perkins, 1969), *Urosporidium crescens* (Perkins, 1971) and *H. lusitanicum* (Azevedo et al., 1985). Cytochemical analysis of the haplosporosomes of *H. lusitanicum* spores (Azevedo, 1984; Azevedo and Corral, 1985) revealed that the internal membrane was partially digested with lipase and therefore contained lipid. The external membrane was strongly positive for glycoprotein, and the matrix indicated the presence of protein. It has been suggested by these investigators and others (Newman et al., 1976; Perkins, 1975) that the haplosporosomes may be used during sporulation or in spore excystment.

Perkins (1968, 1971, 1976) also found haplosporosomes outside of the parasite and postulated that these bodies may be "packets" of lytic enzymes that aid the pathogen in penetration and lysis of juxtaposed host cells. The findings, in the present study, of haplosporosomes outside the MSX plasmodium in areas of host cell and tissue degeneration may be an indication of such a function. The presence of these organelles in areas of lysis of the basal lamina, which the parasite must penetrate in order to enter the systemic circulation, is particularly intriguing. The possible relationship between haplosporosomes and cytopathology in the host deserves further study.

The peripheral orientation and structural polarity of haplosporosomes in close proximity to the plasma membrane of the MSX plasmodium may provide a clue to its function. The highly unusual positioning of this organelle seems to indicate a functional relationship associated with the cell membrane. The alignment of haplosporosomes along the cell membrane may be preparatory to their discharge outside of the cell.

Muller (unpubl.) observed that haplosporosomes tended to be located near the periphery of the cell. Azevedo et al. (1985) also described haplosporosomes located primarily at the periphery of *H. lusitanicum*. The matrix of these bodies appeared to be released into the space between the parasite and the plasmalemma of the host cell. Similar findings were reported by Perkins (1971) in the parasite, *Urosporidium crescens.*

Extreme variation in density of the plasmodial cytoplasm of certain haplosporidan parasites has been described on several occasions (Barrow, 1961; Azevedo and Corral, 1985). Azevedo and Corral (1985) observed the presence of numerous ribosomes in the cytoplasm of *H. lusitanicum* during plasmalemma genesis and offered the hypothesis that the proteins synthesized may become associated with lipids to form lipoproteins necessary for cell membrane formation. The changes in the internal structure of *H. nelsoni* most likely reflect alterations in the physiological state of the parasite. The "dense" plasmodia, rich in ribosomes and haplosporosomes and with normal appearing mitochondria, may typify healthy actively synthesizing parasites. Conversely, the "light" plasmodia, with few ribosomes and haplosporosomes and with structurally disturbed mitochondria, are perhaps degenerating cells, unable to function or develop normally (see Ford, 1988).

The cytopathological and defense-related observations reported in this study suggest that a similar study comparing MSX-mortality resistant with susceptible strains (Ford and Haskin, 1987) may clarify the mechanism that restricts parasites to the gill epithelium in resistant oysters.

ACKNOWLEDGEMENTS

We wish to thank the New Jersey Marine Sciences Consortium (Minigrant Program), the Department of Environmental Science, Rutgers University, and the Northeast Regional Aquaculture Center for helping to fund this study; Dr. Sasha Koulish (City University of New York) for use of his electron microscope facilities; Dr. Tev Airola (Rutgers University) for performing the digital image analysis. This is New Jersey Agricultural Experiment Station Publication No. F-32901-1-88, supported by state funds.

LITERATURE CITED

Azevedo, C. 1984. Ultrastructure of the spore of *Haplosporidium lusitanicum* sp. (Haplosporida, Haplosporidiidae), parasite of a marine mollusc. J. Parasit., 70, 358-371.

Azevedo, C. and L. Corral. 1985. Cytochemical analysis of the haplosporosomes and vesicle-like droplets of *Haplosporidium lusitanicum* (Haplosporida, Haplosporidiidae), parasite of *Helcion pellucidus* (Prosobrancia). J. Invert. Pathol., 46, 281-288.

Azevedo, C., L. Corral and F.O. Perkins. 1985. Ultrastructural observations of spore excystment, plasmodial development and sporoblast formation in

Haplosporidium lusitanicum (Haplosporida, Haplosporidiidae). Z. Parasitenkd., 71, 715-726.

Barrow, J.H. 1961. Observations of a haplosporidian, *Haplosporidium pickfordi* sp. nov. in fresh water snails. Trans. Amer. Microsc. Soc., 80, 319-329.

Bayne, C.J., P.M. Buckley and P.C. Dewan. 1980. Macrophagelike hemocytes of resistant *Biomphalaria glabrata* are cytotoxic for sporocysts of *Schistosoma mansoni in vitro*. J. Parasitol., 66, 413-419.

Burreson, E.M., M.E. Robinson and A. Villalba. 1988. A comparison of paraffin histology and hemolymph analysis for the diagnosis of *Haplosporidium nelsoni* (MSX) in *Crassostrea virginica* (Gmelin). J. Shellfish Res., 7, 19-23.

Cheng, T.C. 1975. Functional morphology and biochemistry of molluscan phagocytes. Ann. N.Y. Acad. Sci., 266, 343-379.

Cheng, T.C. 1981. Bivalves. In: "Invertebrate Blood Cells" (N.A. Ratcliffe, ed.), pp. 233-300. Academic Press, New York.

Cheng, T.C. and A. Cali. 1974. An electron microscope study of the fate of bacteria phagocytized by granulocytes of *Crassostrea virginica*. Contemp. Top. Immunobiol., 4, 25-35.

Cheng, T.C. and D.A. Foley. 1975. Hemolymph cells of the bivalve mollusc *Mercenaria mercenaria*: an electron microscopical study. J. Invert. Pathol., 26, 341-351.

Cheng, T.C. and E. Rifkin. 1970. Cellular reactions in marine molluscs in response to helminth parasitism. In: "A Symposium on Disease of Fishes and Shellfishes" (S.F. Snieszko, ed.), pp. 443-496. Am. Fish. Soc., Washington, D.C.

Cheng, T.C., A. Cali and D.A. Foley. 1974. Cellular reactions in marine pelecypods as a factor influencing endosymbiosis. In: "Symbiosis In the Sea"(W.B. Vernberg, ed.), pp. 61-91. Univ. South Carolina Press, Columbia.

Douglass, W.R. 1977. *Minchinia nelsoni* disease development, host defense reactions, and hemolymph alterations in stocks of oysters (*Crassostrea virginica*) resistant and susceptible to *Minchinia nelsoni*-caused mortality. Ph.D. Dissertation. Rutgers University, New Brunswick, New Jersey.

Douglass, W.R. and H.H. Haskin. 1974. MSX oyster interactions: Leucocyte responses to *Minchinia nelsoni* disease in *Crassostrea virginica*. Proc. Natl. Shellfish. Assoc., 65, 2.

Eble, A.F. 1966. Some observations on the seasonal distribution of selected enzymes in the American oyster as revealed by enzyme histochemistry. Proc. Natl. Shellfish. Assoc., 56, 37-42.

Eble, A.F. and M.R. Tripp. 1968. Enzyme histochemistry of phagosomes in oyster leucocytes. Bull. N.J. Acad. Sci., 13, 93.

Farley, A.C. 1968. *Minchinia nelsoni* (Haplosporida) disease syndrome in the American oyster *Crassostrea virginica*. Protozool., 15, 585-599.

Feng, S.Y., J.S. Feng, C.N. Burke and L.H. Khairallah. 1971. Light and electron microscopy of the leucocytes of *Crassostrea virginica* (Mollusca:Pelecypoda). Z. Zellforsch., 120, 222-245.

Feng, S.Y., J.S. Feng and T. Yamasu. 1977. Roles of *Mytilus coruscus* and *Crassostrea gigas* blood cells in defense and nutrition. Comp. Pathobiol., 3, 31-67.

Foley, D.A. and T.C. Cheng. 1975. A quantitative study of phagocytosis by hemolymph cells of the pelecypods *Crassostrea virginica* and *Mercenaria mercenaria*. J. Invert. Pathol., 25, 189-197.

Foley, D.A. and T.C. Cheng. 1977. Degranulation and other changes of molluscan granulocytes associated with phagocytosis. J. Invert. Pathol., 29, 321-325.

Ford, S.E. 1988. Host-parasite interactions in Eastern oysters selected for resistance to *Haplosporidium nelsoni* (MSX) disease: Survival mechanisms against a natural pathogen. Am. Fish. Soc. Spec. Publ. 18, 206-224.

Ford, S.E. and H.H. Haskin. 1982. History and epizootiology of *Haplosporidium nelsoni* (MSX), an oyster pathogen in Delaware Bay, 1957-1980. J. Invert. Pathol., 40, 118-141.

Ford, S.E. and H.H. Haskin. 1987. Infection and mortality patterns in strains of oysters *Crassostrea virginica* selected for resistance to the parasite *Haplosporidium nelsoni* (MSX). J. Parasit., 73, 368-376.

Ford, S.E. and S.A. Kanaley. 1988. An evaluation of hemolymph diagnosis for the detection of the oyster parasite *Haplosporidium nelsoni* (MSX). J. Shellfish Res., 7, 11-18.

Gezilius, K. 1961. Further studies in the ultrastructure of *Ascariae.* Exp. Cell Res., 23, 300-310.

Haskin, H.H., L.A. Stauber and J.A. Mackin. 1966. *Minchinia nelsoni* n. sp. (Haplosporida, Haplosporidiidae): Causative agent of the Delaware Bay oyster epizootic. Sci., 153, 1414-1416.

Kessel, R.G. 1968. Annulate lamellae. J. Ultrastr. Res., (Suppl.) 10, 1-82.

Koulish, S. 1976. Organization of special parenchymal cells underlying the midgut in some barnacles. J. Exp. Mar. Biol. Ecol., 23, 155-170.

Lackie, A.M. 1980. Invertebrate immunity. Parasitology, 80, 393-412.

Mackin, J.G. 1951. Histopathology of infection of *Crassostrea virginica* (Gmelin) by *Dermocystidium marinum*, Mackin, Owen and Collier. Bull. Mar. Sci. Gulf and Carrib., 1, 72-87.

Mohandus, A. and T.C. Cheng. 1985. An electron microscope study of the structure of lysosomes released from *Mercenaria mercenaria* granulocytes. J. Invert. Pathol., 46, 332-334.

Muller, E. 1967. Electron microscope studies of *Minchinia nelsoni*. Proc. Natl. Shellfish. Assoc., 57, 3.

Myhre, J. 1973. Levels of infection in spat of *Crassostrea virginica* and mehanisms of resistance to the haplosporidian parasite *Minchinia nelsoni*. M.S. Thesis, Rutgers University. New Brunswick, New Jersey.

Myhre, J.L. and H.H. Haskin. 1968. Some observations on the development of early *Minchinia nelsoni* infections in *Crassostrea virginica* and some aspects of the host-parasite relationship. Proc. Natl. Shellfish. Assoc., 58, 7.

Newman, M.W., C.A. Johnson and G.B. Pauley. 1976. *Minchinia*-like haplosporidian parasitizing blue crabs. J. Invert. Pathol., 27, 311-315.

Novikoff, A.B. 1963. Lysosomes in the physiology and pathology of cells: Contributions of staining methods. In: "Lysosomes" (A.V.S. de Reuck and M.P. Cameron, eds.). Little, Brown. Boston, Massachusetts.

Perkins, F.O. 1968. Fine structure of the oyster pathogen *Minchinia nelsoni* (Haplosporida, Haplosporidiidae). J. Invert. Pathol., 10, 287-305.

Perkins, F.O. 1969. Electron microscope studies of sporulation in the oyster pathogen, *Minchinia coastalis* (Sporozoa:Haplospoida). J. Parasitol., 55, 897-920.

Perkins, F.O. 1971. Sporulation in the trematode hyperparasite *Urosporidium crescens* De Turk, 1940 (Haplosporida, Haplosporidiidae)- an electron microscope study. J. Parasitol., 57, 9-23.

Perkins, F.O. 1975. Fine structure of the haplosporidian Kernstab, a persistent intranuclear mitotic apparatus. J. Cell Sci., 18, 327-346.

Perkins, F.O. 1976. Ultrastructure of sporulation in the European flat oyster pathogen, *Marteilia refringens*-Taxonomic implications. J. Protozool., 23, 64-74.

Ratcliffe, N.A. and A.F. Rowley. 1979. Role of hemocytes in defense against biological agents. In: "Insect Hemocytes" (A.P. Gupta, ed.), pp. 331-415. Cambridge Univ. Press, London.

Rosenfield, A., L. Buchanan and G.B. Chapman. 1969. Comparison of the fine structure of spores of three species of *Minchinia* (Haplosporida, Haplosporidiidae) J. Parasitol., 55, 921-941.

Ruddell, C.L. 1971. Elucidation of the nature and function of the granular oyster amebocytes through histochemical studies of normal and traumatized oyster tissues. Histochemie, 26, 98-112.

Smyth, J.D. 1973. Some interface phenomena in parasitic protozoa and platyhelminths. Canad. J. Zool., 51, 367-376.

Yonge, C.M. 1923. Studies on the comparative physiology of digestion. I. The mechanism of feeding, digestion, and assimilation in the lamellibranch *Mya*. British J. Exp. Biol., 1, 15-63.

Yoshino, T.P. and T.C. Cheng. 1978. Snail host-like antigens associated with the surface membranes of *Schistosoma mansoni* miracidia. J. Parasitol., 64, 752-754.

PROTOZOAN DISEASES: CONCLUDING REMARKS

Christian P. Vivarès

Laboratoire de Pathologie Comparée, USTL (URA INRA-CNRS no. 11840)
34060 Montpellier Cedex, France

Although substantial knowledge of protozoan parasites in feral marine animals, ranging from harmless commensals to lethal parasites, already exists, limited consideration has been given to their role in mariculture. Protozoans always constitute a potential threat that mariculture may easily change the natural host-parasite balance. The number of genera of protozoans recorded from marine animals is increasing, and as mariculture is still in its infancy, new species will be introduced. The high prevalences of protozoans in mariculture does not simply mean high mortality rates, but infections can also lead to retarded growth and limited reproduction; this reduces productivity.

Although it may be difficult, every effort should be made to fulfill Koch's postulate to define the disease condition and to demonstrate the role of the etiologic agent.

Dependent on their relationships to the host, Protozoa colonizing marine animals can be assigned to different categories. The first includes free-living ciliates that attack injured or stressed fishes, crustaceans like *Paranophrys carcini* in hemolymph of *Cancer pagurus*, and diatoms in gill of shrimp *Penaeus* spp. Therefore, it is necessary to determine the effects of abiotic factors on the host. The second category consists of ecto- and endocommensals that normally show no clear indication of real pathogenicity and are not known to derive any nutrient from their hosts at any time. The third category includes those protozoans that are on the borderline between commensalism and parasitism, and are potential, sometimes very serious pathogens such as Trichodinids, *Hexamita, Spironucleus,* and paramoebic gill parasites. The infectious population of amoebae may be either local (with increased virulence and decreased host resistance) or exotic (for example, introduced by movements of warm water masses), depending on whether the amoeba persists in the local environment or is sporadically re-established in the susceptible host population.

There are many problems to be considered about these opportunistic pathogens. First, we must study the protozoan *in vitro* to establish possible relationships and the effect of temperature and other physical parameters on the growth of *Paramoeba*. We must also study its morphologic differentiation correlated with the attachment and synthesis of proteases that enable it to attack the tissues of Salmonid fishes. We must study different species of amoebae in the cellular biology field, principally in receptors and ligands, intracellular Ca^{++} concentration, and Ca^{++} flow.

Another aspect is the investigation of the immune response of the fish. Infection with *Labyrinthuloides haliotidis* is lethal to almost all juvenile abalone less than six months old in abalone mariculture. It is capable of using living oyster tissues as a source of nutrients for growth and multiplication. Its artificial culture is possible allowing for the study of its physiology and transmission.

The fourth and largest group includes a wide variety of true parasites depending exclusively on nutrients derived from their hosts. Some of them are well adapted parasites and cause their hosts practically no harm at all. In others, the pathogenicity depends largely on the outcome of the dynamic interaction of the parasite's virulence and the defense mechanisms of its host. Some parasites, notably tissue invading ones (e.g., microsporidans, coccidians), easily establish an infection. Others, like *Amyloodinium* and *Cryptocaryon,* may cause a noteworthy infection only in adverse conditions for the host. Concerning *Amyloodinium*, during the last few years many reports of mortalities in different major species of fishes in mariculture have been made. Some very important contributions were published about its propagation using the host fish or the fish cell line method. Dr. Noga's system may be a useful model to study both the biology and control of protozoan skin parasites of fishes and for quantitatively studying host-parasite interaction at cellular interfaces. This approach could be realized with *Cryptocaryon* from which information has already been obtained on the role of some enzymes that penetrate the host tissues.

The relationships between parasites, their life cycle, and the environment must be studied. An important problem is cross-infection, principally in extensive aquaculture of *Perkinsus* in Gastropod and Bivalve hosts and of *Ichthyobodo* in Cephalopods and Fishes. More experiments are necessary to understand transmission and develop prophylaxis. On the other hand, stress has a strong influence on the health of reared hosts. For example, the causal relationships between specific or general stress conditions and growth, reproduction, invasiveness, or virulence of specific parasitic ciliates of crustaceans requires further investigation. Among endoparasites, some experiments provide evidence that the combined effects of *Trypanosoma* and chronic oil pollution cannot only cause mortality, but can also jeopardize health and reproduction in surviving fish. The quality of water must be evaluated and be under permanent control with automated computing systems.

Infections with endoparasitic protozoans appear generally more slowly; they depend on passive or vector transmission and need a longer time to manifest themselves. Many microsporidian species cause conspicuous lesions in marine fish, destroying musculature and presenting a threat to fish farming. In the chinook salmon, an intracellular microsporidium was described from hematoblastic cells. This infection was associated with an acute anemia in the fish. Often, the transmission has not been very clearly demonstrated. It is necessary to study the biochemistry of these parasites that are among the oldest eucaryotic animals on Earth and to develop therapeutic measures. On the other hand, an understanding of the close metabolic relationships between microsporidians and their hosts must be developed. The evaluation of pathogenic effects is also necessary in coccidian research. The full life cycle and mode of transmission of myxosporeans require clarification before adequate measures to control pathogenic species can be applied. In particular, the Wolf-Markiw hypothesis on the transmission of *Myxobolus cerebralis* requires further study *in vivo* and in cell line culture. Another problem concerns PKD in salmonids. Kent and Hedrick showed that the causative agent, PKX, is a myxosporidan. The infection of *Salmo gairdneri* with spores of *Sphaerospora* spp. suggests that they are the most mature stage of PKX yet observed. For mariculture, it is important to note that the chinook salmon transferred to full strength sea water continue to show PKX cells and typical PKD. The most beneficial aspect of early transfers is to stop complications from bacterial and protozoan infections, which may prevent mortalities associated with PKD.

For protists parasitizing molluscs, some fundamental information is given by electron microscopy in the study of ultrastructure and stages of the life cycle. New parasites of *Ruditapes* and *Crassostrea virginica* presented in this meeting must progress quickly to define a precise taxonomic position of the parasites. Other methods of investigation can be used to analyze *in vitro* the life cycles and the transmission of *Perkinsus* in different species of molluscs. Since it is possible to obtain many parasites, the preparation of DNA permits a molecular taxonomic study. On the other hand, the study of host metabolism and reproduction must be correlated with the pathogenicity of the parasite; a good model is developed with *Haplosporidium* and *Crassostrea*. To understand the metabolism of one parasite, it is necessary to isolate it as with *Martelia* and *Bonamia* and to develop cell culture of molluscs. To help mariculturists quickly identify protozoans, new diagnostic methods, like enzyme immunoassays with monoclonal antibodies, must be applied.

Concerning protozoan diseases in crustaceans, experimental demonstration of the cause and effect relationships between single protozoan species and mortality of Crustacea is needed. Reproducible demonstration of cause and effects, or the lack thereof, is necessary in order to place proper significance on the roles played by various protozoans in the ecology and biology of crustaceans. A good model was developed by Meyer's group with bitter crab disease, which is due to a fatal dinoflagellate infection. The processed meat from the infected Alaskan Tanner crabs is unmarketable. The *in vitro* studies and experimental pathogenesis of this chronic but lethal disease gave, for the first time, important information about a parasitic dinoflagellate disease in invertebrates. Also, little is known concerning protozoan diseases of larval and planktonic crustaceans.

A last point to discuss is the relationship between mariculture and human health. For example, dinoflagellates are known to produce toxins that can cause many digestive perturbations, like ciguaterra from fish infected with *Gambierdiscus*, and PSP (paralytic shellfish poisoning) from molluscs infected with *Dinophysis*. Enzyme immunoassays are already commercially available, but continued investigations of cytotoxicology in cell line culture and *in vivo* must be planned.

Finally, general formulae for the cure of endoparasitic protozoans have to be proposed. Curative baths for ectoparasites are not as easy in salt water as in freshwater. Thus, prevention is presently the most feasible way of control. Controlling environmental parameters can often be a means to prevent disease without the use of chemicals. The improvement of healthy animals in mariculture depends on an increased knowledge of marine animal protozoans and their biology with special attention to their metabolism, which will benefit both mariculture and fundamental research.

METAZOAN PARASITIC DISEASES SECTION

METAZOAN PARASITIC DISEASES: INTRODUCTORY REMARKS

Robin M. Overstreet

Gulf Coast Research Laboratory, P.O. Box 7000,
Ocean Springs, Mississippi 39564-7000, USA

INTRODUCTION

Metazoan parasitic diseases have considerable impact on marine fisheries and aquaculture. Protozoan diseases of finfishes and shellfishes often attract more attention than those caused by metazoans because of the common association of several protozoans with mortalities in aquaculture and the frequent absence of associated knowledge about their life cycles. Metazoan species also cause mortalities in both wild and cultured hosts, but mortality is only one of several ways that metazoans impact natural and cultured stocks. Infections that cause mortalities in wild fishes are often overlooked because the hosts are preyed upon before or soon after they die. Infections with the same metazoan agents in culture systems, however, usually attract immediate attention and are quickly controlled because the organisms can be readily seen and diagnosed, the life histories can be determined or assumed within a relatively short period of time, and the cycles can be interrupted or altered. This relative ease in recognizing metazoan agents also makes metazoan infections especially valuable in interpreting the condition of the environment, the host, and the parasite as well as in assessing potential or actual public health concerns and in understanding host-parasite biology.

ASSOCIATIONS, DISEASE, AND MORTALITY

The strict definition of a "parasite" is inconsequential to most aquaculturists. In fact, from a culturist's point of view, there is little difference among a predator, a fouling organism, a "commensal," and a parasite, if the association adversely affects the stock. As examples, hydroids that have no symbiotic relationship with crustaceans prey on their eggs and larvae in raceways and aquaria, and the polyclad turbellarian *Stylocus ellipticus* preys on the American oyster, *Crassostrea virginica,* especially when barnacles are not available to it. Oysters and other bivalves are also especially vulnerable to a variety of fouling agents and commensal symbionts (e.g., Overstreet,

Pathology in Marine Science

1978). Controlling these and other organisms from harming oysters, other bivalves, crustaceans, or fishes presents the same difficulties for culturists as controlling parasites.

As a part of this symposium, several investigators have reported on metazoan parasites and the diseases, especially pathological alterations, associated with them. Such reports are important to aquaculturists since, unlike medical and veterinary parasitology, the parasitology of marine animals maintained in culture is essentially in its infancy. Any new information contributed represents a step forward toward the control of deleterious parasites and the cure of the diseases associated with them.

INTRODUCED HOSTS AND PARASITES

When either hosts or parasites are introduced into new environments, including culture facilities, pathological responses more intense than expected may develop. In many cases, the introductions initiate poorly adapted host-parasite relationships, resulting in subsequent diseases or epizootics.

When hosts are introduced, they can acquire parasites already established in the habitat. For example, when a stock of the blue mussel, *Mytilus edulis,* was introduced into a farm in Brittany, France, it acquired a significantly higher prevalence of infection with *P. squamatus* than the naturally settling stock of *M. galloprovincialis* (possibly a hybrid produced by that species and *M. edulis*) (Coustau et al., this volume).

A similar intensified host response can develop when the parasite is introduced into a new geographic area. When the swimbladder dracunculoid *Anguillicola crassus* was introduced from China or Japan into Northwestern and Central Europe (Moravec and Taraschewski, 1988), it spread to native eels in the Netherlands where it has apparently caused considerable mortality (van Banning and Haenen, this volume). In its natural host, *Anguilla japonica,* in Japan, two individuals of this blood-feeding nematode typically occupied the eel swimbladder with little associated pathologic response. In contrast, when in either wild or pond-reared *A. anguilla*, the relatively large nematode occurred in abundance in the swimbladder, causing a fibrotic condition and a nonfunctional organ in the smaller survivors (van Banning and Haenen, this volume).

Entire natural host populations can be severely impacted by an introduced parasite (e.g., Van Engel et al., 1966; Andrews, 1980). In the early 1960s, the protozoan *Haplosporidium nelsoni* drastically reduced the oyster population in the lower Chesapeake Bay. At that time, live oysters were transported from the Gulf of Mexico to Virginia where oystermen shucked them, discarding the shells and waste near native oyster beds. Within a year or two, the rhizocephalan barnacle *Loxothylacus panopaei*, previously unknown from the Chesapeake region, was observed there for the first time in the two dominant mud crabs, *Eurypanopeus depressus* and *Rithropanopeus harrisi.* This common castrating parasite from the Gulf of Mexico drastically reduced the populations of those two crab species, primary scavengers of dead oysters, allowing the population of rhizocephalan-resistant *Neopanope texana sayi* to replace them. This altered equilibrium of mud crabs continues to persist today (Andrews, pers. commun.).

When metazoans become established in confined hosts, the potential for disease is increased. This potential exists because of the increased probability for contact between the host and infective agent and because of the host's physically and behaviorally stressed condition. Woods et al. (this volume) presents the leech *Myzobdella lugubris* on *Morone saxatilis* as a good example of an introduction of a metazoan into a confined system. That leech feeds on the blood of fishes, seldom causing a disorder in fish in their natural habitat, optimally a shallow, brackish area (preferably < 15‰) with heavy vegetation. After obtaining blood meals, the leech dislodges, attaches in small groups onto the blue crab, *Callinectes sapidus,* or less frequently on some other invertebrate or hard substratum, and at least on the rear margin of the blue crab's carapace deposits cocoons. No data are available showing that the young leech has a preference for specific fishes. Data, however, suggest that *Ictalurus catus, Mugil cephalus,* and *Fundulus majalis* are the most common hosts in South Carolina, Georgia, and Virginia and that *Paralichthys lethostigma, M. cephalus,* and sometimes *F. grandis* are the most common ones in Mississippi (Sawyer et al., 1975). Perhaps this specificity relates to availability of the fish to the leech rather than selectivity by the leech. In any event, when the leech was available to *M. saxatilis* in intensive culture, infestations caused disease, commonly surpassing 100 and occasionally 300 individuals per fish (Woods et al., this volume).

Careful scrutiny into potential parasitological conditions may not disclose all problems. The study by McGladdery et al. (this volume) provides an excellent example of an unexpected outcome. It reports how metacercariae of *Stephanostomum tenue,* which in the geographic region of their study infected the peritoneal lining of *F. heteroclitus,* also infected rainbow trout in cages. Cercariae released from the mud dog whelk, *Nassarius obsoletus,* localized in the pericardial sac, causing fibrosis and ultimately death in additionally stressed fish. No planning could have predicted this consequence involving a parasite with no apparent previous economic, host health, or human health implications.

In summary, regarding mortalities from metazoan symbionts in aquaculture, the risk, as pointed out by Paperna (1987),is related to the type of facilities and the culture practices. If the agents can be restricted from the system, they cannot accumulate in the host or become established in the facility.

DISSIMILAR EFFECTS FROM RELATED PARASITES

Culturists or other biologists should not necessarily expect members of a specific taxonomic group to have a specific serious effect on a host just because a related parasite severely affected its host. Conversely, the fact that a parasite does not cause a serious pathologic response or condition in one host does not preclude that parasite from having an adverse effect in another host species or even in the same host species from a different age group or under different conditions. These expectations should be considered, especially since the reader already has seen from the above examples that even the same parasite can elicit different responses in different hosts.

Crustaceans serve as hosts for a variety of symbiotic metazoan groups, and members of some of these groups significantly impact stocks (Overstreet, 1983, 1987). Rhizocephalan barnacles serve as examples that cause different effects on their normal hosts, and one member of this influential group, *Loxothylacus panopaei,* has already been discussed above.

Briarosaccus callosus, which infected king crabs in Alaska, caused gonadal sterilization with reduced growth in male blue and golden king crabs, apparently reducing their stocks (Meyers et al., this volume). In other cases, such as infections of *Sacculina granifera* in the commercial portunid sand crab, *Portunus pelagicus,* in Morton Bay, Queensland, Australia, growth was apparently not inhibited (Phillips and Cannon, 1978), even though infections with this castrating barnacle also appeared to have a pronounced effect on the abundance of that crab stock (Bishop and Cannon, 1979).

PUBLIC HEALTH IMPLICATIONS

Whereas the contributions in this volume do not stress parasites that affect human health, several metazoan parasites from aquaculture systems do infect man. When additional animal species are cultured, when the same animals are cultured year after year, and when culturing operations develop in different geographic regions, additional parasites will become established and infect man. As human populations spread around the globe, as people experience different cuisines incorporating raw products or products prepared using different methods that allow infections or contamination by toxic products from parasites, and as more people become immunologically compromised, more people will come into contact and get infected with metazoan and other parasites.

Ponds favor establishment of certain snails and other intermediate hosts, and they also yield the cultured product as an available source of prey, especially for potential avian and mammalian definitive hosts. For example, pond-reared crustaceans are vulnerable to infections with metacercariae of microphallid digeneans (e.g., Heard and Overstreet, 1983), and pond-reared fishes are vulnerable to infections of metacercariae of heterophyids (e.g., Paperna and Overstreet, 1981). A variety of birds and mammals serve as hosts for many members of both of those families, completing the digenean cycles by feeding on the cultured product and simultaneously releasing feces with infective larvae available to the snails. Other digeneans like *Nanophyetus salmincola*, which infects several salmonids as metacercariae, appear to be spreading among human populations (Eastburn et al., 1987). Several other groups of parasites that do or could infect man can also be acquired by cultured animals and transmitted to man in inadequately prepared products. One more example is the juvenile stage of *Eustrongylides* sp. Unlike most of the digeneans indicated above, this nematode is not very specific to its intermediate cultured host and infects a wide variety of cold-blooded vertebrates. It is acquired when those hosts feed on oligochaetes infected from larvae in eggs deposited through birds, such as specific herons. Whereas the worm can penetrate into the peritoneal cavity in man (Guerin et al., 1982), it can also directly affect cultured fish hosts as well as initiate secondary bacterial infections in those fish hosts (e.g., Chung et al., 1985; pers. observ.).

BIOLOGICAL INDICATORS

Also not treated in this volume is the importance of metazoan symbionts as indicators of host biology, parasite biology, and environmental conditions. Pathological response to a metazoan can indicate health state of the host or of the system as well as provide a means to identify the parasite. The presence or absence of specific metazoan symbionts in the natural ecosystem can provide information on pollution or the lack of it, on feeding behavior of the host, and on migration of the host. Similar information, except for that involving migration, also could be obtained in cage-cultured hosts. Perhaps, some of these aspects will be treated in future PAMAQ colloquia.

ACKNOWLEDGEMENT

This study was conducted in cooperation with the U.S. Department of Agriculture, CSRS Grant No. 88-38808-3319 and the U.S. Department of Commerce, NOAA, NMFS, Pub.L.99-659, Project No. 2-IJ-3.

LITERATURE CITED

Andrews, J.D. 1980. A review of introductions of exotic oysters and biological planning for new importations. In: "Marine Fisheries Review" (W. Hobart, ed.), pp. 1-11. National Oceanic and Atmospheric Administration, Seattle, Washington.

Bishop, R.K. and L.R.G. Cannon. 1979. Morbid behaviour of the commercial sand crab, *Portunus pelagicus* (L.), parasitized by *Sacculina granifera* Boschma, 1973 (Cirripedia: Rhizocephala). J. Fish Dis., 2, 131-144.

Chung, H.Y., Y.M. Shen, I.H. Lin and G.H. Kou. 1985. Larval *Eustrongylides* (Nematoda: Dioctophymatidae) occurred in rice-field eels (*Fluta alba*). COA Fish. Ser. No.4, Fish Dis. Res. (VII), 38-44.

Eastburn, R.L., T.R. Fritsche and C.A. Terhune, Jr. 1987. Human intestinal infection with *Nanophyetus salmincola* from salmonid fishes. Am. J. Trop. Med. Hyg., 36, 586-591.

Guerin, P.F., S. Marapendi, L. McGrail, C.L. Moravec and E.L. Schiller. 1982. Intestinal perforation caused by larval *Eustrongylides*-Maryland. Cen. Dis. Con., Morb. Mort. Wk. Rept., 31(23), 383-389.

Heard, R.W. and R.M. Overstreet. 1983. Taxonomy and life histories of two North American species of "*Carneophallus*" (Microphallus) (Digenea: Microphallidae). Proc. Helminthol. Soc. Wash., 50, 170-174.

Moravec, F. and H. Taraschewski. 1988. Revision of the genus *Anguillicola* Yamaguti, 1935 (Nematoda: Anguillicolidae) of the swimbladder of eels, including descriptions of two new species, *A. novaezelandiae* sp. n. and *A. papernai* sp.n.. Fol. Parasitol., 35, 125-146.

Overstreet, R.M. 1978. "Marine Maladies? Worms, Germs, and Other Symbionts from the Northern Gulf of Mexico." Mississippi-Alabama Sea Grant Consortium, Ocean Springs, Mississippi.

Overstreet, R.M. 1983. Metazoan symbionts of crustaceans. In: "The Biology of Crustacea, Vol. 6, Pathobiology" (A. J. Provensano, Jr., ed.), pp. 155- 250. Academic Press, New York.

Overstreet, R.M. 1987. Solving parasite-related problems in cultured Crustacea. Int. J. Parasit., 17, 309-318.

Paperna, I. 1987. Solving parasite-related problems in cultured marine fish. Int. J. Parasit., 17, 327-336.

Paperna, I. and R.M. Overstreet. 1981. Parasites and diseases of mullets (Mugilidae). In: "Aquaculture of Grey Mullets" (O. H. Oren, ed.), pp. 411- 493. International Biological Programme 26, Cambridge Univ. Press, London.

Phillips, W.J. and L.R.G. Cannon. 1978. Ecological observations on the commercial sand crab, *Portunus pelagius* (L.), and its parasite, *Sacculina granifera* Boschma, 1973 (Cirripedia: Rhizocephala). J. Fish Dis., 1, 137-149.

Sawyer, R.T., A.R. Lawler and R.M. Overstreet. 1975. Marine leeches of the eastern United States and the Gulf of Mexico with a key to the species. J. Nat. Hist., 9, 633-667.

Van Engel, W.A., W.A. Dillon, D. Zwerner and D. Eldridge. 1966. *Loxothylacus panopaei* (Cirripedia, Sacculinidae) an introduced parasite on a xanthid crab in Chesapeake Bay, U.S.A. Crustaceana, 10, 111-112.

SYLON (CRUSTACEA: RHIZOCEPHALA) INFECTIONS ON THE SHRIMP IN BRITISH COLUMBIA

Susan M. Bower and James A. Boutillier

Department of Fisheries and Oceans, Biological Sciences Branch,
Pacific Biological Station, Nanaimo, British Columbia V9R 5K6 Canada

ABSTRACT: A species of the rhizocephalan *Sylon* has been found on several species of shrimp from the coastal waters of British Columbia, Canada. The morphology and development of this parasite within crustacean hosts and certain pathologic features resulting from the host-parasite association are reported.

INTRODUCTION

Rhizocephalans are distant relatives of acorn and goose barnacles that parasitize other crustaceans. In this group of parasites is a member of the genus *Sylon,* which parasitizes a variety of species of shrimp. To date, there is only a single recognized species in this genus, *Sylon hippolytes*, and it is thought to occur circumpolar in the northern hemisphere.

The most detailed description of the life cycle of *Sylon* is that by Lutzen (1981). However, even in this excellent contribution there are still a number of speculations. This presentation not only corroborates Lutzen's speculations but also more fully elucidates some aspects of the life cycle of *Sylon*. In addition, the unusual distribution of *Sylon* on shrimp in British Columbia (B.C.) is presented and discussed. The information included herein comes from a series of collection and assessment surveys carried out along the B.C. coast in 1986 through 1988 (Table 1).

OBSERVATIONS AND COMMENTS

The infective stage of *Sylon* is the cyprid larva (Fig. 1). This is the only stage of the animal that has any hint of resemblance to nonparasitic barnacles. The cyprid larvae are approximately 170 μm long. This stage has been observed to swim rapidly with a flicking action of the posterior appendages and to walk on solid surfaces using its antennal appendages with their suction cup-like extremities. In the case of *Sylon*,

Pathology in Marine Science

Table 1. Prevalence[1] of *Sylon* on shrimp (*Pandalus* and *Spirontocaris*) from various locations[2] in British Columbia observed during collection and assessment surveys in 1986 through 1988.

		Shrimp species examined					
Locality	Date	*P. platyceros*	*P. jordani*	*P. tridens*	*P. danae*	*P. goniurus*	*S. holmesi*
7	06/86	677/2735 (81)	1/5	0/114	–[3]	–	–
7	01/87	382/1421	–	10/167	–	–	–
7	06/87	365/2534	–	0/21	–	–	–
7	08/88	256/1007 (14)	0/2	1/130	–	–	–
8	06/87	840/2554	0/119	0/29	–	–	–
8	08/88	251/976 (2)	0/1	0/8	–	–	–
9	06/87	410/2736	0/20	0/27	–	–	–
9	08/88	366/4325 (23)	–	0/195	–	–	–
10	08/88	1/2456	–	0/43	–	–	–
17	10/87	0/9	0/121	–	–	–	0/174
18	10/87	0/1	0/1401	–	17/511	9/500	0/83
19	10/87	–	–	–	1/578	–	–
28	03/87	0/12731	–	–	–	–	–
28	05/87	0/10723	–	–	–	–	–
28	10/87	0/45	9/754	–	–	–	1/54

[1]Prevalence indicated by number of infected shrimp/total examined followed by number of *Sylon* externae examined histologically in parenthesis where applicable.
[2]Localities are indicated as Fisheries Statistical Areas illustrated in Fig. 10.
[3]Not collected.

(continued)

Table 1 (continued)

		Shrimp species examined					
Locality	Date	*P. platyceros*	*P. jordani*	*P. tridens*	*P. danae*	*P. goniurus*	*S. holmesi*
28	11/87	0/9261	–	–	–	–	–
28	03/88	0/7332	–	–	–	–	–
23	06/86	0/1408	–	–	–	–	–
23	10/86	0/1572	–	–	–	–	–
23	12/86	0/1573	–	–	–	–	–
23	08/87	0/1373	–	–	–	–	–
23	05/88	0/1090	–	–	–	–	–
29	10/87	–	–	–	–	–	16/226
124	10/88	–	31/7720 (62)[4]	–	–	–	–

[4]The prevalence data was obtained from a random subsample of the total catch. *Sylon* samples for histological examination were selected from the remainder of the catch.

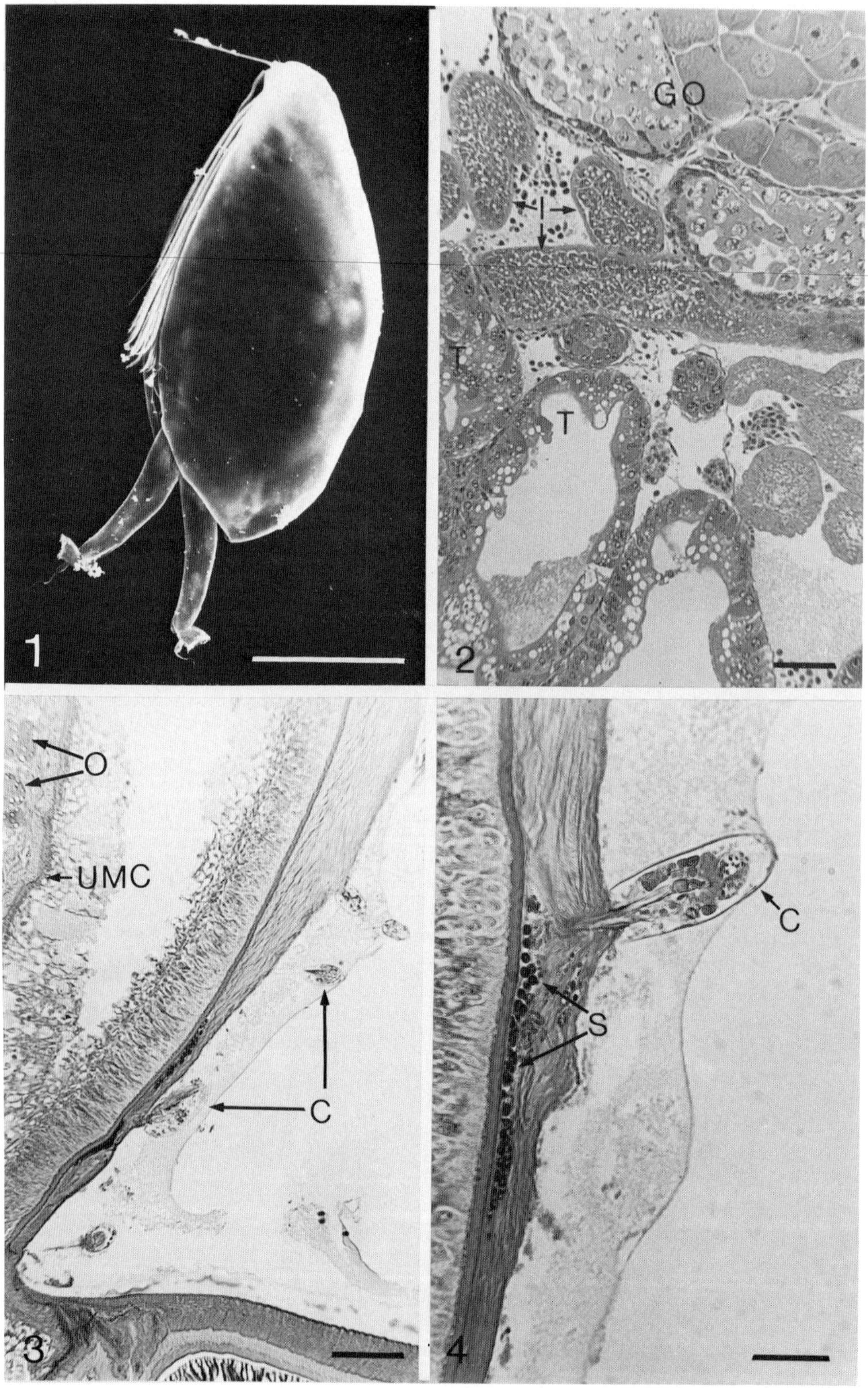
1
GO
I
T
T
2
O
UMC
C
3
C
S
4

the site and method of infection are not known. However, in another rhizocephalan, *Lernaeodiscus porcellanae,* the female cyprid attaches to the gill of the crab host and molts to a permanently attached kentrogon stage (Ritchie and Hoeg, 1981; Hoeg, 1985). Within 60 hr from the time the female cyprid attached to the gill of the crab, the kentrogon will inject a single invasion cell into its host. In *Sylon*, as with other rhizocephalans, the invasion cell (or cells) develops an extensive root system (interna), which penetrates throughout the cephalothorax of the host, even between the organs such as the gonad and digestive gland tubules, and there is virtually no inflammatory response by the host to their presence (Fig. 2). However, as Lutzen (1981) indicated, the internae never seem to invade the tissues of the organs they surround. Eventually the interna extend into the abdominal region of the host between the ventral nerve cord and ventral epidermis. The primordial externa or female gonad of the parasite is usually formed against the cuticle at the posterior edge of the mid-ventral surface of the first pleuron. Yolk deposition begins in the ova within the ovary, a process that may be assisted by a large nurse cell located adjacent to each developing ova. During the next molt of the host, the externa, which is still unfertilized, will emerge.

Some of our work has elucidated the process of fertilization of the *Sylon* ova, which to date had only been speculated upon. The male cyprids attach to the flaccid cuticle of virgin externa with their antennal appendages (Fig. 3). The spermatogonia are inoculated through the antennal appendages into the mantle wall (Fig. 4). The use of antennae in reproduction is unusual for crustaceans. Spermatogonia apparently then migrate through the loose connective tissue and vascular sinuses to the point of attachment of the ovary to mantle wall. After fertilization, the externa will molt, leaving a more compact cuticle. As the externa grows with increased yolk deposition within the ova, the compact cuticle will molt several times and muscular tissue will develop within the mantle wall. In the fertilized externa, a cluster of cuboidal cells is observed at the point of attachment of the ovary and mantle wall (Fig. 5). Boschma (1928) and Lutzen (1981) have designated this structure as the "ganglion." However, the similarity between the cuboidal cells within this structure

---

Fig. 1. Scanning electron micrograph of a cyprid larva from the mantle cavity of a gravid *Sylon* parasitizing *Spirontocaris holmesi*. Bar = 50 μm.

Fig. 2. Histological section through branches of the interna (I) of *Sylon* between the digestive gland tubules (T) and gonad (GO) within the cephalothorax of *Pandalus platyceros*. H&E; Bar = 100 μm.

Fig. 3. Histological section through cyprid larvae (C) adjacent and attached to the flaccid cuticle of a *Sylon* externa via the antennal appendages. This newly emerged externa contains one ovary full of yolk-laden ova (O) and an undeveloped mantle cavity (UMC). H&E; Bar = 100 μm.

Fig. 4. Male cyprid larva (C) with antenna embedded in the flaccid cuticle of *Sylon* externa and in the process of inoculating spermatogonia (S). However, in this instance, the externae may have been previously inoculated with spermatogonia. This would explain the presence of a newly forming compact layer of cuticle below the flaccid cuticle. H&E; Bar = 50 μm.

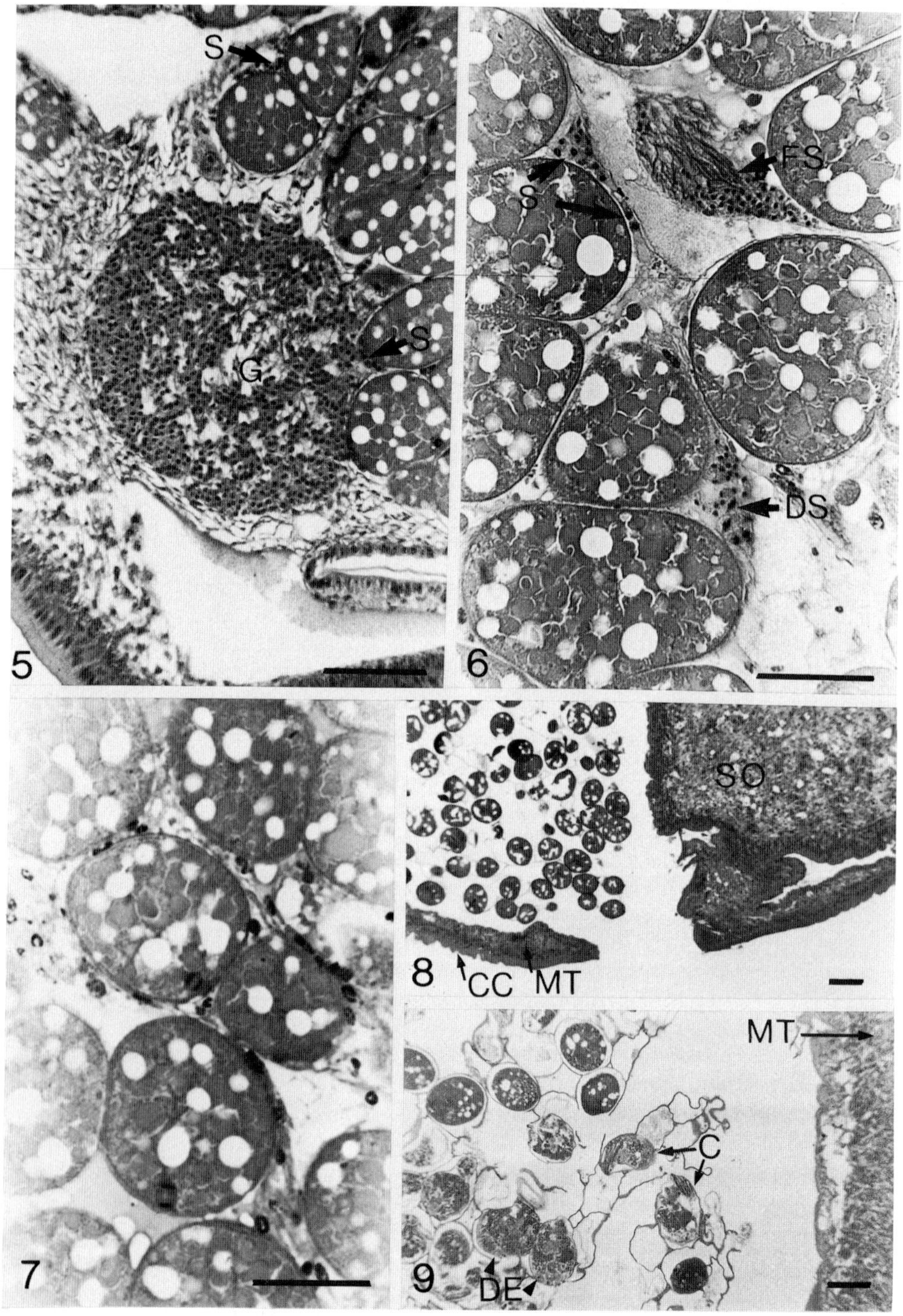
S
S
G
5
FS
S
DS
6
7
SO
8
CC
MT
MT
C
9
DE

and the spermatogonia is noteworthy. Regardless of the identity of this structure, the spermatogonia migrate throughout the ovary where they undergo rapid division, develop into flagellated sperm (Fig. 6), and fertilize the ova. Concurrent with the migration and multiplication of spermatogonia in the ovary, the nurse cells atrophy and appear to disintegrate, leaving only a few connective tissue cells and developing spermatogonia between each ova. Some externae have reached this stage of development without the presence of spermatogonia suggesting that parthenogenesis may occur in some specimens.

The early stages of division of fertilized ova, at least to anaphase, occurs within the ovary (Fig. 7). The mantle wall and the ovary, which had been previously coalescent at the thin cuticular lining of each (UMC, Fig. 3), by this time has split to form the mantle cavity. The fertilized ova are then released into the mantle cavity through the colleteric gland located at the end of the ovary opposite to the point of attachment to the mantle wall.

Soon after the eggs are deposited into the mantle cavity, two pores open in the mantle wall, one on either side of the attachment site of the now spent ovary (Fig. 8). The eggs remain in the mantle cavity embedded in a gelatinous matrix where they will develop directly into cyprid larvae (Fig. 9). The larvae hatch and emerge from the externa through the two pores. This is unlike most other Rhizocephala and nonparasitic barnacles, which undergo a nauplius larval stage prior to the cyprid stage. The spent externa shrivels, and is sloughed off the shrimp leaving a black scar. The host then mounts a massive inflammatory response to the parasite interna, which become melanized and are visible as brown or black lines in the tissues of the prawn.

Many of our findings are similar to those of Lutzen (1981) in terms of the relationship between the parasite and the host. By collecting shrimp of various ages and sex, we have been able to corroborate the observation that all ages and sex of shrimp seem equally susceptible to infection. The majority of the shrimp

---

Fig. 5. Cluster of cuboidal cells (G) at the attachment site of the ovary to the mantle wall of a fertilized externa of *Sylon*, labelled as "ganglion" by other workers but containing cells similar in morphology to the spermatogonia (S). H&E; Bar = 100 μm.

Fig. 6. Spermatogonia (S) undergoing division (DS) and maturation into flagellated sperm (FS) within the ovary and between ova of *Sylon*. H&E; Bar = 50 μm.

Fig. 7. Nuclei of two fertilized ova within the ovary of a *Sylon* externa at the anaphase stage of division. H&E; Bar = 50 μm.

Fig. 8. Histological section through pore in mantle wall located near the attachment site of the spent ovary (SO) of a *Sylon* externa. Note the compact cuticle (CC) on the external surface and the muscular tissue (MT) within the mantle wall, and developing eggs in the mantle cavity. H&E; Bar = 100 μm.

Fig. 9. Histological section through developing eggs (DE) and unhatched cyprid larvae (C) within the mantle cavity of *Sylon* externa. Contractions of the fully developed muscular tissue (MT) in the mantle wall aids in voiding the hatched cyprid larvae from the mantle cavity. H&E; Bar = 100 μm.

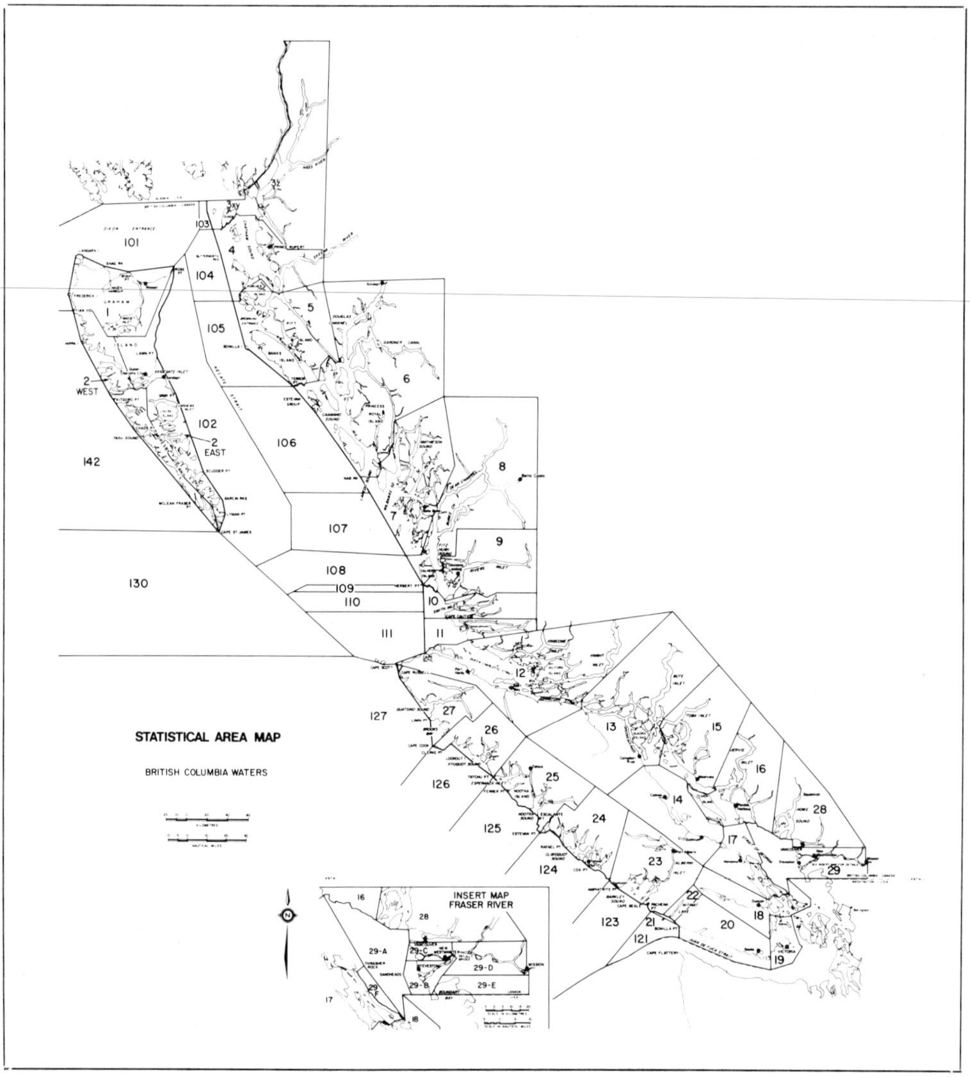

Fig. 10. Map of coastal British Columbia indicating Fisheries Statistical Areas. Table 1 lists prevalence of *Sylon* infection and the numbers and species of shrimp sampled during the 1986 through 1988 surveys.

(*Spirontocaris lilljeborgi*) that Lutzen (1981) examined were female. From his observations, Lutzen speculated that the state of the ovary of the host at the time of infection is largely preserved, but that a considerable degeneration of the ova takes place in ovaries that have reached an advanced stage of ripening. Thus, shrimp parasitized with *Sylon* as a rule are sterilized. Only one shrimp (*Dichelopandalus bonnieri*) carrying numerous embryos of its own has been recorded as being parasitized with *Sylon* (Hoeg and Lutzen, 1985). The same speculation seems to hold for the majority of shrimp species examined during our investigations. However, we have found nine female *Pandalus platyceros* that were berried and infected with *Sylon* at all stages of development. The large size of female *P. platyceros* may account for their ability to remain fecund in some instances. We are currently conducting a more complete analysis of the data and material to examine the ability of *Sylon* to castrate its host.

There does not appear to be any seasonal relationship with the prevalence and stage of infection (Table 1). All stages, from that of early infections (only internae present) to that with shrivelled cyprid-releasing externae, are found in approximately equal prevalence year round. However, the distribution of *Sylon* on the various species of shrimp in B.C. is unusual. *Sylon* was observed in *P. platyceros* at all coastal localities examined in northern B.C. with over 32% infected in some areas. However, none of 47,118 *P. platyceros* from southern B.C. were found infected (Table 1, Fig. 10). Nevertheless, *Sylon* have been observed on other species of shrimp along the entire coast of B.C. (Table 1). In fact, the observation of *Sylon* on *Pandalus goniurus* (statistical area 18, Fig. 10) constitutes a new host record. In an attempt to explain this unusual distribution, we have examined the parasite from the various host species in order to try and identify different species of *Sylon*.

The parasitic stage of *Sylon* on its host has no hard body parts and is pleomorphic with no characters of taxonomic significance for species differentiation. We have examined the superficial ultrastructure of cyprid larvae from four species of hosts (*P. platyceros, Pandalus jordani, Pandalus tridens,* and *Spirontocaris holmesi*) using scanning electron microscopy and have found no differences in the cyprids. Investigations using DNA analysis are currently in progress. Hopefully this technique will give us some insight into *Sylon* species differentiation. However, without sound taxonomic evidence that more than one species of *Sylon* occurs in B.C., we will not be able to support our suspicion that the disjunct distribution of *Sylon* is attributable to species differences and we must look elsewhere for an explanation.

LITERATURE CITED

Boschma, H. 1928. Rhizocephala of the north Atlantic region. Dan. Ingolf Exp., 3, 1-49.

Hoeg, J.T. 1985. Cypris settlement, kentrogon formation and host invasion in the parasitic barnacle *Lernaeodiscus porcellanae* (Muller) (Crustacea:Cirripedia: Rhizocephala). Acta Zool., 66, 1-45.

Hoeg, J. and J. Lutzen. 1985. Crustacea Rhizocephala – Marine invertebrates of Scandinavia, No. 6. Norwegian University Press, Oslo.

Lutzen, J. 1981. Observations on the rhizocephalan barnacle *Sylon hippolytes* M. Sars parasitic on the prawn *Spirontocaris lilljeborgi* (Danielssen). J. Exp. Biol. Ecol., 50, 231-245.

Ritchie, L.W. and J.T. Hoeg. 1981. The life history of *Lernaeodiscus porcellanae* (Cirripedia:Rhizocephala) and co-evolution with its porcellanid host. J. Crust. Biol., 1, 334-347.

INFESTATION OF STRIPED BASS, *MORONE SAXATILIS*, BY THE LEECH *MYZOBDELLA LUGUBRIS*

L.C. Woods, III,[1] M.A. McCarthy,[1] J.N. Kraeuter,[2] and D.R. Sager[3]

[1]Baltimore Gas and Electric, Crane Aquaculture Facility, P.O. Box 1475, Baltimore, Maryland 21203, USA
[2]Rutgers University, Shellfish Research Laboratory, P.O. Box 687, Port Norris, New Jersey 08349, USA
[3]Texas Parks and Wildlife, 4200 Smith School Road, Austin, Texas 78744, USA

ABSTRACT: The estuarine leech *Myzobdella lugubris* has been reported to parasitize a number of fish and invertebrate species (Sawyer et al., 1975). Infestations of striped bass reported in the literature are rare, are usually on fish two years or older, and typically number one leech per fish (Paperna and Zwerner, 1976). We noted heavy leech infestations (81 to 377 per fish) on two year old striped bass reared under intensive culture conditions at the Crane Aquaculture Facility.

INTRODUCTION

Many of the piscicolid leeches belong to the genus *Myzobdella* (see Meyer, 1966). *Myzobdella lugubris* (Fig. 1) is a species that occurs throughout fresh and brackish waters of the United States. It is parasitic on teleosts or commensal on crustaceans and was the first fish leech reported to utilize both a fish and an arthropod in its life history (Daniels and Sawyer, 1975). It appears to parasitize fish for most of the year and then leaves the fish to deposit cocoons during the warmer summer months. *M. lugubris* is reported most frequently on certain brackish water fishes, especially *Mugil cephalus* and *Fundulus* spp., but is known frequently to parasitize a number of freshwater fishes as well, especially species of the genera *Ictalurus* and *Lepomis* (Sawyer et al., 1975). Striped bass 2 years old or older have been parasitized by *M. lugubris* but only rarely, and if so, in low numbers (Paperna and Zwerner, 1976). We observed in November, 1985, a heavy *M. lugubris* infestation (Fig. 2) of 2-year-old striped bass reared at our intensive culture facility. These striped bass are maintained for broodfish research in a 10,000 gal, 20 ft diameter tank receiving oligohaline water from Chesapeake Bay in a single pass open system. An inflow to the tank of 350 gal/min provides two water changeovers per hour. We were

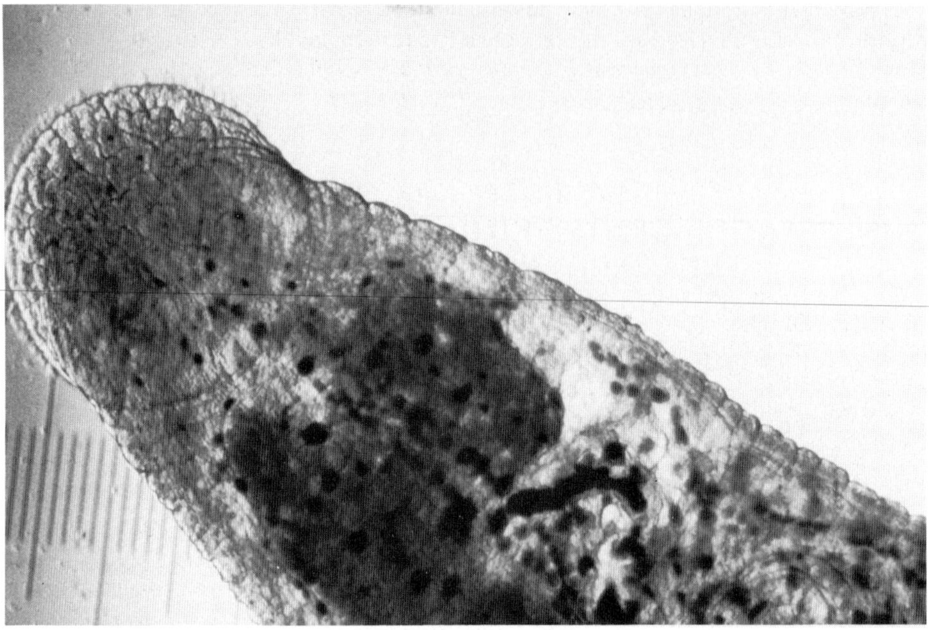

Fig. 1. The piscicolid leech *Myzobdella lugubris.*

Fig. 2. *Myzobdella lugubris* infestation of 2-year-old striped bass.

concerned with the number of leeches attached, far greater than had previously been reported for any species, and the broodfish tissue damaged by *M. lugubris*. This paper describes our attempts to quantify the degree and sites of infestation and means by which the leech can be removed.

MATERIALS AND METHODS

Fifty 2-year-old striped bass were randomly selected from the tank with the leech infestation, anesthetized, weighed to the nearest gram and measured (total length) to the nearest millimeter. Leeches were picked and enumerated from 20 of these fish. Five additional fish were selected for detailed enumeration, which included leech attachment location. The remaining 25 parasitized fish were held for leech control experiments.

Three chemicals were examined for their effects on *M. lugubris*: formalin (37% aqueous formaldehyde); potassium permanganate (99% $KMnO_4$); and Dylox, 80% active ingredient (0,0-dimethyl-1,2,2,2-trichloro-l-hydroxyethyl phosphate). All three were recommended by Herwig (1979) for the control of various leeches on fish. Two preliminary *in vitro* tests with formalin and Dylox were conducted, followed by three *in vivo* tests: the first compared Dylox and formalin, the second compared three different levels of Dylox only, and the final test compared Dylox and potassium permanganate.

In vitro **tests.** For *in vitro* tests, leeches were carefully removed from parasitized fish. In the first test, 80 leeches were utilized, 10 leeches placed in each of eight petri dishes as follows: four Dylox dosage levels (1,5,10,20 ppm); one formalin dosage level (250 ppm); and three control replicates. Leeches were given a 1-hr exposure bath. Each therapeutic agent was added to Chesapeake Bay water held at room temperature (20°C ± 0.5°). After the bath treatment, the leeches were moved to new petri dishes containing room temperature bay water. Leeches were observed for 24 hr post-treatment for signs of distress (detachment from surface of petri dish or rolling up into tight balls) or mortality. In the second *in vitro* test, 360 leeches were used, 10 leeches per petri dish, and 36 petri dishes as follows: 12 control replicates; 12 Dylox dishes, four dosage levels (5,10,15,20 ppm) with three replicates of each dosage; and 12 formalin dishes, four dosage levels (175,200,225,250 ppm) with three replicates of each dosage. Again, 1-hr exposure bath at room temperature was followed by transfer to petri dishes with room temperature bay water. Leeches were evaluated for 24-hr post-treatment for signs of distress or mortality.

In vivo **tests.** In all *in vivo* experiments, therapeutic agents were compared using infested (>100 leeches per individual) striped bass. In each experiment, 27 parasitized fish were utilized, three fish were placed into each of nine 6-ft diameter tanks. Each tank then represented one replicate, with three replicates assigned to each therapeutant tested. The control for all *in vivo* experiments was the population ($n = 300$) of parasitized, untreated striped bass remaining in our larger, 20-ft diameter broodfish research tank.

In the first *in vivo* test, there were two Dylox treatments with dosage levels of 5 and 20 ppm, and one formalin treatment of 250 ppm. After a 1-hr static bath in the

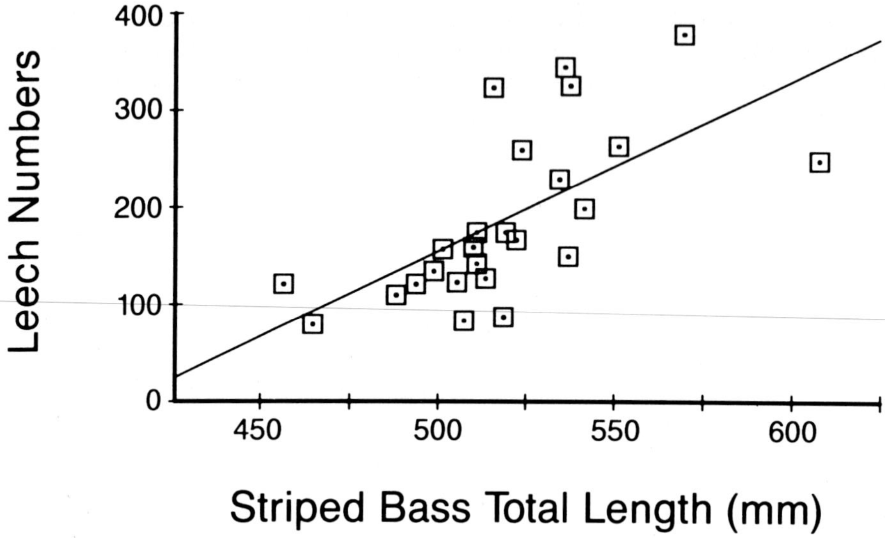

Fig. 3. The relationship between number of leeches and striped bass length.

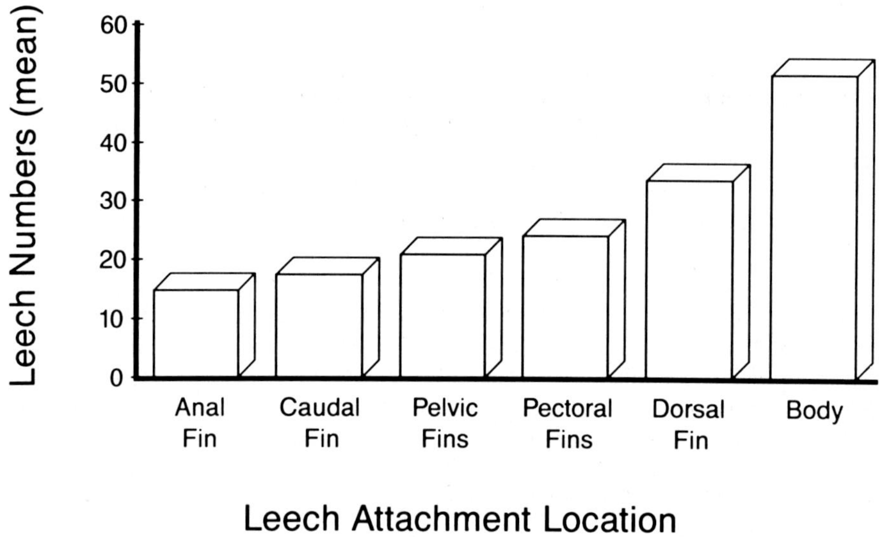

Fig. 4. The number of leeches attached to striped bass body and fin groups.

6-ft tanks, water was turned back on and the chemicals were flushed from the tank. The inflow water from Chesapeake Bay had a temperature of 6 ± 0.5°C. All fish were examined 24 hr after the bath treatment for the presence of leeches.

In the second *in vivo* experiment, formalin was dropped from further consideration. Three Dylox treatments were compared at dosages of 20, 30, and 40 ppm. In addition, the duration of the exposure bath was prolonged from 1 hr to 3 hr. Bay water temperature during the test was 6 ± 0.5°C. All striped bass were examined 24 hr after treatment for the presence of leeches.

In the third and final *in vivo* test, there were two Dylox treatments of 40 ppm, at 11.5°C and 13.5°C, compared with 10 ppm potassium permanganate at 11.5°C. Exposure baths were 3 hr in duration, followed 24 hr later by examinations of the fish to determine the number of leeches remaining attached.

RESULTS AND DISCUSSION

Two-year-old striped bass at our intensive culture facility had an average weight of 1531 g (range 799 to 2397) and length of 511 mm (range 435 to 608). The 25 fish selected for leech counts had a mean weight and length of 1593 g and 518 mm, respectively. With 4654 leeches removed from these 25 fish, the mean number of leeches per fish was 186. There was a relationship between number of leeches attached and the length of the striped bass (Fig. 3). The linear regression of these two variables is described by [$y = -713.1883 + 1.7353x$], with an r value of 0.64. A pattern of leech attachment location (Fig. 4) was observed on the body and fin groups. Excluding the body and the caudal fin, which for the opportunistic *M. lugubris* is difficult to adhere or attach to, we plotted the relationship (Fig. 5) between mean leech number attached to a fin group and the measured surface area (cm^2) of the fin group. The r value for this relationship was 0.97 and the equation is [$y = 5.8126 + 0.285x$].

In vitro **tests.** In the first *in vitro* test, 24 hr post-exposure bath, the leeches were clearly distressed by 10 and 20 ppm Dylox and 250 ppm formalin. Most leeches were found detached and balled-up in these particular treatments. No significant mortality was noted. In the second *in vitro* experiment, 24 hr post-exposure bath, neither Dylox nor formalin was effective in killing leeches; however, Dylox at 15 and 20 ppm and formalin at 250 ppm was effective in causing detachment from the petri dishes.

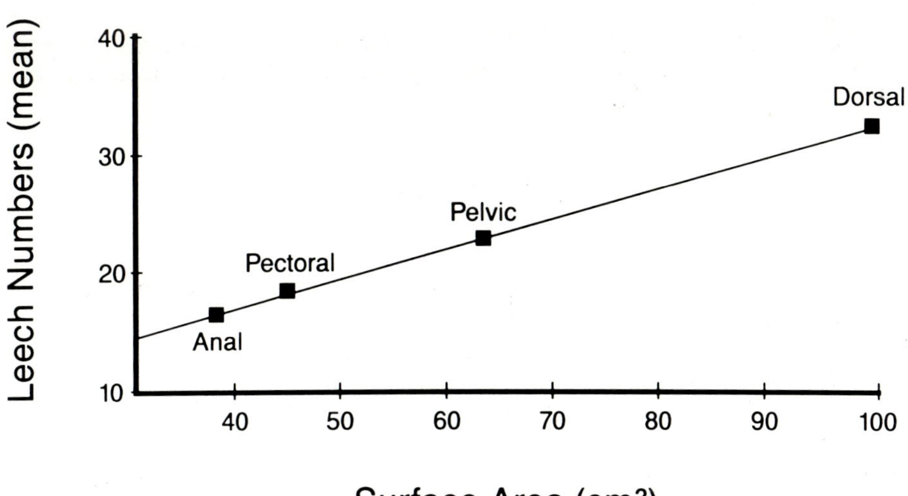

Fig. 5. The relationship between number of leeches and available surface area of fin groups.

Limited mortality did occur at 250 ppm formalin (< 2%) and at the two highest Dylox concentrations (7%).

In vivo **tests.** Heavy leech infestations persisted for 24 hr after the first *in vivo* test with no observed stress to the attached leeches or reduction in number of leeches on the striped bass. Water temperatures were much cooler in the fish tanks receiving Chesapeake Bay water directly (6°C) than in the Petri dishes held at room temperature *in vitro* (20°C). The second *in vivo* experiment, also conducted in 6°C, showed similar results to the first, with no apparent reduction in the number of *M. lugubris* attached to the striped bass. Statistically significant results were obtained in the third *in vivo* test where water temperatures were elevated above 10°C. No reduction in attached leech numbers were observed in any of the three potassium permanganate replicates (mean number leeches remaining = 189) even with the relatively high permanganate level (10 ppm), which caused the death of one fish. The Dylox treatments, on the other hand, were effective at causing nearly total detachment by *M. lugubris* from our 2-year-old striped bass (mean number leeches remaining for both treatments was < 1). Of the 18 parasitized striped bass used in the two Dylox treatments, 10 had no leeches remaining.

M. lugubris has been reported to parasitize more than two dozen species of wild fish in fresh and brackish waters. However, we found no reports of *M. lugubris* parasitism for any species of the magnitude we observed, and we report the first incidence of *M. lugubris* parasitizing cultured striped bass. Adult *M. lugubris* are very poor swimmers (Sawyer et al., 1975) and appear to utilize available surface area of body and especially the fins for sites of attachment. To date, and since 1985, only striped bass 2-years-old or older have been found with *M. lugubris* attached. For control of *M. lugubris*, we found 40 ppm Dylox with temperatures above 10°C to be quite effective.

ACKNOWLEDGEMENTS

We are grateful to Donald Klemm for the positive identification of *Myzobdella lugubris*. We also thank John Woiwode for his assistance with figures and statistical analysis as well as his critical review of the paper.

LITERATURE CITED

Daniels, B.A. and R.T. Sawyer. 1975. The biology of the leech *Myzobdella lugubris* infesting blue crabs and catfish. Biol. Bull., 148, 193-198.

Herwig, N. 1979. "Handbook of Drugs and Chemicals Used in the Treatment of Fish Diseases." Charles C. Thomas, Springfield, Illinois.

Meyer, F.P. 1966. Parasites of freshwater fishes. U.S. Fish and Wildlife Service, Leaflet FDL-5.

Paperna, I. and D.E. Zwerner. 1976. Parasites and diseases of striped bass, *Morone saxatilis* (Walbaum), from the lower Chesapeake Bay. J. Fish. Biol., 9, 267-281.

Sawyer, R.T., A.R. Lawler and R.M. Overstreet. 1975. Marine leeches of the eastern United States and Gulf of Mexico with a key to the species. J. Nat. Hist., 9, 633-667.

MICROCOTYLE CHRYSOPHRII (MONOGENEA: POLYOPISTHOCOTYLEA), A PATHOGEN FOR CULTURED AND WILD GILTHEAD SEABREAM, *SPARUS AURATA*

M. Faisal[1*] and E.A. Imam[2]

[1]Department of Avian and Aquatic Animal Medicine, Faculty of Veterinary Medicine, University of Alexandria, Egypt
[2]Faculty of Veterinary Medicine, Cairo University, Cairo, Egypt

ABSTRACT: Hyperinfestation by the gill monogenean *Microcotyle chrysophrii* caused mass mortality in juveniles of the gilthead seabream *Sparus aurata* cultured in holding tanks in Alexandria, Egypt. Most infested fish showed respiratory manifestations, swam near the water surface, and sometimes tried to jump out of the water.

In the same geographical area, *M. chrysophrii* was found parasitizing the wild population of *S. aurata* in the Mediterranean Sea throughout the year with a maximum prevalence of 80% and mean intensity of 30 ± 8 in spring. When the intensity exceeded 10 parasites/fish, the condition of the host was adversely affected. This might be attributed to the hematophagous feeding nature of *M. chrysophrii* as well as the impairment of gill respiration and excretory functions by the relatively large polyopisthocotylean. Whether this monogenean adversely affects feral populations is not known.

*Present address: Department of Microbiology and Immunology (Factor 12-538), School of Medicine, University of California at Los Angeles, Los Angeles, California 90024, USA

INTRODUCTION

Reports of mortalities caused by monogeneans have increased with the expansion of mariculture projects (Rohde, 1984). Two cases of heavy mortalities caused by monogeneans have been described in cultured gilthead seabream, *Sparus aurata*. The first was reported by Paperna et al. (1977) as well as Paperna and Baudouy-Laurencin (1979) by the monopisthocotylea *Furnestinia echensis* in tank-cultured *S. aurata*. The second one, caused by the polyopisthocotylean *Polylabris tubicirrus* in *S. aurata* reared in raceways, was reported by Silan et al. (1985).

In the present study, we report a third case of heavy mortalities (75) in tank-reared juvenile *S. aurata* caused by the monogenean *Microcotyle chrysophrii* (Polyopisthocotylea). The same trematode exerts adverse effects on the condition factor of infested wild *S. aurata*.

MATERIALS AND METHODS

Samples of *S. aurata* were collected at monthly intervals from the Mediterranean Sea at Alexandria, Egypt, for 1 year (January-December, 1986). Moreover, during March, 1987, severe respiratory manifestations associated with 78% mortality were observed among 1.5 year old *S. aurata* (240 $\pm$ 15 g; 22.5 $\pm$ 7.4 cm) in El-Nozha fish farm (Alexandria). The fish were reared in circular tanks (3 m^3) receiving salty ground water (36% salinity, 10 mg O_2/l). Moribund fish swam near the water surface with widely opened mouths and accelerated opercular movement. Some fish tried to jump out of the water. Two hundred gill smears of moribund fish revealed hyperinfestation with monogeneans.

Identification of the parasite was accomplished on fresh compressed individuals (65 worms) from the gills according to the techniques described elsewhere (Bullock and Bresch, 1985). The compressed worms were fixed in glycerin alcohol (1:4) and examined microscopically. Illustration of the parasite was drawn with the aid of a camera lucida.

Infested fish, free from any other internal or external parasites, were divided according to the number of worms/fish into four groups (1-10, 11-20, 21-40, > 40 worms/fish). The condition factor for each fish was calculated according to the formula:

$$\text{condition factor} = \frac{\text{weight (g)}}{L^3\ \text{(cm)}}$$

The means and standard deviation of each group (27 fish each) were calculated and compared with those of noninfested fish using student's t-test.

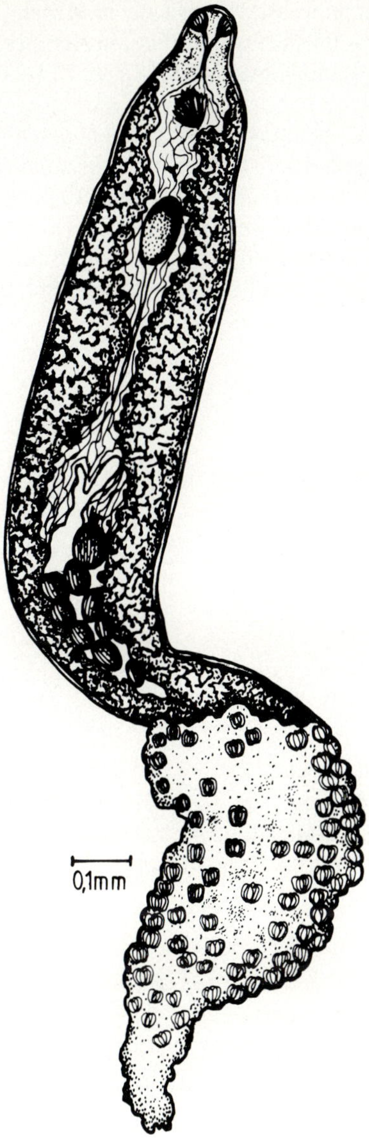

Fig. 1. *Microcotyle chrysophrii* from the gills of *Sparus aurata*.

RESULTS

Gill smears of *S. aurata* showed white to yellowish actively motile, relatively large (1-4 mm long) trematodes. The morphological characteristics of this trematode (Fig. 1) resemble completely those of the monogenean *Microcotyle chrysophrii*.

M. chrysophrii was found on the gills of *S. aurata* caught from the Mediterranean Sea at Alexandria during the entire year. The number of infected fish, as well as the number of parasites per fish, increased drastically in March (80%, 30 worms/fish), maintained high levels (57.1-60%, 6.4 worms/fish) in summer, then decreased in fall (27-35%, 4.4-6.2 worms/fish) and winter (42.9-48.8%, 12.1-21.2 worms/fish) (Table 1, Fig. 2).

Comparison of the condition factors of different groups of *S. aurata* infested with different numbers of parasites/fish against *M. chrysophrii*-free fish, revealed a significant suppression in those harboring more than 10 parasites/fish ($p < 0.01$). *S. aurata* with < 10 *M. chrysophrii*/fish were not affected (Fig. 3). During March, 1987, the parasite induced a 75% mortality (6.5 ton loss) of juvenile *S. aurata* that were reared in tanks. *M. chrysophrii* could be seen in large numbers (100-780 worms/fish) on the gill wet smears of moribund *S. aurata*.

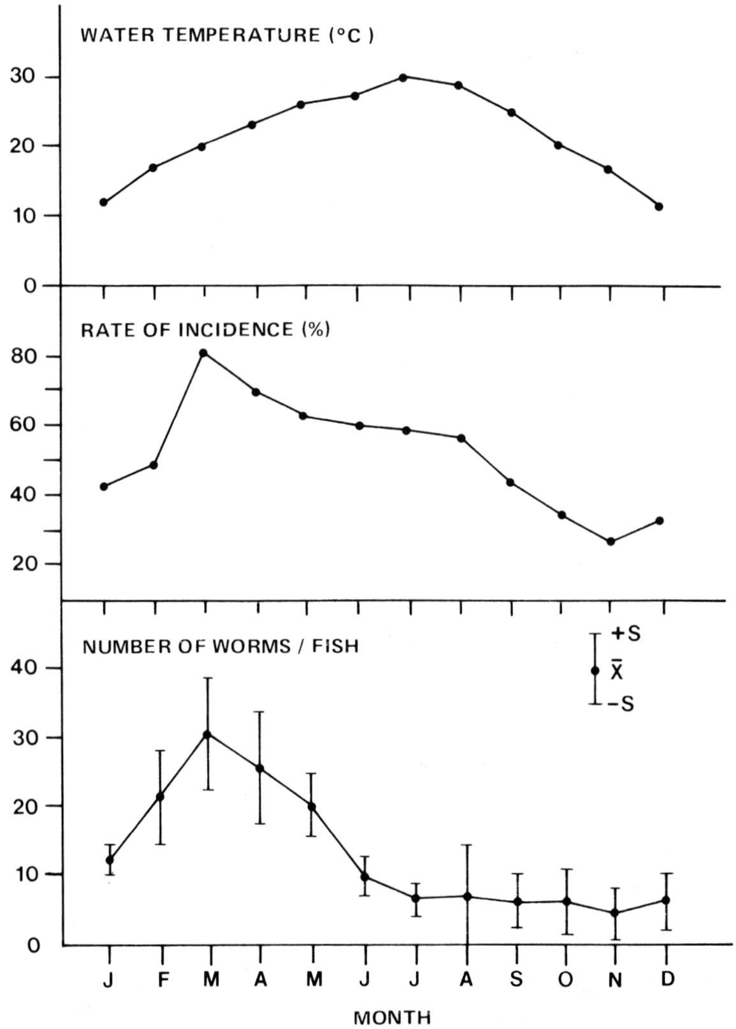

Fig. 2. Monthly prevalence and intensity of *M. chrysophrii* infestation.

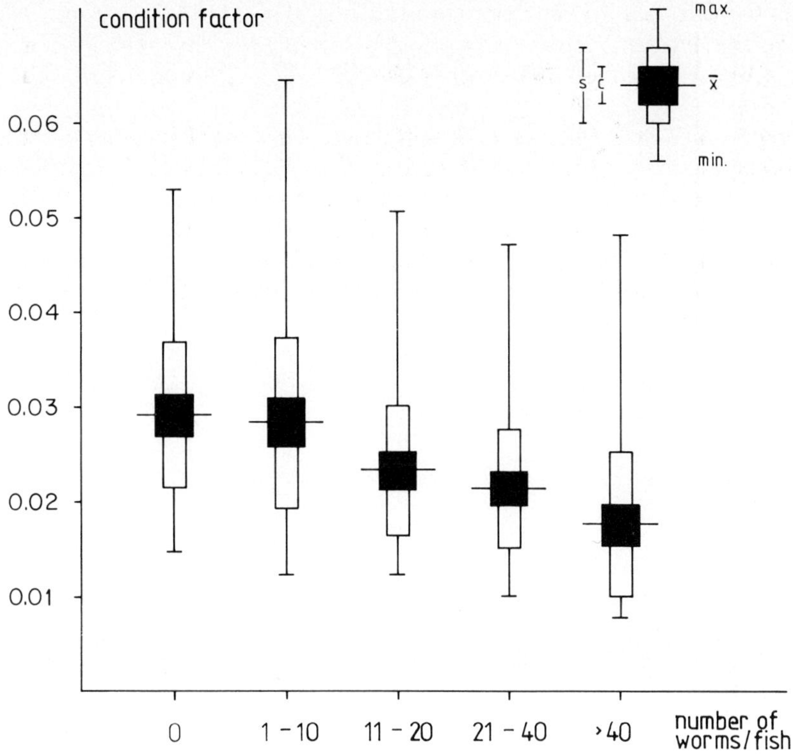

Fig. 3. Effect of *M. chrysophrii* infestation on the condition factor of sparus aurata. Data are expressed as means (X) ± standard deviation(s). Confidence limit (C) as well as maximum (max) and minimum (min) value of each group are shown in the figure.

DISCUSSION

Formerly, *M. chrysophrii* was considered to be a harmless monogenean occupying the gills of the gilthead seabream *Sparus aurata* (Paperna, 1987). The present study records *M. chrysophrii* as the cause of 75% mortality of cultured *S. aurata*. Under mariculture conditions, the susceptibility of *S. aurata* to monogenean infestation seems to increase. Moreover, monogeneans that are normally occupying and specific to other fish species were able to induce mortalities of *S. aurata* under culture conditions. For example, *Polylabris tubicirrus* as well as *Microcotyle sargi,* usually parasites of *Diplodus* spp. (Pisces: Spariidae), caused losses in tank and raceway reared *S. aurata* in France (Silan et al., 1985) and Corsica (C. Maillard, pers. commun. to Paperna, 1987). From our results, as well as other studies (Paperna et al., 1977; Paperna and Baudouy-Laurencin, 1979; Silan et al., 1985), it is apparent that monogeneans represent a threat for the gilthead seabream in confined areas. Whether this is due to a lowering of defense mechanisms under the stressful conditions of mariculture or to the possibility that overcrowding increases the access of larval monogeneans to gill tissue are questions requiring further investigation.

Table 1. The monthly incidence of *Microcotyle chrysophrii* among *Sparus aurata*.

Month	Average Water Temp. °C	Number of fish examined	Number of infested fish	% of infestation	Number of parasites/fish
January	11.9	205	88	42.9	12.1 ± 2
February	16.9	129	63	48.8	21.2 ± 6.7
March	20.0	129	104	80.6	30.5 ± 8
April	23.1	131	92	70.2	25.4 ± 7.9
May	26.3	145	92	63.5	20.1 ± 4.5
June	27.5	205	123	60	9.6 ± 2.7
July	30.1	114	67	58.7	6.4 ± 2.3
August	20.1	105	60	57.1	6.8 ± 7.6
September	25.3	134	59	44	5.9 ± 3.8
October	20.4	116	41	35.3	6.2 ± 4.7
November	16.9	125	34	27.2	4.4 ± 3.7
December	11.8	214	71	33.2	6.1 ± 4.1

Other microtylids have been implicated in deaths of cultured fish such as the yellowtail *Seriola quinqueradiata* in Japan by *Axine heterocerca* (Kubota and Takakuwa, 1963) and *Siganus luridus* in Eilat by *Allobivagina* spp. (Paperna et al., 1984). Microcotyle-infested fish showed severe anemia (Eto et al., 1976).

Our data also show an adverse effect of infestation on the condition factor of the host, which strongly suggest that *M. chrysophrii* (> 10 worms/fish) as the primary cause for weight loss. This may be attributed to the hematophagous nature of *Microcotyle* spp. (Halton, 1976). Moreover, the polyopisthocotylean monogeneans possess large complex opisthohaptors bearing several pairs of clamps that attach firmly to the surface of the gill lamellae. Rohde (1984) has indicated that such monogeneans impair gill functions, e.g., excretion, osmoregulation, and irritate the branchial epithelium.

M. chrysophrii was found in the wild population of *S. aurata* throughout the year. No other published data on the prevalence of this monogenean in the Egyptian coastal water could be found.

ACKNOWLEDGEMENT

The authors express their thanks to Prof. W.J. Hargis, Jr. for the critical review of the manuscript.

LITERATURE CITED

Bullock, A.M. and H. Bresch. 1985. Labor-Methoden, Toxizitätsteste mit Fischen. In: "Grundlagen der Fischpathologie" (R.J. Roberts and H.J. Schlotfeldt, eds.), pp. 324-367. Verlag Paul Parey, Berlin.

Eto, A., S. Sakamoto, M. Fujii and Y. Yone. 1976. Studies on a anaemia of yellowtail parasitized by a trematode, *Axine* (*Heteraxine*) *heterocera*. Rep. Fish. Res. Lab. Kyushu Univ., 3, 45-51.

Halton, D.W. 1976. *Diclidophora merlangi*: Sloughing and renewal of hematin cells. Exp. Parasit., 40, 41-47.

Kubota, S.S. and M. Takakuwa. 1963. Studies on the diseases of marine cultured fishes. I. General description and preliminary discussion of fish diseases in Mie prefecture. J. Fac. Fish. Prefect. Univ. Mie-Tsu, 6, 107-124.

Paperna, I. 1987. Solving parasite-related problems in cultured marine fish. Int. J. Parasitol., 17, 327-336.

Paperna, I. and F. Baudouy-Laurencin. 1979. Parasitic infections of the sea bass *Dicentrachus labrax* and the gilthead seabream, *Sparus aurata*, in mariculture facilities in France. Aquaculture, 16, 173-175.

Paperna, I., A. Colorni, G.W. Kissil and H. Gordin. 1977. Diseases of *Sparus aurata* in marine culture at Eilat. Aquaculture, 10, 195-213.

Paperna, I., A. Diamant and R.M. Overstreet. 1984. Monogenean infestations and mortality in wild and cultured Red Sea fishes. Helgoland. Meeresuntersuch., 37, 445-462.

Rohde, K. 1984. Disease caused by metazoans: Helminths. In: "Diseases of Marine Animals," vol. IV (1): Introduction, Pisces (O. Kinne, ed.), pp. 210- 229. Biologische Anstalt Helgoland, Hamburg.

Silan, P., P. Cabral and C. Maillard. 1985. Enlargement of the host range of *Polylabris tubicirrus* (Monogenea, Polyopisthocotylea) under fish farming conditions. Aquaculture, 47, 267-270.

PROSORHYNCHUS SQUAMATUS (TREMATODA) PARASITOSIS IN THE *MYTILUS EDULIS–MYTILUS GALLOPROVINCIALIS* COMPLEX: SPECIFICITY AND HOST-PARASITE RELATIONSHIPS

C. Coustau,[1] C. Combes,[2] C. Maillard,[1] F. Renaud,[1] and B. Delay[3]

[1]Laboratoire de Parasitologie comparée, U.R.A. C.N.R.S. 166, and
[3] Institut des Sciences de l'évolution, U.R.A. C.N.R.S. 87;
Université de Montpellier, Pl. E. Bataillon, 34000 Montpellier cedex, France
[2]Centre de Biologie et d'Ecologie tropicales et méditerranéennes, U.R.A. C.N.R.S. 166, Université de Perpignan, Ave de Villeneuve, 66025 Perpignan, France

ABSTRACT: The parasitization of the mussels *Mytilus edulis, M. galloprovincialis*, and a natural hybrid of these two species by the bucephalid trematode *Prosorhynchus squamatus* from the northwestern French coast is being reported. The pathology of the infections is described. The principal syndromes include atrophy of host tissues, and hence reduction in market value, and parasitic castration.

Prevalence data indicate that *P. squamatus* is more commonly a parasite of *M. edulis* than of hybrids. The consequences of this parasitosis on individuals and populations led to research on pathogenesis of *P. squamatus* in the *M. edulis-M. galloprovincialis* complex. It is suggested that characterization of the susceptibility of host populations relative to genetic structure should render possible the selection of resistant populations that possess the commercial features of *M. edulis.*

INTRODUCTION

In aquaculture, improved profitability requires better knowledge of the risks to which stocks are exposed, especially since these are often kept under conditions of high density or in a confined space. Fish farms are generally propitious environments for the development of parasites, resulting in reduced productivity. Good knowledge of the biology of parasite populations in stocks can make it possible to understand the determinism of parasite infection and host-parasite relationships and then to plan effective and lasting control.

Pathology in Marine Science

At a given stage in its life cycle, every parasite species develops in a host belonging to one or more species, hence the concept of parasite specificity (Euzet and Combes, 1980). These mechanisms of association and interaction in parasite systems can be approached by studying the different responses to potential host genomes. A good case for this approach is that of a parasite, specific to a host taxon, which is itself able to hybridize with a non-host taxon. In this case, the parasite is confronted with receptive genomes, nonreceptive genomes, and all the hybrid genomes.

An example of the above is the mussel *Mytilus edulis* raised on floating cultures and on bouchots on the Atlantic coast of France. This lamellibranch is parasitized by the larval stage of a trematode, which we identified as *Prosorhynchus squamatus* (Odhner, 1905, Bucephalidae). The life cycle of members of the Bucephalidae involves three successive hosts: lamellibranch molluscs (first intermediate host), teleost fishes (second intermediate host), and ichthyophagous teleost fishes (definitive host). The sporocyst and cercarial stages of *P. squamatus* have only been reported to date in *Mytilus edulis* (Chubrik, 1966; Matthews, 1972) and not in any other mollusc. However, on the Atlantic coast of Europe there is sympatry between *M. edulis* and another bivalve, *M. galloprovincialis*. Genetic differentiation between *M. edulis* and *M. galloprovincialis* is small (Skibinski et al., 1980; Gosling, 1984; Edwards and Skibinski, 1987) and hybridization has been experimentally verified for the first and second generations (Lubet et al., 1984). In addition, natural hybrids have been found in syntopic zones (Skibinski et al., 1978; Skibinski and Beardmore, 1979; Gosling and Wilkins, 1981; Skibinski et al., 1983).

Thus, in Britanny, France, *P. squamatus* is confronted not only with *M. edulis* but also with individuals that display different degrees of intergradation between *M. edulis* and *M. galloprovincialis*. In this special situation, we effected a double approach to the biology of host-parasite relationships: (1) on the scale of the individual in order to understand the mechanisms of the action of the parasite on its host, and (2) on a population scale, to evaluate parasite specificity and the sensitivity of hosts.

The first results of the study are described here.

MATERIALS AND METHODS

Host. The mussels examined were all from a floating suspended culture on the coast of the island of Groix, France (Fig. 1). The parasitological study compared two distinct populations occurring in this farm: *M. edulis* imported as brood, and mussels from local natural settlement. The *M. edulis* reared for sale are collected in the spring and summer in the mussel breeding sector at Noirmoutier, France (Fig. 1). They are transferred in summer (July and August) to lines at the farm of Groix.

The population from natural settlement on nonexploited ropes at the Groix rearing contains individuals that display all the intermediate morphological characters between *M. edulis* and *M. galloprovincialis* and are considered to be hybrids.

Parasite. The trematode was detected by dissection of the mussels under a binocular microscope. It is found in the tissue as branched sporocysts containing cercariae (Fig. 3a). The latter were exposed by rupturing sporocysts and examined under a light microscope. Observation of the cercaria permitted their identification as

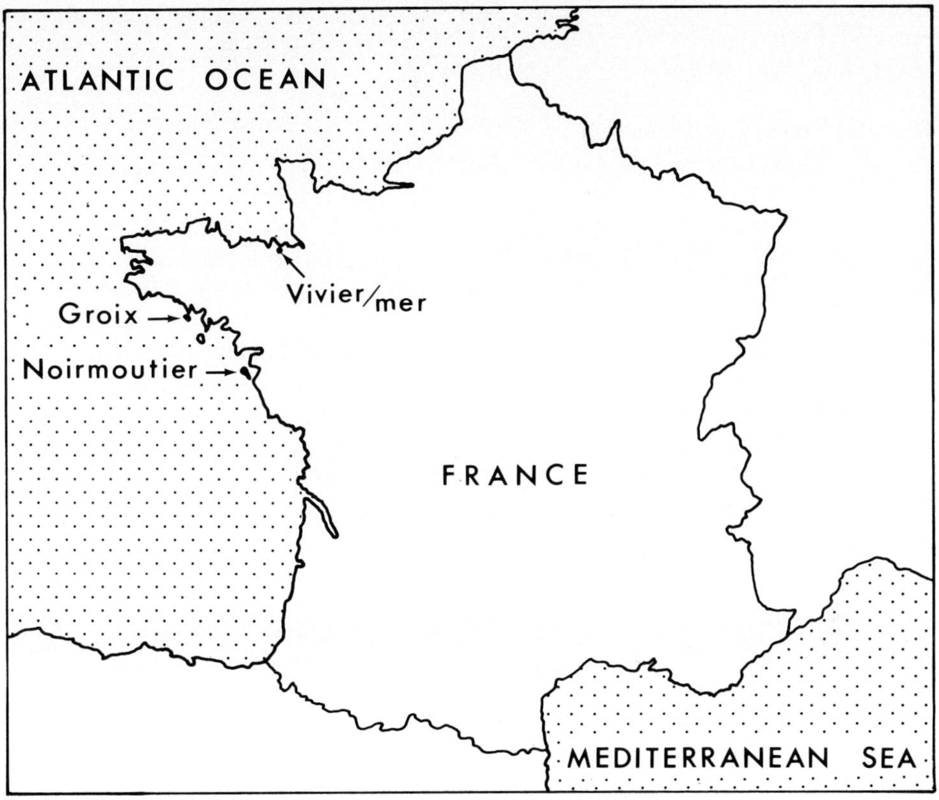

Fig. 1. Location of sampling sites.

Prosorhynchus squamatus according to Matthews' (1972) revised definition. For histological observations, the tissues were fixed with Bouin-Hollande solution and stained with Prenant-Gabe's trichrome: Groat's hematoxylin, eosin, and vert lumière.

For this study, samplings were carried out in both populations (Fig. 2) from March to August and in October and December, 1987. Samples were all taken at a depth of between 3 and 5 meters. Prevalence (percentage of parasitized individuals) was determined each month.

RESULTS AND DISCUSSION

Study of host-parasite relationships on an individual scale; pathogenic effect. Branched sporocysts of *P. squamatus* were observed in the mantle, the visceral mass, and at the surface of the digestive gland; in bivalves of the genus *Mytilus* these are the sites of development of the genitalia (Lubet, 1959). We compared healthy and parasitized specimens in order to evaluate the effect of the parasite on the tissue of the host. In healthy individuals (Fig. 3b, c) examined during the reproduction period (April, 1987), the mantle was formed of reserve tissue in which numerous genital follicles had formed. In parasitized individuals sampled at the same time (Fig. 3d), the reserve tissue did not contain genital follicles but included well-developed sporocysts with cercariae at different stages of development. Thus, our observations revealed that in all cases there was a reduction in reproductive capacity that could lead to total castration of parasitized individuals.

Investigation on population scale.

Seasonal changes of parasitism. Synchronous seasonal changes of prevalence was observed in the two syntopic populations (*M. edulis* and the local population) with maximum prevalence in May, June, and July (Fig. 2). In addition, the results of dissections carried out at different times of the year suggest that there is an annual development cycle of *P. squamatus* in the first intermediate host. Specifically, the following features were observed: (1) immature sporocysts containing undeveloped cercariae in March-April; (2) well-developed sporocysts containing developed cercariae in June-July; (3) degenerated and lysed sporocysts in August; and (4) absence of sporocysts in October and December.

Comparison of parasitism between host populations. The results revealed that there are significant differences between prevalence in *M. edulis* and in the local population in Groix. The latter displayed less prevalence each month than *M. edulis* (Fig. 2).

What can account for these differences between syntopic populations? In our analysis, all the individuals in the two host populations were sampled at the same depth and on neighboring ropes. We, therefore, consider that the probabilities of contact between *P. squamatus* miracidia and the different types of mussels were identical.

It was then necessary to check that the greater prevalence in *M. edulis* was not related to infection at the site of collection at Noirmoutier. For this, in May, 1987, we examined 96 *M. edulis* from the same collection but reared at Vivier sur Mer (northern Brittany) (Fig. 1). Prevalence was only 2% in contrast with 35% at Groix (Fig. 2). The great majority of infection of *M. edulis*, therefore, took place at Groix and not at Noirmoutier.

The difference in prevalence observed between *M. edulis* and mussels forming the local population at Groix should therefore be attributed to the genetic structure of the host population rather than to ecological factors. Indeed, the local population

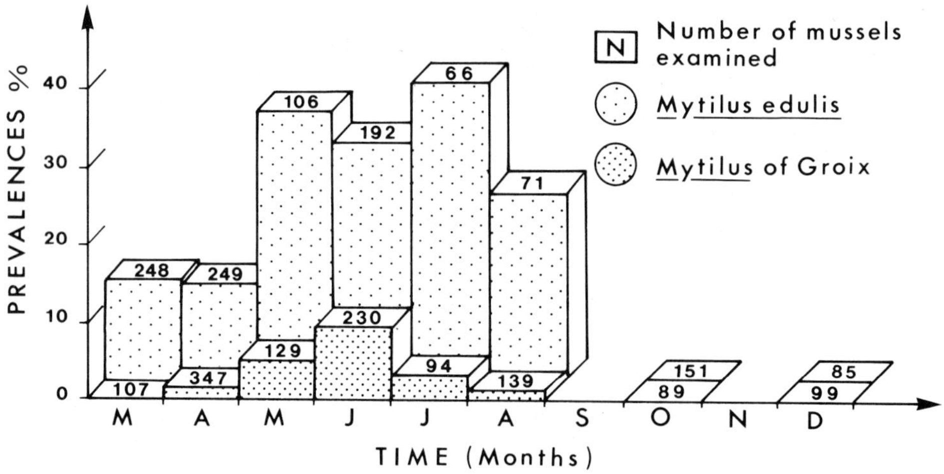

Fig. 2. Prevalence of *Prosorhynchus squamatus* in two mussel populations.

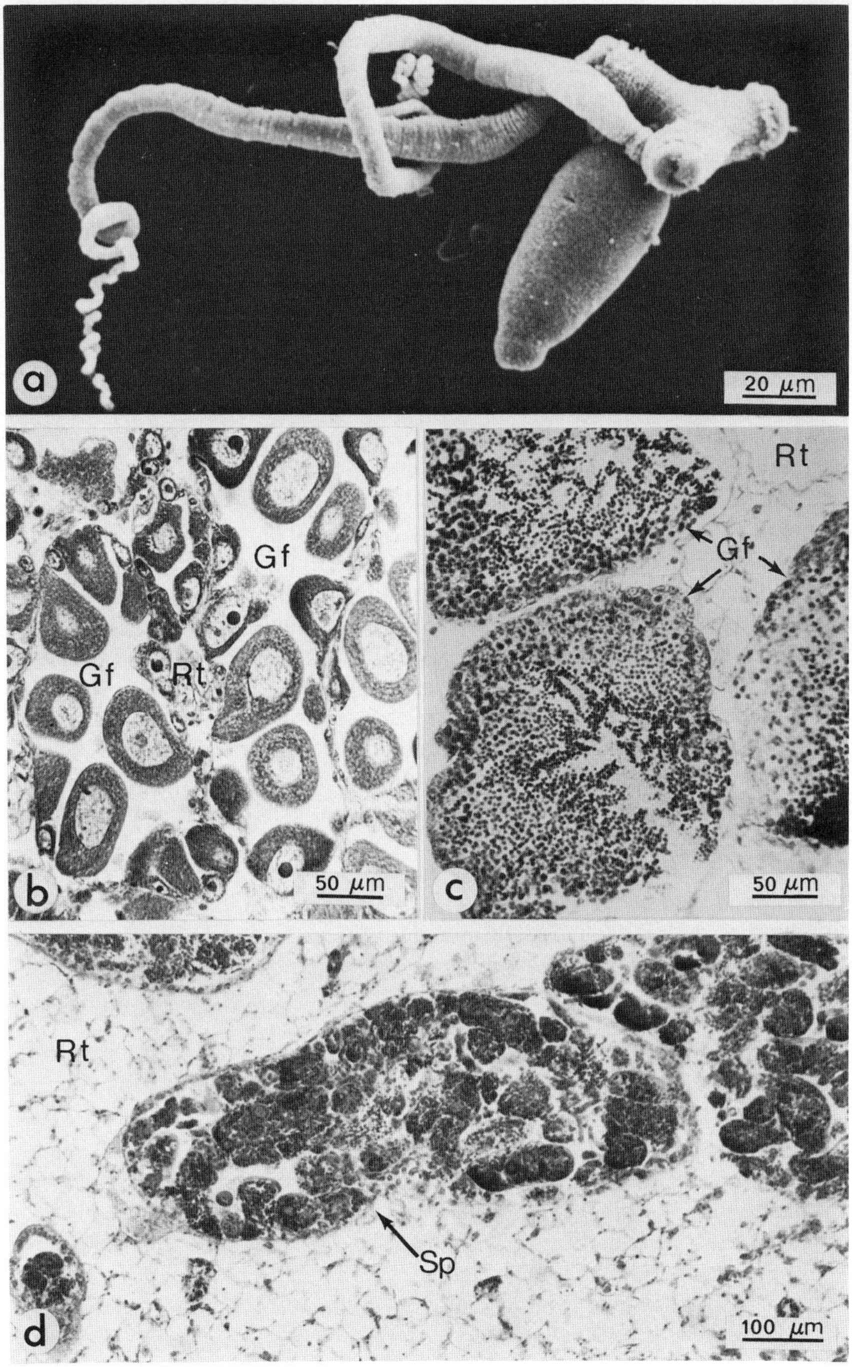

Fig. 3. (a) Scanning electron micrograph of the typical cercaria of *Prosorhynchus squamatus*, with its trilobed tail stem. (b, c) Normal histology of mature female (b) and male (c) mantle of *Mytilus edulis*. Gf, genital follicle; Rt, reserve tissue. (d) Mantle of *Mytilus edulis* infected by branched sporocysts (Sp) of *Prosorhynchus squamatus*. Notice absence of genital follicles in reserve tissue (Rt).

consisted of a set of individuals that displayed hybrid morphological characteristics of the taxons *M. edulis* (genome receptive to the parasite) and *M. galloprovincialis* (nonreceptive genome). This local population is, therefore, the result of a genetic mix, which must be the cause of differential sensitivity in the compositional genomes. We, therefore, propose the hypothesis that the more the genetic structure of an individual is of the *M. galloprovincialis* type, the less it is likely to be infected by this parasite. This is in agreement with the results on the prevalence displayed by the monogenean *Diplozoon gracile*, a parasite of *Barbus meridionalis* and of hybrids of *B. barbus* and *B. meridionalis* (Le Brun, 1988).

CONCLUSIONS

The trematode *P. squamatus* seems appropriate for its own use the resources allocated for reproduction in its intermediate host, resulting in the castration of this host by the definition of Malek and Cheng (1974). Two aspects of the consequences of this phenomenon can be envisaged:

(1) From the standpoint of aquaculture, the commercial value of parasitized individuals decreases because of poor condition caused by atrophy of mantle tissue (gonads + reserve tissue). In addition, it is probable that this parasitosis leads to a death rate that has not yet been assessed. With the knowledge that *M. edulis* possesses better commercial features than *M. galloprovincialis*, our study should thus make it possible to envisage the search for natural populations that are not affected by the trematode *P. squamatus* but have the economic features of *M. edulis*.

(2) From the viewpoint of evolutionary biology, the parasites are a source of selective forces (castration, mortality) that affect the fitness of host individuals; they participate in the evolutionary process of populations. In the hybrid-zone that we studied, the pressure exerted by *P. squamatus* differed according to the genetic characteristics of the mussel populations (*M. edulis* and local mussels). This special situation thus forms an excellent study model for the understanding of the role of the parasite in the mechanisms that govern the evolution of host populations.

The study that we propose to develop comprises two levels of perception:

(1) Intrapopulation: within the taxon *M. edulis*, which displays the highest prevalence (38%), might not the selective forces induced by the parasite be the cause of modifications in the genetic structure of populations by selection of the less receptive genomes?

(2) Interpopulation: in an analysis of the genetic and demographic population structures of three taxa (*M. edulis, M. galloprovincialis*, and their hybrids) on the British coast, Skibinski (1983) and Gardner and Skibinski (1988) compared in identical age classes the relative proportions of the different taxa in the environment. These authors observed a systematic diminution of *M. edulis* as a function of time and thus proposed that differential viability is acting in the *M. edulis-M. galloprovincialis* complex. These results thus lead us to envisage the role of parasitism in the evolution of the potential local host populations in the hybridization zone at Groix.

ACKNOWLEDGEMENTS

We wish to acknowledge the assistance of M.H. Omnes and D. Blateau in the collection of samples. We are indebted to La Trinité-sur-mer station of IFREMER for the generous use of the facilities of the Shellfish Laboratory. This work was supported by a grant from IFREMER.

LITERATURE CITED

Chubrik, G.M. 1966. Fauna and ecology of larval trematodes parasites in molluscs of Barents and White seas. Life cycle of parasitic worms of north seas. Trudy. Akad. Nauk. SSSR., 10, 78-158. (In Russian).

Edwards, C.A. and D.O.F. Skibinski. 1987. Genetic variation of mitochondrial DNA in mussel (*Mytilus edulis* and *M. galloprovincialis*) populations from South West England and South Wales. Mar. Biol., 94, 547- 556.

Euzet, L. and C. Combes. 1980. Les problèmes de l'espèce chez les animaux parasites. In: "Les Problèmes de l'Espèce dans le Règne Animal" 3, Mem. Soc. Zool. Fr., 40, 239-285.

Gardner, J.P.A. and D.O.F. Skibinski. 1988. Historical and size dependent genetic variation in hybrid mussel populations. Heredity, 61, 93-105.

Gosling, E.M. 1984. The systematic status of *Mytilus galloprovincialis* in Western Europe: a review. Malacologia, 25, 551-568.

Gosling, E.M. and N.P. Wilkins. 1981. Ecological genetics of the mussels *Mytilus edulis* and *M. galloprovincialis* on Irish Coasts. Mar. Ecol. Prog. Ser., 4, 221-227.

LeBrun, N. 1988. Specificité des relations hôte-parasites. Le modèle Cyprinidae-*Diplozoon*. Thèse de Doctorat, U.S.T.L., Montpellier, 168 pp.

Lubet, P. 1959. Recherches sur le cycle sexuel et l'émission des gamètes chez les Mytilidés et les Pectinidés. Rev. Trav. Inst. Pèches Marit., 23, 389-548.

Lubet, P., G. Prunus, M. Masson and D. Bucaille. 1984. Recherches experimentales sur l'hybridation de *Mytilus edulis* L. et. *M. galloprovincialis* Lmk. (Mollusques Lamellibranches). Bull. Soc. Zool. Fr., 87-98.

Malek, E.A. and T.C. Cheng. 1974. "Medical and Economic Malacology," Academic Press, New York.

Matthews, R.A. 1972. The life cycle of *Prosorhynchus crucibulum* (Rudolphi, 1819) Odhner, 1905, and a comparison of its cercaria with that of *Prosorhynchus squamatus* Odhner, 1905. Parasitology, 66, 133-164.

Skibinski, D.O.F. 1983. Natural selection in hybrid mussel populations. In: "Protein polymorphism: Adaptative and Taxonomique Significance" (G.S. Oxford and D. Rollinson, eds.), Systematics Assoc. Spec. Vol. 24. Academic Press, London.

Skibinski, D.O.F. and J.A. Beardmore. 1979. A genetic study of intergradation between *Mytilus edulis* and *Mytilus galloprovincialis*. Experientia, 35, 1442-1444.

Skibinski, D.O.F., M. Ahmad and J.A. Beardmore. 1978. Genetic evidence for naturally occurring hybrids between *Mytilus edulis* and *Mytilus galloprovincialis*. Evolution, 32, 354-364.

Skibinski, D.O.F., T.F. Cross and M. Ahmad. 1980. Electrophoretic investigation of systematic relationships in the marine mussels *Modiolus modiolus* L., *Mytilus edulis* L., and *Mytilus galloprovincialis* Lmk. (Mytilidae; Mollusca). Biol. J. Linn. Soc., 13, 65-73.

Skibinski, D.O.F., J.A. Beardmore and T.F. Cross. 1983. Aspects of the population genetics of *Mytilus* (Mytilidae; Mollusca) in the British Isles. Biol. J. Linn. Soc., 19, 137-183.

EXPERIMENTAL INFECTION OF JUVENILES GREY MULLET, *LIZA RAMADA*, BY METACERCARIA OF *LABRATREMA MINIMUS* (DIGENEA: BUCEPHALIDAE): ULTRASTRUCTURAL ANALYSIS OF THE HOST-PARASITE RELATIONSHIP

E. Faliex

Marine Biology Laboratory, University of Perpignan, Ave. de Villeneuve,
F- 66025 Perpignan Cedex, France

ABSTRACT: The ultrastructural study of the relationship between *Labratrema minimus* metacercaria and the liver of juvenile grey mullet, *Liza ramada*, shows: (1) the laying down of three layers forming the cyst wall; (2) the evolution of the three layers constituting the fully developed cyst wall; and (3) the passing of lipid droplets through the inner layer from the nearest hepatocytes to the cyst cavity. Thus, the metacercarial cyst is of mixed origin, parasitic and hepatic. This cyst wall protects the parasite against possible reactions from the host. The solidification of the cyst wall by compaction involves the appearance of numerous desmosomes. Also, the wall permits passing of nutritive reserves that can be utilized by the parasite.

INTRODUCTION

Different species of teleosts of cultural value, such as the grey mullet or sea bream, are commonly infected by metacercariae of trematodes when they live in Mediterranean lagoons (Biagianti and Maillard, 1986; Maillard, pers. comm.). Indeed, Maillard (1976) has shown that these larval distomes commonly occur in these lagoons and their study is of considerable interest because of the pathogenic consequences induced in the young stages of those teleosts, which are the second intermediate hosts in the life cycles of these trematodes (Maillard et al., 1980).

The biological impact study of these trematode metacercariae consists of analyzing the cytological modifications observed in the liver of the second intermediate hosts, at the interface between the fish host and the parasite. Firstly, we have observed the effect of *Labratrema minimus* (Digenea: Bucephalidae) metacercaria in the liver of naturally infected silversides, *Atherina boyeri* (Faliex and Biagianti, 1987). Secondly, we present here a study of the effect of *Labratrema minimus* metacercaria on the liver of juvenile grey mullet (*Liza ramada*), after

Pathology in Marine Science

moderate experimental infection. This work was aimed primarily at analyzing hepatic disturbances linked with the encystment and maintenance of *Labratrema minimus* metacercaria in the host organ at the ultrastructural level. A similar study has been reported by Higgins (1980) for another bucephalid, *Bucephalus haimeanus*.

MATERIALS AND METHODS

Juvenile grey mullet were caught in the Mediterranean Sea and kept in captivity in the laboratory for 6 months. Infections were made by exposing these fish to the cercariae of *Labratrema minimus*. Samples were prepared regularly during a 7-month period post-experimental infection. The fish livers were processed for electron microscopy by the method described by Bruslé (1983).

RESULTS

Presented here is information pertaining to: (1) the laying down of the three layers forming the cyst wall; (2) the development of the fully developed cyst wall; and (3) the constitution of nutritive reserves benefit of the parasite.

Laying down cyst wall layers (Fig. 1). When the parasite enters the host's liver, the first stage in the formation of the cyst wall, at one day post-infection, is the constitution of the inner layer composed of noncellular, loose, and thin granular material secreted directly by the parasite (Higgins et al., 1977).

Between the inner layer and the parasite there is a cyst cavity and, on the other side of this layer, there are the electron-dense hepatocytes that become flattened as a result of compression. These hepatocytes contain numerous clear vacuoles, mitochondria with dense matrices, and some residual bodies. Some of them show alterations like swollen mitochondria with clear matrices.

Subsequently, the cyst wall is further developed involving the appearance of two new layers. This development occurs from the 1st through the 15th day post-infection.

The degenerative and necrotic cells, representing the closest hepatocytes that had been disrupted during the penetration by cercariae, constitute the middle layer, which appears electron dense and very vacuolated in some areas.

Other hepatocytes, which had been flattened as a result of compression by the parasite, and some fibroblasts constitute the outer layer. The inner, middle, and outer layers are characteristic of the fully developed cyst wall.

Further development of cyst wall (Fig. 2). After the formation of the three layers, the fully developed cyst wall thickens until the 7th month after infection. Furthermore, solidification and compaction of the wall occurs.

The inner layer becomes thicker as a result of the constant deposition of granular material produced by the parasite's tegument until the 4th month after infection. Furthermore, this layer becomes compacted and appears to be composed of four sublayers at 7 months post-infection: (1) the first is irregularly compacted; (2)

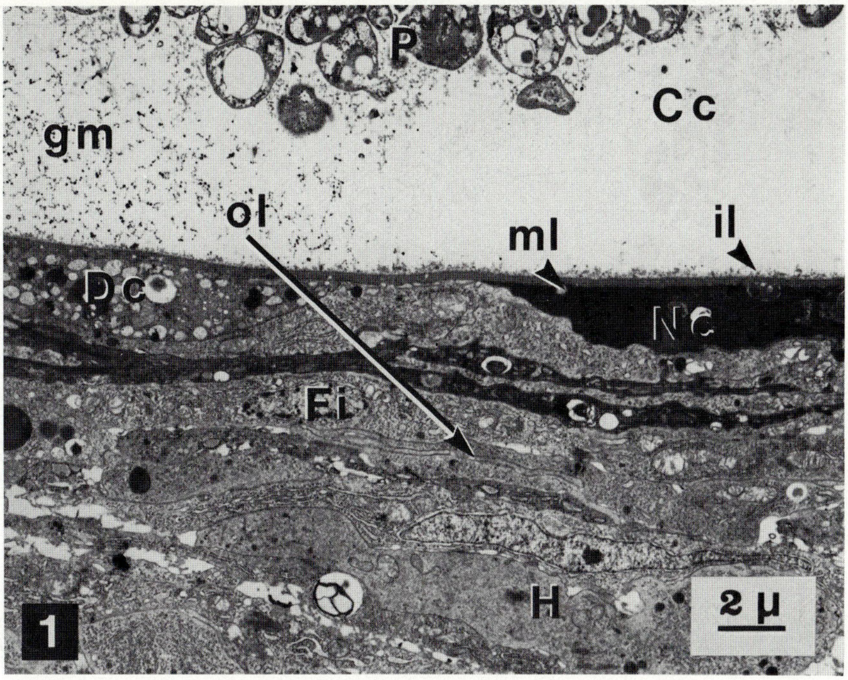

Fig. 1. Laying down of the three layers forming the cyst wall. Cc, Cyst cavity; Dc, Degenerative cell; Fi, Fibroblast; gm, granular material; H, Hepatocyte; il, inner layer; ml, middle layer; Nc, Necrosed cell; ol, outer layer; P, Parasite. 4,500X.

the second, very compacted; (3) the third, less compacted; and (4) the last one is very compacted.

The middle layer becomes thicker and appears electron dense and is very vacuolated in some areas. This layer is now formed of very compacted degenerative and necrotic cells.

Simultaneous to the others, the outer layer continues to develop and becomes composed of extremely flattened hepatocytes that are closely associated. It appears as a superposition of membranes that are kept close together by numerous desmosomes, which aid to the cohesion and solidification of this layer. At some places elongate fibroblasts occur.

Since the 2nd month post-infection, between the outer layer and the underlying hepatocytes, a brittle zone appears and clearly separates the cyst wall from the host's hepatic tissue.

The underlying hepatocytes are normal in shape or slightly elongate. Some portray alterations such as hypertrophied vacuolated mitochondria and residual bodies.

Constitution of nutritive reserves that benefit the parasite (Fig. 3). There is an accumulation of lipid droplets in the cyst cavity near the inner layer at 3 days post-infection. These lipid droplets are of hepatic origin, and most of these have passed through the inner layer from the nearest hepatocytes to the cyst cavity. Although evidence is still not available, these lipids could be used by the parasite.

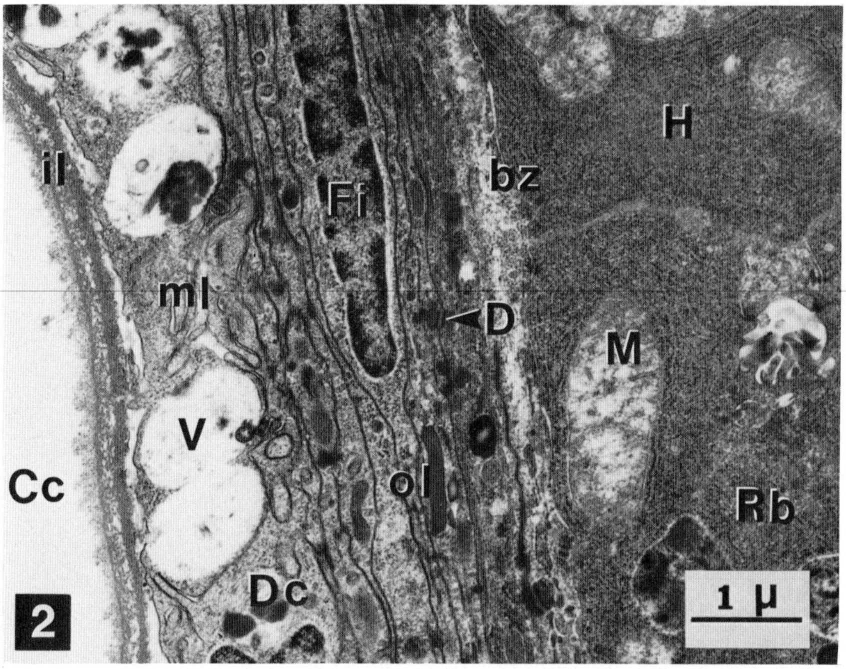

Fig. 2. Development of fully developed cyst wall. bz, brittleness zone; Cc, Cyst cavity; D, Desmosome; Dc, Degenerative cell; Fi, Fibroblast; H, Hepatocyte; il, inner layer; M, Mitochondria; ml, middle layer; ol, outer layer; Rb, Residuals bodies; V, Vacuole. 15,000X.

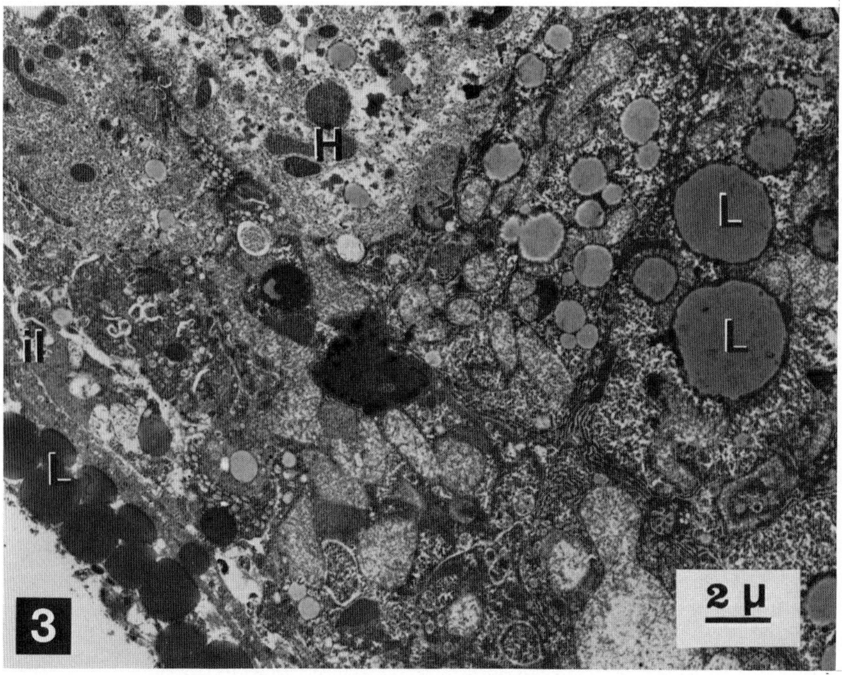

Fig. 3. Occurrence of lipid droplets in the metacercarial cyst. H, Hepatocyte; il, inner layer; L, lipid droplets. 4,500X.

DISCUSSION AND CONCLUSIONS

After experimental infection of juvenile grey mullet by *Labratrema minimus*, we have described the formation of three metacercarial cyst layers, the inner, middle, and outer layers, during a 7-month period.

First, we have demonstrated the mixed origin, parasitic and hepatic, of the cyst wall, which protects the parasite against a possible inflammatory reaction from the host. This protection is possibly realized as a result of using host hepatocytes in the formation of the cyst wall.

Secondly, we have described the further development of the three layers forming the fully developed cyst wall, involving the thickening and solidification of these layers by compaction and appearance of numerous desmosomes.

Finally, we have shown the occurrence of lipid droplets in the metacercarial cyst contributed by host hepatocytes. These may serve as nutritive reserves that may benefit the parasite.

This work represents the first stage in the study of the biological impact of this distomatosis on fishes of commercial value. It would appear interesting to go on by studying two special points: (1) the possible existence of inhibition mechanisms directed against the penetration of the parasite at the cercarial stage in the definitive host, *Dicentrarchus labrax*, and (2) the degeneration of *Labratrema minimus* metacercariae, which is sometimes observed in the liver of the second intermediate host, to see how the parasite can be eliminated from the host's hepatic tissue.

LITERATURE CITED

Biagianti, S. and C. Maillard. 1986. Lesions ultrastructurales induites par un Trematode, *Labratrema minimus* (Stossich, 1887) (Bucephalidae) dans le foie de *Liza ramada* et *Liza aurata* (Mugilidae). Europ. Aquacult. Soc., Spec. Publ. No. 9, pp. 153-170, Bredene, Belgium.

Bruslé, S. 1983. Contribution to the sexuality of a Hermaphroditic teleost, *Serranus hepatus* L. J. Fish Biol., 22, 283-292.

Faliex, E. and S. Biagianti. 1987. Metacercarial infection of marine fish by *Labratrema minimus* (Digenea, Bucephalidae): Histocytological analysis of host-parasite relationship. Aquaculture, 67, 229-232.

Higgins, J.C. 1980. Formation of the cyst wall and related changes in the structure of the tegument of *Bucephalus haimeanus* during its metamorphosis from the cercarial to the metacercarial stage. Parasitology, 81, 47-59.

Higgins, J.C., D.E. Wright and R.A. Matthews. 1977. The ultrastructure and histochemistry of the cyst wall of *Bucephalus haimeanus* (Lacaze-Duthiers, 1854). Parasitology, 75, 207-214.

Maillard, C. 1976. Distomatoses de poissons en milieu lagunaire. Thèse Doct. d'Etat USTL, Montpellier, France.

Maillard, C., A. Lambert and A. Raibaut. 1980. Nouvelle forme de distomatose larvaire. Etude d'un Trématode pathogène pour les alevins de Daurade (*Sparus aurata* L. 1758) en écloserie. C.R. Acad. Sci. Paris, Sér. D., 290, 535-538.

THE EFFECTS OF *STEPHANOSTOMUM TENUE* (DIGENEA: ACANTHOCOLPIDAE) ON MARINE AQUACULTURE OF THE RAINBOW TROUT, *SALMO GAIRDNERI*.

S.E. McGladdery[1], L. Murphy[2], B.D. Hicks[3], and S.K. Wagner[1]

[1]Department of Fisheries and Oceans, Gulf Fisheries Centre, P.O. Box 5030, Moncton, New Brunswick, E1C 9B6, Canada.
[2]Department of Fisheries and Oceans (Gulf Region), P.O. Box 1236, Charlottetown, Prince Edward Island, C1A 7M8, Canada.
[3]Pacific Aqua Foods, Suite 350, 601 West Cordova Street, Vancouver, British Columbia, V6B 1G1, Canada.

ABSTRACT: Following mass mortalities of maricultured rainbow trout (*Salmo gairdneri*) as a result of infection by metacercariae of *Stephanostomum tenue* (Digenea: Acanthocolpidae), prevalence and abundance of infection were recorded from rainbow trout held in four cages kept 0.3 m, 1.8 m, 4.9 m, and 7.3 m from the estuary floor, between May and November, 1986. Infections were first noted at the end of June in trout from the three cages suspended less than 5 m from the bottom. Infections did not appear in the deeper water cage until late July. Abundance of infection increased steadily until mid-August in the two cages held less than 2 m from the bottom. The cages held in water over 4 m from the bottom maintained consistently low levels of infection (less than 2 metacercariae per fish) from late June until the end of the experiment. Cercarial release from the first intermediate host, the mud dog whelk (*Nassarius obsoletus*), was restricted to the period when water temperatures exceeded approximately 16-17°C. No correlation was found between cercarial release, salinity, or dissolved oxygen. The pathological effects of this parasite on its accidental trout host and the natural life cycle of *S. tenue* in the affected estuary system were also investigated.

INTRODUCTION

In June, 1985, mass mortalities of rainbow trout, *Salmo gairdneri*, were reported from an experimental aquaculture site operated by the Canadian Department of Fisheries and Oceans, off northern Prince Edward Island. Similar losses were also reported by various private aquaculturists from other sites around Prince Edward

Island. The organism responsible for these mortalities was identified as the metacercarial stage of *Stephanostomum tenue* (Digenea: Acanthocolpidae), which was found encysted in the pericardial cavity of the trout. This is the first record of *S. tenue* in Canadian Atlantic waters.

Martin (1939) detailed the life cycle of *S. tenue* from Woods Hole, Massachusetts. The first intermediate host was the mud dog whelk, *Nassarius obsoletus* (= *Nassa obsoleta*), the second intermediate host was the silverside, *Menidia menidia*, and the definitive hosts were experimentally infected puffer fish, *Spheroides maculatus*, and naturally infected striped bass, *Morone saxatilis* (= *Roccus lineatus*). Of these, the only hosts common to all affected sea-cage sites were the mud dog whelks. An intensive survey of the fish and gastropods present in each affected estuary was initiated to determine not only the source of infection to the rainbow trout, but also the other hosts involved in the *S. tenue* life cycle. The question of whether or not infection by *S. tenue* could be controlled or avoided within affected estuaries was also investigated, since the prime sites for marine aquaculture on Prince Edward Island are the shallow, sheltered, mud-bottom estuaries, which harbor *S. tenue*.

MATERIALS AND METHODS

Host survey. Before it could be determined whether or not rainbow trout infection by *S. tenue* could be controlled by removal of one or more hosts involved in its life cycle, the natural host species had to be identified. Samples of gastropods and fish were collected from three sites around Prince Edward Island that had been subject to rainbow trout mortalities, between June and November, 1985. Gastropod samples were collected by scuba divers from around the sea-cage sites and fish were collected by trapnet fixed gear, designed to sample pelagic and benthic fish.

Challenge infections. Two challenge infections were run: one using uninfected rainbow trout and infected mud dog whelks in August, 1986, and another using American eel fed infected and uninfected trout hearts in August, 1987. This was in order to determine whether or not the specimens found in the rainbow trout were identical to those found in the wild hosts and, if they were, how long it took for development of the metacercarial cysts.

Exposure of uninfected trout to infected whelks. A total of 28 uninfected rainbow trout, which were collected from an unaffected aquaculture site, were exposed for 21 days to 150 mud dog whelks collected from the Bideford Estuary (thereby giving at least 10 infected whelks, assuming 6-7% infection at that location [see below]). Samples of four fish were removed and examined daily for the first 4 days and thereafter two fish were removed every 2-3 days up to day 16 post-exposure. The remaining two trout were left for an additional 5 days before being examined.

Eels fed infected and uninfected trout hearts. The second challenge infection involved feeding 13 eels, which were collected from Malpeque Bay, a total of 55 trout hearts, each of which contained at least two metacercarial cysts. An additional 14 eels were fed uninfected trout hearts. Six days post-feeding, five eels from each group were examined for adult *S. tenue*. The remaining eels were examined 13 days post-feeding.

Histopathology. Subsamples of one or two fish were taken from the first challenge experiment (see above) and were used for histological examination. This was in order to determine why encystment of the metacercariae in the pericardial cavity had provoked the high mortalities observed. The route by which the metacercariae entered the pericardial cavity was also examined.

Life span of free swimming cercariae. Cercariae collected from mud dog whelks in July, 1986, were held in filtered estuarine sea water for a maximum of 75 days at 4°C in order to determine their life span. The infective viability of the cercariae, however, was not determined from this experiment.

Sea-cage location experiments. In 1985, rainbow trout were collected from the same areas surveyed for gastropod and fish hosts of *S. tenue*. In 1986, however, all examinations concentrated on the four experimental sea-cages managed by the Department of Fisheries and Oceans, Canada, in the Bideford Estuary off Malpeque Bay, northern Prince Edward Island (Fig. 1).

These cages were designed to investigate whether or not infection by *S. tenue* could be controlled by avoidance. Prevalence and abundance of *S. tenue* were recorded weekly from trout held in cages with 0.3 m (cage A), 1.8 m (cage B), 4.9 m (cage C), and 7.3 m (cage D) clearance between the bottom of the cage and the estuary floor (Fig. 2). Clearance was chosen as a primary criterion for avoidance of infection owing to the benthic nature of the first intermediate host (*N. obsoletus*) and the reported adhesion of *S. tenue* cercariae to the substrate (Martin, 1939). The clearances chosen for this experiment were designed to span the range available at

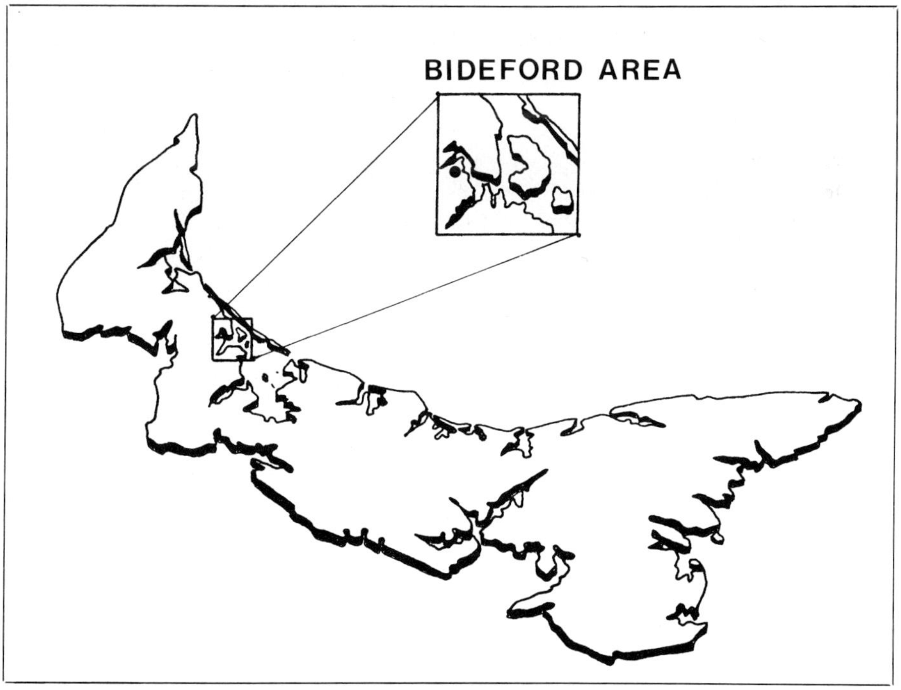

Fig. 1. Map of Prince Edward Island, Canada, showing the location of the Bideford Estuary study area.

manageable sites (those sites easily accessed, even in bad weather) within the estuary.

Each cage consisted of a 4.3 m^2 surface frame from which a 2.8 cm^2 mesh net, measuring 3.7 m x 3.7 m x 3.0 m was suspended. The top of each cage was covered by netting to reduce losses to birds. Approximately 1700 rainbow trout, averaging 17.5 cm in total length, were introduced to each cage. The fish were fed dry commercial feed twice daily in order to be consistent with the management practices of private aquaculturists. Bottom, midwater, and surface temperatures, salinity, and dissolved oxygen were recorded daily. Any rainbow trout mortalities were removed and examined.

Weekly samples of rainbow trout consisted of 10 fish, examined fresh, from each of the four cages. The gills, alimentary tract, peritoneal cavity, and pericardial cavity were examined for metacercarial encystment.

RESULTS AND DISCUSSION

Host survey. The three gastropod species most commonly found in the estuaries where rainbow trout were subject to *S. tenue* infection were the periwinkle (*Littorina littorea*), the dog whelk (*Thais lapillus*), and the mud dog whelk (*Nassarius obsoletus*). Only *N. obsoletus* was found to be infected by *S. tenue* rediae, which were first observed on 11 July, 1986. Of the 435 *N. obsoletus* examined between July and September, 1986, 33 were found to be infected, giving an overall prevalence of 7%. This is much higher than the prevalence of 0.4% reported by Martin (1939) in *N. obsoletus* (=*Nassa obsoleta*) from Woods Hole, Massachusetts. No infections were found, however, in additional samples of 100 *N. obsoletus* each, collected in

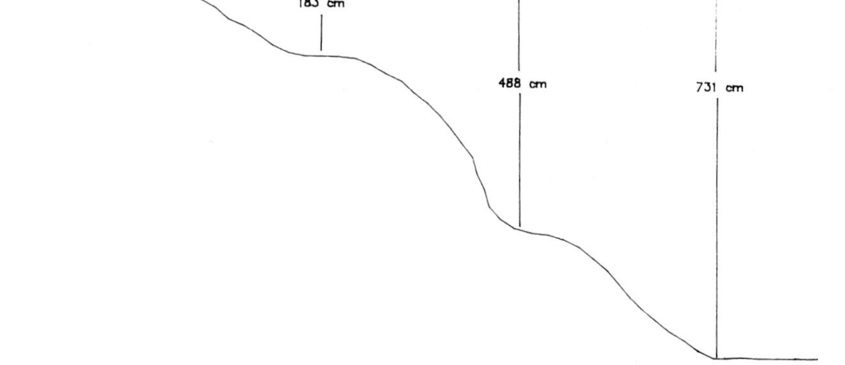

Fig. 2. Diagrammatic representation of the different clearances between the bottom of the four sea-cages and the estuary floor.

November, 1985 and 1986, indicating that *S. tenue* rediae do not appear to overwinter in the first intermediate host.

Of all the fish species examined (26 *Pseudopleuronectes americanus*, 13 *Tautoglabrous adspersus*, 5 *Menidia menidia*, 46 *Anguilla rostrata*, 53 *Fundulus heteroclitous*, 10 *Microgadus tomcod*, and 1 *Gadus morhua*) only mummichog (*F. heteroclitous*) were found to be infected by *S. tenue* metacercariae and only American eel (*A. rostrata*) were found to be infected by adult *S. tenue*. No striped bass (*Morone saxatilis*) were caught in the estuaries surveyed.

Of the 53 mummichog examined, 34 (64%) were infected with a range of 1-23 *S. tenue* metacercariae and, of the 46 *A. rostrata* examined, 10 (22%) were infected with a range of 2-15 adult *S. tenue*.

All the rainbow trout examined were from aquaculture sites, since there are no wild rainbow trout populations in Prince Edward Island. Of the 29 trout examined in 1985, 28 were infected with a range of 3-64 *S. tenue* metacercariae. The cysts were all located in the pericardial cavity of the trout, in direct contrast to the body cavity location of *S. tenue* metacercariae in mummichog.

Challenge infections.

Exposure of uninfected trout to infected whelks. In order to verify whether or not the difference in site of encystment indicated that the metacercariae from the trout and mummichog were different species, uninfected rainbow trout were exposed to mud dog whelks from the Bideford Estuary. The first metacercarial cysts were found in the pericardial cavity 9 days post-exposure, confirming that the cercariae being released in the estuary were responsible for the trout pericardial infections. This also provided a development time of approximately 9 days for the appearance of the metacercarial cyst in the trout, at ambient water temperatures (17° - 20°C).

Eels fed infected and uninfected trout hearts. The second challenge infection, designed to determine whether or not the metacercariae in the trout hearts were the same species as those found in the American eel, was inconclusive.

The only eels available at the time of the experiment were from just outside the Bideford Estuary and both control and experimentally infected eels were found to harbor adult *S. tenue* in the rectum. Slight differences in the stage of egg development in the adult worms could not be used to determine whether or not they were derived from the infected trout hearts.

Histopathology. Results from 11 of the 28 trout used for the challenge infection between uninfected rainbow trout and infected mud dog whelk were examined histologically. The results from these examinations are shown in Figures 3 and 4. There was a thick fibrogranulomatous layer of inflammation produced by the host in response to the presence of the metacercarial cyst, even though the cyst itself did not actually penetrate deeper than the surface of the cardiac musculature. Degenerating metacercariae were encysted in granulomas, which were surrounded by this pericardial inflammatory tissue (Figs. 3, 4). Cellular elements of this pericarditis included melanomacrophages, neutrophils, and lymphocytes embedded in a matrix of fibroblasts. This inflammation extended anteriorly along the bulbous arteriosus, diminishing in severity at the distal reaches of the ventral aorta. Degenerating larvae were seen to provoke a much more severe inflammatory response than non-degenerating larvae (Fig. 4). Most cysts were found on or around the bulbous arteriosus, and some were even found lying loose in the pericardial cavity fluids. The actual pathologic effect of the metacercarial infection did not appear to be directly

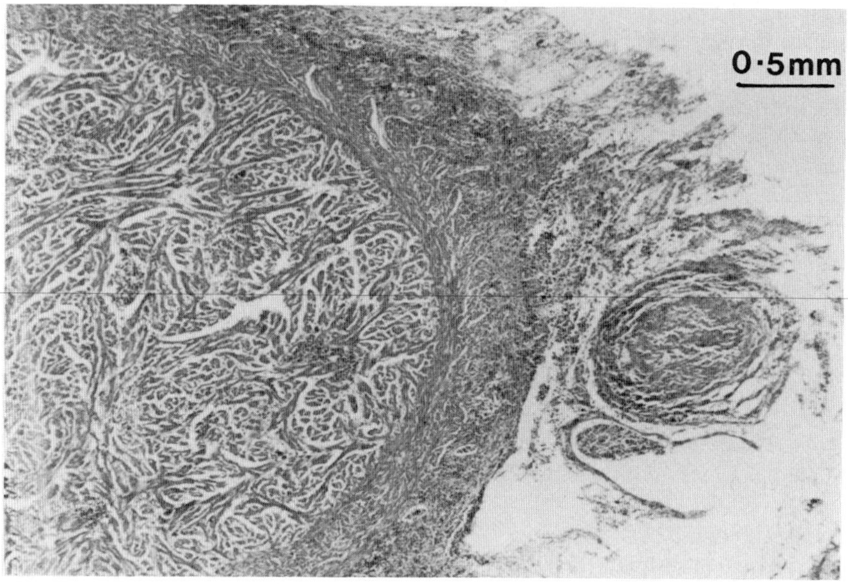

Fig. 3. Extensive pericardial inflammation and fibroplasia associated with a degenerating metacercaria. H & E stain; 200X.

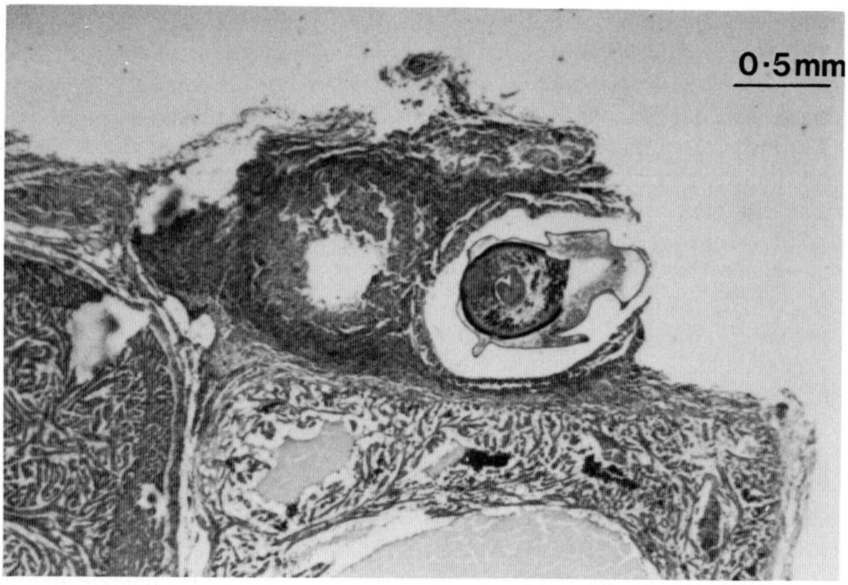

Fig. 4. Comparative effects of pericardial inflammatory response around the bulbous arteriosus to a degenerate and non-degenerate metacercaria. H & E stain; 200X.

related to damage caused by the larva itself, but appeared to be due mostly to the host tissue response of a fibrogranulomatous sheath laid down around each metacercarial cyst. The deposition of this tissue on the surface of the heart, and especially around the bulbous arteriosus, is believed to impede the ability of the fish to compensate for increased oxygen demands by expanding its cardiac volume. This cardiac dysfunction would have been especially serious in the summer of 1985, when water temperatures broke 50 year old records, and could explain why mortalities were not quite as heavy in 1986, when water temperatures were not as high as in the previous year. The low oxygen concentrations in such warm water in conjunction with the decreased ability of the trout to adapt its cardiac volume accordingly led, inevitably, to the heavy mortalities encountered. The connective tissue response also explains why some fish were killed with only one cyst in the pericardial cavity, while others appeared to survive much heavier infections. Those trout that had a single cyst surrounded by a strong connective tissue sheath at the base of the bulbous arteriosus would not be able to expand cardiac volume to compensate for increased oxygen demands as easily as a fish with more cysts all located on a less elastic tissue of the heart, such as the ventricle.

The route of invasion within the trout was found to be via the posterior buccal cavity wall, where a strong inflammatory response to the penetration of metacercariae was observed. This indicated that the trout take these larvae into the buccal cavity passively, and that they are not actively ingested, as is the case with infection of the mummichog (pers. observ.) and silverside (Martin, 1939). Since the cultured trout were fed commercial pellet food and showed no evidence (in the form of stomach contents) of active feeding on microscopic food, such as *S. tenue* cercariae, this passive entrance into the buccal cavity must have occurred during the normal process of respiration. This indicated that the cercariae were not spatially restricted to the estuary floor, as would be assumed for larvae that are released from benthic gastropods, such as the mud dog whelk, and that have been demonstrated to use an adhesive gland in their tails for adhering to the substrate (Martin, 1939). It was, therefore, concluded that *S. tenue* cercariae were distributed throughout at least the lower portions of the estuary water-column and might subsequently be using the cages themselves as a substrate. No larvae were isolated, however, from sea-cage scrapings.

It should be noted that, since *S. tenue* is more or less harmless in its natural intermediate host, it would have been impossible to predict that it would be pathogenic in an accidental host. Instances of parasites producing aberrant effects in accidental hosts are rare, but not unknown (e.g., *Paralaphostrongylus tenuis*, a parasite of white-tailed deer, *Odocoileus virginianus*, that produces fatal disorientation in the moose, *Alces alces* (Anderson, 1964)). With this in mind, it may be worthwhile for aquaculturists who are planning to expand into new or unexplored environments to set out test cages with a "dispensible" number of fish in order to test for parasites (or other phenomena) that may adversely affect their stock species while appearing quite harmless in the natural environment.

Life span of free swimming cercariae. Free swimming cercariae were kept alive in filtered 4°C sea water for a period of 75 days. This indicated that they could survive winter water temperatures (above freezing); however, their transmission or infective viability over this holding period was not investigated.

Sea-cage location experiment. Having established the source of infection of the caged rainbow trout and the life cycle of *S. tenue* in the affected estuaries of Prince Edward Island, the next question that had to be answered was whether or not infections could

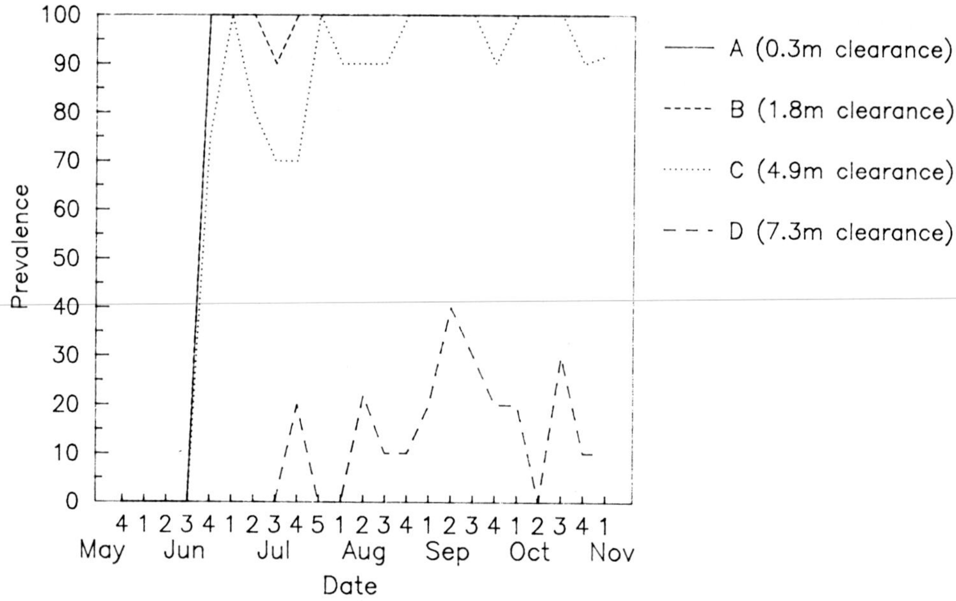

Fig. 5. Prevalence of *Stephanostomum tenue* metacercariae in rainbow trout collected weekly between May and November, 1986, from each of the four experimental cages.

be eradicated. Removal of either the first intermediate host (the mud dog whelk) or the definitive host (the eel) in order to break the life cycle was impractical because of the immense numbers of the former and the importance of the latter as a commercial fishery in adjacent waters. The possibility of avoiding exposure of the trout to infective cercariae was therefore examined.

It was known that cercariae were present in the water column; however, since all the sea cages affected in 1985 were in shallow water (approximately 4 m), it could not be determined how far up the water column the cercariae penetrated. In 1986, four experimental sea cages were placed at various locations throughout the Bideford Estuary in order to provide four different bottom clearances at low tide (Fig. 2).

The prevalence of *S. tenue* metacercariae in the trout sampled weekly from each cage is shown in Figure 5. Metacercariae first appeared in the trout in the last week in June, increasing from 0% prevalence in the previous samples to 100% in cages A and B, and 75% in cage C within 1 week. These prevalences of 75-100% remained high until the end of the experiment, in November. No infections were found in the trout from cage D until the last week in July, when the prevalence of *S. tenue* increased from 0-20%. The highest prevalence found in trout from cage D was 40%, in the 2nd week of September, and throughout most of the summer (11 of the 19 weeks over which infections were monitored) prevalences of metacercariae in trout from cage D were significantly lower ($P < .05$) than those in fish from the other three cages. This clearly indicated that, although cercariae were present in the water column up to 7.3 m away from the estuary floor, they were significantly less prevalent than in the water column less than 5 m away from the estuary floor.

There was an inverse correlation between abundance of *S. tenue* metacercariae and distance between the bottom of the cage and the estuary floor (Fig. 6). The highest numbers of metacercariae were found in trout from cage A, with 30 cm clearance between cage and estuary floor (maximum number of metacercariae = 75), while the lowest numbers of larvae were found in the trout from cage D over 7 m from the estuary floor (maximum number of metacercariae = 1).

From these results, it was concluded that although infection by *S. tenue* metacercariae could not be completely avoided within the Bideford Estuary, exposure to the infective cercariae could be reduced to a more or less negligible level (i.e., prevalences of less than 40%, and an abundance of 1) by placing the sea-cages in water with over 7 m clearance from the estuary floor.

Dissolved oxygen, salinity, turbidity, and water temperature were measured to see whether or not any of these physical parameters could be correlated to metacercarial infections in the trout. The only physical parameter that showed any degree of variation over the duration of the experiment was water temperature (Fig. 7). It ranged from 11.5°C in May to a maximum of 21.9°C in the first week of August, and a minimum of 3.3°C in November. The abrupt change from 0% to 75-100% prevalence of *S. tenue* in trout from cages A, B, and C in late June corresponded to a bottom temperature increase from 12.6-13.8°C in the second week of June to 16.0-16.5°C in the third week. Bottom temperatures did not increase past 16.0°C until the first week of July at cage site D. Allowing a minimum development time of approximately 10 days (derived from a 9-day development time at 21.0°C [see above]), this would mean that cercariae were first released from the mud dog whelk when the ambient water temperature at the estuary floor was 16.0°C at cage C, 16.8°C at cage B, and 17.1°C at cage A. The bottom temperature at site D at the same time was 15.6°C. Ten days prior to the first appearance of *S. tenue* in the rainbow trout from cage D, however, the bottom temperature was 16.2°C.

From these results it appears that cercarial release may be triggered by water temperatures of 16.0°C and higher. Samples of approximately 100 mud dog whelk

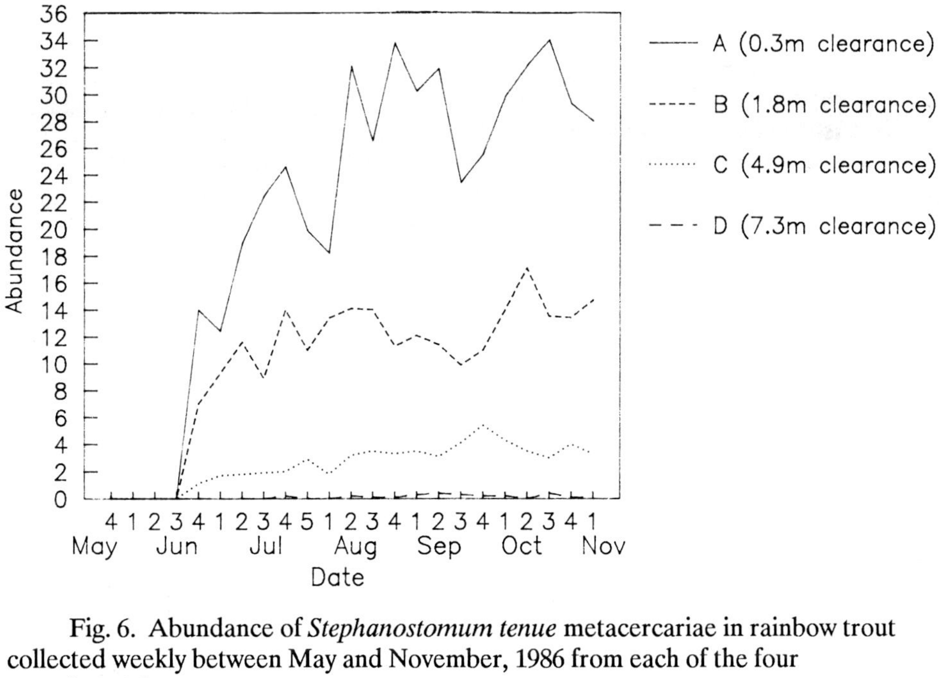

Fig. 6. Abundance of *Stephanostomum tenue* metacercariae in rainbow trout collected weekly between May and November, 1986 from each of the four experimental cages.

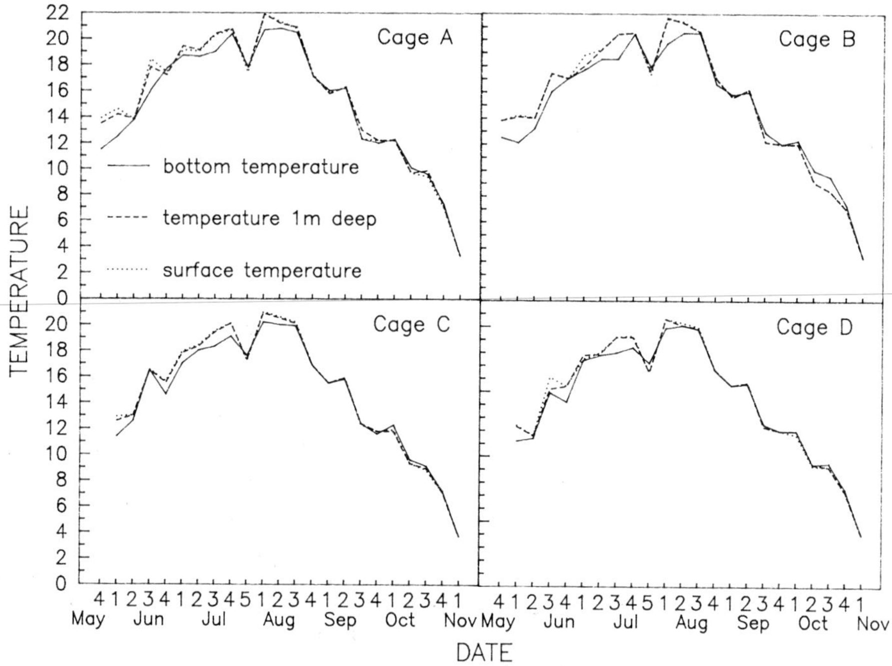

Fig. 7. Water temperatures taken from the bottom, 1 m below the surface and at the surface at each of the four cage sites.

collected in November, 1985 and November, 1986 showed no infections by *S. tenue* rediae. Thus, it appears that the infections that first appeared in the trout in June must have come from mud dog whelks infected by *S. tenue* miracidia in spring. This is possible, since American eel were found to be infected in late September and are present in the estuary over winter, lying dormant, buried in the mud, until spring. This proximity to the mud dog whelk over winter and in spring would certainly facilitate transmission of the miracidial stage and repetition of the life cycle.

Eels move out of the Bideford Estuary into Malpeque Bay in late spring, thereby reducing the efficiency of transmission of the miracidial stage to the estuarine mud dog whelk. The majority of mud dog whelk infections must, therefore, occur in spring. Unfortunately, mud dog whelks were not monitored regularly over spring, so this theory remains to be tested.

CONCLUSIONS

The rainbow trout, *Salmo gairdneri*, is an accidental host of *Stephanostomum tenue* metacercariae. These metacercariae normally infect mummichog (*Fundulus heteroclitous*) or silverside (*Menidia menidia*) (Martin, 1939), encysting harmlessly in the body cavity. In trout, however, *S. tenue* metacercariae encyst in the pericardial cavity where they provoke a severe host immune response that can be fatal under certain conditions, such as unusually warm water and low oxygen concentrations. Control or removal of the definitive host, the American eel, *Anguilla rostrata*, is not possible because it forms a commercially important fishery in the affected areas.

Control of the first intermediate host, the mud dog whelk, *Nassarius obsoletus*, was not possible either, since it was present in too large numbers. Infection levels in trout were found to be controlled by placing the sea-cages in water with over 7 m clearance between the bottom of the cage and the estuary floor. Although this did not avoid infestation completely, it did reduce the levels of infection (prevalences <40%; abundance = 1) significantly, compared to those found in trout kept in cages placed in water with less than 2 m clearance (prevalences >90%; abundances ranging from 7 to 33.8). Trout placed in a cage with 4.9 m clearance showed intermediate levels of infection.

ACKNOWLEDGEMENTS

The authors wish to acknowledge the assistance of Mr. J. MacLeod in providing samples from his private aquaculture site and the staff of the Ellerslie Research Station for maintenance of the caged trout and collection of molluscan and fish samples. Financial support for this research was provided by the National Research Council of Canada through the Trout Growers Association of Prince Edward Island and by the Department of Fisheries and Oceans, Canada.

LITERATURE CITED

Anderson, R.C. 1964. Neurologic diseases in moose infected experimentally with *Paralaphostrongylus tenuis* from white-tailed deer. Path. Vet., 1, 289-322.

Martin, W.E. 1939. Studies on the trematodes of Woods Hole II. The life cycle of *Stephanostomum tenue* (Linton). Biol. Bull., 77, 65-73.

EFFECTS OF THE SWIMBLADDER NEMATODE *ANGUILLICOLA CRASSUS* IN WILD AND FARMED EEL, *ANGUILLA ANGUILLA*

P. van Banning[1] and O.L.M. Haenen[2]

[1]Netherlands Institute for Fishery Investigations, P.O.B. 68, 1970 AB IJmuiden, The Netherlands
[2]Central Veterinary Institute, P.O.B. 65, 8200 AB Lelystad, The Netherlands

ABSTRACT: The nonindigenous swimbladder nematode *Anguillicola crassus* has been introduced into the inland waters of the Netherlands, probably via importation of live European eel, *Anguilla anguilla*, for commercial purposes. The nematode has spread very quickly through the Dutch wild eel populations and by 1987 it had a prevalence of 80-100% for most inland waters. The high infection levels, together with increasing mortalities in infected cultured eel, has lead to questions about the effects of this nematode on eel. Wild young eel (length: 6-15 cm) from a fresh water lake and farmed young eel (8-10 cm and 25-40 cm) were studied by direct observations of squash preparations and by histology of the swimbladder and other organs. It was observed that eel can be infected very soon after immigration of the elvers into fresh water. The nematode may even reach the adult phase in such small eels, adapting their size to the restricted lumen of the small swimbladder.

The observed effects on the swimbladder wall were dilation of blood vessels, inflammation, and rupture. The swimbladder of eels surviving these acute stages have fibrotic walls and often show adhesion to the surrounding organs. Such nonfunctional swimbladders restrict survival and reproductive potential of the nematode.

INTRODUCTION

The increasing demand for eel for commercial and aquaculture purposes in western Europe has lead to world-wide importation of live eel into some countries, including the Netherlands. Such a situation carries the risk of introducing nonindigenous parasites and pathogens into fish farms and the environment. This is demonstrated by recently made observations of parasites of Asiatic and North American origin in wild and farmed eel in Europe. Køie (1987) described in Denmark three parasites introduced from Asia: the monogeneans *Pseudodactylogyrus*

Pathology in Marine Science

Fig. 1. Swimbladder taken out of an eel and opened to show the burden of *A. crassus* adults in the lumen.

anguillae and *P. bini*, and the nematode *Anguillicola crassa*. The latter has now been renamed *A. crassus* by Moravec and Taraschewski (1988). Both this nonindigenous eel parasite (Neumann, 1985; Peters and Hartmann, 1986) and the North American acanthocephalan *Paratenuisentis ambiguus* (see Taraschewski et al., 1987) have recently been recorded in Germany. In the Netherlands, the introduction and widespread settlement of *A. crassus* in the wild eel populations was most obvious (Fig. 1). Although *A. crassus* was first recorded in 1985 (van Banning et al., 1985), its widespread distribution in different eel populations in the Netherlands at that time suggests that introduction could have already taken place 2 or 3 years earlier. The prevalence increased rapidly in 1986 and 1987, up to levels of 94% (Dekker and van Willigen, 1987).

The introduction of *A. crassus* into the Netherlands became a matter of concern because of the predicted impact on the condition and survival of wild and farmed eel. Although *A. crassus* does not cause serious effects in its original host, the Japanese eel, *Anguilla japonica*, it has been reported that the parasite causes pathological effects on the European eel, *Anguilla anguilla*, in eel farming in Japan (Egusa, 1979). Furthermore, it was observed that an increased mortality rate of 10-20% and a retardation of growth of 20-30% could occur in infected farm eels (Liewes and Schaminee-Main, 1987). Although most of the questions concerning the effects of *A. crassus* on eel are directed to the fresh water phase in order to examine the parasite's biological cycle and mode of infection, it can also affect marine eel farming as these are stocked with fresh water eel, which can already be infected with *A. crassus*.

Wild eels of 6-15 cm length were selected to gain insight into the primary infection process and prevalence in small eel. Further, it was expected that this length class would offer a better possibility for studying the first effects of the infection. Young eels of 6-7 cm length have just left the marine elver stage and recently entered the fresh water environment. It can be assumed that such small eels have not been previously infected with *A. crassus* and are free from any effect of the infection because this nematode is not infectious in the marine environment. Therefore, the pathogenesis of the nematode on the eel can be studied from the very initial period of stay following initial entry into the freshwater environment. For comparison of the effects, a study was also made of farmed eel.

MATERIALS AND METHODS

Sampling and examination of wild eels. Wild eels, *Anguilla anguilla*, with *A. crassus* infection were obtained in June and October, 1987, with fine-mesh beam trawling from the IJsselmeer, a fresh water lake situated in the northern part of the Netherlands. For practical and histological reasons, a maximum length of 15 cm of the eel was sampled. The eels were transferred alive to the laboratory. Within 20 hr after capture, the effects of *A. crassus* infection were studied by macroscopical observation and squash preparations of the swimbladder and other internal organs for the presence and the condition of the parasites (abnormal = dead or not well developed; normal = alive and active). For this purpose the eels (N = 25) were anaesthetized with 2-phenoxy-ethanol (0.2-0.5 ml/liter). For histological studies, other samples (N = 20, distributed over all length ranges) of the same catches were taken. The eels were fixed in 10% buffered formalin, paraffin wax embedded, sectioned transversally at 4 μm, and stained with haematoxylin and eosin.

Sampling and examination of farmed eels. Samples of farmed eel, *Anguilla anguilla*, were provided by a fresh water eel farm and a marine eel farm (salinity 30 g/kg) in the Netherlands. The samples were taken at random (length class 8-10 cm) or selected (length class 25-40 cm) as eels infected with *A. crassus* by the eel grower. The farmed eel of 8-10 cm length had been artificially infected 5 months before at the freshwater farm. Farmed eels could have had a different infection history and other stress conditions compared to wild eels. So, for farmed eels other parameters were also taken into account, such as secondary bacterial infections. For the bacteriological study, swabs from liver, spleen, and kidney were taken and grown on TSA- and HIS-agar plates at 22°C. When growth was observed, the bacteria were differentiated using morphological and biochemical tests according to Cowan (1977) and Bergey (1984).

Larval stages. For this study, the larval stages of the nematode were defined according to data from wild eel (Kuwahara et al., 1974; Hirose et al., 1976; Puqin and Yuru, 1980) and from experimental observations (De Charleroy et al., 1988a, b; Haenen et al., 1988). The following life cycle was proposed: within the egg, larval stage 1 (L_I) develops, which molts into larval stage 2 (L_{II}), still covered by the egg sheath; L_{II} leaves the eel, hatches in the water and is eaten by the first intermediate host (fresh water copepods) in which the L_{II} molts to L_{III}; the infected copepod is eaten by the eel and L_{III} migrates from the intestine towards the swimbladder wall, where it molts to L_{IV}; ultimately, L_{IV} migrates to the swimbladder lumen and becomes pre-adult and then adult.

Table 1. Observations of presence of *Anguillicola crassus* in squash preparations of organs of young eel, sampled in June, 1987 in the IJsselmeer. Prevalence of infection: 80%. Prevalence of swimbladders with inflammatory reactions: 60%. No presence of fibrotic swimbladders.

Length of eel in cm	Number of L_{III} in stomach, liver, intestine	Number of L_{III} / L_{IV} in swimbladder wall	Number of worms in swimbladder lumen		Presence of L_I / L_{II} in swimbladder lumen	Presence of inflammatory reactions and dilations of blood vessels
			pre-adult	adult		
6.5	0	0	0	0	–	–
7.0	0	0	0	0	–	–
7.4	1	0	0	0	–	–
7.5	0	0	0	0	–	–
7.7	0	0	0	0	–	–
8.5	0	3	1	2	+	+
8.8	0	4	2	0	–	+
9.0	0	0	1	1	+	+
9.0	0	0	0	1	–	+
9.5	0	0	0	2	+	+
9.5	2	1	0	1	–	–
9.5	0	4	1	1	–	+
9.8	2	4	0	2	+	+
9.8	0	2	0	4	+	+

\+ = present
– = not present

(continued)

Table 1 (continued)

Length of eel in cm	Number of L_{III} in stomach, liver, intestine	Number of L_{III} / L_{IV} in swimbladder wall	Number of worms in swimbladder lumen		Presence of L_I / L_{II} in swimbladder lumen	Presence of inflammatory reactions and dilations of blood vessels
			pre-adult	adult		
9.8	0	1	0	3	+	+
9.8	0	1	0	4	+	+
10.2	0	8	1	1	–	+
10.3	0	1	1	2	+	–
10.5	0	1	2	0	–	–
11.2	0	0	0	0	–	–
11.2	0	2	1	2	–	+
11.3	0	1	1	2	+	+
12.1	1	0	0	3	+	–
12.6	1	2	0	2	+	+
16.5	1	1	0	2	+	+

\+ = present
– = not present

RESULTS

Wild eel. *Anguillicola crassus* was found to be already abundant in the small eels sampled in June. In this month, such small eel have lived in the freshwater environment for only a short period (approximately 6-8 weeks), which can thus be also considered as the infection period. In view of such a short infection period, it was remarkable that *A. crassus* already showed a prevalence of 80% in June (Table 1). Furthermore, all stages of the nematode were already present in the swimbladder of the young eel from 8.5 cm upwards, including egg producing adults. The adults in these small eel were in the range of 3-22 mm length (Table 2).

Table 2. Lengths in mm of different stages of *Anguillicola crassus* as observed in young eel of 6-15 cm, sampled in June, 1987 from the IJsselmeer.

stage	n	length min. - max. range	length x ± SD
L_{IV} in swimbladder	36	0.5 - 2.8	1.1 ± 0.7
pre-adult	10	3 - 6	3.9 ± 1.1
adult female	18	6 - 22	10.1 ± 3.6
adult male	19	3 - 8	5.3 ± 1.8

The presence of *A. crassus* could be detected with squash preparations showing L_{III} in stomach, intestine, swimbladder, liver, and kidney, and L_{IV}, adults, L_I, and L_{II} in the swimbladder. The latter organ is considered to be the target organ for the effects of the activities of *A. crassus*. Acute inflammatory reactions, including dilation of blood vessels, formation of connective tissue, and rupture of the swimbladder were observed (Tables 1, 3). The presence of inflammatory reactions and connective tissue formation in the swimbladder wall of the small eel were clearly linked to the presence of *A. crassus*, especially when adults are present together with L_I and L_{II}. In the initial infection period (represented by the June sampling) the total infection prevalence in young eel was 80%, 60% showed inflammatory reactions in the swimbladder wall but thickened swimbladders were not detected macroscopically (Table 1). In the subsequent period (represented by the October sampling) these figures changed to 100%, 28%, and 80%, respectively (Table 3). In the squash preparations, eels with thickened or occluded swimbladders had fewer L_{III}, L_{IV}, and adults of *A. crassus*. In these adults, nonfunctional gonads were shown by the absence of eggs and L_I. Therefore, it seems that fibrotic thickened swimbladders are not suitable for further settlement and survival of *A. crassus*.

The presence of larvae, pre-adults, and adults of *A. crassus* was also detected in histological eel sections (Table 4). In both sampling periods (June and October) the first occurrence of adult parasites was found in 11 cm long eel. The digestive system of the worms was filled with eel erythrocytes, indicating that they had actively fed on the blood of the eel. No parasites were found in sections from eels measuring 7-9 cm.

Table 3. Observations of presence of *Anguillicola crassus* in squash preparations of organs of young eel sampled in October, 1987 in the IJsselmeer. Prevalence of infection: 100%. Prevalence of swimbladders with inflammatory reactions: 28%. Prevalence of eels with fibrotic swimbladder: 80%.

length of eel in cm	number of L_{III} in stomach, liver, intestine		number of L_{III} / L_{IV} in swimbladder wall		number of worms in swimbladder lumen				presence of L_I / L_{II} in swimbladder	presence of inflammatory reactions	presence of fibrotic swimbladder wall
					pre-adult		adult				
	A	N	A	N	A	N	A	N			
7.3	0	0	1	3	0	0	1	0	–	+	+
8.4	1	0	1	0	0	0	0	0	–	–	+
8.6	0	0	3	1	0	0	0	0	–	+	–
8.7	1	0	0	3	0	0	0	0	–	–	+
8.9	0	0	4	0	0	0	0	0	–	–	+
9.1	0	0	4	4	0	0	1	2	–	+	+
9.8	0	0	2	1	0	0	0	0	–	–	+
9.8	0	0	1	2	0	0	0	0	–	–	+
10.2	0	0	1	2	0	0	1	0	–	–	+
10.2	0	0	0	0	0	0	2	0	–	–	+
10.3	0	0	1	0	0	0	0	0	–	–	+
10.4	0	1	3	0	0	0	0	0	–	–	+

A = abnormal (dead or not well developed)
N = normal
+ = present
– = not present

(continued)

Table 3 (continued)

length of eel in cm	number of L_{III} in stomach, liver, intestine		number of L_{III} / L_{IV} in swimbladder wall		number of worms in swimbladder lumen				presence of L_I / L_{II} in swimbladder	presence of inflammatory reactions	presence of fibrotic swimbladder wall
					pre-adult		adult				
	A	N	A	N	A	N	A	N			
10.5	0	1	0	1	1	0	0	0	−	−	+
10.5	0	0	1	1	0	0	2	0	−	−	+
10.5	0	0	1	0	0	0	0	0	−	−	+
10.8	0	0	1	2	0	0	2	0	−	−	+
10.9	0	0	0	18	0	0	0	3	+	+	−
11.1	1	0	0	0	0	0	0	0	−	−	+
11.3	0	0	1	0	0	0	0	0	−	−	+
11.3	0	0	3	2	0	0	0	0	−	−	+
11.8	0	0	4	0	0	0	2	0	−	−	+
11.9	0	0	0	6	0	0	0	3	+	+	−
12.0	0	0	4	1	0	0	3	0	−	−	+
12.3	0	0	0	6	0	1	0	3	+	+	−
13.0	0	0	0	2	0	3	0	6	+	+	−

A = abnormal (dead or not well developed)
N = normal
+ = present
− = not present

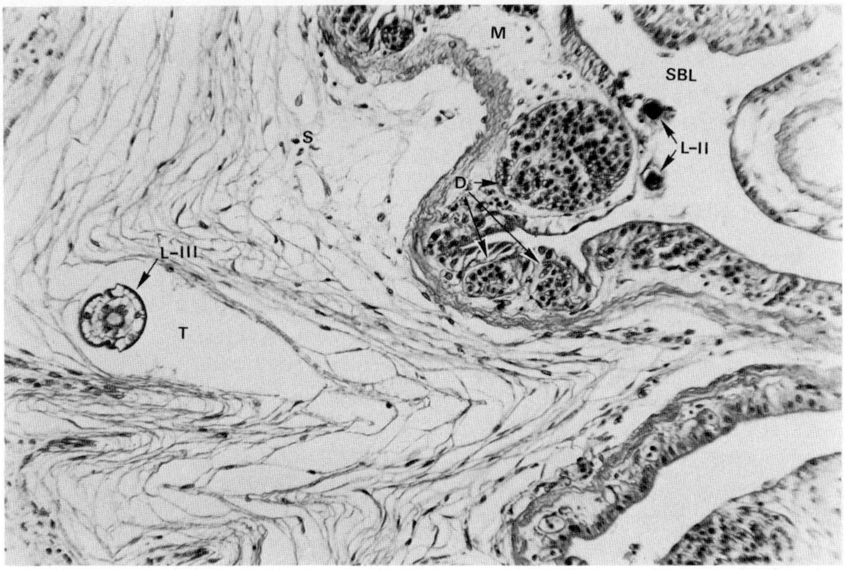

Fig. 2. Cross section of the swimbladder of wild eel (12 cm in length), infected with *Anguillicola crassus*. L_{II} larvae are attached to the thickened swimbladder wall. The mucosa (M) shows dilation of blood vessels (D). An L_{III} larva can be found in a tunnel (T) within the subserosa (S). H&E; 109X. SBL = swimbladder lumen.

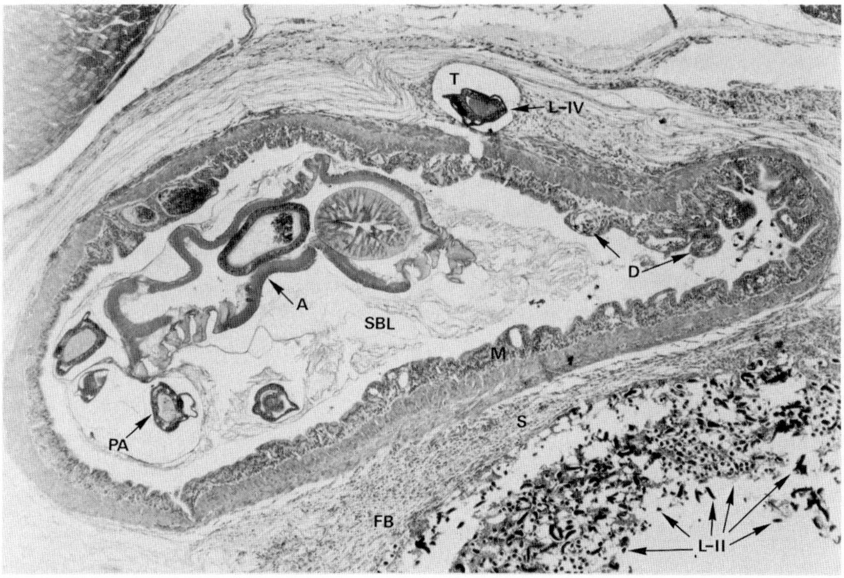

Fig. 3. Cross section of swimbladder of wild eel (14 cm in length) infected with *Anguillicola crassus*. L_{II} larvae are numerous in the fibrotic bulb (FB) of the subserosa (S) of the swimbladder. An L_{IV} larva has formed a tunnel (T) in this subserosa. Dilation of blood vessels (D) is seen in the mucosa (M) of the swimbladder wall. Pre-adult (PA) and adult (A) parasites can be found in the swimbladder lumen (SBL). H&E; 21.8X.

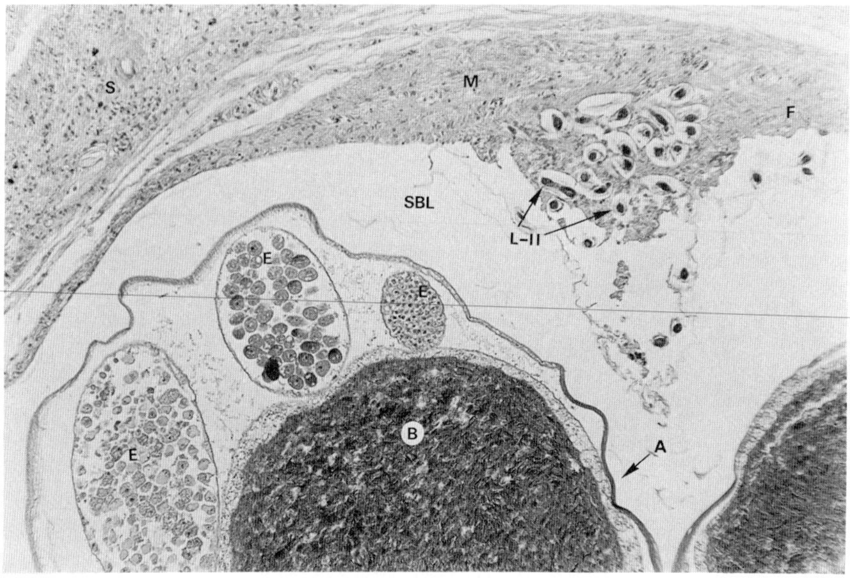

Fig. 4. Cross section of an *Anguillicola crassus* adult female (A) situated in the swimbladder lumen (SBL) of wild eel (12 cm in length). Different egg stages (E) can be seen. The parasite's gut is filled with erythrocytes of the eel (B). L_{II} larvae can be seen within the mucosa (M) of the swimbladder wall causing fibrosis (F). Fibrosis of the serosa (S) can also be observed. H&E; 54.5X.

The histological findings as to the effects of the parasite on swimbladder and organs of young eel are summarized in Figures 2, 3, 4 and Table 4. The following microscopical lesions were found: inflammatory reactions; tunnels in the swimbladder wall; dilation of blood vessels of the swimbladder wall; fibrosis and fibrotic conglomerates of the swimbladder, gut, and other organs; stenosis of intestine by shrinkage of scar tissue; and enteritis.

A few cases of larval presence (L_I and L_{III}) in muscular tissue were observed in squash preparations and histological sections, but most of these larvae were dead or encapsulated. Because of the rarity and the inactivated situation of such larvae in muscular tissue, they were considered as cases showing aberrant migration and, therefore, not representative of the normal behavior of *A. crassus* larvae within eel.

Farmed eel. Farmed eel heavily infected with *A. crassus* show loss of apetite, abnormal behavior by hanging near the surface, the presence of open skin ulcers in the caudal part of the belly, and a red and swollen anus. The swimbladders of such diseased eels always contain *A. crassus* adults with high production of eggs, L_I, and L_{II}. Macroscopically, their swimbladders frequently show hemorraghic characteristics and ruptures, resulting in the presence of free adult worms in the body cavity. In general, *A. crassus* infection strongly favors (secondary) bacterial infections in farmed eel. Bacterial typing showed that the infections were caused by *Aeromonas hydrophila*, and, additionally, in the marine farmed eel, by *Pseudomonas* spp. Such bacteria are to be considered as general opportunistic pathogens that enter the eel through the lesions caused by *A. crassus*. These secondary bacterial infections can result in an increase of mortality up to 10-20% in eel farms (Liewes and

Schaminee-Main, 1987). Farmed eel that survive the *A. crassus* infection can also show a thickened fibrotic swimbladder as observed in wild eel.

Histological sections of farmed eel showed the following:

(1) The first group consisted of eels 8-10 cm long from a freshwater farm. These had been fed 5 months previously with copepods infected with *A. crassus* larvae. Pre-adult and adult parasites were found in the swimbladder lumen. These small eels also showed L_{II} in the intestinal lumen and L_{III} were found in the fibrotic swimbladder wall and in the swimbladder lumen.

(2) The second group consisted of eels 25 cm or more in length, originating from a marine farm. These eels must have already been infected for a long time, at least since their introduction from fresh water for growth in the marine farm. L_{IV} was found in the swimbladder wall. Pre-adults and adults occurred in the swimbladder lumen. An adherent, fibrotic complex of swimbladder, kidney, and intestine had formed, filling up the body cavity. L_{II} were found in this complex. In one eel, L_{II} were observed in the ventral abdominal musculature.

DISCUSSION AND CONCLUSIONS

As Egusa (1979) has stated, *A. crassus* causes pathological changes in the European eel. The infection route and life cycle were defined by Hirose et al. (1976), Kuwahara et al. (1974), Puqin and Yuru (1980), De Charleroy et al. (1988a,b), and Haenen et al. (1988). Our results support and enhance the findings presented in the earlier studies.

The observed length of adult worms ranges of 3-22 mm in young eel (Table 2). This differs from those observed in larger eel by Kuwahara et al. (1974) and Taraschewski et al. (1987), which measured 20.5-71.5 mm and 5.7-36.7 mm, respectively. This indicates the ability of *A. crassus* to adapt the size of the adult stage to the available space in the swimbladder.

In eel, L_{III} were found in the intestinal wall, body cavity, swimbladder wall, and swimbladder lumen, which confirms the hypothesis that there is direct migration of L_{III} from the intestinal lumen to the swimbladder. L_{IV} were found in the swimbladder wall. Pre-adults and adults were found only in the swimbladder lumen.

A severe reaction to the parasite in the swimbladder was found both in wild and farmed eel: tunnel formation, acute inflammation with dilation of blood vessels, and ultimately fibrosis of the swimbladder wall. Finally, the swimbladder adheres totally to surrounding organs, such as kidney and intestine. In this complex, L_{II} tend to migrate and may get stuck, or lost, as in the case of incidental findings of L_{II} in the muscle. Furthermore, it seems that this phase of migration of L_{II}, with the accompanying irritation and inflammation of the tissues surrounding the swimbladder, is the most harmful phase of the parasite in the eel host. Whether eels with heavy fibrosis of the swimbladder cannot be reinfected or are less suitable for survival of the parasite needs further study.

Lesions in European eel caused by *A. crassus* appear to be much more severe than in the Japanese eel. In addition, aberrant migration routes have been found in the European eel. This may reflect the fact that European eel is not a natural host for *A. crassus*.

Secondary bacteriological infections were found in farmed eel. These were absent or very rare in wild eel, but it must be kept in mind that the dead or very

Table 4. Histological findings of presence of *Anguillicola crassus* in sections of young eel, sampled in 1987 from the IJsselmeer and from eel farms.

	WILD EEL										FARMED EEL	
											f.w. farm	sea farm
	J	O	J	O	J	O	J	O	J	O		
(cm)	7-8	7-8	9-10	9-10	11-12	11-12	13-14	13-14	15-16	15-16	8-10	> 25
L_{I} present	–		–	–	*	*	*	–	–	–	*	*
L_{II} in s.b. wall/lumen	–	–	–	–	*	*	*	*	*	*	*	*
L_{II} in body cavity	–	–	–	–	–	–	*	*	*	–	–	*
L_{III} in intest. wall/lumen	–	–	–	*	–	*	–	–	–	*	*	*
L_{III} in body cavity	–	–	–	*	–	–	–	–	–	–	*	–
L_{III} in s.b. wall/lumen	–	–	–	–	*	–	*	*	–	–	*	–
L_{IV} in s.b. wall	–	–	–	–	–	–	*	*	–	–	–	*
Pre-adult in s.b. lumen	–	–	–	*	*	*	*	*	–	–	*	*
Tunnels in s.b. wall	–	–	–	–	*	*	*	*	*	*	*	*
Inflammation s.b. wall	–	–	–	–	*	*	*	*	*	*	–	*
Dilation bloodv. s.b. wall	–	–	–	–	*	*	*	*	*	*	–	*
Fibrosis of s.b. wall	–	–	–	–	*	*	*	*	*	*	*	*
Fibrosis of intestine	–	–	–	*	*	*	*	*	–	*	–	*
Fibrosis of kidney	–	–	–	*	*	*	*	*	–	*	–	*
Bacterial infection	–°	–°	–°	–°	–°	–°	–°	*	–°	–°	*	*

The results of bacteriological tests are given in the bottom row for reference.
* = positive, – = no indication; ° = not tested; s.b. = swimbladder
f.w. = freshwater; J = June, 1987; O = October, 1987

diseased eels are not caught by fishing, whereas in eel farms they are. It is thus difficult to compare the losses in farmed and wild eels that are primarily due to *A. crassus*. Despite this, it will be interesting to monitor the pathological effects of the parasite in wild and farmed eels over the coming years. Furthermore, the prediction of a prevalence stabilization of the infection as a result of an increasing number of eels with thickened swimbladders in eel populations should be checked.

A. crassus is causing severe pathological effects in both wild and farmed European eel, *Anguilla anguilla*, in the Netherlands. In farmed eel, risk factors may facilitate a secondary bacterial sepsis of the *A. crassus*-infected eel. The general changes are acute inflammation followed by fibrosis of the swimbladder, which ultimately become incorporated into surrounding tissues and become nonfunctional. A survey of the wild stocks over the coming years is needed to determine the long-term effect of the introduced *A. crassus* on the recruitment and condition of young European eel.

LITERATURE CITED

Bergey, D.H. 1984. "Bergey's Manual of Systematic Bacteriology," Vol. 1 (N.R. Krieg and J.G. Holt, eds.). Williams and Wilkins, Baltimore.

Cowan, S.T. 1977. "Manual for the Identification of Medical Bacteria." Cambridge Univ. Press, Cambridge, England.

De Charleroy, D., K. Thomas and C. Belpaire. 1988a. Notes on the biology and diagnostical methods of *Anguillicola crassa*, a swimbladder nematode in the European eel (*Anguilla anguilla* L.). Rept. Zool. Instit. Univ. Leuven, Belgium.

De Charleroy, D., L. Grisez, K. Thomas, C. Belpaire and F. Ollevier. 1988b. The life cycle of *Anguillicola crassa* (Kuwahara, Niimi and Itagaki). Proc. F.H.S./A.F.S. Intl. Fish Health Conference, Vancouver (Abstract).

Dekker, W. and J. van Willigen. 1987. Short note on the distribution and abundance of *Anguillicola* in the Netherlands. Euro. Inland Fisher. Adv. Comm. (FAO), Report of Eel Working Group, Bristol, April, 1987.

Egusa, S. 1979. Notes on the culture of the European eel (*Anguilla anguilla*) in Japanese eel farming ponds. In: "Eel Research and Management" (F. Thurow, ed.). Rap. Proc. Verb. Reun. Cons. Int. Exp. Mer, 174, 51-58.

Haenen, O.L.M., L. Grisez, D. De Charleroy, C. Belpaire and P. Ollevier. 1988. Artificial infection of the European eel, *Anguilla anguilla* L., with L_{III} larvae of *Anguillicola crassus* (Nematoda, Dracunculoidea). Proc. F.H.S./A.F.S. Intl. Fish Health Conference, Vancouver (Abstract).

Hirose, H., T. Sekino and S. Egusa. 1976. Notes on the egg deposition, larval migration and intermediate host of the nematode *Anguillicola crassa* parasitic in the swimbladder of eels. Fish Pathol., 11, 27-31 (In Japanese; English summary).

Køie, M. 1987. Eel parasites in natural waters and aquaculture in Denmark. In: "Parasites and Diseases in Natural Waters and Aquaculture in Nordic Countries" (A. Stenmark and G. Malmberg, eds.). Zoo-tax, Naturhistoriska Riksmuseet, Stockholm, Sweden.

Kuwahara, A., A. Miimi and H. Itagaki. 1974. Studies on a nematode parasitic in the air bladder of the eel. I. Description of *Anguillicola crassa* n.sp. (Philometridae, Anguillicolidae). Japan. J. Parasit., 23, 275-279.

Liewes, E.W. and S. Schaminee-Main. 1987. Onderzoek naar de effecten van de parasiet *Anguillicola crassa* op de ontwikkeling van de paling (*Anguilla anguilla*) in een zoutwater palingmesterij. Intern Report Texvis BV, Den Berg, Texel, The Netherlands.

Moravec, F. and H. Taraschewski. 1988. Revision of the genus *Anguillicola* Yamaguti, 1935 (Nematoda: Anguillicolidae) of the swimbladder of eels, including descriptions of two new species, *A. novaezelandiae* sp.n. and *A. papernai* sp.m. Folia Parasitol., in press.

Neumann, W. 1985. Schwimmblasenparasit *Anguillicola* bei Aalen. Fischer und Teichwirt, 11, 322.

Peters, G. and F. Hartmann. 1986. *Anguillicola*, a parasitic nematode of the swimbladder spreading among eel populations in Europe. Dis. Aquat. Organ., 1, 229-230.

Puqin, W. and Z. Yuru. 1980. Observations on the life history of *Anguillicola globiceps* (Nematoda:Anguillicolidae). Acta Zool. Sin., 26, 243-249.

Taraschewski, H., F. Moravec, T. Lamah and K. Anders. 1987. Distribution and morphology of two helminths recently introduced into European eel populations: *Anguillicola crassus* (Nematoda,Dracunculoidea) and *Paratenuisentis ambiguus* (Acanthocephala, Tenuisentidae). Dis. Aquat. Organ., 3, 167-176.

van Banning, P., W. Heermans and J.A. van Willigen. 1985. *Anguillicola crassa*, een nieuwe aalparasiet in Nederlandse wateren. Visserij, 38, 237-240.

METAZOAN PARASITIC DISEASES: CONCLUDING REMARKS

Claude Combes

Centre de Biologie et d'Ecologie Tropicale et Méditerranéenne,
Université de Perpignan, France

INTRODUCTION

By definition, a parasite feeds at the expense of its host, which implies a damage, even if this does not result in obvious pathology. This damage results in a statistical decrease in the fecundity of the host, since it must assign a part of its resources to its defense mechanisms (Price, 1975) and to the restoration of injured tissues. Furthermore, the parasitized animal's foraging activities, chances of mating, production of offspring, and ability to protect the young are also reduced. Anderson and May (1978) even consider that an organism must decrease the instantaneous growth rate of the host population to deserve the qualification of parasite. If the host's offspring are affected in quantity or even in quality, the chances for the parasite to come across the living environment it needs for survival are reduced. It can be assumed that this issue is especially crucial when the parasite has a narrow specificity, which is frequently the situation.

For the host-parasite system to persist through time, it is essential that the pathologic alterations caused by the parasite do not entail an exaggerated decrease in the host population. We must draw attention to the populational aspect of this question. It does not matter if the pathologic changes in some individual host is severe (even total castration) if the pathology at the population level remains unimportant. Regarding the problem we are facing, it is not the pathologic picture in the individual that counts but the populational pathology, strongly influenced by the prevalence, the abundance (according to the definitions of Margolis et al., 1982), and the distribution (aggregated or not) of the parasites. Incidentally, we must realize that a similar question arises in prey-predator systems. While they subtract a part of the prey population, the population of predators must preserve a certain equilibrium in order to assure the future of their descendants. It is well known that if a balance is not maintained, the predator population faces cyclic or definitive collapse.

If the above reasoning is sound, we can expect that the selective pressures exerted on host-parasite systems should provoke a reduction as pronounced as possible of the pathology caused by the parasite on the host. It is obvious that the selective pressures act both on the parasite and the host genome so that the diminution of the pathologic changes will be the result at the same time of a true decrease of the pathogenic effect of the parasite and of the development of the host defense mechanisms. Logically, the parasites should not be pathogenic, or at least very mildly so. Unfortunately, in the case of human health and for economically important animals, this is not the case. So the question is, "Under what circumstances does a parasite become a pathogen?", or more straightforward, "When is a parasite a pathogen?"

SIX PATHOLOGY RULES

If the hypothesis of a selection of minimal pathogenicity does not seem to apply to most observed situations, especially in marine aquaculture, we believe that six possible explanations exist. These are not exclusive from each other, as we will see later.

Hypothesis is erroneous because the reasoning is erroneous. The hypothesis, as formulated, supposes that the pathogenesis directed towards the host is definitely unfavorable to the fate of the parasite. However, in quite a number of cases, the pathology of the host is favorable, or even necessary, to the completion of the parasite's life cycle. This is often true, as May and Anderson (1983) note in the case of "microparasites" such as viruses and bacteria, but also in the case of certain metazoan parasites. This is the case when a host harbors the infective stage of a parasite that must be ingested by a different host in which another stage of the parasite will develop. This process, long known and for which the global term of "favorisation" has been proposed (Combes, 1980), intervenes during the course of trematode and cestode life cycles in which the definitive host becomes infected by consuming a host harboring metacercariae or cysticerci, respectively. A beautiful example is provided by McGladdery (this volume). The metacercariae of *Stephanostomum tenue* provoke in the fish host a pericardiac pathology that can easily be suspected to diminish the escape performances of the fish, making them more vulnerable to predators. Other spectacular cases have been reported in aquatic environments. For example, Maillard (1976) demonstrated that the metacercariae of another trematode, *Cainocreadium labracis*, encyst in the fin muscles of the second intermediate host, *Potamoschistus microps*, which serves as prey of the definitive host, *Dicentrarchus labrax*.

The case of *Diplostomum spathaceum* is also classical. The metacercariae become established in the eyes of the fish intermediate host, which suffer a decrease of visual acuity (Erasmus, 1958) and exhibit weakened escape reactions in the presence of the piscivorous birds that serve as definitive hosts. *Ornithodiplostomum ptychocheilus*, the metacercariae of which encyst in the brain, also provoke alterations in fish behavior, which favor transmission of the parasite (Radabaugh, 1980).

Curtis (1987) demonstrated that a pathological change in host behavior can also be observed in tramsmission processes not involving predation. A snail from American estuaries, *Ilyanassa obsoleta*, migrates towards unusual environments (the beaches and sandbars) when it is infected by the trematode *Gynaecotyla adunca*, and field observations suggest that it is indeed a favorisation process; the displacement of the parasitized molluscs increases the probability of encystment of the cercariae in crustacean hosts (the cercariae have no tail and are unable to migrate large distances). Curtis has remarked, "... since parasitized snails are virtually all sterile, any adaptive value in altered host behavior goes to the parasite." It is, therefore, clear that in marine aquaculture, as in any other situation (either natural or artificial), the damage caused by parasites in intermediate hosts is sometimes the logical result of the process that selected them.

Incidentally, permit me to mention that Crofton (1971) considered certain pathogenic effects caused by heavy burdens of parasites (related to the aggregative distribution of parasites) have an adaptive value as well, i.e., provoking the death of these heavily infected hosts and regulating the parasite population. The effectiveness of such a phenomenon has not been demonstrated often, but it illustrates very well the distinction made previously between individual and populational pathology. If it does occur, the precise mechanism of such a selection remains to be clarified.

Host species exhibit mechanisms that compensate impact of parasitism. Having accepted the concept that the pathology caused by a parasite reduces host fecundity, which, in the long term, can be disadvantageous for transmission, we need to examine this phenomenon further. Certain host species possess very high fecundity levels that are normally adjusted by regulatory mechanisms. In this case, parasitism, even if it reduces the number of offspring of the host, will have no effect on the demography of the population. For instance, Euzet (pers. commun.) informed me that in the Etang de Thau (South of France), the prevalence of *Sacculina carcini* (which causes total castration of crabs) often exceed 50%. The fecundity of noninfected crabs is, therefore, sufficient to compensate for the nonreproducing status of half the population. It must be stated that the negative impact of parasitism is just one population reduction factor to which the host species adjusts as a result of its high reproductive potential.

It would be interesting to determine if parasitism, by exerting pressures, can be a driving force in the evolution of the reproductive capacity of the host. If this is the case, the enhanced number of offspring can be favorable to the parasite, which can augment the pressures, and so on. This process is very similar to that of immunity, which evolves under the pressure of the parasites, which, in turn, evolves to adapt to immunity, until a balance is reached.

We must acknowledge that, in the present status of the knowledge, it is not possible to assess if the selective pressures exerted by parasites actually are responsible for influencing host fecundity. Since Hamilton (1980) has suggested that "pressure of parasites that are short-lived and rapid-evolving compared to the hosts they attack could be an evolutionary factor sufficiently general to account for sex wherever it exists," a possible role of parasites in influencing the evolution of host fecundity, in addition to the evolution of sexuality, cannot be discarded. This difficult question has received little attention. However, increasing knowledge on the roles played by host genome evolution and by parasite genome evolution in the achievement of host-parasite systems would be of equal interest for basic and applied purposes.

Host-parasite system too young for selective pressures to be efficient. Futuyma and Kim (1987) consider that "the phylogenies of hosts and parasites show little congruence at any taxonomic level ... lateral transfer has been a major feature of parasite evolution."

Several recent investigations demonstrate that, in certain lineages, parasites are remarkably faithful to their hosts for long durations. This "parallel evolution" is known as Farenholz's A rule ("Parasite phylogeny mirrors host phylogeny"). Clear demonstrations of this rule have recently been provided by Durette-Desset (1986), who used paleobiogeography of hosts, and by Hafner and Nadler (1988), based on phenetic analysis. In the case of parasites of marine organisms, the demonstration of Renaud (1988) of fish cestodes, based on isoenzymatic studies, can be mentioned.

Even if the above reveal that the opinion of Futuyma and Kim (1987) must be somehow "blended," it remains that lateral transfers (designated as "captures" by Chabaud (1975), Quentin (1971), and others) have played a marked role in the evolution of parasites. These transfers occur in nature when evolutionary shifts result in the sharing of niches of certain species with others, or simply at the change of geographical areas.

Animal transportations by man during the last thousand years, especially during the last decade, have multiplied the opportunities for the parasites to engage in lateral transfers; either an uninfected host is transported into an area where a potential (i.e., candidate to transfer) parasite exists, or a parasite is introduced with its host into an area where an uninfected but related host exists. Lightner (this volume) is correct in calling attention on the danger of "jumbo-transports" that multiply the risks.

Reasoning, as well as observations, lead to the assumption that a young host-parasite system (i.e., resulting from a recent capture) can be extremely harmful for the host, since selective pressures have not had enough time to reduce the pathogenic effect. The example of sleeping sickness is well known. In west Africa, man is afflicted by a "variety" of trypanosomes that cause a chronic disease probably acquired long ago. In east Africa, an almost indistinguishable parasite is responsible for an acute and very serious disease and is considered to have been acquired recently (Baker, 1974). Sometimes it is possible to witness such an evolution towards a moderate pathogenicity. This is the case for some viruses. A striking example being that of the myxomatosis virus, which, introduced in Europe in the 1950s, has selected in 25-30 years in favor of increasingly less pathogenic strains (May and Anderson, 1983). Rau (1982) even writes "parasite-associated pathology and host-death are frequently considered to be little more than temporary evolutionary maladaptations which natural selection will tend to minimize as the host-parasite relationship matures."

In marine aquaculture, it is not surprising that the sudden introduction of molluscs, crustaceans, or fishes into new environments is often accompanied by explosive parasitoses (in the broad sense). Everything can go well if a providential narrow specificity hampers lateral transfer, but in the opposite case, the worst can happen. We see a good illustration of that in the contribution by van Banning and Haenen (this volume) who demonstrate that the nematode *Anguillicola crassus*, initially a nonharmful parasite of the Japanese eel, *Anguilla japonica*, with which it was imported, has become highly pathogenic to the European eel, *Anguilla anguilla*. Instead of using the swimbladder pneumatic duct to leave the host (as they do in their normal host), this nematode larva makes its way through the tissues!

Host-parasite system is loosely specific. We have interpreted the evolution towards limited pathogenicity as a result of pressures that favor the less pathogenic parasites. This implies, on the part of these individuals, a precise adaptation to the morphology, physiology, and defense mechanisms of the host. For instance, in the case of

Anguillicola crassus (see above), finding the pneumatic duct as a path to exit from the host is a behavioral adaptation that lessens pathogenesis. It is sure that symmetrically (but on a scale always difficult to estimate), the host itself acquires adaptations to the parasite. When these reciprocal selective pressures are demonstrated, they constitute a pattern of coevolution.

It seems that the best possible equilibrium can be selected only if the system comprises one host species and one parasite species (or perhaps several ones but very closely related to each other). This question of the role of specificity in the evolution of host-parasite systems has been analyzed in detail by Euzet and Combes (1980).

It can be expected that a negative correlation exists between specificity and pathogenicity. More specifically, the resulting pathology should be more pronounced in certain species of the host spectrum, probably the more recently acquired ones if we consider the previous paragraph. We think that such cases do exist. For instance, *Schistosoma japonicum*, a parasite with a broad host spectrum, is more pathogenic to man than are other members of the genus *Schistosoma*, which are more host-specific. Moreover, it seems to be more harmful in man, a recently acquired host, than in many other mammals in endemic areas.

It is likely that similar examples could be found among parasites of marine animals. If this were verified, it could be assumed that an oioxenic (Euzet and Combes, 1980) or stenoxenic specificity is associated with limited pathogenicity, while a euryxenic specificity is associated with a more pronounced pathogenicity.

Hybridizations have provoked lack of resistant genes in certain individual hosts. We must recall two very recent results concerning the parasitism in hybrid organisms. The first was reported by Sage et al. (1986). These authors demonstrated that in a zone (Germany) where the mice *Mus musculus musculus* and *Mus musculus domesticus* hybridize, the individuals trapped in the hybridization zone are significantly more parasitized by nematodes than the ones caught in zones where only the parent mice live. The authors emphasized the fact that no peculiarity of the environment can account for an increase of nematode transmission. From the Sage et al. data, we have drawn Figure 1A.

The second was reported by Dupont and Crivelli (1988). This concerns freshwater fishes of the genera *Rutilus* and *Alburnus*, and their natural hybrids, in Greece. As previously (but in an even more spectacular way), it was observed that the hybrids are significantly more parasitized. Figures 1B and 1C have been constructed from the authors' data. In the papers by Sage et al. (1986) and Dupont and Crivelli (1988), the authors report a strong aggregative distribution of the population of parasites in the population of hosts.

Sage et al. (1986) propose a convincing explanation. They suppose that "resistance genes" (on which nothing is known to date) are situated at different loci so that among individuals originating from crosses between hybrids and back-crosses certain ones would be totally lacking resistance genes. This explanation accounts for the increase of both prevalence and aggregation.

In view of the above, we think that the risk must be considered in introducing hybrids, supposedly more resistant than their parents, to certain diseases. The results reviewed above show that, in particular cases, the hybrids are more susceptible to pathogens.

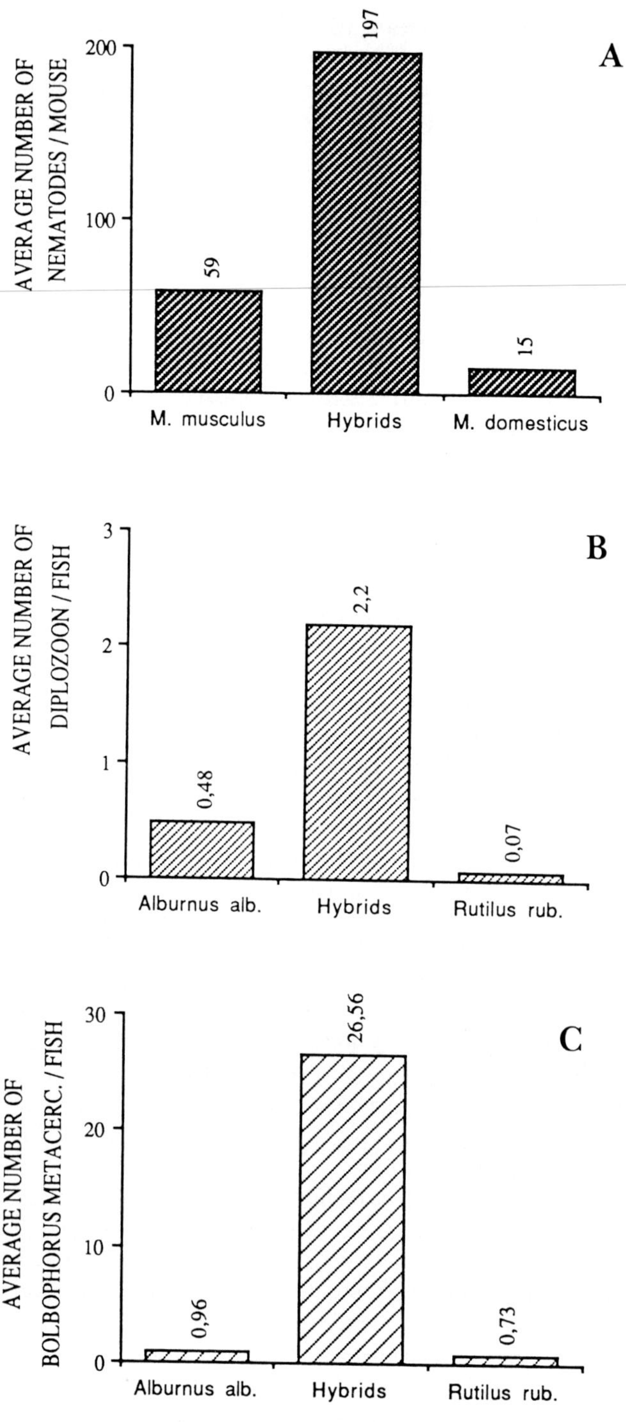

Fig. 1. Influence of hybridization on parasitism. A. Graph drawn from the data of Sage et al. (1986). B and C. Graph drawn from the data of Dupont and Crivelli (1988).

Environmental conditions are not ones in which selection operated. Selection of adaptive traits is only of value for the environment in which the selective pressure operated. This remains true for the selection of the characters (of the parasite but also of the host) that lead to the equilibrium between the members of host-parasite systems.

In breeding conditions, and more particularly in aquaculture, the environments in which the parasite and the host are kept are often quite different from the natural milieu (Euzet and Raibaut, 1985). The most evident change concerns the probability of encounter between the infecting stages of parasites and their targets, since the dispersion of the parasites is limited while the density of hosts is increased. Silan and Maillard (1986) have demonstrated that the parasitism of *Dicentrarchus labrax* by gill monogeneans never reaches lethal levels in natural conditions, but that the threshold is rapidly crossed in farms. Silan and Maillard (1986) also note that, in breeding conditions, distribution of monogeneans in the fish population becomes strongly aggregated, which enhances pathologic alterations. Cheng (this volume) shows the negative influence that alterations in ionic composition of water can have upon the defense mechanisms of *Crassostrea virginica*.

A typical example of the influence of the environment is reported by Raibaut (1985). Natural populations of the copepod *Ergasilus lizae*, a parasite of grey mullets, are regulated when the fish host migrates into the open sea, the parasite being unable to survive in high salinities. If the fish are permanently maintained in brackish water for breeding purposes, the copepod populations are no longer regulated and the outbreak of parasites provokes disease.

Renaud (pers. comm.) emphasizes that the stocks, which are exploited in aquaculture, often arise from a very limited number of progenitors, which causes an impoverishment of the gene pool when compared with wild populations. This certainly increases their vulnerability. Everybody knows the disasters that sometimes followed the use in agriculture of cultivars that have lost all genetic variability.

Of course, changes are not always negative and, on the contrary, it may happen that breeding conditions prevent the development of certain parasites. In the open environment where marine aquaculture is usually performed, these cases are likely to be exceptions.

One must keep in mind that in the conditions of marine aquaculture, even if a novel host-parasite system is not created, the chances given to the parasite to cause an exaggerated pathology can be considerably increased by elevations of parasite burdens resulting from better transmission.

DISCUSSION AND CONCLUSIONS

When the conditions we have enumerated to explain serious pathologies caused by metazoan parasites are carefully analyzed, it is striking that these rules of pathology are often met in the conditions of marine aquaculture. Two or more often apply in a single farm.

The first (natural selection of pathology) applies to animals that are intermediate hosts in heteroxenic life cycles, or, more generally, every time pathologic changes favor transmission. This can be the case for any of the marine animals that are bred: molluscs, crustaceans, and fishes.

The second rule (non-selection of a limited pathology for demographic reasons) also concerns the genetic characteristics of the hosts. It is clear that it is not in the interest of the farmer to choose species that "accept" a marked pathology when confronted with parasites. This would combine the inconveniences of a high mortality and of a reduction of the commercial value.

The third rule (youth of the host-parasite system) applies more especially as farmers often import allochtonous eggs or larvae. One of the major measures is to avoid the formation of novel host-parasite systems as a consequence of hazardous introductions. The risk of importing parasites to which local host races or species will be more receptive is equal to the risk of importing exotic races of species of hosts that will become infected by local parasites.

The fourth rule (importance of specificity) applies when the host spectrum of a parasite is broad; not only is the parasite likely to infect various species of hosts but also the resulting pathology can be important. Strict control is necessary to prevent the importation of such parasites to farms.

The fifth rule (susceptibility of hybrids) has not been thoroughly investigated in any marine environment, but it must be kept in mind when supposedly resistant hybrids are selected.

Finally, the sixth rule (environment favoring the parasite) is the most widely accepted since, by definition, the breeding conditions diverge more or less from nature. As soon as a species of commercial interest is bred, changes in the occurrence or frequency of parasites are likely to appear, whatever the sense (increase or decrease). It seems that one can often observe a replacement of many but less pathogenic species by a lesser number of more pathogenic ones. Only a thorough study of the transmission patterns will permit the selection of "ecological" measures to reduce the level of transmission.

The six rules we have stated illustrate the reasons why the pathogenicity of a parasite can be numerous and varied. As far as the theoretical selection for minimal pathology is concerned, the fact is that the most perfect coevolution process cannot overcome certain limits. The best proof is probably provided by the vector arthropods; their active role in transmission is so essential that any reduction of their activity can only be prejudicial to the parasite. Nevertheless, very accurate studies (Rowland and Boersma, 1988) have demonstrated that anopheline mosquitoes harboring Plasmodium have a flight ability that can be reduced by one third. This reduction is probably "the best compromise" possible for the parasite, between the necessity of feeding at the prejudice of the mosquito and the necessity of permitting it to fly.

The questions we have evoked are not specific to marine aquaculture; nevertheless, it is clear that the control of pathogens in the aquatic environment is especially difficult. One of the main obstacles to prevention lies in the difficulty of isolating the breeding places from contaminations or re-contaminations.

If protecting marine aquaculture from parasitism seems indeed a challenge, we would like to stress how the notions of genetics and selection have been present in our analysis. This probably means that studies pertaining to genetic structure of individuals, populations, or species constitute the main hope to modify or even manipulate these structures, and hence to permit improvements.

Immunology, with the perspective of vaccine produced by genetic engineering, and therapy, with the discovery of new molecules, should help displace equilibriums in favor of hosts. Whatever the progress may be, research on factors of the environment that influence transmission will always remain extremely important to achieve the optimum conditions of aquaculture.

ACKNOWLEDGEMENTS

I am very grateful to Professor T.C. Cheng, Medical University of South Carolina; Professor L. Euzet and Professor A. Raibaut, Université de Montpellier; Professor R. Barbault, Ecole Normale Supérieure de Paris; and Dr. F. Renaud, Centre National de la Recherche Scientifique, for their valuable comments and suggestions.

LITERATURE CITED

Anderson, R.M. and R. M. May. 1978. Regulation and stability of host-parasite population interactions. I. Regulatory processes. J. Ani. Ecol., 47, 219-249.

Baker, J.R. 1974. Epidemiology of African sleeping sickness. In: "Trypanosomiasis and Leishmaniasis", pp. 29-50. Elsevier. Excerpta Medica. North Holland, Amsterdam.

Chabaud, A. 1975. Spécificité parasitaire chez les Nématodes parasites de Vertébrés. In: "Traité de Zoologie, 4, Némathelminthes" (Grassé, ed.), pp. 548-557. Masson, Paris.

Combes, C. 1980. Les méchanismes de recrutement chez les Métazoaires parasites et leur interprétation en termes de stratégies démographiques. Vie et Milieu, 30, 55-63.

Crofton, H.D. 1971. A quantitative approach to parasitism. Parasitology, 62, 179-194.

Curtis, L.A. 1987. Vertical distribution of an estuarine snail altered by a parasite. Science, 235, 1509-1511.

Dupont, F. and A. Crivelli. 1988. Do parasites confer a disadvantage to hybrids? A case study of *Alburnus alburnus* x *Rutilus rubilio*, a natural hybrid of Lake Mikri Prespa, Northern Greece. Oecologia, 75, 587-592.

Durette-Desset, M.-C. 1986. Trichostrongyloid nematodes and their vertebrate hosts: reconstruction of the phylogeny of a parasitic group. Adv. Parasitol., 24, 239-306.

Erasmus, D.A. 1958. Studies on the morphology, biology and development of a strigeid cercaria (*Cercaria* x Baylis, 1930). Parasitology, 48, 312-335.

Euzet, L. and C. Combes. 1980. Les problèmes de l'espèce chez les animaux parasites. In: "Les Problèmes de l'Espèce dans le Règne Animal". Mém. Soc. Zool. France, 40, 239-285.

Euzet, L. and A. Raibaut. 1985. Les maladies parasitaires en pisciculture marine. Symbioses, 17, 51-68.

Futuyma, D.J. and J. Kim. 1987. Phylogeny and coevolution. Science, 237, 441-442.

Hafner, M.S. and S.A. Nadler. 1988. Phylogenetic trees support the evolution of parasites and their hosts. Nature, 332, 258-259.

Hamilton, W.D. 1980. Sex versus non-sex parasite. Oikos, 35, 282-290.

Maillard, C. 1976. Distomatoses de Poissons en milieu lagunaire. Thèse Université de Montpellier, France. 383 pp.

Margolis, L., G.W. Esch, J.C. Holmes, A.M. Kuris and G.A. Schad. 1982. The use of ecological terms in parasitology. J. Parasitol., 68, 131-133.

May, R.M. and R.M. Anderson. 1983. Epidemiology and genetics in the coevolution of parasites and hosts. Proc. Roy. Soc. London, B 219, 281- 313.

Price, E.W. 1975. "Evolutionary Strategies of Parasitic Insects and Mites." Plenum, New York.

Quentin, J.-C. 1971. Sur les modalités d'évolution chez quelques lignées d'Helminthes de Rongeurs Muroidea. Cahiers O.R.S.T.O.M., sér. Entomol. Méd. Parasitol., 9, 103-176.

Radabaugh, D.C. 1980. Changes in minnow, *Pimephales promelas* Rafinesque, schooling behavior associated with infection of brain-encysted larvae of the fluke *Ornithodiplostomum ptychocheilus*. J. Fish Biol., 16, 621-628.

Raibaut, A. 1985. Les cycles évolutifs des Copépode parasites et les modalités de l'infestation. Année Biol., 24, 233-274.

Rowland, M. and E. Boersma. 1988. Changes in the spontaneous flight activity of the mosquito *Anopheles stephensi* by parasitization with the rodent malaria *Plasmodium yoelii*. Parasitology, 97, 221-227.

Rau, M.E. 1982. Behavioural pathology and parasite transmission: A speculative synthesis. In: "Aspects of Parasitology" (Meerovitch, ed.), pp. 335-360. The Institute of Parasitology, McGill University, Montreal.

Renaud, F. 1988. Biologie et évolution des populations d'Helminthes parasites. Le modèle Helminthes-Téléostéens. Thèse Université de Montpellier, France. 136 pp.

Sage, R.D., D. Heyneman, K.C. Lim and A.C. Wilson. 1986. Wormy mice in a hybrid zone. Nature, 324, 60-63.

Silan, P. and C. Maillard. 1986. Modalités de l'infestation par *Diplectanum aequans*, Monogène ectoparasite de *Dicentrachus labrax* en aquiculture. Eléments d'épidémiologie et de prophylaxie. In: "Pathology in Marine Aquaculture" (Vivarès, Bonami, and Jaspers, eds.), pp. 139-152. European Aquaculture Society, Special Publication No. 9, Bredene, Belgium.

TOXICOLOGICAL PATHOLOGY SECTION

TOXICOLOGICAL PATHOLOGY: INTRODUCTORY REMARKS

John A. Couch

Environmental Research Laboratory, U.S. Environmental Protection Agency, Gulf Breeze, Florida 32561 USA

Toxicological pathology, as a discipline, deals with the effects of toxins and toxicants in causing disease or influencing disease processes in organisms. As a hybrid science, toxicological pathology utilizes concepts, methods, and tools from both toxicology and pathology.

Toxicological pathology is of key importance in understanding diseases that may affect culture of marine species. Once infectious agents or nutritional deficiencies have been eliminated as possible disease causes in marine culture, attention must be turned to possible chemical agents as etiologic. Central to this concept is the fact that there can be no healthy culture of marine species without high water quality, which, of course, depends upon presence or absence of chemical toxins and toxicants, as well as normal ranges for pH, O_2, salinity, and temperature.

The water quality principle applies to culture of marine species both in natural water bodies and in artificial cultures systems. Thus, there is a great need for a better understanding of those lesions and diseases that may be caused by waterborne toxins and toxicants.

The role of toxicological pathology in marine aquaculture is to determine the following: (1) causes of specific tissue and cellular effects by specific chemicals, i.e., linkage of cause and effect; (2) identification, characterization, and understanding of lesion and disease syndromes linked to toxins or toxicants; (3) resolution of etiological agents, e.g., biological, infectious agents versus chemical agents, in disease syndromes; (4) an understanding of interactions, enhancements, inhibitions, or other relationships between and among biological and chemical injurious agents; and (5) prescription, following an understanding of a disease process linked to chemical agents, for a preventive or therapeutic treatment of the condition.

In the last 20 years the published literature on toxicological pathology of aquatic species has grown considerably (see reviews by Couch, 1975, 1988; Meyer and Hendricks, 1982; Hinton and Couch, 1984; Couch and Harshbarger, 1985).

The identification of lesions that are biomarkers or indicators of specific or general exposures to either single chemical agents or mixtures of chemical agents is an important area of present research. These biomarker lesions are occasionally identified by laborious efforts in investigative studies resulting from the need to solve practical problems, but some of these lesions have been carefully, experimentally induced and their biomarker significance ascertained experimentally, under controlled conditions. Once the lesion is identified as a biomarker of a general or specific exposure in aquatic animals, its presence in samples of animals is indicative of contamination.

Three examples of possible biomarker or indicator lesions in fishes are: (1) Hepatocellular carcinoma – indicative of hepatocarcinogen exposure, e.g., aflatoxins, N-nitroso compounds; (2) fatty change and hepatocellular necrosis – indicative of certain volatile organics, PCBs, and some hepatotoxicants, such as carbon tetrachloride; and (3) Spongiosis hepatis – indicative of exposures to N-nitroso compounds and, perhaps, other hepatotoxicants.

The most relevant example of these biomarker lesions is spongiosis hepatis. Spongiosis hepatis (SH) is a lesion heretofore reported in the liver of certain mammals and fishes following experimental and possibly inadvertent exposure to hepatotoxicant chemicals. Bannasch et al. (1981) first reported and identified the lesion as a distinct entity in rats experimentally exposed to N-nitroso compounds. Since that report, the same lesion type has been induced experimentally in several species of fishes with chemical carcinogens (Couch and Courtney, 1987) and in feral fishes from contaminated habitats (Myers et al., 1987). SH has its origin in the perisinusoidal cell (cell of Ito) in the space of Disse, and progresses from small reticulated foci to larger foci and eventually to pericytomatous form in the liver of the sheepshead minnow following experimental exposure to diethylnitrosamine. Partially similar lesion development has also been reported for medaka, guppy, and feral flatfishes following exact or suspected exposure to various chemical carcinogens. The identification of the cell or origin, analysis of SH lesion progression, and ultrastructural characterization have permitted reasonable confidence that the lesion is homologous in mammals and fishes and probably represents a specific indicator of exposure to certain chemical carcinogens, either in the laboratory or in the field. At present, there is no contrary evidence that suggests that infectious agents or innate causes may be responsible for the lesion in mammals or fishes.

LITERATURE CITED

Bannasch, P., M. Bloch and H. Zerban. 1981. Spongiosis hepatis: Specific changes of the perisinusoidal liver cells induced in rats by N- nitrosomorpholine. Lab. Invest., 44, 252-264.

Couch, J.A. 1975. Histopathological effects of pesticides and related chemicals on the liver of fishes. In: "Pathology of Fishes" (Ribelin and Migaki, eds.), pp. 559-584. Univ. Wisconsin Press, Madison.

Couch, J.A. 1988. Role of pathobiology in experimental marine biology and ecology. J. Exp. Mar. Biol. Ecol., 118, 1-6.

Couch, J.A. and L. Courtney. 1987. N-nitrosodiethylamine-induced hepatocarcinogenesis in estuarine sheepshead minnow (*Cyprinodon variegatus*): Neoplasms and related lesions compared with mammalian lesions. J. Natl. Canc. Inst., 79, 292-321.

Couch, J.A. and J. Harshbarger. 1985. Effects of carcinogenic agents on aquatic animals: An experimental and environmental overview. Environ. Carcinogenesis Revs., 3, 63-105.

Hinton, D. and J. Couch. 1984. Pathobiological measures of marine pollution effects. In: "Concepts in Marine Pollution Measurements" (H.H. White, ed.), pp. 7-32. Maryland Sea Grant Publication UM-SG-TS-84-03, College Park.

Meyers, T.R. and J. Hendricks. 1982. A summary of tissue lesions in aquatic animals induced by controlled exposures to environmental contaminants, chemotherapeutic agents, and potential carcinogens. Mar. Fish. Revs., 44, 1-17.

Myers, M., L. Rhodes and B.M. Cain. 1987. Pathologic anatomy and patterns at occurrence of hepatic neoplasms putative preneoplastic lesions, and other idiopathic hepatic conditions in English sole (*Parophrys vetulus*) from Puget Sound, Washington. J. Natl. Canc. Inst., 78, 333-363.

QUALITATIVE AND QUANTITATIVE ASSESSMENT OF THE RESPONSE OF MILKFISH, *CHANOS CHANOS*, FRY TO LOW-LEVEL COPPER EXPOSURE

Helmut Segner[1] and Thomas Braunbeck[2]

[1]Zoologie II (Physiologie), Universität Karlsruhe, Kaiserstr. 12, D 7500 Karlsruhe, Federal Republic of Germany
[2]Zoologie I (Morphologie/Ökologie), Universität Heidelberg, Im Neuenheimer Feld 230, D 6900 Heidelberg, Federal Republic of Germany

ABSTRACT: The response of liver histological structure in milkfish, *Chanos chanos*, fry subjected to environmental concentrations of copper, which were without overt effects on organismic performance parameters of the fish, was investigated. The fry were exposed to 20 and 100 μg/liter of waterborne copper under the regimen of two diets (live food vs. artificial food) over an experimental period of 27 days. The effects of copper contamination were assessed by analysis of growth and survival, of whole body copper concentrations, and of liver histological alterations.

Dry weight increase and survival rates of the fry were significantly influenced by the kind of diet, but not by copper exposure. Body copper concentrations were affected by the environmental concentration of copper, by diet, and by ontogenetic changes in body weight. Morphometrical analysis of liver structure showed that both exposure to 20 μg Cu/liter and 100 μg Cu/liter induced a significant increase in the number and size of hepatic lysosomes. In fry exposed to 100 μg Cu/liter, an additional significant enlargement of liver glycogen fields could be measured. However, copper contamination did not evoke pathological alterations. Whereas lysosomal proliferation may be interpreted as a specific detoxification response of the fry to the enhanced copper burdens, the liver glycogen increase may be regarded as a general stress response.

The results of the study show that histology is capable of revealing sensitively and selectively even subtle effects of environmentally induced changes in fish. The crucial point is to predict the ecological significance of structural alterations evoked by low-dose exposure to toxins and xenobiotic compounds.

INTRODUCTION

The assessment of environmental impact on the organism, regardless whether natural or man-made, is an important challenge. For the evaluation of pollution effects on fish, criteria commonly accepted are organismic responses such as survival, growth, reproduction, spawning, and hatching success. Since adaptation to environmental change is a hierarchical process involving different levels of biological organization (e.g., Sprague, 1971; Vogt, 1987), macroscopically overt signs of toxicity are almost consistently preceded by changes at the organ, tissue, cellular, and molecular levels. The ability to assess qualitative and quantitative changes occurring prior to organismic changes can provide an "early warning system" (Moore, 1985).

A method of proven value in the assessment of environmental impact at the organ and cellular level is histology (e.g., Hinton et al., 1973, 1978; Meyers and Hendricks, 1985; Moore, 1985; Vogt, 1987; Braunbeck et al., 1989a,b). Particularly in small fishes, whose minute organs are not suited for physiological analyses, histology is a valuable approach. Nevertheless, as pointed out by Johnson and Bergman (1984), the value of histology in aquatic toxicology has yet to be fully established. For instance, structural alterations occurring under conditions of long-term, low-concentration exposure (a situation that is typical of many environmental contamination problems) may be difficult to recognize because of the normal variability of histological structures and to the action of additional factors that superimpose the toxicant effects.

The present study was focused on this problem. Two questions were to be answered:

(1) Will the histological effects of low concentrations of a toxicant be detectable among the considerable normal variability of liver histological structure (sensitivity of the method)? In order to increase the precision of histological analysis, morphometrical techniques were included in the study, as recently suggested by Johnson and Bergman (1984) and Hinton et al. (1987). Morphometry may also be powerful for the determination of dose-response relationships based on morphological criteria (Hinton et al., 1984).

(2) Will it be possible to distinguish liver alterations induced by a toxicant from the concurrent impact of other factors (selectivity of the method)? In the present study, additional factors to be considered include an endogeneous one, ontogenetic development: milkfish fry completed metamorphosis within the experimental period. The second factor is an exogeneous one: food quality. Both variables are known to induce changes in the physiological condition and the liver structure of teleosts (e.g., Mamcarz, 1984; Barni et al., 1985; Segner and Juario, 1986).

Milkfish, *Chanos chanos*, was selected as the experimental animal since there is prolonged experience in the normal and altered histology of this species (e.g., Juario and Storch, 1983; Segner and Juario, 1986; Storch and Juario, 1983; Storch et al., 1983, 1984; Segner et al., 1984, 1987). Such knowledge is a prerequisite to successfully employ histology in environmental research (Sprague, 1971; Braunbeck et al., 1989b). Copper was selected as the toxic agent because of its frequent use for aquatic vegetation control in fish culture systems; usually it is applied in concentrations between 0.3-2.0 mg/liter (Nor, 1987). Since copper is an essential trace element, the organism is equipped with mechanisms to regulate its metabolism. Thus, the animal may be expected to react with a broader range of adaptive responses than after exposure to a non-essential compound.

MATERIAL AND METHODS

Animals. Milkfish fry, freshly caught along the shore of Hamtik, Antique, The Philippines, and approximately 20 days old, were transferred to the SEAFDEC (Southeast Asian Fisheries Development Center) laboratories at Tigbauan, Iloilo, The Philippines. Mean initial dry weight of the fry was 1.77 ± 0.40 mg; mean standard length was 14.3 ± 0.4 mm. The fry were randomly distributed to glass aquaria at a stocking density of 6 fry/liter. The aquaria were mildly aerated and provided with filtered seawater (60 μm mesh). The experiment was conducted at ambient water temperature (25.6 ± 0.72°C at 6 a.m., and 29.6 ± 0.86°C at 6 p.m.), at a salinity of 32‰ and at a pH of 8.0 ± 0.13. Dissolved oxygen remained at 6.2 ± 0.3 mg/liter throughout the experiment, ammonia levels were at 0.6 ± 0.3 mg/liter. Background values of the experimental water for copper amounted to 5 ± 1.2 μg Cu/liter.

Experimental design. After an acclimation period of 1 day, the larvae were subjected to the following treatments: (1) live food (1 day-old *Artemia* nauplii, China strain), no addition of copper; (2) live food + 20 μg Cu/liter added (0.31 μmol Cu/liter; nominal concentration); (3) live food + 100 μg Cu/liter added (1.56 μmol Cu/liter; nominal concentration); (4) artifial food (SEAFDEC formula), no addition of copper; (5) artificial food + 20 μg Cu/liter added (0.31 μmol Cu/liter; nominal concentration); and (6) artificial food + 100 μg Cu/liter (1.56 μmol Cu/liter; nominal concentration).

The proximate analysis of the diets is given in Table 1. The artificial food was offered *ad libitum*, the nauplii were maintained at a density of 1-2/ml.

Copper was added to the aquarium water as $CuSO_4$. A semi-static system was used for the experiment, with the water being changed and the $CuSO_4$ addition being renewed every morning. In a preliminary 65-day experiment, it was assured that exposure of milkfish fry to 100 μg Cu/liter does not exert significant effects on survival or growth of the fry.

The experiment was conducted over 27 days. The day when the treatments started is referred to as day 0.

Table 1. Proximate analysis of the diets (Duray and Bagarinao, 1984).

Diet	Moisture (%)	Crude Protein (%)	Crude Fiber (%)	Crude Fat (%)	Ash (%)	Nitrogen-Free Extract (%)
Artificial diet	9.6	40.7	2.5	9.3	14.2	33.2
Artemia nauplii (China strain)	17.0	47.3	5.0	12.0	21.4	14.3

Light microscopy. At days 10, 20, and 27 of the experiment, livers of 4 specimens from each treatment were immersed in Bouin's fluid for 24 hr, afterwards dehydrated and embedded in paraffin; 7 μm sections were stained with hematoxylin-eosin.

Histochemistry. For detection of glycogen, the osmium-induced argentaffin reaction was used (Singh, 1964). Semithin sections of araldite-embedded specimens were incubated in an alkaline 1% solution of silver diamine for 1.5 hr at 60°C.

Transmission electron microscopy. At days 10, 20, and 27 of the experiment, livers of four specimens from each treatment were dissected and fixed for 2 hr in cold 2.5% glutardialdehyde in 0.1 M Soerensen's phosphate buffer, pH 7.1. After rinsing the tissues several times in buffer, they were postfixed in 1% osmium tetroxide in Soerensen's buffer, pH 7.1. For further processing, samples were stored in 70% ethanol and transferred to West Germany. After dehydration and embedding in araldite, semithin and ultrathin sections were prepared with glass knives. Semithin sections (0.5 μm) were stained with methylene blue-azur II. The ultrathin sections were mounted on Formvar-coated copper grids, stained with uranyl acetate and lead citrate prior to examination under the electron microscope (Zeiss EM 9 S2) at the Zoological Institute of Heidelberg.

Morphometry. To conduct the morphometric analysis, three light micrographs and six electron micrographs were taken of each individual fish fixed in glutaraldehyde. Three levels of magnification were employed: for estimation of the relative volume of the hepatocytes, the number and volume density of hepatocellular nuclei in light micrographs were determined. The light micrographs were taken at a final magnification of 1,100x from 0.5 μm semithin sections stained with methylene blue-azur II. For estimation of the volume density of glycogen fields, electron micrographs were randomly recorded from ultrathin sections at a final magnification of 7,200x. Morphometric measurements on mitochondria and dense bodies were performed on electron micrographs at a final magnification of 18,000x.

Morphometric measurements were carried out using the point counting method as described by Weibel (1979). The volume density V_{vi} of a structure was estimated by placing a lattice of P_T testpoints on a micrograph. The number of test points P_i enclosed within profiles of the structure investigated were counted and the fraction $P_{pi} = P_i/P_T$ was determined:

$$V_{vi} = P_{pi}$$

Test Points falling on extrahepatocellular structures (bile canaliculi, sinusoids, etc.) were subtracted from the total number of test points. To determine the volume densities of nuclei (V_{vnh}), mitochondria (V_{vmito}) and dense bodies (V_{vdb}), a test point lattice with 391 systematically spaced points (distance d = 10 mm) was used, for estimating the volume density of glycogen fields (V_{vglyc}) a test point lattice of 96 points (d = 20 mm) was used.

Numerical densities N_{vi} of hepatocyte nuclei N_{vnh} and of dense bodies N_{vdb} were calculated following the equation:

$$N_{vi} = (1/b) \times [(N_{vAi})^{3/2} / (V_{vi})^{1/2}]$$

with N_{vAi} as the number of actually counted profiles divided by the test area, with V_{vi} as the volume density of the structure investigated and with b as a shape-dependent coefficient. Since both nuclei and dense bodies were considered to be of circular outline, this coefficient was assumed to be 1.38 (Weibel, 1979).

Atomic absorption spectroscopy. Whole body concentrations of copper were measured. At days 20 and 27 of the experiment, ten fry were sampled from each treatment. They were briefly rinsed with 1% nitric acid (supra pure grade) to remove all copper attached only superficially. Then the fry were lyophilized and the dry weight was determined. The weighed samples were transferred to pre-weighed Eppendorf reaction tubes and hydrolyzed with 65% nitric acid (supra pure grade). After a 6 hr pre-digestion at room temperature, digestion was completed over 24 hr at 95°C. The digested samples were filled with distilled water to 1000 μl, weighed, and immediately analyzed by flameless atomic absorption spectroscopy (Perkin Elmer 5000, HGA 500) using a graphite furnace and the l'Vov platform technique (Hinderberger et al., 1981) without matrix modification. Measurements were controlled by parallel processing and analysis of the reference standard "bovine liver No. 1577" from the U.S. National Bureau of Standards.

	number of measurements	certified value (μg/g dry weight)	own results (μg/g dry weight)
copper	8	193 ± 10	187 ± 7

Statistics. Differences between means from growth and copper measurements were evaluated using Student's t test. Mean values from morphometric measurements were compared using the non-parametric Wilcoxon-Mann-Whitney U test (Sachs, 1984). The level of significance adopted throughout the study was $P < 0.05$.

RESULTS

Growth. Growth results are presented in Table 2. The kind of food (live food vs. dry food) had a highly significant influence both on dry weight increase and mortality. The artificial diet not only resulted in retarded weight gain of the animals but also in enhanced mortality. A significant effect of copper exposure on weight increase and survival, however, could not be detected, although growth appeared to be slightly depressed in the groups contaminated with 100 μg Cu/liter.

Copper contents of fry. Results on body copper concentrations, total body copper contents, and accumulation factors of the fry are given in Table 3.

Background values for copper, as measured in the control groups, were significantly different after feeding live food or dry food. Contamination with 20 μg Cu/liter and 100 μg Cu/liter significantly increased the body copper concentrations, both in fry fed live food and fry fed artificial food. The accumulation factors did not display a linear increase of copper accumulation when shifting environmental concentration from 20 μg Cu/liter to 100 μg Cu/liter. A diet-related modification of copper accumulation was especially obvious for fish exposed to 100 μg Cu/liter, with the accumulation factor for copper differing by the factor two between live food and dry food groups (Table 3).

A reverse relationship seems to exist between copper concentration and body weight (Fig. 1): copper concentrations of the fry decreased (live food groups) or stagnated (dry food groups) between day 20 and 27, whereas body weight increased

Table 2. Growth data of milkfish, *Chanos chanos*, fry.

	Live food			Artificial food		
	control	+ 20 µg Cu/l	+ 100 µg Cu/l	control	+ 20 µg Cu/l	+ 100 µg Cu/l
			day 0 of the experiment			
body dry weight (mg) (mean ± SD)			1.77 ± 0.40			
			day 20 of the experiment			
body dry weight (mg) (mean ± SD)	12.11[a)] ± 0.93	12.53[a)] ± 1.99	11.66[a)] ± 2.32	5.03[b)] ± 0.30	5.32[b)] ± 0.41	4.95[b)] ± 0.77
			day 27 of the experiment			
body dry weight (mg) (mean ± SD)	21.93[c)] ± 5.06	20.47[c)] ± 5.49	19.40[c)] ± 5.23	6.91[d)] ± 1.46	6.97[d)] ± 1.24	6.52[d)] ± 1.87
cumulative mortality (%)	9	8	12	25	29	27

From each treatment, 10 specimens were measured at each sampling date. Treatments with different superscripts are significantly ($P < 0.05$; Student's t test) different. SD = standard deviation.

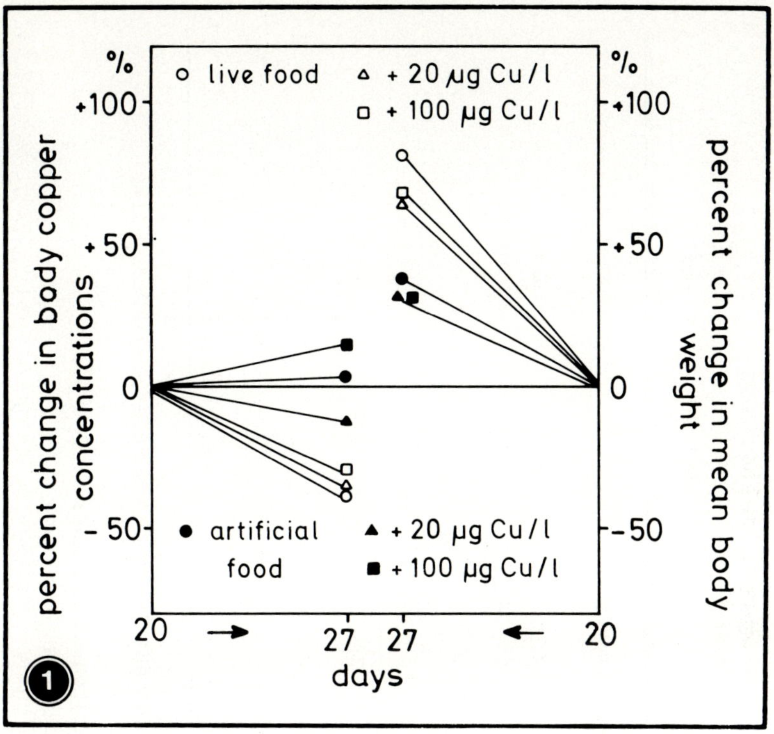

Fig. 1. Comparison of the percentile change in mean body copper concentrations of milkfish fry between day 20 and day 27 of the experiment (left side of the diagram) and percentile change of mean body dry weight of the fry (right side of the diagram).

substantially within the same period. In line with this, there was no reduction in the absolute amount of copper found per individual, which is presented as "total body copper burden" in Table 3.

Liver histology. After a 10 day-feeding period, clear-cut differences in liver histology were evident between the fry reared on live food and those reared on dry food. In paraffin sections, the fry receiving the artificial diet displayed large hepatocytes, with extended unstained areas (Fig. 2, arrows), which, as proven by histochemistry and electron microscopy, represented glycogen deposits. Basophilic structures, mainly corresponding to the rough endoplasmic reticulum (RER) at the ultrastructural level, were restricted to smaller areas around the nuclei and along the cell borders. Fry fed live food, on the other hand, possessed significantly smaller glycogen fields (volume density of 0.264 ± 0.023, compared to 0.381 ± 0.045 in the dry food group), but the basophilic areas were extended (Fig. 3, arrows). These corresponded to a strong, stack-like development of the RER in the electron micrographs.

Copper contamination effects were evident with respect to dense bodies (presumably lysosomes) and to glycogen storage. The livers of fry exposed to 20 and 100 µg Cu/liter showed a conspicuous increase in the number of dense bodies. Exposure to 100 µg Cu/liter led to an additional significant augmentation of glycogen storage in the liver parenchymal cells (Fig. 8). Pathological alterations of liver cell structure as a consequence of the copper contamination could not be observed.

Table 3. Copper contents of milkfish, *Chanos chanos*, fry.

	Live food			Artificial food		
	control	+ 20 µg Cu/l	+ 100 µg Cu/l	control	+ 20 µg Cu/l	+ 100 µg Cu/l
			day 20 of the experiment			
whole body copper concentration (µg Cu/g dry weight) (mean ± SD)	6.79[a,e)] ± 1.38	14.60[b)] ± 2.59	57.16[c)] ± 11.63	14.10[b)] ± 1.82	28.37[d)] ± 3.65	54.40[c)] ± 10.78
total body copper burden (µg)	0.06	0.12	0.45	0.05	0.09	0.17
accumulation factor		107	84		101	37

n = 8-10 for each sample. Results with different superscripts are significantly ($P < 0.05$, Student's t test) different. The accumulation factor is computed according to

$$\frac{\text{(copper concentration in the contaminated group)}}{\text{(copper concentration in the control group)} \times \text{copper}_{water}}$$

SD = standard deviation.

(continued)

Table 3 (continued)

	Live food			Artificial food		
	control	+ 20 µg Cu/l	+ 100 µg Cu/l	control	+ 20 µg Cu/l	+ 100 µg Cu/l
			day 27 of the experiment			
whole body copper concentration (µg Cu/g dry weight) (mean ± SD)	4.13[a)] ± 1.56	8.97[e)] ± 2.40	39.00[f)] ± 7.26	14.51[b)] ± 3.08	25.34[d)] ± 8.83	62.41[c)] ± 16.33
total body copper burden (µg) (percent increase when compared to day 20)	0.06 (0)	0.12 (0)	0.51 (13)	0.06 (20)	0.11 (22)	0.25 (47)
accumulation factor		109	94		87	43

n = 8-10 for each sample. Results with different superscripts are significantly ($P < 0.05$, Student's t test) different. The accumulation factor is computed according to

$$\frac{\text{(copper concentration in the contaminated group)}}{\text{(copper concentration in the control group)} \times \text{copper}_{\text{water}}}$$

SD = standard deviation.

At day 20 of the experiment, i.e., at the end of the metamorphosis period, drastic changes in liver histology, which were virtually identical in all experimental groups, could be noted (Figs. 5, 8): hepatocellular size was reduced if compared to day 10 (Fig. 4), and the whole cytoplasm stained basophilic. Glycogen fields had almost disappeared (Figs. 4, 5, 8). The existence of shrunken, very dark hepatocytes with irregularly shaped nuclei (Fig. 5, arrow) gave rise to an enhanced intraparenchymal variability in cellular morphology. At the ultrastructural level, the hepatocytes were characterized by a loss of the regular stack-like organization and a partial fragmentation of the RER. The cisternae were evenly distributed throughout the cytoplasm.

After 27 days of feeding, liver structure, as indicated by the development of the glycogen volume density (Fig. 8), had partially recovered from the profound changes as observed at day 20. Moreover, inter-group differences resulting from diet and copper exposure had been re-established. Significant differences could be determined in hepatocellular size between the six treatments, as indicated by the number of hepatocytes/cm^3 liver tissue (Fig. 9). The mean cell volume of uncontaminated fry fed live food was computed to be 1951 μm^3 (Table 4), which is significantly lower than the value obtained for fry fed the artificial diet (2431 μm^3). Exposure to 100 μg Cu/liter resulted in a significant increase of hepatocellular size in either dietary treatment (Fig. 9, Table 4). These differences in cell volume were accompanied by corresponding changes in nuclear volume density (Fig. 9) and nucleus/cell ratios (Table 4).

Fig. 2. 7 μm paraffin section of the liver of milkfish fry fed artificial diet for 10 days. Hematoxylin-eosin stain. Most of hepatocellular cytoplasm is occupied by unstained glycogen fields (arrows).

Fig. 3. 7 μm paraffin section of the liver of milkfish fry fed live food for 10 days. Hematoxylin-eosin stain. Unstained areas are reduced when compared to fry fed artificial diet (Fig. 2). Enlarged dark staining basophilic parts within the cytoplasm of the hepatocytes (arrows) represent RER at the ultrastructural level.

Fig. 4. 0.5 μm semithin section of the liver of milkfish fry fed live food and exposed to 100 μg Cu/liter for 10 days. Argentaffin stain. The hepatocytes are characterized by well-developed perinuclear and peripheral layers of RER and extended dark-staining glycogen fields.

Fig. 5. 0.5 μm semithin section of the liver of milkfish fry fed live food and exposed to 100 μg Cu/liter for 20 days. Argentaffin stain. Cell size is reduced, glycogen has disappeared, parenchymal heterogeneity (i.e., variation in cell size, shape, and basophilia) is increased. Arrow: Dark, shrunken hepatocyte.

Fig. 6. Electron micrograph of the liver of milkfish fry fed live food and exposed to 100 μg Cu/liter for 27 days. Cell organization and glycogen content is partly re-established when compared to day 20. Number of prominent dense bodies is enhanced.

Fig. 7. 0.5 μm semithin section of the liver of milkfish fry fed live food and exposed to 20 μg Cu/liter for 27 days. Argentaffin stain. Even in light micrographs, numerous dense bodies are obvious.

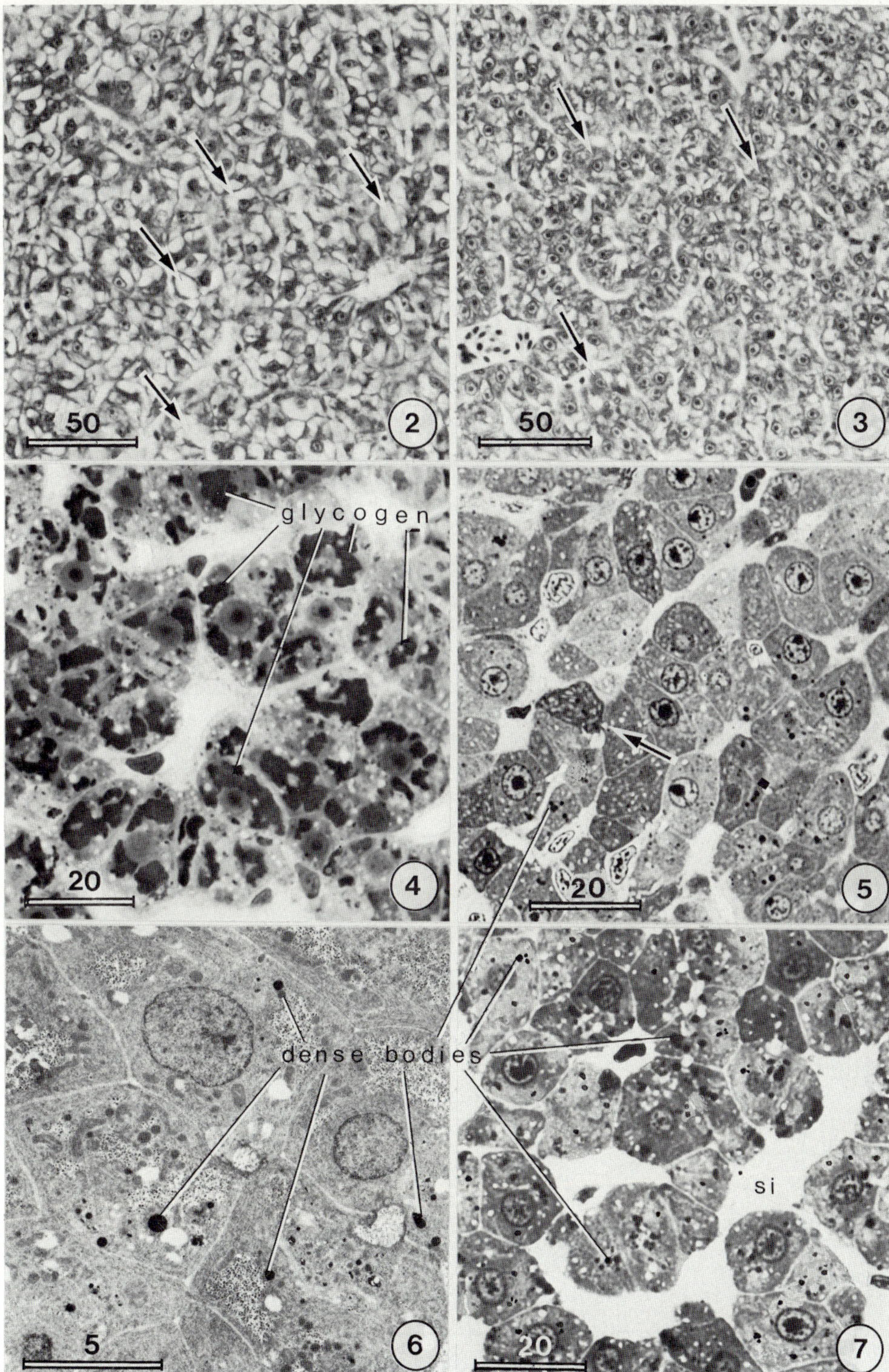
50
2
50
3
glycogen
20
4
20
5
dense bodies
5
6
si
20
7

The variation of cell size between the test groups was paralleled by variation of the glycogen volume densities (Fig. 9, Table 4). Whereas in control fry fed live food only 8.3% of the cytoplasmic volume was occupied by glycogen deposits, these amounted to 12.6% in fish fed live food and contaminated with 100 μg Cu/liter, 18.1% in control fish fed the artificial diet and 22.1% in a fish fed the artificial diet and contaminated with 100 μg Cu/liter.

The RER, which had been seriously disturbed after 20 days of the experiment, partly re-established its stack-like organization after 27 days in any experimental treatment. Although the RER was more developed in the livers of fry receiving live food, the difference compared to the dry food group was not as pronounced as after 10 days of the experiment.

Mitochondrial volume density was significantly affected by dietary treatment, but not by copper exposure (Fig. 9). Since no obvious differences of mitochondrial size between live and dry food-reared specimens were evident by qualitative examination in the electron microscope, the observed difference of mitochondrial volume density between the two treatments must be caused by an increased number of mitochondria in the live food groups.

The kind of food also exerted a significant influence on the volume density of the dense bodies (lower values in the dry food group) but not on the lysosomal numerical density (Fig. 9). In consequence, the mean volume of a single dense body in the liver cells of control fish fed with live food could be computed as 0.41 μm^3, whereas it only amounted to 0.20 μm^3 in the dry food group. Copper exposure consistently enhanced the volume density and the number of lysosomal elements

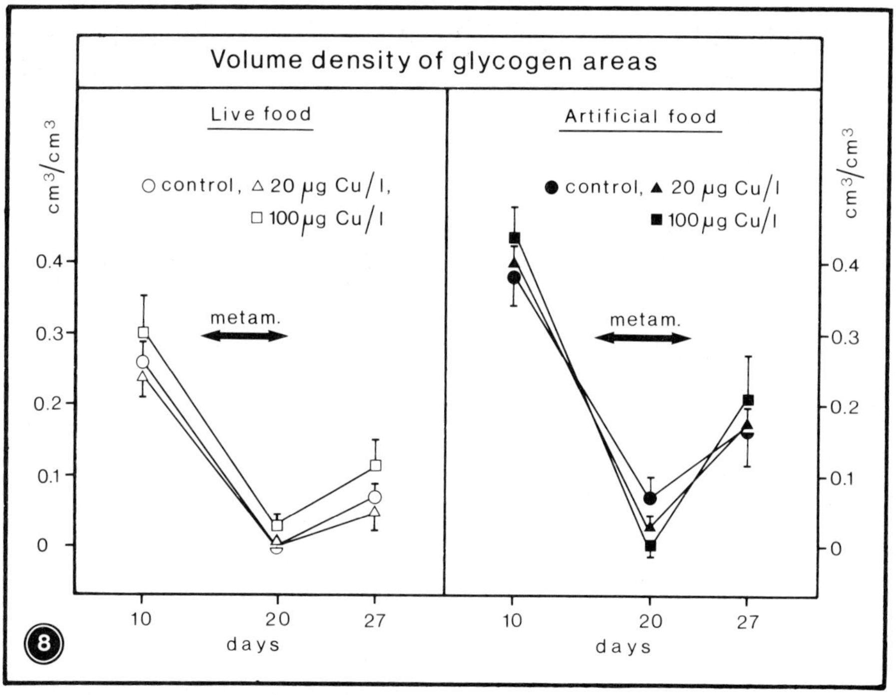

Fig. 8. Morphometrical evaluation of the volume density of glycogen areas within the liver tissue (cm^3/cm^3) of milkfish fry between day 10 and day 27 of the experiment. Abbreviation: "metam." = metamorphosis.

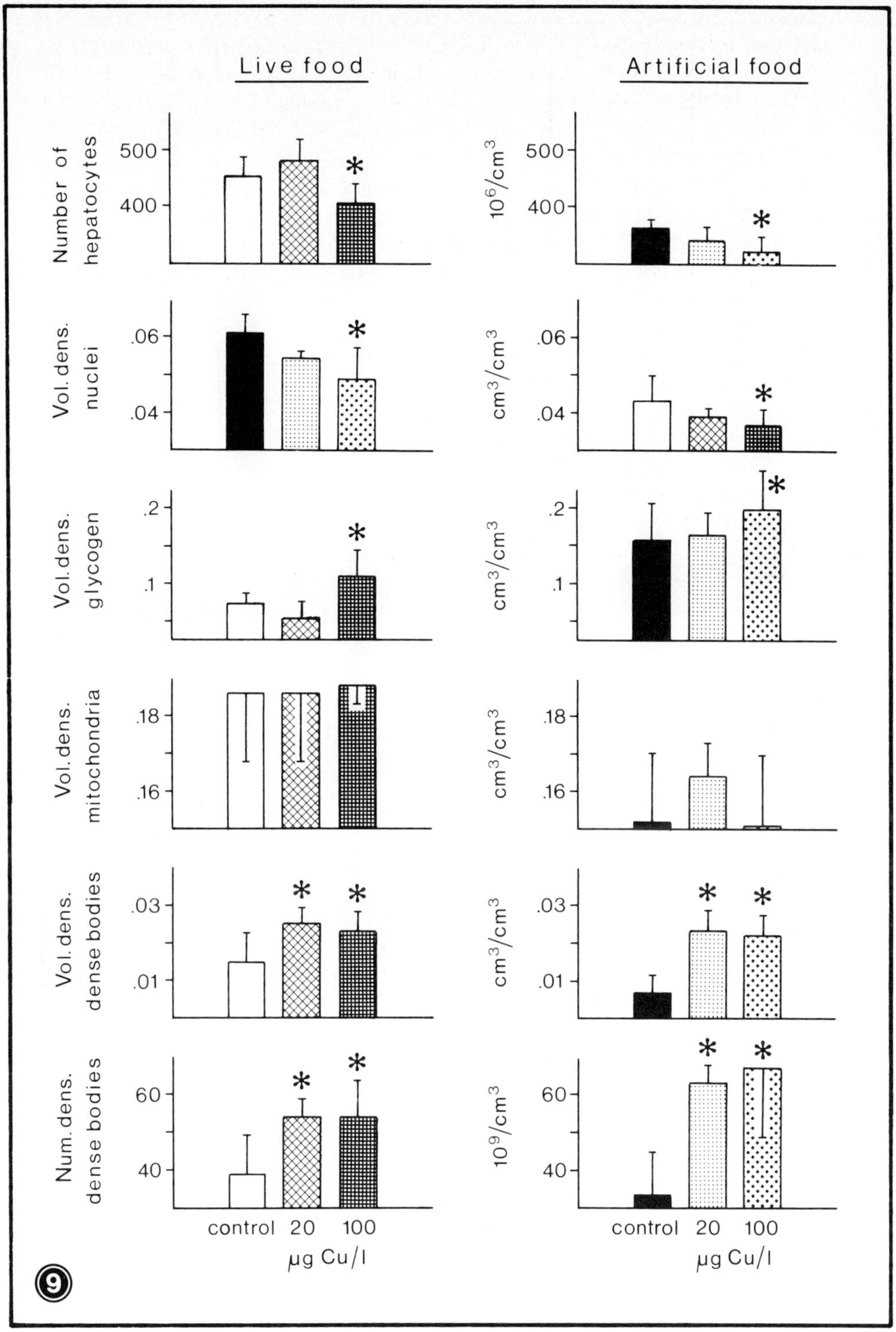

Fig. 9. Morphometrical description of liver structure of milkfish fry after 27 days of experiment. Asterisks indicate significant differences ($p < 0.05$).

Table 4. Average cell data of hepatocytes from milkfish fry after 27 days of experiment. Data are expressed in μm³ (Weibel, 1979).

	Live food			Artificial food		
	control	+ 20 μg Cu/l	+ 100 μg Cu/l	control	+ 20 μg Cu/l	+ 100 μg Cu/l
hepatocyte volume (μm^3)	1951	1833	2173	2431	2580	2708
nucleus volume (μm^3)	135	112	121	119	114	113
nucleus/cell ratio	0.069	0.061	0.055	0.049	0.044	0.041
volume of glycogen fields (μm^3)	162	110	274	439	478	597

The percentage of total liver occupied by hepatocytes amounted to 89.01%. The volume of a single hepatocyte was computed from the numerical density of hepatocyte nuclei assuming that each hepatocyte is mononuclear (Weibel, 1979).

under both feeding regimens (Figs. 6, 7, 9). In contrast to the observations on the glycogen contents, copper-induced increase of dense bodies was already significant in fish exposed to 20 μg Cu/liter.

DISCUSSION AND CONCLUSIONS

Although the toxicological effects of copper on fish are fairly well documented, the variability of the results reported is considerable (e.g., Hodson et al., 1979; Hellawell, 1986; Mance, 1987). In many cases, a negative effect of copper exposure on fish growth has been observed (for references see Collvin, 1985). Larval fish, when exposed to copper from the embryonic stage onwards, show reduced growth even with low concentrations of the metal. McKim et al. (1978) reported that a contamination with copper concentrations between 31.7-43.5 μg Cu/liter from the embryo up to 30-60 days post-hatch is sufficient to reduce the standing crop of rainbow trout, brook trout, brown trout, lake trout, white sucker, herring, and smallmouth bass. Northern pike were found to be more resistant, showing growth retardation only after exposure to 104.1 μg Cu/liter. When exposing fathead minnow to copper from the embryo stage up to 28 days post-hatch, Scudder et al. (1988) obtained a significant reduction of weight and length for all copper concentrations tested (61 through 621 μg Cu/liter).

For our study with milkfish fry, we selected copper concentrations that, as we established in range-finding experiments, remain without observable effects on growth of the fish during this developmental stage. This lack of response at the organismic level may partly be accounted for by species-specific differences in copper sensitivity, but also by the fact that the fry was not exposed to copper from the egg-stage onward.

Body copper stores of the animals were significantly elevated by exposure to waterborne copper. The fact that the accumulation factors were slightly lower at the higher environmental copper concentrations may indicate the onset of regulatory mechanisms. Chernoff and Dooley (1979) also observed in *Fundulus heteroclitus* decreasing accumulation factors with increasing concentration of metals in the water. According to these authors, fish apparently do not expend energy to regulate metal burdens below a certain threshold; therefore, fish exposed to a low environmental metal concentration accumulate the metal at a comparatively higher rate than fish exposed to a high environmental metal concentration. Interestingly enough, regulation of copper accumulation seems to depend not only on the external concentration of copper but also on the nutritional status of the fry as related to the different dietary regimens as applied in this study. The importance of the nutritional status of the fish for uptake, metabolism, and excretion of heavy metals was already pointed out by Dixon and Hilton (1985), Segner and Storch (1985), and Segner (1987). In the present study, the observation of an enhanced accumulation factor of copper in fry fed live food may be explained by a particularly intensive metabolism induced by the live diet as it is suggested by, e.g., the high growth rate in this group.

A decrease of contaminant body concentrations with increasing fish size as obtained in the present study with milkfish fry was already demonstrated by other authors (Chernoff and Dooley, 1979; Milner, 1982; Newman and Mitz, 1988). This effect may be due to the inverse relationship between fish size and percentile weight of viscera; since smaller fish contain a higher percentage of visceral tissue than larger fish, and since the viscera usually contain significantly higher metal burdens than does somatic muscle tissue (e.g., Chernoff and Dooley, 1979), the negative allometric development of viscera to whole body during ontogeny could account at least partly for the decrease or stagnation of body copper concentrations in milkfish fry between days 20 and 27 of the experiment. However, the onset of additional regulatory processes in juvenile fish after metamorphosis cannot be excluded when considering the percentile increase in whole body copper burdens between days 20 and 27 (Table 3).

Analysis of fish copper concentrations showed that although copper exposure remained without effect on growth of the fry, it affected whole body copper stores of the fry in a dose-dependent manner. The next step in the analysis of the consequences of copper exposure was the investigation of possible alterations at the tissue and cellular level. We approached this problem by investigating histology and cytology of the liver, an organ playing a central role in copper and overall metabolism of the fish.

In the present study, the influence of each experimental factor tested could be clearly separated from the action of the other factors. Copper exposure evoked a response of the liver, which was clearly isolated from the diet- and metamorphosis-related hepatic changes. Both diet and copper influenced lysosomal and glycogen volume density; however, by means of morphometrical evaluation, the two factors could be clearly discriminated. Diet additionally evoked changes in nuclear volume, mitochondrial volume density, and morphology of the RER, which were not affected by copper contamination. Metamorphosis, as a profound physiological disturbance, superposed both the influence of copper and of nutrition leading to a more or less uniform liver morphology in all treatments.

The finding that certain structural alterations in the liver or definite combinations of alterations could be specifically ascribed to the influence of one of the environmental factors is in contrast to the widespread belief that the liver reacts only in an unspecific way (Couch, 1975; Sindermann, 1980). An unspecific reaction, of course, can be expected if, as, e.g., shown by Segner (1987), the combined action of several factors creates a new quality of the enviromental impact, which is different from the isolated action of each of the relavant factors. However, basically, there is no reason to expect only unspecific structural alterations in the fish liver. Mammalian liver pathologists, who have a similarly limited set of liver reaction patterns at the electron microscopical level, are nevertheless able to diagnose a wide variety of liver diseases and metabolic disorders. The major problem encountered by the fish liver pathologists is the still restricted amount of ultrastructural data available.

Copper-related structural alterations of the liver revealed a dose-response relationship, irrespective of whether the fry was fed live or dry food, with a significant increase in the number and size of dense bodies after exposure to 20 μg Cu/liter and 100 μg Cu/liter, and an additional enlargement of hepatocellular glycogen fields after exposure to 100 μg Cu/liter. Pathological changes, i.e., alterations of liver histology and cytology beyond its normal range, could be observed in no case.

The dense bodies noted in milkfish hepatocytes are most likely lysosomes, although we have not performed cytochemical stainings on acid phosphatase. Despite the fact that copper accumulation in fish liver induces cytoplasmic metallothionein synthesis (e.g., Dixon and Sprague, 1981; McCarter et al., 1982), the bulk of the metal is deposited in the lysosomal fraction of the hepatocytes (up to 90%; Weis, 1984). In line with this, a proliferation of lysosomes has been observed repeatedly in copper-contaminated fish (Weis et al., 1986; Bunton et al., 1987; Lanno et al., 1987). As pointed out by Laurén and McDonald (1987), detoxification of copper within the fish liver is not primarily achieved by complexation with metallothioneins. Our results, as well as those of Weis (1984), Weis et al. (1986), Bunton et al. (1987), and Lanno et al. (1987), indicate that fish detoxify hepatic copper by deposition within lysosomes. The sequestration of copper within these organelles obviously protects the other cell components from the deleterious effects of elevated levels of the essential trace element copper. This measure appears to be very effective because, apart from acute studies (Baker, 1969), usually pathological lesions have not been reported for copper-contaminated fish (Weis et al., 1986; Lanno et al., 1987; this study). Only Bunton et al. (1987) obtained degenerative features in the liver of *Morone americana*, but only in individuals with tremendously high copper concentrations (up to 1,000 μg Cu/g wet weight).

The enhanced hepatic glycogen deposition in consequence of exposure to 100 μg Cu/liter indicates an alteration of carbohydrate metabolism, possibly as a cortisol-mediated general stress response (cf. Peters et al., 1980; Larsson et al., 1985). For copper, an increase of plasma cortisol in fish has been reported by Donaldson and Dye (1975). Hanke et al. (1983) and Yousri and Hanke (1985) pointed out that the pollution stress response of fish may be biphasic: a first phase under conditions of low-level or short-term exposure is characterized by elevated liver glycogen values as a result of intensive cortisol-mediated hepatic gluconeogenesis; the second phase, typical of prolonged exposure or high concentrations of the pollutant, is characterized by reduced liver glycogen levels. According to this interpretation, milkfish fry exposed to 100 μg Cu/liter are still within the first phase of environmental stress.

This study was undertaken to test the sensitivity and selectivity of histological and morphometrical methods in detecting and diagnosing internal responses of fish occurring at toxicant concentrations that induce no overt signs of intoxication at the organismic level. Our results demonstrate that histological analysis can detect even subtle cellular alterations that are still within the range of normal plasticity, and histological analysis can separate toxicant-related effects from other, simultanously acting influences. Thus, histology is effective in establishing cause-response relationships. Moreover, as evident, e.g., from the findings on the lysosomes, it can cast light on the "mode of action" (Sprague, 1971) of the toxicant. In our opinion, the limiting factor in the use of histology and cytology in aquatic toxicology is neither selectivity nor sensitivity of the method, but problems encountered with the interpretation of the physiological and ecological significance of alterations observed (Malins, 1982; Johnson and Bergman, 1984). As discussed, e.g., by Sprague (1971), Bayne (1985), and Vogt (1987), each change occurring at one level of biological organization (in the present study the organ and cellular level) finds its significance on the next higher level of biological organization, in this case the organismic level.

In Figure 10, possible responses of an organism to an altered environment as well as their consequences for the survival potential of the organism are shown. Environment-induced changes in the functional state of an organism may well be within the normal range of adaptation of an organism and are tolerated indefinitely, leading to no reduction of fitness or survival potential of the fish (Rosenthal and Alderdice, 1976; Bayne, 1985). More serious changes, although they may be resisted for some time, ultimately result in the death of the organism. Between these two possibilities there is a zone, where the organism can compensate the environmental impact, although at increasing metabolic cost. At a first stage of intoxication, symptoms of a selective response dominate (Wilson, 1980), which are not necessarily deleterious to the well-being of the fish. By means of these physiological or structural changes, however, the organism may be better able to compete with the new environmental condition (Bayne, 1985). With increasing concentrations of the

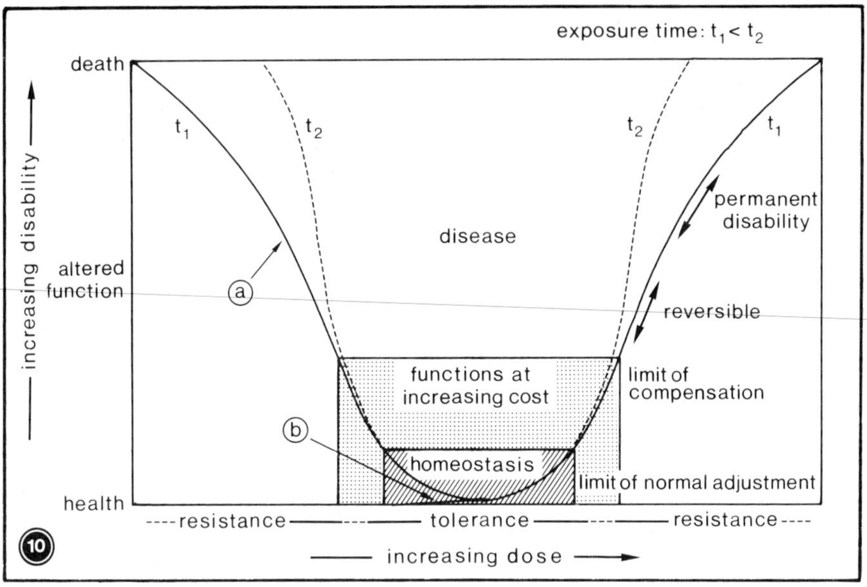

Fig. 10. Schematic presentation of possible responses of an organism subjected to the impact of an essential (line a) or non-essential (line b) compound (adapted from Wilson, 1980).

contaminant, non-selective responses requiring increased energy expenditure appear. In the present study, lysosomal proliferation may represent a specific response to copper exposure, whereas the augmentation of liver glycogen values, as observed in fish contaminated with 100 μg Cu/liter, may be indicative of a general stress response (Peters et al., 1980; Bayne, 1985).

The copper-related structural and concomitant functional alterations (lysosomal proliferation, alteration of carbohydrate metabolism) do not appear to be sufficient to increase the metabolic costs to such an extent that growth or survival of the fry are significantly reduced within the experimental period (although for the groups exposed to 100 μg Cu/liter a slight reduction was indicated). The assumption that the copper exposure remains without significant effects on the functional state of the organism may be further substantiated by the observation that, although body copper concentrations differed significantly between contaminated fry fed live food or dry food, the effects of copper on the liver structure were virtually identical in both groups. This suggests that the upper limit of the fry's capacity to manage and compensate enhanced body copper burdens was not reached.

The observation that the copper exposure remains without significant extra cost for milkfish fry is only true for the combination and range of environmental variables set in this study and for the exposure period of 27 days. Collvin (1985) has shown in an excellent way for perch, *Perca fluviatilis*, that even when the fish is acclimated to copper exposure at a given set of environmental variables, the acclimation may get lost when one of the environmental factors change. Thus, a minute alteration, which is definitely without adverse consequences for the well-being of the animal under given environmental conditions, may become of primary importance for survival as soon as an additional environmental change occurs. Furthermore, the importance of the interaction of concentration and time factor, as visualized in Figure 10 by the two lines

t_1 and t_2, has to be emphasized. The effect of a toxicant on the fish (under constant environmental conditions) is the result of contaminant concentration and exposure time. Considering the fact that a persistant exposure to low contaminant concentrations is the typical situation for the fish fauna in many surface water bodies, considerable additional work dealing with the detection and understanding of the relevance of "small" effects (i.e., effects with no obvious harmful action in the present situation) has to be done.

LITERATURE CITED

Baker, J.T.P. 1969. Histological and electron microscopical observations on copper poisoning in the winter flounder (*Pseudopleuronectes americanus*). J. Fish. Res. Bd. Canada, 26, 2785-2793.

Barni, S., G. Bernocchi and G. Gerzelli. 1985. Morphohistochemical changes in hepatocytes during the life cycle of the European eel. Tiss. Cell, 17, 97- 109.

Bayne, B.L. 1985. General introduction. In: "The Effects of Stress and Pollution on Marine Animals" (B.L. Bayne et al., eds.), pp. x-xvi. Praeger Scientific, New York.

Braunbeck, T., R. Nagel and V. Storch. 1989a. Sex-specific reaction of liver ultrastructure in zebra fish (*Brachydanio rerio*) after prolonged sublethal exposure to 4-nitrophenol. Aquat. Toxicol., 14, 185-202.

Braunbeck, T., V. Storch and H. Bresch. 1989b. Species-specific reaction of liver ultrastructure in zebrafish, *Brachydanio rerio*, and trout, *Salmo gairdneri*, after prolonged exposure to 4-chloroaniline. Arch. Environ. Contam. Toxicol., in press.

Bunton, T.E., S.M. Baksi, S.G. George and J.M. Frazier. 1987. Abnormal hepatic copper storage in a teleost fish (*Morone americana*). Vet. Pathol., 24, 515-524.

Chernoff, B. and J.K. Dooley. 1979. Heavy metals in relation to the biology of the mummichog, *Fundulus heteroclitus*. J. Fish Biol., 24, 309-328.

Couch, J.A. 1975. Histopathological effects of pesticides and related chemicals on the liver of fishes. In: "The Pathology of Fishes" (W.E. Ribelin and G. Migaki, eds.), pp. 559-584. University of Wisconsin Press, Madison.

Collvin, L. 1985. The effect of copper on growth, food consumption and food conversion of perch, *Perca fluviatilis*, offered maximal food rations. Aquat. Toxicol., 6, 105-114.

Dixon, D.G. and J.W. Hilton. 1985. Effects of available dietary carbohydrate and water temperature on the chronic toxicity of waterborne copper to rainbow trout (*Salmo gairdneri*) Can. J. Fish. Aquat. Sci., 42, 1007-1013.

Dixon, D.G. and J.B. Sprague. 1981. Copper bioaccumulation and hepatoprotein synthesis during acclimation to copper by juvenile rainbow trout. Aquat. Toxicol., 1, 69-81.

Donaldson, E.M. and H.M. Dye. 1975. Corticosteroid concentrations in sockeye salmon (*Oncorhynchus nerka*) exposed to low concentrations of copper. J. Fish. Res. Bd. Canada, 32, 533-539.

Duray, M. and T. Bagarinao. 1984. Weaning of hatchery-bred milkfish larvae from live food to artificial diets. Aquaculture, 41, 325-332.

Hanke, W., G. Gluth, H. Bubel and R. Müller. 1983. Physiological changes in carp induced by pollution. Ecotox. Environ. Safety, 7, 229-241.

Hellawell, J.M. 1986. "Biological Indicators of Freshwater Pollution and Environmental Management." Elsevier Applied Science Publishers, London and New York.

Hinderberger, E.J., M.L. Kaiser and S.R. Koirtyohan. 1981. Furnace atomic absorption analysis of biological samples using the l'Vov platform and matrix modification. Atom. Spectrosc., 2, 1-7.

Hinton, D.E., M.W. Kendall and B.B. Silver. 1973. Use of histologic and histochemical assessments in the prognosis of the effects of aquatic pollutants. In: "Biological Methods for the Assessment of Water Quality", pp. 194-208. Am. Soc. Test. Mat. ASTM STP 528, Washington, D.C.

Hinton, D.E., J.E. Klaunig and M.M. Lipsky. 1978. PCB-induced alterations in teleost liver: a model for environmental disease in fish. Mar. Fish. Rev., 65, 47-50.

Hinton, D.E., R.C. Lantz and J.A. Hampton. 1984. Effect of age and exposure to a carcinogen on the structure of the Medaka liver: a morphometric study. Natl. Cancer Inst. Monogr., 65, 239-249.

Hinton, D.E., R.C. Lantz, J.A. Hampton, P.R. McCuskey and R.S. McCuskey. 1987. Normal versus abnormal structure: considerations in morphologic response of teleosts to pollutants. Environ. Health Persp., 71, 139-146.

Hodson, P.V., U. Borgmann and H. Shear. 1979. Toxicity of copper to aquatic biota. In: "Copper in the Environment" (J.O. Nriagu, ed.), pp. 308-372. John Wiley, New York.

Johnson, R.D. and H.L. Bergman. 1984. Use of histopathology in aquatic toxicology: a critique. In: "Contaminant Effects on Fisheries" (V.W. Cairns, P.V. Hodson and J.O. Nriagu, eds.), pp. 19-36. John Wiley, New York.

Juario, J.V. and V. Storch. 1983. Biological evaluation of phytoplankton (*Chlorella* sp., *Tetraselmis* sp., and *Isochrysis galbana*) as food for milkfish, *Chanos chanos*, fry. Aquaculture, 40, 193-198.

Lanno, R.P., B. Hicks and J.W. Hilton. 1987. Histological observations on intrahepatocytic copper-containing granules in rainbow trout reared on diets containing elevated levels of copper. Aquat. Toxicol., 10, 251-263.

Larsson, A., C. Haux and M.L. Sjöbeck. 1985. Fish physiology and metal pollution: Results and experiences from laboratory and field studies. Ecotox. Environ. Safety, 9, 250-281.

Laurén, D.J. and D.G. McDonald. 1987. Acclimation to copper by rainbow trout, *Salmo gairdneri:* Biochemistry. Can. J. Fish. Aquat. Sci., 44, 105-111.

McCarter, J.A., A.T. Matheson, M. Roch, R.W. Olafson and J.T. Buckley. 1982. Chronic exposure of coho salmon to sublethal concentrations of copper. II. Distribution of copper between high- and low-molecular weight proteins in liver cytosol and the possible role of metallothionein in detoxification. Comp. Biochem. Physiol., 72C, 21-26.

McKim, J.M., J.G. Eaton and G.W. Holcombe. 1978. Metal toxicity to embryos and larvae of eight species of freshwater fish. II. Copper. Bull. Environ. Contam. Toxicol., 19, 608-616.

Malins, D.C. 1982. Alterations in the cellular and subcellular structure of marine teleosts and invertebrates exposed to petroleum in the laboratory and the field: a critical review. Can. J. Fish. Aquat. Sci., 39, 877-889.

Mamcarz, A. 1984. Rearing of coregonid fishes (Coregonidae) in illuminated lake cages. III. Phenotypic variability in body size of *Coregonus peled* Gmel. during the first year. Aquaculture, 40, 241-250.

Mance, G. 1987. "Pollution Threat of Heavy Metals in Aquatic Environments." Elsevier Applied Science Publishers, London and New York.

Meyers, T.R. and J.D. Hendricks. 1985. Histopathology. In: "Fundamentals of Aquatic Toxicology" (G.M. Rand and S.R. Petrocelli, eds.), pp. 283-331. Hemisphere, Washington.

Milner, N.J. 1982. The accumulation of zinc by O-group plaice, *Pleuronectes platessa* (L.), from high concentrations in seawater and food. J. Fish Biol., 21, 325-336.

Moore, M.N. 1985. Cellular responses to pollutants. Mar. Pollut. Bull., 16, 134-139.

Newman, M.C. and S.V. Mitz. 1988. Size dependence of zinc elimination and uptake from water by mosquitofish *Gambusia affinis* (Baird and Girard). Aquat. Toxicol., 12, 17-32.

Nor, Y.N. 1987. Ecotoxicity of copper to aquatic biota: A review. Environ. Res., 43, 274-282.

Peters. G., H. Delventhal and H. Klinger. 1980. Physiological and morphological effects of social stress in the eel (*Anguilla anguilla* L.). Arch. FischWiss., 30, 157-180.

Rosenthal, H. and D.F. Alderdice. 1976. Sublethal effects of environmental stressors, natural and pollutional, on marine fish eggs and larvae. J. Fish. Res. Bd. Canada, 33, 2047-2065.

Sachs, L. 1984. "Angewandte Statistik," 6th ed. Springer, Berlin-Heidelberg-New York.

Scudder, B.C., J.L. Carter and H.V. Leland. 1988. Effects of copper on development of the fathead minnow, *Pimephales promelas* Rafinesque. Aquat. Toxicol., 12, 107-124.

Segner, H. 1987. Response of fed and starved roach, *Rutilus rutilus*, to sublethal copper contamination. J. Fish Biol., 30, 423-427.

Segner, H. and J.V. Juario. 1986. Histological observations on the rearing of milkfish, *Chanos chanos*, fry using different diets. J. Appl. Ichthyol., 2, 162-173.

Segner, H. and V. Storch. 1985. Influence of water-borne iron on the liver of *Poecilia reticulata* (Peters 1859). J. Appl. Ichthyol., 1, 39-47.

Segner, H., B. Orejana-Acosta and J.V. Juario. 1984. The dietary value of *Brachionus plicatilis* grown on three different species of phytoplankton on the ultrastructure of the hepatocytes of *Chanos chanos* (Forskal) fry. Aquaculture, 42, 109-115.

Segner, H., P. Burkhardt, E.M. Avila, J.V. Juario and V. Storch. 1987. Nutrition-related histopathology of the intestine of milkfish *Chanos chanos* fry. Dis. Aquat. Org., 2, 99-107.

Sindermann, C.J. 1980. The use of pathological effects of pollutants in marine environmental monitoring programs. Rapp. P.-v. Réun. Cons. Int. Explor. Mer, 179, 129-134.

Singh, I. 1964. A modification of the Masson-Hamperl method for staining of argentaffin cells. Anat. Anz., 115, 81-82.

Sprague, J.B. 1971. Measurement of pollutant toxicity to fish. III. Sublethal effects and "safe" concentrations. Water Res., 5, 245-266.

Storch, V. and J.V. Juario. 1983. The effect of starvation and subsequent feeding on the hepatocytes of *Chanos chanos* (Forskal) fingerlings and fry. J. Fish Biol., 23, 95-103.

Storch, V. W. Stählin and J.V. Juario. 1983. Effects of different diets on the ultrastructure of hepatocytes of *Chanos chanos* fry: an electron microscopic and morphometric analysis. Mar. Biol., 74, 101-104.

Storch, V., H. Segner, M.N. Duray and J.V. Juario. 1984. Influence of nutrition on the hepatocytes of milkfish, *Chanos chanos* (Chanidae). Zool. Anz., 213, 151-160.

Vogt, G. 1987. Monitoring of environmental pollutants such as pesticides in prawn aquaculture by histological diagnosis. Aquaculture, 67, 157-164.

Weibel, E.R. 1979. "Stereological Methods. Practical Methods for Biological Morphometry," Vol. 1. Academic Press, New York.

Weis, P. 1984. Metallothionein and metal tolerance in the killifish, *Fundulus heteroclitus*. Mar. Environ. Res., 14, 153-166.

Weis, P., J.D. Bogden and E.C. Enslee. 1986. Hg- and Cu-induced hepatocellular changes in the mummichog, *Fundulus heteroclitus*. Env. Health Perspect., 65, 167-173.

Wilson, K.W. 1980. Monitoring dose-response relationships. Rapp. P.-v. Réun. Cons. Int. Explor. Mer, 179, 333-338.

Yousri, R. and W. Hanke. 1985. The effects of pentachlorphenol, phenol and other pollutants on the liver of carp (*Cyprinus carpio* L.). Comp. Biochem. Physiol., 82C, 283-290.

EFFECTS OF TOXICANTS ON CERTAIN FUNCTIONS OF THE LYMPHORETICULAR SYSTEM OF FISH

B.A. Weeks, J.E. Warinner, E.S. Mathews, and A. Wishkovsky*

Division of Chemistry and Toxicology, Virginia Institute of Marine Science, School of Marine Science, The College of William and Mary, Gloucester Point, Virginia 23062, USA

ABSTRACT: Spot (*Leiostomus xanthurus*) and hogchoker (*Trinectes maculatus*) from the Elizabeth River, Virginia, which is heavily contaminated with polynuclear aromatic hydrocarbons (PAHs), exhibit greatly reduced phagocytic and chemotactic activity of kidney macrophages. In addition, kidney macrophages from Elizabeth River spot were markedly deficient in generating a zymosan-induced chemiluminescent (CL) response, as were macrophages from York River spot exposed in the laboratory to Elizabeth River sediments. The CL response of hogchoker and oyster toadfish (*Opsanus tau*) macrophages was significantly depressed by exposure to 400 μg/liter tributyltin (TBT), while croaker (*Micropogonias undulatus*) CL activity was depressed at 40 μg/liter TBT. These assay techniques are proposed as potential bioindicators of environmental pollution.

INTRODUCTION

Interest in the effect of environmental stress on the health of fish has increased in recent years. While acutely toxic substances often cause death quickly, sublethal concentrations of the same substances may have a chronically debilitating effect that lowers resistance to disease (Sindermann, 1983). In fish, as in higher vertebrates, the lymphoreticular system is an important defense mechanism because of its capacity to destroy foreign substances. Since macrophages comprise an important part of the immune system, it is of interest to determine the effect of toxic substances on the normal immune functions of macrophages.

---

*Present address: Department of Medicine, School of Veterinary Medicine, University of California, Davis, California 95616, USA

The waters of the Elizabeth River, Virginia, are known to be among the most heavily contaminated in the Chesapeake Bay region. The Elizabeth River has been the site of civilian and military shipbuilding, shoreside commerce, and associated manufacturing and processing industry for hundreds of years. Concentrations of polynuclear aromatic hydrocarbons (PAHs) have been found to exceed thousands of parts per million in the sediments of some areas of the river (Bieri et al., 1986). Many of these substances are known carcinogens or mutagens. Biological data derived from the collection and analysis of organisms collected at these sites reveal that the natural community has been impacted and that the impacts are greatest in the areas where the PAHs are in highest concentration (Huggett et al., 1987; Hargis and Zwerner, 1988).

Several aspects of immune function were examined in species of bottom-feeding fish captured in the Elizabeth River or exposed to toxicants in the laboratory. It is hypothesized that alterations in immune activity levels might characterize the health status of fish that have been exposed to chronic environmental contamination.

MATERIALS AND METHODS

Fish. Spot (*Leiostomus xanthurus*) and hogchoker (*Trinectes maculatus*) were collected by trawl net from the reach of the Elizabeth River most contaminated with PAHs and from the York River, which is relatively uncontaminated by PAHs. In addition, croaker (*Micropogonias undulatus*) and oyster toadfish (*Opsanus tau*) used for laboratory exposure to toxicants were collected from the York River. Fish were allowed to acclimate several days in flowing estuarine water before use.

Collection of macrophages (Weeks and Warinner, 1986). Suspensions of kidney cells from toadfish, hogchoker, and croaker were prepared by macerating kidney tissue in a glass homogenizer. Cell preparations contained on the average 40% macrophages, as identified by differential staining. Purified macrophage suspensions from spot and hogchoker were prepared by passage of the kidney cell preparation through Percoll density gradients. Greater than 85% of the cells in the resultant macrophage fraction were macrophages, with the remainder consisting of lymphocytes and red blood cells.

Phagocytosis assay (Weeks and Warinner, 1984). Purified kidney macrophages from spot and hogchoker captured in the Elizabeth River or in clean waters were prepared in monolayers on glass coverslips. Formalin-killed *Escherichia coli* suspension was added to the coverslips. At 30-min intervals for 150 min, coverslips were fixed and stained with Wright's stain. Phagocytosis was evaluated microscopically by counting the number of macrophages containing intracellular bacteria and expressing the result as percent phagocytosis. Differences were evaluated by the Student's t test.

Chemotaxis assay (Weeks et al., 1986). Kidney macrophages were isolated from Elizabeth River and York River spot and hogchoker and purified. Chemotactic activity was measured in a double-chambered apparatus in which macrophages were allowed to migrate from the upper chamber through a Nuclepore filter membrane to the lower chamber containing *E. coli* as the chemotactic stimulus. Membranes were removed at 30-min intervals, fixed, stained with Wright's stain, and mounted on slides.

The number of macrophages on each side of the filter was determined microscopically and the results expressed as percent chemotaxis. Differences in the means were evaluated by the Student's t test.

Chemiluminescence assay (Warinner et al., 1988; Wishkovsky et al., in press). Chemiluminescence (CL) was measured in a liquid scintillation counter in mini-vials containing macrophages, zymosan stimulus, and luminol to amplify the CL response. Using macrophages from Elizabeth River and York River spot, the time-course of the CL response was determined by counting the vials at 1-min intervals. In a second experiment, the effect of tributyltin (TBT) on the CL response was evaluated using kidney cell suspensions from toadfish, hogchoker, and croaker. TBT solution was added to vials containing cell suspension, zymosan, and luminol. Data were recorded as counts per minute at the peak response and analyzed by one-way analysis of variance.

RESULTS

Figure 1 shows the effects of the Elizabeth River environment on phagocytosis by macrophages from spot and hogchoker. Phagocytic activity increased with time, reaching maximum values within 120 min. Peak levels of phagocytosis were reduced significantly ($P < .05$) in both species of fish obtained from the Elizabeth River. The phagocytic activity of macrophages from Elizabeth River hogchoker returned to control levels after the fish were held in clean water for 2-3 weeks (data not shown).

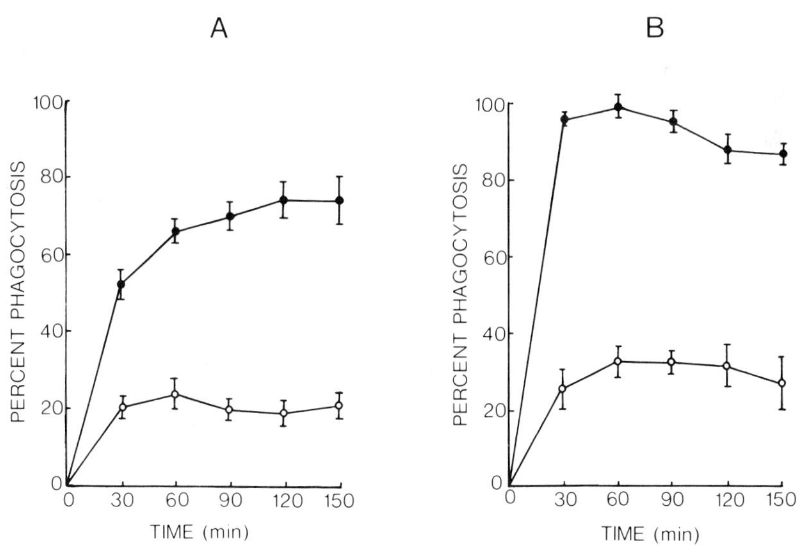

Fig. 1. Macrophage phagocytosis in (A) spot and (B) hogchoker from the Elizabeth River and from control waters. ● control fish; o Elizabeth River fish. Each value represents the mean ± SEM of 4 experiments using 5-10 fish per experiment (Weeks and Warinner, 1984). © Elsevier Applied Science Publishers Ltd., England, 1984.

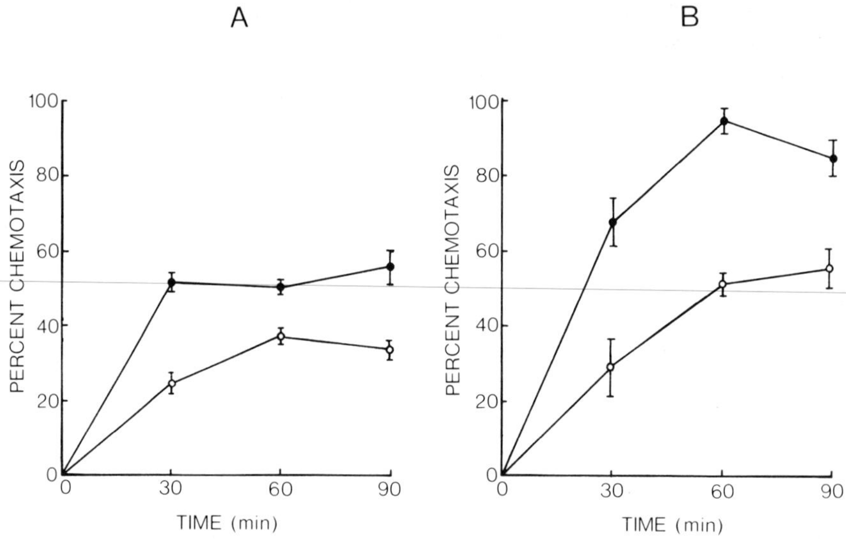

Figure 2. Macrophage chemotaxis in (A) spot and (B) hogchoker from the Elizabeth River and the York River. ● control fish; o Elizabeth River fish. Each value represents the mean ± SEM of 4 experiments using 6-10 fish per experiment (Weeks et al., 1986). © The Fisheries Society of the British Isles, 1986.

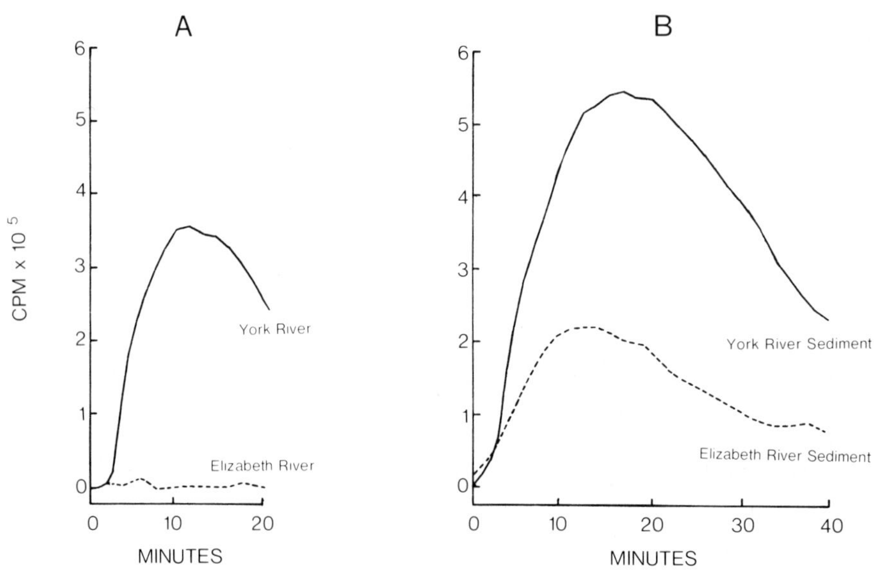

Figure 3. (A) Zymosan-stimulated CL response of macrophages from spot captured in the Elizabeth River and in the York River; (B) Zymosan-stimulated CL response of macrophages from York River spot exposed in the laboratory to Elizabeth River sediment (Warinner et al., 1988). © Elsevier Applied Science Publishers Ltd., England, 1988.

Chemotactic activity of spot and hogchoker kidney macrophages is shown in Figure 2. Maximum values were reached within 90 min, but activity was reduced significantly ($P < .05$) in the Elizabeth River fish.

Figure 3(A) shows the zymosan-stimulated CL response of purified macrophage suspensions from spot captured in the Elizabeth River and in the York River. Macrophages from Elizabeth River fish showed no measurable CL response. Figure 3(B) show a similar decrease in the zymosan-stimulated CL response of macrophages from York River fish exposed in the laboratory to Elizabeth River sediment. The peak amplitude of cells from fish exposed to contaminated sediments was greatly reduced as compared to controls.

The effect of TBT on the chemiluminescence activity of kidney cell suspensions from toadfish, hogchoker and croaker is shown in Table 1. In toadfish and hogchoker, a significant reduction from the control was observed at 400 μg/liter TBT, while in croaker a significant reduction in the response occurred at 40 μg/liter and 400 μg/liter TBT.

Table 1. Effect of TBT on zymosan-stimulated chemiluminescence: peak amplitude/thousand macrophages.

Source of Macro-phages[1]	TBT (μg/liter) 0	0.04	0.4	4	40	400
Toadfish (n=10)	9.9±3.1[2,3]	9.0±2.4	8.9±2.3	8.7±2.0	10.3±2.1	1.2±0.7*[4]
Hogchoker (n=8)	12.9±3.1	13.2±3.5	13.1±3.1	11.0±2.6	8.3±2.4	1.6±0.6*
Croaker (n=7)	16.8±4.2	16.1±4.0	17.1±3.9	14.8±3.2	1.9±1.1*	0.4±0.1*

[1]Kidney cell suspensions were prepared from individual toadfish kidneys or from kidneys pooled from 6-11 hogchoker or 4-8 croaker.
[2]Cpm in thousands.
[3]Results are expressed as the mean ± SEM.
[4]Significant differences from the zero level are indicated by *($P < 0.001$).

DISCUSSION

These results demonstrate that three functional aspects of macrophage activity (phagocytosis, chemotaxis, and chemiluminescence) were markedly reduced in

Elizabeth River fish as compared to clean water controls. The immunotoxic effects may be reversible, as indicated by the return to control levels of phagocytic activity following depuration of Elizabeth River fish. Chemical analyses have shown a gradient of concentrations of PAHs in Elizabeth River sediments, with a maximum in the reaches in which fish specimens were caught (Huggett et al., 1987). It is likely that these toxicants contribute to the reduced macrophage activity, but the effects of age, nutrition, season, and the presence of other environmental toxicants need to be evaluated.

Continuing studies of other fish species should be performed to clarify the immunological status of fish resident in polluted waters. If the results reported here apply to a wide variety of species, these techniques may make it possible to monitor fish health on a broad scale and thereby assess the degree of environmental pollution.

ACKNOWLEDGEMENT

This research was supported by the Virginia Graduate Marine Science Consortium. VIMS Contribution No. 1513.

LITERATURE CITED

Bieri, R.H., C. Hein, R.J. Huggett, P. Shou, H. Slone, C.L. Smith and C.W. Su. 1986. Polycyclic aromatic hydrocarbons in surface sediments from the Elizabeth River subestuary. Int. J. Environ. Anal. Chem., 26, 97-113.

Hargis, W.J., Jr. and D.E. Zwerner. 1988. Some histologic gill lesions of several estuarine finfishes related to exposure to contaminated sediments: A preliminary report. In: "Understanding the Estuary" (M.P. Lynch and E.C. Krome, eds.). Chesapeake Research Consortium, Gloucester Point, Virginia.

Huggett, R.J., M.E. Bender and M.A. Unger. 1987. Polynuclear aromatic hydrocarbons in the Elizabeth River, Virginia. In: "Fate and Effects of Sediment-Bound Chemicals in Aquatic Systems" (K.L. Dickson, A.W. Maki and W.A. Brungs, eds.). Pergamon Press, New York.

Sindermann, C.J. 1983. An examination of some relationships between pollution and disease. Rapp. P-v. Riun. Cons. Int. Explor. Mer., 182, 37-43.

Warinner, J.E., E.S. Mathews and B.A. Weeks. 1988. Preliminary investigations of the chemiluminescent response in normal and pollutant-exposed fish. Mar. Environ. Res., 24, 281-284.

Weeks, B.A. and J.E. Warinner. 1984. Effects of toxic chemicals on macrophage phagocytosis in two estuarine fishes. Mar. Environ. Res., 14, 327-335.

Weeks, B.A. and J.E. Warinner. 1986. Functional evaluation of macrophages in fish from a polluted estuary. Vet. Immunol. Immunopathol., 12, 313-320.

Weeks, B.A., J.E. Warinner, P.L. Mason and D.S. McGinnis. 1986. Influence of toxic chemicals on the chemotactic response of fish macrophages. J. Fish Biol., 28, 653-658.

Wishkovsky, A., E.S. Mathews and B.A. Weeks. Effect of tributyltin on the chemiluminescent response of phagocytes from three species of estuarine fish. Arch. Environ. Contam. Toxicol., in press.

TERATOLOGICAL AND NEOPLASTIC DISEASES SECTION

SEASONAL PREVALENCE AND INTENSITY AND DISEASE PROGRESSION OF NEOPLASIA IN SOFT SHELL CLAMS, *MYA ARENARIA*, FROM THE SHREWSBURY RIVER, NEW JERSEY[1]

Bruce J. Barber

Virginia Institute of Marine Science, School of Marine Science,
College of William and Mary, Gloucester Point, Virginia 23062, USA

ABSTRACT: Soft shell clams, *Mya arenaria*, from the Shrewsbury River, New Jersey, were examined from September, 1986, to August, 1987, for the prevalence (proportion of individuals that were affected) and intensity (proportion of blood cells that were neoplastic) of sarcomatous neoplasia. Monthly samples consisted of 100 individuals obtained from commercial fishermen. In addition, 27 diseased and 27 non-diseased clams were individually marked, replanted, and examined at 2-week intervals between May and August, 1987, to monitor disease progression. All diagnoses were performed on fixed and stained blood preparations. Prevalence (2-19%) and intensity (< 1-64%) varied seasonally and tended to be inversely related. Maximum mean intensities, which occurred in October and April, were associated with low prevalences. As intensity subsequently decreased, prevalence increased, reaching maxima in December and May. In the transplanted clams, mortality was significantly ($P < 0.001$) greater in the diseased group (93%) than in the non-diseased group (48%). Four clams (15%) that were initially diagnosed as not having the disease developed the disease and four clams (15%) initially diagnosed as having the disease lost the disease. All "reversals" occurred at intensities below 23%. At intensities above 23%, the disease always progressed and resulted in death. Time to death was significantly ($P < 0.01$) and inversely related to disease intensity. Considering the relatively low prevalence found in this population, it is unlikely that neoplastic disease has a major impact on the soft shell clam fishery of New Jersey.

---

[1]Contribution No. 1516 from the Virginia Institute of Marine Science, College of William and Mary, Gloucester Point, Virginia 23062.

Pathology in Marine Science

INTRODUCTION

The existence of sarcomatous neoplasia in bivalve molluscs was first described in *Mytilus edulis* from Oregon by Farley (1969). Since then, neoplastic diseases have been found in *M. edulis* populations in Great Britain (Lowe and Moore, 1978; Green and Alderman, 1983), Denmark (Rasmussen, 1986), and British Columbia (Cosson-Mannevy et al., 1984). Neoplastic disease has been reported in *Mya arenaria* from Maine (Brown et al., 1977; Yevich and Barszcz, 1977), Rhode Island (Brown et al., 1977; Cooper et al., 1982a), Massachusetts (Reinisch et al., 1984), Connecticut (Brousseau, 1987), and Maryland (Farley et al., 1986). Other species in which neoplasia has been found include *Ostrea lurida* from Oregon (Mix et al., 1977), *Ostrea edulis* from France (Balouet et al., 1986), *Crassostrea virginica* from Virginia (Frierman, 1976), and *Cerastoderma edule* from France (Poder and Auffret, 1986) and Ireland (Twomey and Mulcahy, 1984).

The disease is characterized by abnormally large, atypical cells, presumed to be hemocytes, having a high mitotic rate, DNA content, and a nucleus:cytoplasm ratio double that of normal hemocytes (Farley, 1969; Lowe and Moore, 1978; Cooper et al., 1982b; Rasmussen, 1986). In histological sections, vesicular connective tissue is infiltrated and replaced to varying degrees by the abnormal cells (anaplastic in appearance). Nuclei of neoplastic cells are round or lobulate, have diffuse chromatin and one to several nucleoli, and are 2-4 times larger than those of normal hemocytes. Cytoplasm of neoplastic cells is basophilic and agranular (Farley, 1969; Lowe and Moore, 1978; Cooper et al., 1982b). *In vivo*, neoplastic cells are considerably larger than normal hemocytes, lack pseudopodia, and do not aggregate (Cooper et al., 1982b).

The physiological effects and resultant ecological consequences of this disease on bivalve populations are poorly understood. *M. arenaria* having neoplastic disease exhibited a 20-50% reduction in growth and a greater mortality than clams not having the disease (Appeldoorn et al., 1979). In laboratory studies, neoplastic disease has been found to be progressive and lethal in both *M. arenaria* (Cooper et al., 1982a; Farley et al., 1986) and *M. edulis* (Elston et al., 1988).

Since neoplastic disease is capable of causing mortality, potential impacts on bivalve populations having high prevalences of this disease should be considered. For example, Farley et al. (1986) reported that neoplastic disease was non-existant in Maryland soft shell clams (*M. arenaria*) prior to 1979, but is now epizootic in some locations. Recently, prevalences up to 90% have been found, with concurrent mortalities (Farley et al., unpubl.).

This study was undertaken to determine the impact of neoplastic disease on the soft shell clam, *Mya arenaria*, fishery of New Jersey. Seasonal prevalence of the disease in New Jersey was examined for the first time. Intensity was determined as the proportion of neoplastic to total blood cells (actual counts) rather than the more subjective staging previously employed. In addition, disease progression was monitored in individual clams maintained *in situ*.

MATERIALS AND METHODS

Seasonal prevalence and intensity. Commercial harvesting of soft shell clams in New Jersey is confined primarily to the Shrewsbury River, Navesink River, and Sandy Hook Bay region (Fig. 1). For this study, clams were obtained directly from fishermen while they were working at low tide. Fifty clams were obtained from each of two sites within the estuary on a monthly basis for a year, beginning in September, 1986. Water temperature (°C) and salinity (‰) were recorded at the time of collection to the nearest 0.5 unit using a thermometer and refractometer, respectively.

Clams were returned to the laboratory and placed in a recirculating sea water system prior to analysis the following day. Neoplastic disease was diagnosed using the histocytological procedures developed by Cooper et al. (1982b) and Farley et al. (1986). A 0.5 ml sample of blood from the anterior adductor muscle was drawn into a hypodermic syringe containing 2 ml of filtered (0.45 μm) sea water. The blood-sea water mixture was emptied into a settling chamber over a microscope slide that had been coated with poly-L-lysine. Blood cells were allowed to settle for 1 hr. Chambers and fluid were then removed and the blood cells were fixed in methanol and stained with a modified Wright's stain technique (eosin Y and methylene blue). Preparations were allowed to air dry before being permanently covered.

The fixed, stained slides were examined (200X) for the presence of neoplastic cells. Prevalence of the disease was calculated as the percentage of a sample (N=50)

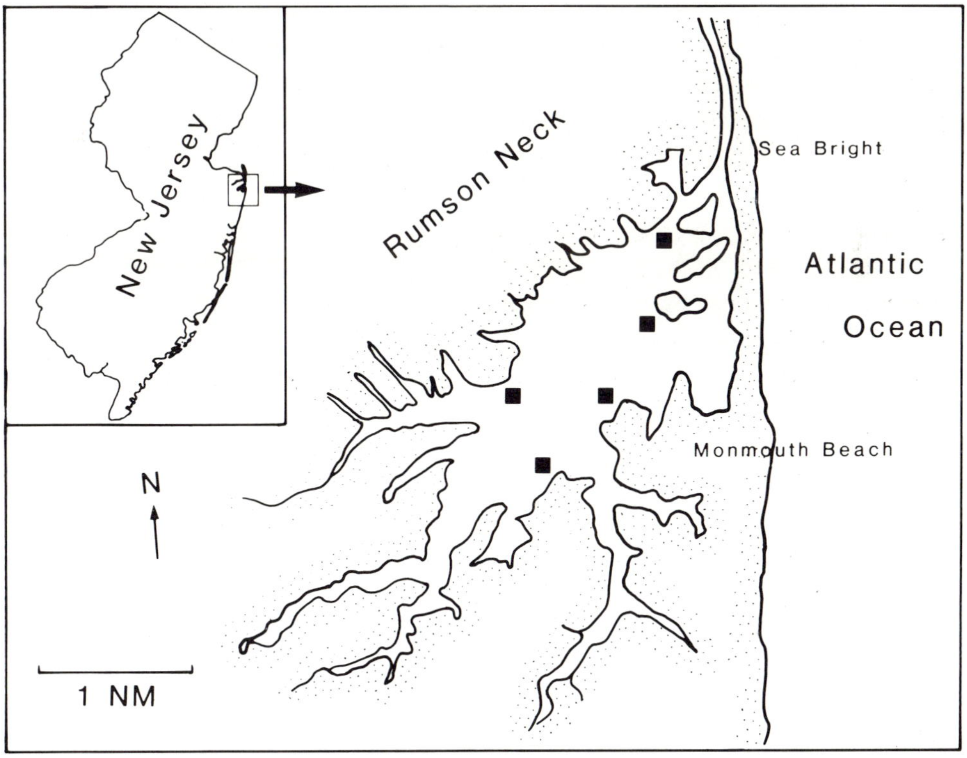

Fig. 1. Map of Shrewsbury River showing locations of sites from which soft shell clams, *Mya arenaria*, were collected.

having neoplastic cells. Prevalences at the two sites were compared using the testing procedure for differences between two proportions (Zar, 1974). Intensity of the disease was calculated as the percentage of total cells that were neoplastic, based on actual cell counts in up to 16 fields. Mean intensity of infected clams was then calculated for each sample.

Disease progression. To monitor disease progression, a separate sample of *M. arenaria* was screened for disease in May, 1987, until 27 diseased clams and 27 non-diseased clams were obtained. These were individually numbered and transplanted to an intertidal site in lower Delaware Bay where they were allowed to rebury inside a wire mesh cage. This site was chosen because of its ease of access, proximity to the laboratory, and suitability (a sparse natural population exists here). As the bleeding technique has been shown to be non-destructive (Brown et al., 1977; Cooper et al., 1982a), the clams were resampled every 2 weeks through August, 1987. Disease intensity was determined for those having abnormal cells. The relationship between disease intensity and time to death was examined by averaging the time to death (in 2-week intervals) for clams in each of five categories of initial intensity: 0-20%, 21-40%, 41-60%, 61-80%, and 81-100%.

RESULTS

Seasonal prevalence and intensity. Over the course of the study clams were collected from five separate sites within the Shrewsbury River (Fig. 1). However, the prevalence of neoplastic clams was never significantly different ($P > 0.05$) between the two sites sampled on any date. Thus all clams were considered to belong to a homogenous population and the two samples collected each month were combined for consideration of seasonal prevalence and intensity. The shell length of clams used in this study ranged from 35.5 to 82.3 mm. No samples were obtained in January and February, 1987, because of ice and water quality problems.

Table 1. Water temperature (°C) and salinity (‰) of Shrewsbury River, New Jersey at times of collection of soft shell clams.

Date	Temperature	Salinity
29 September 1986	21.0	25.0
27 October 1986	14.0	24.5
25 November 1986	7.5	19.0
23 December 1986	3.0	16.0
23 March 1987	7.0	17.5
30 April 1987	12.0	19.0
29 May 1987	16.0	23.0
30 June 1987	25.0	24.0
30 July 1987	27.0	23.0
31 August 1987	23.0	24.0

Water temperature in the Shrewsbury River ranged from 3.0°C in December, 1986, to 27.0°C in July, 1987 (Table 1). Salinity also exhibited a seasonal cycle, ranging from 16 ‰ in December, 1986, to 23-25 ‰ between May and October, 1987 (Table 1).

Prevalence of neoplastic disease ranged from 2-19% (Fig. 2). In most months prevalence was below 10%, but was 15% in December, 1986; 12% in March, 1987; and 19% in May, 1987. Thus, peak prevalences occurred in conjunction with minimum water temperature and salinity (December-March, 1987) and in May, 1987, after spawning had been completed.

Mean intensity of diseased clams ranged from <1-64% (Fig. 2). Intensity in individual clams ranged from 0.04-98.6%. Distinct maxima in intensity occurred in October, 1986, and April, 1987 (both 64%). In each instance, a sharp decrease in intensity was recorded the following month. In general, disease intensity was related to disease prevalence in that peak intensities occurred in conjunction with low prevalences and were followed (within the next month or two) by increasing prevalence.

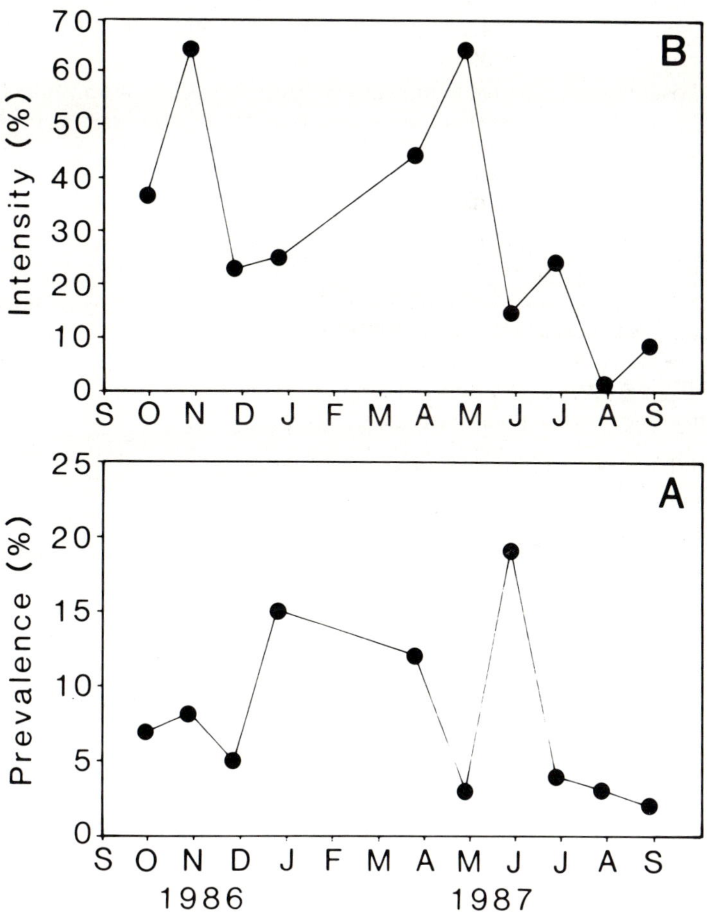

Fig. 2. Seasonal prevalence (A) and intensity (B) of neoplasia in soft shell clams, *Mya arenaria*, from the Shrewsbury River, New Jersey.

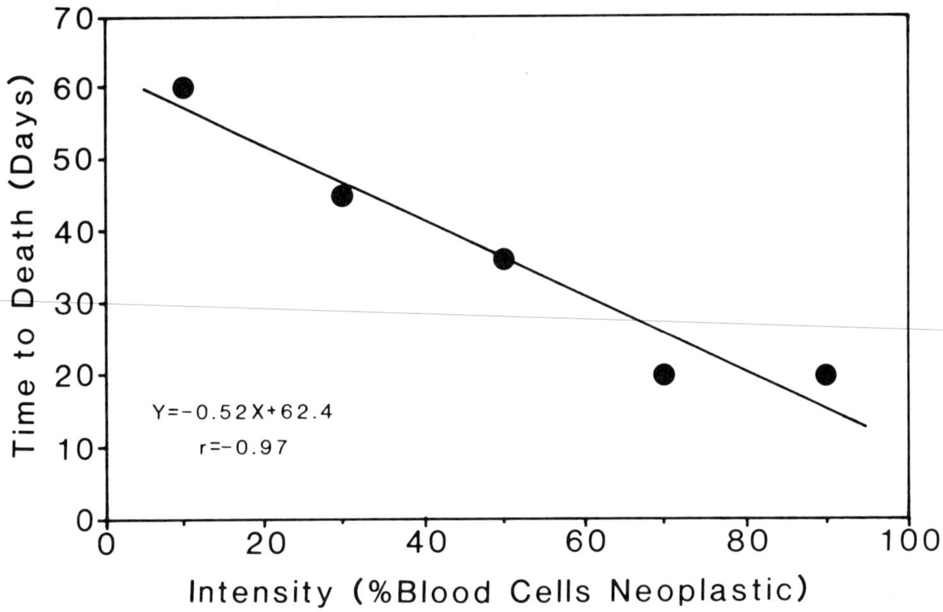

Fig. 3. Relationship between intensity of neoplastic disease and time to death for soft shell clams, *Mya arenaria*, from the Shrewsbury River, New Jersey. Points are means of all times falling within each 20% class of intensity.

Disease progression. Within the group that was initially disease free, 48% (13/27) died over the course of the 3-month experiment. Four clams (15%) initially diagnosed as being disease-free, developed the disease. Of these four, one died within 6 weeks, one subsequently lost the disease, and two remained chronically diseased until termination of the experiment. In all four of these clams, disease intensity never exceeded 4.4%.

Within the group that was initially diseased, 93% (25/27) died over the study period. Mortality was significantly greater ($P < 0.001$) in the diseased group than in the group not having the disease. In most clams, disease intensity steadily increased until death occurred. However, four clams (15%) that had the disease at the beginning of the experiment lost the disease. Of these four, two subsequently died and two remained alive at the end of the experiment. Of the four clams that lost the disease, one had an initial intensity of 23% and the other three had intensities less than 2%.

Among the clams that were initially diagnosed as having the disease, there was a significant ($P < 0.01$) inverse correlation between infection intensity and time to death (Fig. 3). That is, the greater the disease intensity, the shorter the life expectancy.

DISCUSSION

Seasonal prevalence and intensity. Neoplastic disease exhibited a seasonal cycle in both prevalence and intensity in the soft shell clam, *M. arenaria*, population of the Shrewsbury River, New Jersey. Considering the relatively small size of the estuary

(Fig. 1), it is not surprising that no differences in disease prevalence were found between sites sampled. Peak prevalences occurred in December (15%) and May (19%), and minimum prevalences (< 10%) were found in April and from June to November. Disease intensity was approximately inversely related to prevalence in that intensity maxima (64%), which occurred in October and April, were associated with low prevalences. Subsequently, as intensity decreased, prevalence increased, reaching maxima in December (15%) and May (19%). This suggests that a dynamic interaction between prevalence, intensity, and mortality may exist within this clam population. The number of individuals infected (prevalence) increases followed by a gradual increase in intensity. Mortality of heavily diseased clams may then lead to lower prevalence as well as lower mean intensity. A similar cycle of increasing prevalence followed by increasing intensity, leading to considerable mortality, followed by reduced prevalence occurs in oysters, *Crassostrea virginica*, infected by the endoparasite MSX (*Haplosporidium nelsoni*) in Delaware Bay (Ford, 1985).

Seasonal fluctuations in the prevalence of neoplastic disease have been reported previously for bivalve molluscs in other locations. *M. edulis* from two of three sites in Denmark had prevalences of neoplasia that were lowest in the spring (Rasmussen, 1986). Similarly, two out of three populations of *M. edulis* from British Columbia, Canada, had peak prevalences during the winter (Cosson-Mannevy et al., 1984). Prevalence of neoplasia has been examined in *M. arenaria* from Rhode Island (Cooper et al., 1982a), Connecticut (Brousseau, 1987) and Maryland (Farley et al., 1986). In all locations, prevalence was greatest sometime between October and April and lowest between June and September. Thus a fairly universal characteristic of neoplastic disease is a seasonality in prevalence, with lowest prevalences occurring in the spring and summer and highest prevalences occurring in the fall and winter. In addition, this study and that of Cooper et al. (1982a) found a second prevalence peak in May.

The fairly consistent seasonal component suggests that environmental factors have a role in the dynamics of this disease. In this study, maximal prevalences were found during the winter, when salinity and temperature were lowest for the year. This may be the result of an alteration in the availability or activity of etiological agents, a reduction in disease virility, or increase in host response (allowing more infected individuals to resist mortality in spite of being diseased), or a reduction in the activity of predators that may selectively prey on diseased individuals (Ford, 1985; Newell and Barber, 1988). The prevalence maxima found in May in this study and that of Cooper et al. (1982a) occurred during the spawning period when bivalves are generally in a state of negative energy balance (Newell and Barber, 1988). Even though temperature, salinity, and reproductive state may have an influence, much remains to be learned about the factors controlling the outbreak, prevalence, and tolerance of this disease.

Disease progression. Previous studies have established that neoplastic disease results in the death of bivalves. Frierman (1976) noted that neoplasms were associated with mortality in oysters, *C. virginica*. Appeldoorn et al. (1979) found that mortality was greater (not significantly) in diseased soft shell clams than in non-diseased clams that had been cross-transplanted between clean and oil-impacted sites. Farley et al. (1986) maintained 10 diseased *M. arenaria* in aquaria without food at 10°C and found that in all cases disease intensity increased and mortality resulted within 6 months. Cooper et al. (1982a) found that disease intensity increased and resulted in the death of 50% of diseased *M. arenaria* maintained in a flowing sea water system. Elston et al.

(1988) collected 40 *M. edulis* from a natural population and maintained them in flowing sea water (temperature 9-12°C; salinity 30 ‰) for 128 days. Twenty individuals developed the disease, 12 of which died, and 6 of the 8 remaining had advanced cases of the disease at the end of the experiment. The present study was the first to examine the progression of neoplastic disease *in situ*, where clams were protected from predators, but in an otherwise natural environment. Over the course of 3 months, mortality (93%) was significantly greater in the diseased group than in the control group (45%). The progressive nature of the disease was illustrated by the significant inverse correlation between disease intensity and time to death (Fig. 3).

Although it is clear that neoplastic disease is progressive and lethal in most cases, reversals of disease status and apparent cases of remission have been reported. In the study of Cooper et al. (1982a), the disease remained chronic (stable, low level) in 40% and decreased or was lost in 10% of the diseased clams, and was acquired in 10% of the previously non-diseased clams. Twenty percent of the mussels examined by Elston et al. (1988) exhibited an ability to respond to the disease such that remission resulted. This appeared to result from the entrapment of abnormal cells in an extracellular matrix secreted by normal hemocytes. In the present study, disease reversals occurred in 15% of both groups. In all cases, the loss of neoplasia in diseased clams and the initial appearance of neoplasia occurred at intensities less than 23%. In a few cases the disease remained chronic, and this also occurred at intensities below 23%. Thus it appears that bivalves are capable of tolerating and even recovering from this disease, at least at relatively low intensities. However, once a certain intensity is exceeded (25% in the present study), the disease progresses and results in death.

Effect on fishery. Any commercial fishery having a high prevalence might be subject to impact by this disease, considering its progressive and lethal nature. Prevalences of neoplasia of 2-19% (with a yearly average of 8%) in *M. arenaria* from the Shrewsbury River were relatively low compared to other bivalve populations in which prevalences as high as 90% (Reinisch et al., 1984; Farley et al., unpubl.), 42-65% (Farley et al., 1986), 64% (Brown et al., 1977), and 43% (Cooper et al., 1982a) were found.

A very rough estimate of mortality caused by neoplastic disease in the Shrewsbury River soft shell clam population can be calculated as follows. Using the mortality figures from the transplant study, mortality in the diseased group was 45% higher than in the control group. Prevalence of neoplasia ranged from 2-19% on a monthly basis, with a yearly average of 8%. Resultant mortality might therefore be expected to range from 0.9-8.6% on a monthly basis, with a yearly mortality of 3.6%. Thus, even though neoplastic disease is epizootic (>1%) in the Shrewsbury River, it does not appear to have a major impact on the soft shell clam fishery of New Jersey.

ACKNOWLEDGEMENTS

I would like to thank D. Owens (New Jersey Bureau of Shellfish Control) for assistance with clam collection, J. Cortelyou for laboratory assistance, and S. Ford for encouragement and manuscript review. This project was supported by a grant provided by the New Jersey Department of Higher Education and administered by the New Jersey Marine Sciences Consortium.

LITERATURE CITED

Appeldoorn, R.S., R.S. Brown and C.W. Brown. 1979. A three month growth and mortality study of normal and neoplastic *Mya arenaria* cross-transplanted between clean and oil-impacted areas. Proc. Nat. Shellfish. Assoc., 69, 192.

Balouet, G., M. Poder, A. Cahour and M. Auffret. 1986. Proliferative hemocytic condition in European flat oysters (*Ostrea edulis*) from Breton Coasts: A 6-year survey. J. Invert. Pathol., 48, 208-215.

Brousseau, D.J. 1987. Seasonal aspects of sarcomatous neoplasia in *Mya arenaria* (soft-shell clam) from Long Island Sound. J. Invert. Pathol., 50, 269-276.

Brown, R.S., R.E. Wolke, S.B. Saila and C.W. Brown. 1977. Prevalence of neoplasia in 10 New England populations of the soft-shell clam (*Mya arenaria*). Ann. N.Y. Acad. Sci., 298, 522-534.

Cooper, K.R., R.S. Brown and P.W. Chang. 1982a. The course and mortality of a hematopoietic neoplasm in the soft shell clam, *Mya arenaria*. J. Invert. Pathol., 39, 149-157.

Cooper, K.R., R.S. Brown and P.W. Chang. 1982b. Accuracy of blood cytological screening techniques for the diagnosis of a possible hematopoietic neoplasm in the bivalve mollusc, *Mya arenaria*. J. Invert. Pathol., 39, 281-289.

Cosson-Mannevy, M.A., C.S. Wong and W.J. Cretney. 1984. Putative neoplastic disorders in mussels (*Mytilus edulis*) from southern Vancouver Island waters, British Columbia. J. Invert. Pathol., 44, 151-160.

Elston, R.A., M.L. Kent and A.S. Drum. 1988. Progression, lethality and remission of hemic neoplasia in the bay mussel *Mytilus edulis*. Dis. Aquat. Organ., 4, 135-142.

Farley, C.A. 1969. Sarcomatoid proliferative disease in a wild population of blue mussels (*Mytilus edulis*). J. Nat. Cancer Inst., 43, 509-516.

Farley, C.A., S.V. Otto and C.L. Reinisch. 1986. New occurrence of epizootic sarcoma in Chesapeake Bay soft shell clams, *Mya arenaria*. Fish. Bull., 84, 851-857.

Ford, S.E. 1985. Chronic infections of *Haplosporidium nelsoni* (MSX) in the oyster *Crassostrea virginica*. J. Invert. Pathol., 45, 94-107.

Frierman, M.E. 1976. Occurrence of hematopoietic neoplasms in Virginia oysters (*Crassostrea virginica*). Mar. Fish. Rev., 38, 34-36.

Green, M. and D.J. Alderman. 1983. Neoplasia in *Mytilus edulis* L. from United Kingdom waters. Aquaculture, 30, 1-10.

Lowe, D.M. and M.N. Moore. 1978. Cytology and quantitative cytochemistry of a proliferative atypical hemocytic condition in *Mytilus edulis* (Bivalvia, Mollusca). J. Nat. Cancer Inst., 60, 1455-1459.

Mix, M.C., H.J. Pribble, R.T. Riley and S.P. Tomasovic. 1977. Neoplastic disease in bivalve mollusks from Oregon estuaries with emphasis on research on proliferative disorders in Yaquina Bay oysters. Ann. N.Y. Acad. Sci., 298, 356-373.

Newell, R.I.E. and B.J. Barber. 1988. A physiological approach to the study of bivalve molluscan diseases. Am. Fish. Soc. Spec. Publ., 18, 269-280.

Poder, M. and M. Auffret. 1986. Sarcomatous lesion in the cockle *Cerastoderma edule* I. Morphology and population survey in Brittany, France. Aquaculture, 58, 1-8.

Rasmussen, L.P.D. 1986. Occurrence, prevalence and seasonality of neoplasia in the marine mussel *Mytilus edulis* from three sites in Denmark, Mar. Biol., 92, 59-64.
Reinisch, C.L., A.M. Charles and A.M. Stone. 1984. Epizootic neoplasia in soft shell clams collected from New Bedford Harbor. Hazard. Waste, 1, 73-81.
Twomey, E. and M.F. Mulcahy. 1984. A proliferative disorder of possible hemic origin in the common cockle, *Cerastoderma edule*. J. Invert. Pathol., 44, 109-111.
Yevich, P.P. and C.A. Barszcz. 1977. Neoplasia in soft-shell clams (*Mya arenaria*) collected from oil-impacted sites. Ann. N.Y. Acad. Sci., 298, 409-426.
Zar, J.H. 1974. "Biostatistical Analysis." Prentice-Hall, Englewood Cliffs, New Jersey.

ULTRASTRUCTURAL OBSERVATIONS ON MELANIZED SPERM IN DEVELOPING AND FULLY FORMED SPERMATOPHORES OF MALE SHRIMP, *PENAEUS VANNAMEI*

W.J. Dougherty and M.M. Dougherty

Marine Biomedical Research Program, Medical University of South Carolina, Charleston, South Carolina 29412, USA

ABSTRACT: Developing spermatophores in the medial vas deferens and black spermatophores in the terminal ampules of pond-cultured *Penaeus vannamei* males were prepared for light and electron microscopical study. As in previous studies, black spermatophores frequently contained mixtures of normal appearing unistellate sperm, abnormal spikeless sperm, and degenerating sperm, with oval pigment droplets occurring in the nuclei of both normal appearing unistellate sperm and of abnormal spikeless sperm. Oval pigment droplets also occurred among sperm in the extracellular matrix of spermatophores. Developing spermatophores in both the proximal and distal segments of the medial vas deferens similarly contained normal appearing unistellate sperm and abnormal spikeless sperm, each with oval intranuclear pigment droplets. Testicular sperm were unistellate just prior to their exit from the testicular lobes into the proximal vas deferens. No intranuclear pigment droplets were encountered in testicular sperm. Sperm melanization may occur as sperm leave the proximal vas deferens and enter the medial vas deferens. The mechanism by which sperm become melanized remains to be determined. No bacteria and no hemocytes were encountered in either light or electron microscopical preparations of developing or fully formed melanized spermatophores.

INTRODUCTION

Pond-raised male penaeid shrimp are known to become afflicted with a condition referred to as "black spermatophore." This results in diminished male breeding capacity. The etiology of this condition is poorly understood although several causes have been suggested, including bacterial infection and mechanical trauma (Chamberlain et al., 1983). The report of Harris and Sandifer (1986) noted a high frequency of occurrence of the condition in *Macrobrachium rosenbergii* males that were repetitively electroejaculated. Symptoms of the disease in *Penaeus setiferus,*

P. stylirostris, P. vannamei, and *M. rosenbergii* are similar and include the presence of darkened spermatophores, ranging in color from tan to black, within terminal ampules of the vas deferens. Most workers presumed that the dark coloration is due to the abnormal accumulation of the pigment melanin in the spermatophores. Dougherty and Dougherty (1989) have recently presented histochemical evidence supporting the view that the pigment is melanin. They also showed that the pigment occurred within the nuclei of penaeid sperm and that many sperm in black spermatophores were abnormally spikeless.

In order to better understand this condition and to determine whether sperm melanization and spikeless sperm occur not only in fully formed spermatophores but also in spermatophores forming within the vas deferens, the reproductive tracts of *P. vannamei* males exhibiting black spermatophores within their terminal ampules were examined by light and electron microscopy.

MATERIALS AND METHODS

Reproductive tracts containing black spermatophores in the terminal ampullae were dissected from male *P. vannamei* held at the Waddell Mariculture Center, Blufton, South Carolina, and were prepared for microscopical study by immersion for 2 to 18 hr at room temperature in a fixative consisting of 6% glutaraldehyde in 0.1 M cacodylate buffer, pH 7.2, diluted with sea water. This fixative is modified from that used previously in studies of black spermatophores (Dougherty and Dougherty, 1989). Several normal-appearing, i.e., nonpigmented, spermatophores were collected from *P. vannamei* and *P. setiferus* males and were prepared in the same manner for microscopical study. While in aldehyde fixative, segments of the vas deferens and

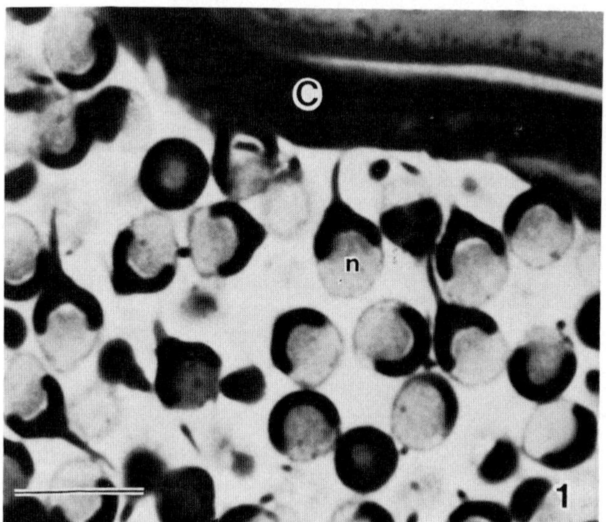

Fig. 1. Light photomicrograph of a semi-thin section of a normal spermatophore of *P. vannamei*. Note sperm with single spikes and oval bodies. No pigment occurs within normal sperm nuclei nor in the extracellular matrix among sperm. Sperm nucleus (n). Spermatophore capsule (C). Bar = 10 μm.

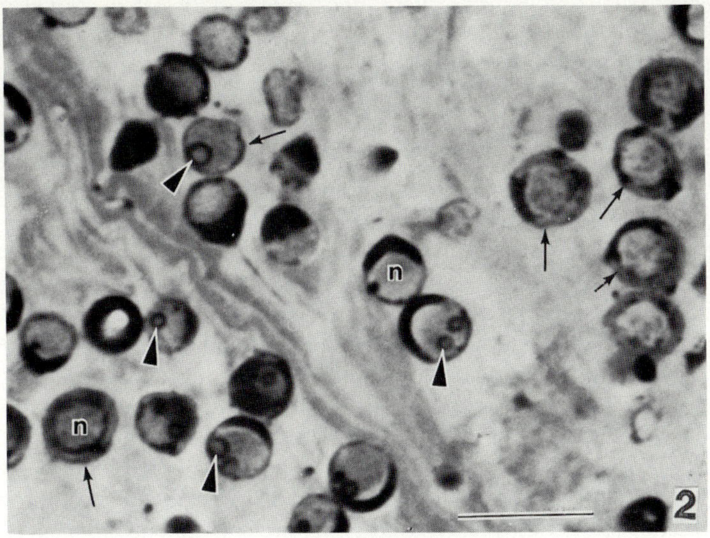

Fig. 2. Light photomicrograph of a semi-thin section of a developing spermatophore stained with toluidine blue obtained from the medial vas deferens of an animal with the black spermatophore condition. This image illustrates spikeless (arrows) sperm, some of which contain pigment droplets (arrowheads) within their nuclei (n). Bar = 10 μm.

spermatophores were sliced into smaller segments. After washing in several changes of 0.1 M cacodylate-buffered sea water, several tissue samples were post-fixed in 1% OsO_4 in 0.1 M cacodylate buffer, pH 7.2, at 4°C for 1 to 2 hr. Samples were then rinsed in cold sea water several times, dehydrated in an ascending series of acetone, and infiltrated and embedded in low viscosity epoxy resin (Spurr, 1969).

Semi-thin sections ($\approx$ 1 μm) were cut with glass knives and were examined and photographed in a brightfield microscope either unstained or following staining with a solution of toluidine blue. Ultrathin sections were cut with a diamond knife mounted in a Sorvall MT-2 ultramicrotome, picked up on copper grids, stained sequentially with uranyl acetate and lead citrate, and examined in a Philips 301 transmission electron microscope operating at 60 kV.

OBSERVATIONS

The normal penaeid spermatophore consists of a sperm mass surrounded by a chitinous capsule (Fig. 1). The sperm mass normally contains many unistellate sperm embedded in a non-cellular extracellular matrix.

As in a previous study (Dougherty and Dougherty, 1989), light microscopical study of black spermatophores revealed varying numbers of aggregated and individual yellow-brown to black polymorphic pigment droplets, measuring 2 to 400 μm in diameter, in spermatophore capsules. These droplets were not encountered in normal spermatophores. Severly afflicted spermatophores contained mostly degenerate sperm. Other less severely afflicted spermatophores contained mixtures of normal appearing unistellate sperm and abnormal spikeless sperm (Fig. 2). Oval, yellow-brown to black pigment droplets, measuring 1 to 2.5 μm in diameter, were encountered not only in the extracellular matrix of the sperm masses but also in the nuclei of sperm (Fig. 2).

Light and electron microscopical study of developing spermatophores in the medial vas deferens revealed both unistellate sperm and spikeless sperm containing intranuclear melanin droplets (Figs. 3, 4). Intranuclear droplets were bound by a single membrane, exhibited fairly homogeneous internal contents, and were frequently enmeshed in strands of intranuclear chromatin threads. Spikeless sperm, in addition to containing oval electron dense intranuclear droplets, exhibited arrays of small, electron dense particles packed in the region of the sperm where one would expect to have encountered the spike (Fig. 4), suggesting that these particles play a role in sperm spike depolymerization. The chemical nature of these small particles is not known. No bacteria and no hemocytes were encountered in developing spermatophores within the segments of the medial vas deferens.

Degenerating spermatozoa in fully formed spermatophores exhibited no spikes and no pigment droplets in the remains of the sperm cell bodies (Fig. 5).

Light and electron microscope study of the distal region of several testicular lobes of animals afflicted with black spermatophore revealed only normal appearing unistellate sperm. No testicular sperm exhibited intranuclear melanin droplets.

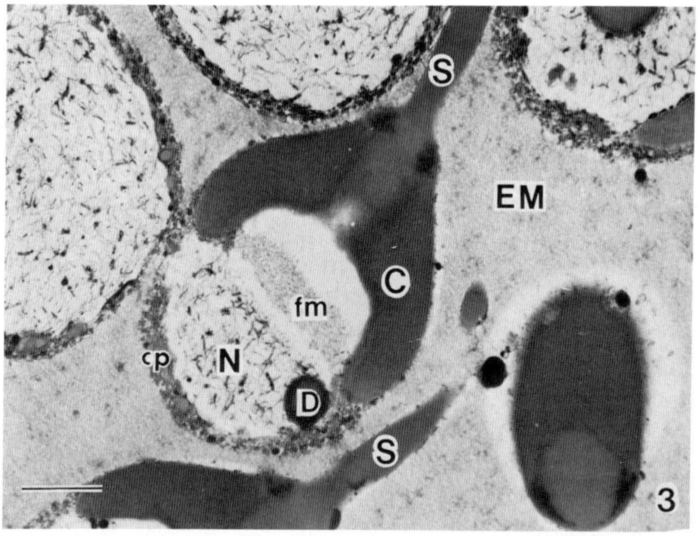

Fig. 3. Electron micrograph of spiked sperm in a developing spermatophore. Sperm consist of a spike (S) and hemispherical cap (C) of moderate electron density, a nucleus (N) containing a network of chromatin threads, a filamentous meshwork (fm) between the nucleus and hemispherical cap, and a posterior hemispherical rim of cytoplasmic particles (cp). The nuclear compartment contains an electron dense droplet (D) corresponding in size and shape to the pigmented droplets visualized in light microscope images of melanized sperm. Extracellular matrix (EM). Bar = 1 μm.

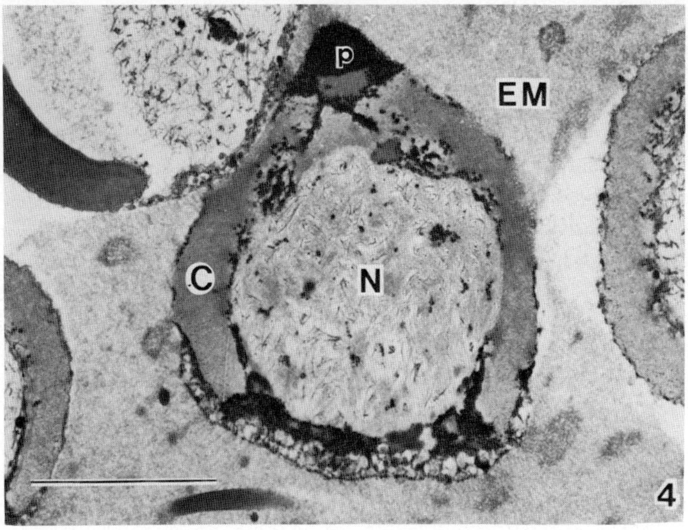

Fig. 4. Electron micrograph of spikeless sperm in a developing spermatophore. The hemispherical cap (C) exhibits lowered electron density and aggregates of small, electron dense particles (p) occur in the area of the sperm cell where one would normally expect to encounter a spike. Other small electron dense particles occur within the matrix of the hemispherical cap. Nucleus (N). Extracellular matrix (EM). Bar = 1 μm.

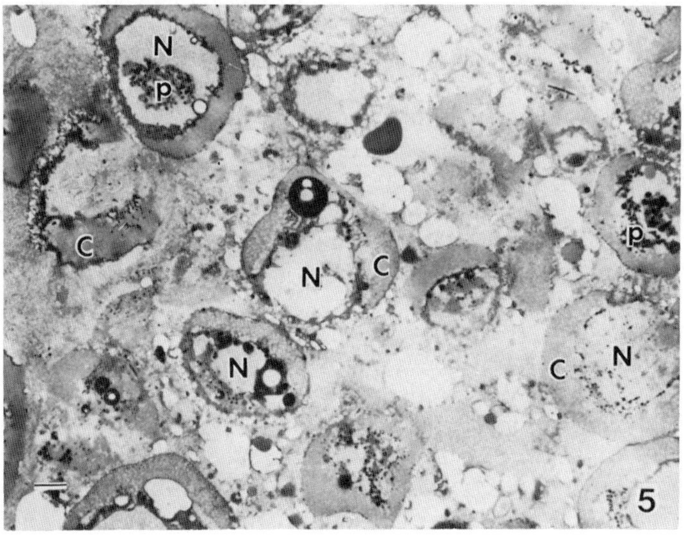

Fig. 5. Electron micrograph illustrating sperm with anamalous structure, presumably in process of degeneration, in a fully formed black spermatophore. No sperm in this field exhibit spikes. Many contain small, electron dense particles (p) within and at the perimeter of their nuclear (N) compartments and within the hemispherical cap material (C). Chromatin threads are no longer apparent in nuclei of disintegrating sperm. Bar = 1 μm.

DISCUSSION AND CONCLUSIONS

Our observations indicate that normal appearing unistellate sperm and abnormal spikeless sperm, each containing intranuclear melanin droplets, occur both in developing spermatophores within the medial vas deferens and in fully formed spermatophores within terminal ampules of *P. vannamei* males afflicted with the black spermatophore condition. Previous studies have demonstrated that the intranuclear droplets are melanin in nature (Dougherty and Dougherty, 1989). Our inability to find melanized unistellate sperm within testicular lobes of animals afflicted with the black spermatophore condition suggests that melanization of sperm is initiated somewhere along the reproductive tract, most probably in the proximal segment of the medial vas deferens. Reduced fecundity in this species probably follows the depolymerization of sperm spikes and degeneration of sperm.

Various suggestions have been made to account for the black spermatophore condition in shrimp, including bacterial or fungal infection and mechanical or electrical trauma incurred during repetitive spermatophore ejaculation. To these we might add nutritional deficiency, aging, and autoimmunity.

Chamberlain et al. (1983) reported the occurrence of *Pseudomonas* sp. in melanized spermatophores of the shrimp species they studied. They were unable, however, to produce the melanized condition in animals inoculated with bacterial culture broth. Talbot et al. (1988) noted the darkening of *P. setiferus* spermatophores 35 days after the introduction of wild-caught animals into pond culture. The presence of numerous bacteria in the sperm masses of these animals was not detected, however, until 48 days after introducing the shrimp into pond culture. In the present study, no bacteria were observed by light or electron microscopy in the vicinity of melanin pigment droplets in the sperm or sperm mass matrices of developing or fully formed melanized spermatophores.

Onset of reproductive tract melanization may follow repetitive manual or electrical ejaculation of spermatophores both in penaeid (Chamberlain et al., 1983; Sandifer et al., 1984) and caridean shrimp (Harris and Sandifer, 1986). These traumatic treatments do not offer complete or fundamental explanations of the cause of the condition in view of our observations that the melanized condition appeared spontaneously in *P. vannamei* males not previously ejaculated by manual or electrical means. Spontaneous onset of the melanized condition may have its etiology in a nutritional deficiency or in age-related changes in pond-raised shrimp. Nutritional deficiency, however, is not likely to be the complete explanation of the onset of this condition in view of the reported adverse effects of mechanical and electrical trauma on apparently nutritionally sufficient animals. The correlation of the melanized condition with age-related changes in shrimp remains to be explored.

Melanization occurs in many invertebrates as a response to injury or to infection and frequently involves the coordinated responses of hemocytes in the detection and encapsulation of foreign objects with the subsequent deposition of a melanin capsule around the isolated mass (Salt, 1965; Nappi, 1975; Bauchau, 1981; Lightner, 1985; Söderhäll and Smith, 1986). While hemocytic melanization appears to be a common defense response in many arthropods, a humoral melanization mechanism appears to occur intracellularly (Bradley and Nayar, 1985) or in the hemolymph in several arthropod species, without the direct participation of hemocytes (Poinar and Leutenegger, 1971; Poinar, 1973; Gupta, 1986). Söderhäll and Smith (1986) discount humoral melanization and suggest that polyphenoloxidase, the enzyme that catalyzes the formation of melanin and occurs in hemolymph, is of hemocyte origin. It is difficult to invoke hemocytic origins for the intracellular polyphenoloxidase of mosquito Malpighian tubules (Bradley and Nayar, 1985). It is interesting that no bacteria, no hemocytes, and no melanin capsules were encountered in either light or electron microscope preparations of developing or fully formed melanized spermatophores. In the absence of data that would implicate bacterial or fungal invaders, mechanical or electrical trauma, or nutritional deficiencies as primary elicitors of melanization of *P. vannamei* sperm, one must consider the possibility that the sperm themselves, or macromolecular constituents of their plasma membranes, being nonself entities, might activate constitutive polyphenoloxidase of the spermatophore, such enzyme activation resulting in melanized sperm. Malek and Bawab (1971) reported the occurrence of a "tanning" oxidase, presumably polyphenoloxidase, in the spermatophore of the shrimp *Penaeus trisulcatus* while polyphenoloxidase activity has been detected histochemically and biochemically also in normal appearing spermatophores of *M. rosenbergii* and *P. setiferus* (Dougherty, unpubl.). The possible role of this or other nonhemocytic enzymes in the pathological melanization of sperm and spermatophores of shrimps remains to be explored.

ACKNOWLEDGEMENTS

This work was supported in part by a grant from the South Carolina Sea Grant consortium, Grant No. NA85AA-D-SG121, Am.5. Ms. Susan Harris of the South Carolina Department of Wildlife and Marine Resources made this serious and extensive problem known to us and Mr. Steve Hopkins of the Waddell Mariculture Center, Bluffton, South Carolina, very generously made melanized specimens available for study.

LITERATURE CITED

Bauchau, A.G. 1981. Crustaceans. In: "Invertebrate Blood Cells" (N.A. Ratcliffe and A.F. Rowley, eds.), vol. 2, pp. 385-420. Academic Press, London.

Bradley, T.J. and J.K. Nayar. 1985. Intracellular melanization of the larvae of *Dirofilaria immitis* in the Malpighian tubules of the mosquito, *Aedes sollicitans*. J. Invert. Pathol., 45, 339-345.

Chamberlain, G.W., S.K. Johnson and D.H. Lewis. 1983. Swelling and melanization of the male reproductive system of captive adult penaeid shrimp. J. World Maricul. Soc., 14, 135-136.

Dougherty, W.J. and M.M. Dougherty. 1989. Electron microscopic and histochemical observations on melanized sperm and spermatophores of pond-cultured shrimp, *Penaeus vannamei*. J. Invert. Pathol., in press.

Gupta, A.P. 1986. Arthropod immunocytes: identification, structure, functions, and analogies to the functions of vertebrate B- and T-lymphocytes. In: "Hemocytic and Humoral Immunity in Arthropods" (A.P. Gupta, ed.), pp. 3-59. John Wiley & Sons, New York.

Harris, S.E.G. and P.A. Sandifer. 1986. Sperm production and the effects of electrically induced spermatophore expulsion in the prawn *Macrobrachium rosenbergii* (DeMan). J. Crust. Biol., 6, 633-647.

Lightner, D.V. 1985. A review of the diseases of cultured penaeid shrimps and prawns with emphasis on recent discoveries and developments. Proc. First Internatl. Conf. on Culture of Penaeid Prawns/Shrimps, Iloilo City, Philippines (Y. Taki, J.H. Primavera and J.A. Llobrera, eds.), pp. 79-103. Aquaculture Department, Southeast Asian Fisheries Development Center.

Malek, S.R.A. and F.M. Bawab. 1971. Tanning in the spermatophore of a crustacean (*Penaeus trisulcatus*). Experientia, 27, 1098.

Nappi, A.J. 1975. Parasite encapsulation in insects. In: "Invertebrate Immunity" (K. Maramorsch and R.F. Shope, eds.), pp. 293-325. Academic Press, New York.

Poinar, G.O., Jr. 1973. The mosquito-filarial nematode association: pathology and defense mechanisms. WHO Bull., 107, 1-4.

Poinar, G.O., Jr. and R. Leutenegger. 1971. Ultrastructural investigation of the melanization process in *Culex pipiens* (Culicidae) in response to a nematode. J. Ultrastruct. Res., 36, 149-158.

Salt, G. 1965. Experimental studies in insect parasitism. XIII. The haemocyte reaction of a caterpillar to eggs of its habitual parasite. Proc. Roy. Soc. Lond., B, 162, 303-318.

Sandifer, P.A., A.L. Lawrence, S.G. Harris, G.W. Chamberlain, A.D. Stokes and W.A. Bray. 1984. Electrical stimulation of spermatophore expulsion in marine shrimp, *Penaeus* spp. Aquaculture, 41, 181-187.

Söderhäll, K. and V.J. Smith. 1986. Prophenoloxidase-activating cascade as a recognition and defense system in arthropods. In: "Hemocytic and Humoral Immunity in Arthropods" (A.P. Gupta, ed.), pp. 251-285. John Wiley & Sons, New York.

Spurr, A.R. 1969. A low viscosity epoxy resin embedding medium for electron microscopy. J. Ultrastruct. Res., 26, 31-43.

Talbot, P., D. Howard, T. Lee, W. Li, H. Ro, J. Trujillo and A. Lawrence. 1988. Unexpected rapid destruction of sperm in captive penaeid shrimp. Abstracts of 19th Ann. World Aquacult. Soc. Meeting, p. 75.

EPIDEMIOLOGY/EPIZOOTIOLOGY SECTION

EPIDEMIOLOGY/EPIZOOTIOLOGY: A BASIS FOR CONTROL OF DISEASE IN MARICULTURE

A.H. McVicar

DAFS Marine Laboratory, P.O. Box 101, Victoria Road, Aberdeen, Scotland

INTRODUCTION

Epidemiology, which involves investigations on the incidence, patterns of spread, and control of disease, is one of the oldest areas of medical and veterinary science. It is clearly beyond the scope of introductory comments of a symposium session to deal with this broad topic in any detail, so it is the intention of this presentation to consider some of the main principles and problem areas that are of practical relevance to disease studies in mariculture. Emphasis is placed on teleosts because of the author's personal experience and to complement the concluding comments by van Banning (this volume), which center largely on shellfish. Key features identified for particular consideration are the concept of fish farm disease as a complex of conditions, the role of husbandry (environment) in disease, and the determination of disease rates.

The basic methods of study and principles involved are equally applicable to animals and man and hence the distinction frequently made between epidemiology (pertaining to human populations) and epizootiology (pertaining to animal populations) is somewhat artificial. Epidemiology is the term used in preference here. Central to most studies is the concept of dealing with the total disease pattern within a whole population rather than in individual members. Such investigations are directed primarily on the interaction and consequences of the factors that combine to produce an observed disease pattern and only subsequently on the more focused studies of the individual components such as the disease agent, host resistance, and environment. In practice, it is through study and understanding of the roles of these influencing factors and their underlying inter-relationships in causing disease that the best strategies for control and prevention of disease are often developed. Klontz (1985) noted that epidemiological concepts have proved to be very reliable and powerful tools in containing and even eradicating diseases in humans and animals and advocated their use in fish disease diagnosis. Consequently, in an applied field, such as disease control in mariculture, epidemiologically-based studies should play an important and early role.

Disease complexes. Fish disease does not normally occur in isolation in a "pure" form in populations grown in natural or semi-natural conditions such as fish farms. The disease pattern observed is partially the result of a potentially wide and variable range of influencing and associated factors (including other diseases) and of secondary consequences of the disease, each of which may also be subject to high variability. The more chronic a disease, the greater the chance of additional factors having an important role. Disease, as usually observed in a fish farm, may therefore be considered to be a complex of conditions, resulting from the modification of the original primary disease by these factors. As these factors may change from farm site to farm site, and even at different times within the same site, different outbreaks of the same disease will seldom have identical signs. Frequently, individual farm sites or geographical areas may have their own characteristic pattern of a disease, which may substantially differ from other areas or sites. A consequence for disease epidemiological studies in mariculture is that it is thus obligatory to consider the wider aspect of a disease throughout its distribution range to elucidate local, seasonal, and other variable effects and not to depend on single or limited case studies.

The practicing fish farmer is primarily concerned with the total disease complex as this is what he observes killing or reducing the quality of his stock and consequently is what he wishes to prevent or treat. Much of the practical basis for dealing with mariculture diseases centers on the control of the associated or secondary aspects of a disease problem as this can often reduce or even eliminate the more serious disease symptoms and consequences and allow the farm stock to tolerate the disease. For example, while it is recognized that outbreaks of trichodinosis and vibriosis in fish, shellfish, and other species are often associated with problems of stress, handling, crowding, and generally bad husbandry within the mariculture system (Lom, 1970; Pearse, 1972; Austin and Austin, 1987), adequate control has been achieved for many serious outbreaks of these diseases through use of chemotherapeutic or antibiotic treatments. This is, therefore, an acceptable and valuable strategy for the epidemiologist to pursue. While the primary cause of the disease remains uncontrolled, the improvement in the disease situation may only be temporary owing to the continual recurrence of the same (or new) secondary and associated problems. In addition, the frequent prevention or treatment of secondary problems is often expensive and stressful to stock. Consequently, it should remain a priority for the epidemiologist to identify and control primary causes of disease whenever possible.

In practice, finding the primary cause of disease can often prove very difficult, and the scientific literature abounds with controversy on the interpretation of what is primary, secondary, and concomitant disease within a particular disease complex. In Norwegian farmed Atlantic salmon, "Hitra disease" or "cold water vibriosis" is generally acknowledged to be a complex of disease. Nutrition, husbandry, and the bacterium *Vibrio salmonicida* are prominently involved, but their relationships to each other and to the total disease pattern has not yet been fully elucidated (Egidius et al., 1981; Poppe et al., 1985). Similarly, with Pancreas Disease (PD) of farmed Atlantic salmon, infectious and nutritional primary etiologies have been proposed (McVicar, 1987a; Ferguson et al., 1986). The disease was first recognized in Scotland in 1976 (Munro et al., 1984) and has since been found in sea farmed salmon throughout Europe and also in the western USA. Detailed studies throughout the Scottish industry have indicated a wide range of associated pathologies with substantial variations between different farm sites and the same farm in different years. The lack of consistency of occurrence of many of these pathologies in most PD outbreaks has indicated those features that were not integral parts of the primary disease and were probably due to other locally associated problems. In common with many preliminary studies on fish diseases, early studies on this condition were complicated by having to evaluate the significance and relationship of the wide range of pathologies, many of which were subsequently shown to occur late in the disease process and provided little information on the primary cause of the disease. However, by returning to the basic principles of epidemiology of the disease on fish farms, circumstantial evidence as to the primary nature of PD was obtained; after the first outbreak on a farm the disease persists through transfer between year classes, transfer of the disease occurs between adjacent farm sites, and survivors appear to have acquired resistance. These key features of the epidemiology suggested the possibility of infection as a primary cause (McVicar, 1987a) and gave direction to experimental studies focused in this area, which currently suggest that PD is primarily caused by a virus (McVicar and Munro, 1988).

Host agent environment. The host-causative agent-environment model is commonly used in epidemiological studies to explain patterns observed for both infectious and noninfectious diseases and this is also a familiar concept in fish disease studies. With naturally-occurring disease, a steady state is apparent when each aspect has the same proportional effect and a balanced relationship then exists between host and disease for extended periods. However, a change in any of the components (e.g., host resistance capability, transmission ability, virulence of an infective agent, or environmental parameters) may affect the steady state and lead to increasing or decreasing levels of disease. Frequently, the appearance of new disease in a population or new host species has serious implications as this balance and steady state has not been established and the disease epidemiology may be unpredictable. This is clearly seen with *Bonamia* in *Ostrea edulis* (van Banning, 1985) and the eel swim bladder nematodes *Anguillicola crassa* in Europe (van Banning and Haenen, this volume). There are also many parallels between this situation and any typical mariculture system where both the fish and its environment are being manipulated by man to produce a product. These actions may often have disease consequences. This was clearly apparent with the disease "Black Patch Necrosis" in juvenile Dover sole, *Solea solea*, being tank-reared in Scotland. The disease appeared at approximately 3 months post-metamorphosis, could cause in excess of 90% mortality, and could not be satisfactorily treated with antibiotics and chemotherapy (McVicar and White, 1982). Although an infectious agent was suspected from the epidemiological pattern, it was not initially identifiable from among the background bacterial load of the fish. Attention was therefore directed at the third component of the model, the environment within rearing tanks. Examination of such parameters as stocking densities, water exchange and quality, light intensity, and feeding practices revealed no correlations but tank substrate was experimentally shown to be a significant element in the disease process. The presence of sand on tank bottoms not only cured the disease but also prevented its occurrence, even when the infective agent was present (McVicar, 1987b). It was suggested that sole require the mildly abrasive action of the sand and possibly also the cover to reduce stress to prevent the disease. This case graphically indicates that the range of variables present in a mariculture system must not lie outside the tolerance limits of the animal being grown. Similar conclusions have been reached for other species of fish. For example, Baudin Laurencin et al. (1985) considered that "summer pathology" in marine cultured rainbow trout was associated with unfavorable environment, specifically high temperatures and salinity, reducing adaptability of the fish. Prevention methods suggested included choosing the best period for transfer of fish to sea water and dietary manipulation. It is worth bearing in mind that most fish and shellfish species currently being cultivated are little developed beyond their wild counterparts, which have evolved characteristics to accommodate the variables encountered in their natural range. For some species (such as sole) this may be very narrow and significant elements of their natural environment (e.g., sand) may be required in the mariculture system.

An extension of this concept is the identification and alteration of key factors in the model equation by the epidemiologist that would permit the balance between host and disease to be shifted in favor of the host. Many such factors lie in aspects of husbandry and it is frequently stated that most fish farm diseases are caused either directly or indirectly by the fish farmer. In these cases the fish farmer has shifted the balance of the relationship in favor of the disease. Theoretically, it should be possible to redress this balance or even to tip the advantage in favor of the host. In practice, while many of these factors contributing to disease in mariculture are obvious (e.g., overstocking, fouling of nets or tanks, bad handling of stock), they cannot always be easily prevented or altered for economic or practical reasons. Frequently, in mariculture there is a finely judged compromise between the requirements of economic productivity and what the species being cultured can tolerate. It is then that the skill of the fish farmer becomes paramount; the good farmers judge correctly; the bad or inexperienced, wrongly.

Prevalence and incidence. A key feature in any epidemiological study is knowledge of the disease prevalence (level of disease found at a given time) and incidence (rate of occurrence of new cases over a certain period of time). Information on both is essential before the dynamics and effects of a disease in a population can be determined. A high disease prevalence in a population may indicate an epidemic, a commonly repeated transient condition, or a disease that has little effect on its host and accumulates in the population with time. Conversely, a low disease prevalence can exist for several reasons: because the disease is rare, because the disease is transient, or because affected individuals in the population are rapidly removed by mortality. Therefore, in contrast to the widespread popular impression that commonly occurring diseases are most significant, it is important to note that, without epidemic situations, it is only those diseases showing low prevalence that could be most pathogenic and show highest incidences. In extensive natural or semi-natural culture systems, direct observation of mortality is notoriously difficult and it is often necessary to obtain disease or mortality rates by other means.

Prevalence levels are relatively easy to determine from whole or subsamples of populations, but their accuracy is dependent on the accuracy of diagnosis and the representativeness of any subsamples. Both of these areas would warrant extensive discussion beyond that possible here, but several points are of particular relevance to epidemiological studies in mariculture. For infective disease, direct diagnosis by isolation of the agent is preferable but not always possible because of practical considerations (e.g., lack of suitable facilities in the field) or inability to culture agents (e.g., because of fastidious growth requirements). In such cases, diagnosis is dependent on pathological or physiological parameters, both of which frequently suffer from lack of specificity of cause, lack of sensitivity through delay in their appearance after the event, and variability in their response. Subsampling of populations is potentially a major source of error in prevalence studies, and many examples of problems could be given for almost any type of study. In cage culture of fish, particularly in the large volume cages that are being increasingly used in salmon farming, it is difficult to obtain regular subsamples without the labor intensive and stock stressful operation of lifting nets. Farmers are naturally reluctant to do so and unless carefully instructed would frequently resort to attracting fish with food. Such practices will clearly bias samples towards feeding fish and could consequently give a totally false disease picture.

Before incidence can be measured, data on two parameters are normally required: the size of the population at risk and the time course or rate of occurrence of the disease. In captive mariculture populations, knowledge of population size is usually available, especially at the time of input, although this may become less certain with time owing to difficulties in obtaining accurate mortality figures and loss resulting from predation and escape. The more extensive the mariculture system, the greater the difficulties. In released populations, e.g., oyster deposition on to the sea bed and salmon ranching, these problems are compounded. When management of totally free living populations is involved, at best only estimates of the population size can be obtained. The time course of a disease incident can be determined from frequent samples of prevalence levels; however, in practice, available resources and sampling constraints often make it difficult to achieve ideal sample size and frequency. Extreme caution should be exercised when attempts are made to obtain disease rates from experimental situations. Unless all significant influencing factors are precisely duplicated in the experimental system, the results may bear little relationship to the actual rates in the field. For example, with a disease that causes significant debilitation, predation on affected stocks may be a significant cause of actual death in a natural situation, but its absence in experimental trials may result in protracted morbidity of individuals.

When a sample or subsample is taken at a point in time from a population (as illustrated in Fig. 1), a contribution is obtained from the residue of diseased and non-diseased stock in each cohort (e.g., year class or size group) present. The prevalence of the disease is the total infected residue in the sample as a proportion of the total population residue, and the disease incidence is the slope of the interface line between non-infected and infected residual parts of the population. The disease residue is the result of the dynamic relationship between the number of new disease cases and the loss of disease cases. Measurement of this relationship cannot be obtained directly from a single sample. Data must be transferred from the vertical (point in time) axis to the horizontal (rate) axis. Transfer can be achieved if a time index was obtained of the progress of the disease within the diseased population residue. A practical illustration is provided by *Ichthyophonus hoferi.* This fungal pathogen has been implicated in high mortality rates in some fish populations in North American and European North Atlantic waters including fish farms (McVicar, 1982). Field and experimental evidence indicated the disease was lethal to plaice but, as already indicated, true mortality rates could not be deduced from tank trials. However, it was observed that the disease is highly antigenic and that all infected plaice ultimately produce antibody (McVicar, 1982). In Figure 2, line ac represents the disease residue in a sample, ab the proportion of diseased fish with antibody to *Ichthyophonus*, and bc the proportion without antibody. Because of the strong immunogenicity of the disease, it can be stated that affected fish without antibody had

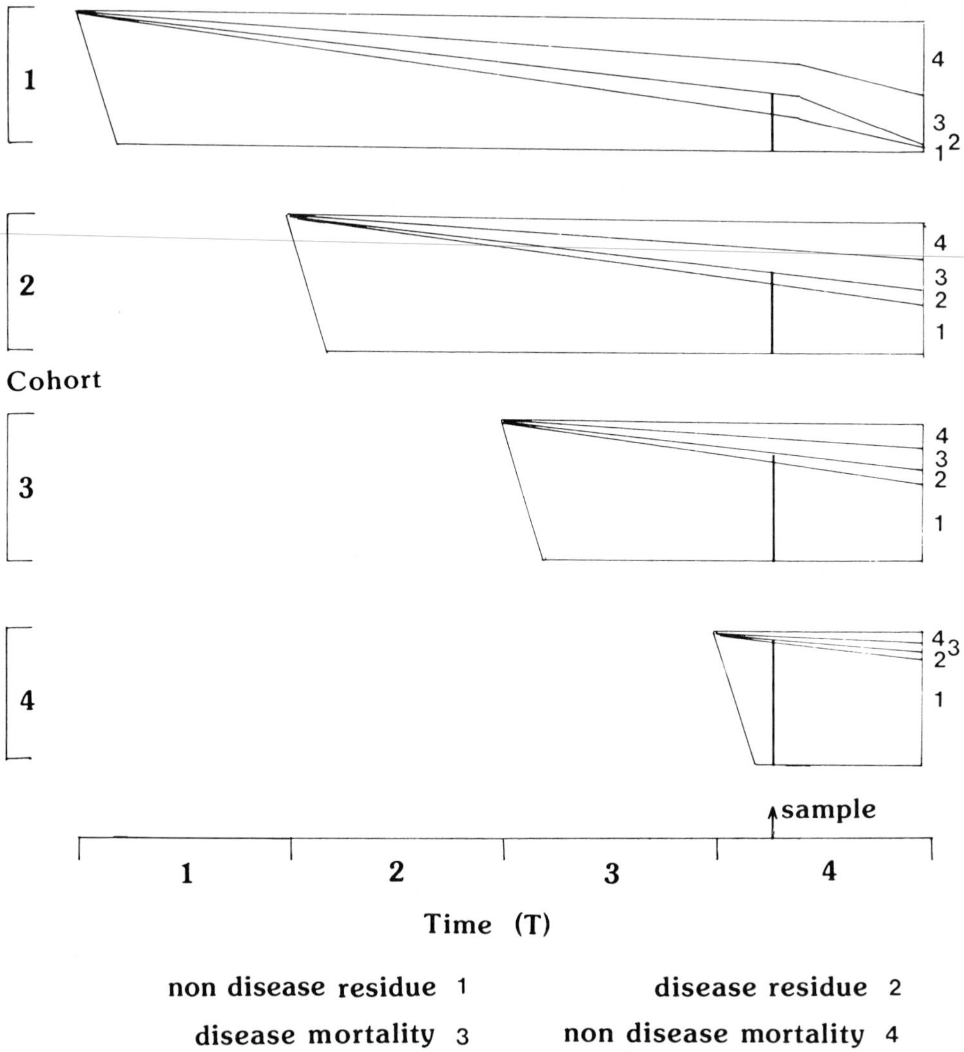

Fig. 1. Schematic diagram of a population consisting of four cohorts with steady incidence of a lethal disease and constant mortality rates (disease and non-disease induced) until senescence. A sample taken at a point in time has contributions from the residual populations as indicated.

become infected within the time period necessary to produce detectable antibody. Experimental work has shown that several species of fish produce detectable antibody to *Ichthyophonus* within 36 days at temperatures similar to those of natural conditions. It is possible that some variation may exist between antibody production rates in experimental and natural conditions, but there is sufficient knowledge of fish immunological systems to indicate that this is unlikely to be significant on this occasion. On the time axis (T) cd in Figure 2 can therefore be calculated as 36 and determined on a proportional basis:

$$\frac{bc}{ab} = \frac{cd}{de}, \qquad de = \frac{ab}{bc} \times 36$$

Using actual *Ichthyophonus* prevalence data for north Scotland waters

$$de = \frac{41.7}{58.3} \times 36 = 26 \text{ days}$$

Survival time after infection (ce) = cd + de = 26 + 36 = 62 days. The mean prevalence of *Ichthyophonus* off north Scotland is 9.35% and the annual incidence or mortality rate can be calculated as:

$$\frac{365}{62} \times 9.35 = 55.04\%$$

Any other recognizable time marker of the stage of development of a disease, e.g., parasite development stage or a particular pathological feature, could be similarly used in any species of animal. Although knowledge of population size is not required, some information on population limits, immigration, and emigration is desirable.

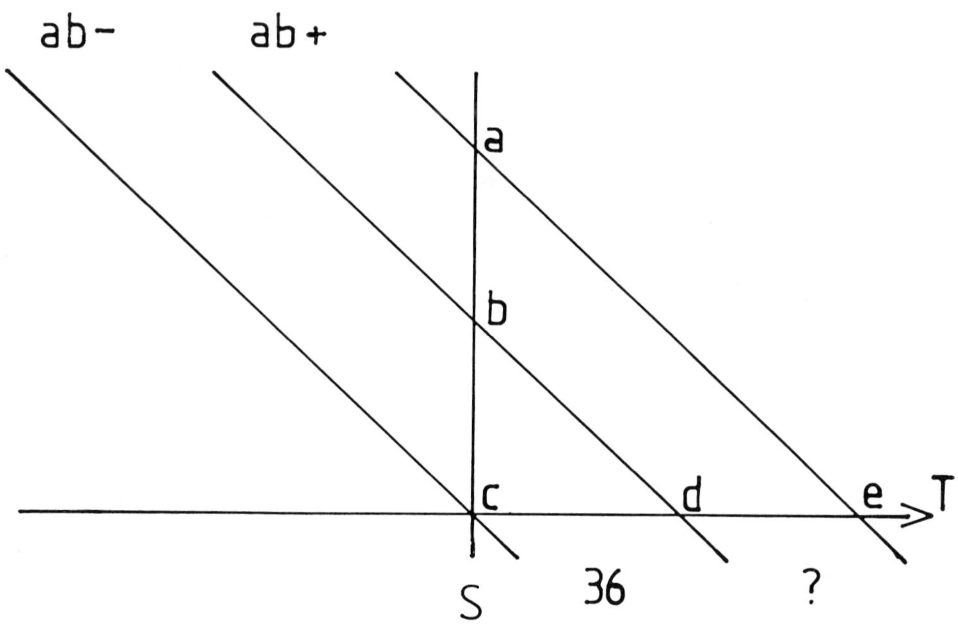

Fig. 2. Schematic diagram of the diseased residue in a population with constant incidence and mortality rate. Development of antibody at a point during the disease progression is shown. ac = diseased residue available to a sample s; cb = portion without antibody; ab = portion with antibody; T = time; S = sample.

SUMMARY

Epidemiological studies are central to the development of methods for the control and prevention of diseases in mariculture. Although disease agents may differ, the host disease relationships and their response to specific cultivation and environmental conditions show many similarities so that a common solution to problems is often possible through the discipline of epidemiology. In the fish farm environment, the disease pathology caused by a particular etiological agent is typically modified or masked by locally associated secondary and accompanying problems including other diseases sometimes to the extent that total disease symptoms are site or area specific. While control of associated problems can permit tolerance of many diseases, failure to identify and control primary etiologies may lead to persistent recurrence of problems, particularly as many mariculture techniques used may be close to the tolerance limits of the animal being grown. For the dynamics and effects of a disease to be determined, information on its prevalence and incidence is required, a task that becomes increasingly difficult the more extensive the mariculture system used.

LITERATURE CITED

Austin, B. and D.A. Austin. 1987. "Bacterial Fish Pathogens: Disease in Farmed and Wild Fish." Ellis Horwood Limited, Chichester, England.

Baudin Laurencin, F., J.F. Aldrin, J.L. Messager and G. Tixerant. 1985. "Summer pathology" in marine cultured rainbow trout. In: "Fish and Shellfish Pathology" (A.E. Ellis, ed.), pp. 211-221. Academic Press, London.

Egidius, E., K. Andersen, E. Clausen and J. Raa. 1981. Cold water vibrosis or "Hitra disease" in Norwegian salmonid farming. J. Fish Dis., 4, 353-354.

Ferguson, H.W., R.J. Roberts, R.H. Richards, R.O. Collins and D.A. Rice. 1986. Severe degenerative cardiomyopathy associated with pancreas disease in Atlantic salmon, *Salmo salar* L. J. Fish Dis., 20, 95-98.

Klontz, G.W. 1985. Diagnostic methods in fish diseases: present status and needs. In: "Fish and Shellfish Pathology" (A.E. Ellis, ed.), pp. 3-10. Academic Press, London.

Lom, J. 1970. Protozoa causing diseases in marine fishes. In: "A Symposium on Diseases of Fishes and Shellfishes" (S.F. Snieszko, ed.), pp. 101-123. American Fisheries Society, Washington.

McVicar, A.H. 1982. *Ichthyophonus* infections of fish. In: "Microbial Diseases of Fish" (R.J. Roberts, ed.), pp. 243-269. Academic Press, London.

McVicar, A.H. 1987a. Pancreas disease of farmed Atlantic salmon, *Salmo salar*, in Scotland: epidemiology and early pathology. Aquaculture, 67, 71-78.

McVicar, A.H. 1987b. Black patch necrosis of the skin of *Solea solea* (L.): the role of sand in prophylaxis and treatment. J. Fish. Dis., 10, 59-63.

McVicar, A.H. and A.L.S. Munro. 1988. Pancreas disease in farmed Atlantic salmon: evidence for an infectious aetiology. ICES CM 1988/F:6, 1-6 (mimeo).

McVicar, A.H. and P.G. White. 1982. The prevention and cure of an infectious disease in cultivated juvenile Dover sole *Solea solea* (L.). Aquaculture, 26, 213-222.

Munro, A.L.S., A.E. Ellis, A.H. McVicar, H.A. McLay and E.A. Needham. 1984. An exocrine pancreas disease of farmed Atlantic salmon in Scotland. Helgol. Meeresunters, 37, 571-586.

Pearse, L. 1972. A note on a marine trichodinid ciliate parasite on the skin of captive flatfish. Aquaculture, 1, 261-266.

Poppe, T.T., T. Hastein and R. Salte. 1985. "Hitra disease" (Haemorrhagic syndrome) in Norwegian salmon farming: present status. In: "Fish and Shellfish Pathology" (A.E. Ellis, ed.), pp. 223-229. Academic Press, London.

van Banning, P. 1985. Control of *Bonamia* in Dutch oyster culture. In: "Fish and Shellfish Pathology" (A.E. Ellis, ed.), pp. 393-396. Academic Press, London.

DISEASES OF FIVE COMMERCIAL CRAB SPECIES IN SOUTHEAST ALASKA

Theodore R. Meyers, Sally Short, and William Eaton[1]

Alaska Department of Fish and Game, Fisheries Rehabilitation Enhancement and Development Division, Fish Pathology Section, P.O. Box 3-2000, Juneau, Alaska 99802, USA

ABSTRACT: A continuous survey of parasitic and other diseases of economically important crustaceans in Alaskan waters has revealed several diseases. The crabs examined were three species of king crabs, Tanner crabs, and Dungeness crabs. In the king crabs, a rhizocephalan parasite, a proliferative disorder, an intracellular prokaryotic agent, and a ciliate infection are being reported. In the Dungeness crabs, intranuclear inclusions exist, while in the Tanner crabs, the bitter crab disease occurs. Histopathologic descriptions of the above-mentioned parasitic and other diseases are provided.

INTRODUCTION

Three king crab species, Tanner crabs (*Chionoecetes bairdi*), and Dungeness crabs (*Cancer magister*) have supported a multi-million dollar commercial fishing industry in southeast Alaska. Population surveys of these crab species performed by the Alaska Department of Fish and Game (ADF&G) Commercial Fisheries and Fisheries Rehabilitation Enhancement and Development (FRED) Divisions and the University of Alaska at Juneau have identified six diseases. Although information on four of these entities is still quite preliminary regarding their significance, the other two conditions have been studied extensively and show great potential to negatively impact certain crab populations, especially when commercial fishing is an added population stressor. The purpose of this report is to identify indigenous diseases of commercially important crab species within Alaska that may significantly affect the management of these fisheries. As this study continues, other diseases will most likely emerge.

---

[1]Joint appointment with the University of Alaska Southeast, School of Fisheries and Ocean Sciences, 11120 Glacier Highway, Juneau, Alaska 99801.

MATERIALS AND METHODS

All crabs were collected using standard commercial crab pots for each respective species and maintained in flowing sea water for varying periods of time until necropsies could be performed. Major organs were preserved in Helley's fixative (substituting $ZnCl_2$ for $HgCl_2$) and processed using standard histological procedures with hematoxylin and eosin staining.

Stains and procedures used for Tanner crab hematology have been described previously (Meyers et al., 1987). Tissues processed for transmission electron microscopy (TEM) were taken from Helley's fluid and post-fixed in 1% osmium tetroxide (OsO_4) in 0.1 M cacodylate buffer (pH 7.2) and placed in 70% ethyl alcohol for storage. Tissue samples were embedded in Spurr's or Medcast resins, thin sectioned, and stained with uranyl acetate and lead citrate. Sections were examined in a Philips 200 TEM operated at 40 kV. Scanning electron microscopy and TEM on Tanner crab materials were performed as described in Meyers et al. (1987).

RESULTS

King crab diseases

Rhizocephala. All king crabs were initially collected for gross and histopathological studies of parasitism by the rhizocephalan barnacle *Briarosaccus callosus*. Parasitized crabs were easily recognized by the large sausage-shaped externa of the parasite exposed by lifting the abdominal apron (Fig. 1). Occasionally, crabs were found with only a protruding leathery stalk beneath the flap where an externa had once been attached (Fig. 1). Necropsies were performed on 67 blue king crabs (*Paralithodes platypus*) composed of 26 unparasitized and 41 parasitized individuals collected from Muir and Adams Inlets of Glacier Bay, Alaska, in early March, 1984. In May, 1984, 45 golden king crabs (*Lithodes aequispina*), including 28 parasitized individuals, were examined from Lynn Canal near Haines. Necropsies were also done on 45 red king crabs (*P. camtschatica*) from waters in the Juneau area and included 21 parasitized animals. Internal gross observations in all king crab species indicated atrophy or absence of gonads in parasitized hosts and marked replacement of hepatopancreas by the bright green rootlets of the barnacle comprising the interna (Fig. 2). Parasite rootlets penetrated connective tissues and

---

Fig. 1. *In situ* detached externa (E) of *Briarosaccus callosus* infecting a red king crab from Southeast Alaska. Note stalk attachment site (arrow). (Photograph courtesy of T. Koeneman.)

Fig. 2. Gross appearance of internal rootlet mass (R) of *Briarosaccus callosus* infecting a blue king crab from Southeast Alaska. Hepatopancreas (H) and a nerve tract (arrow) of the thoracic ganglion.

Fig. 3. Squash preparation of living *Briarosaccus callosus* rootlets from a blue king crab demonstrating the internal green peripheral granules. 608X.

Fig. 4. Histologic appearance of a *Briarosaccus callosus* rootlet (R) that has invaded the neuropile of a nerve tract from the thoracic ganglion in a blue king crab. H & E; 608X.

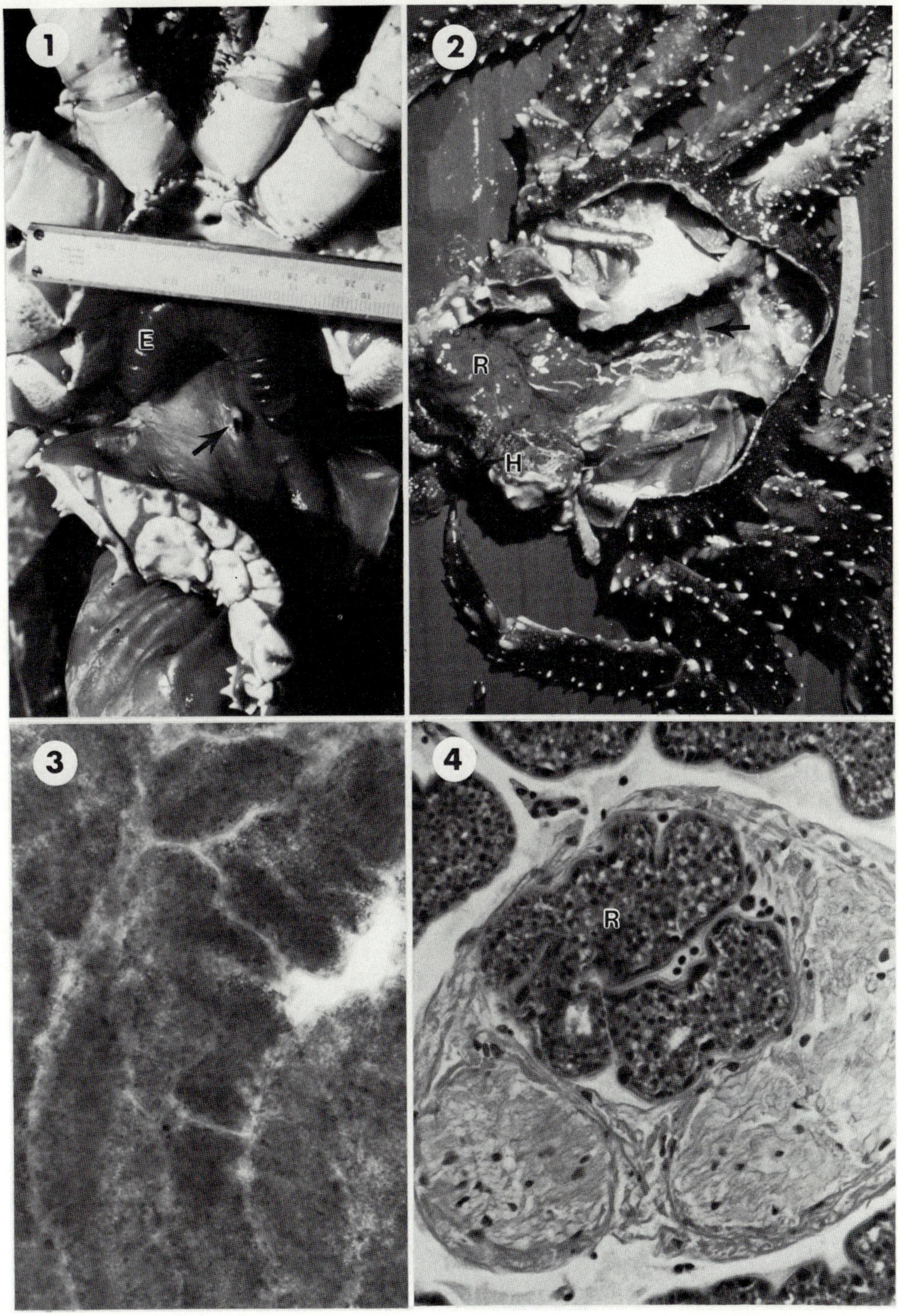
1
E
2
R
H
3
4
R

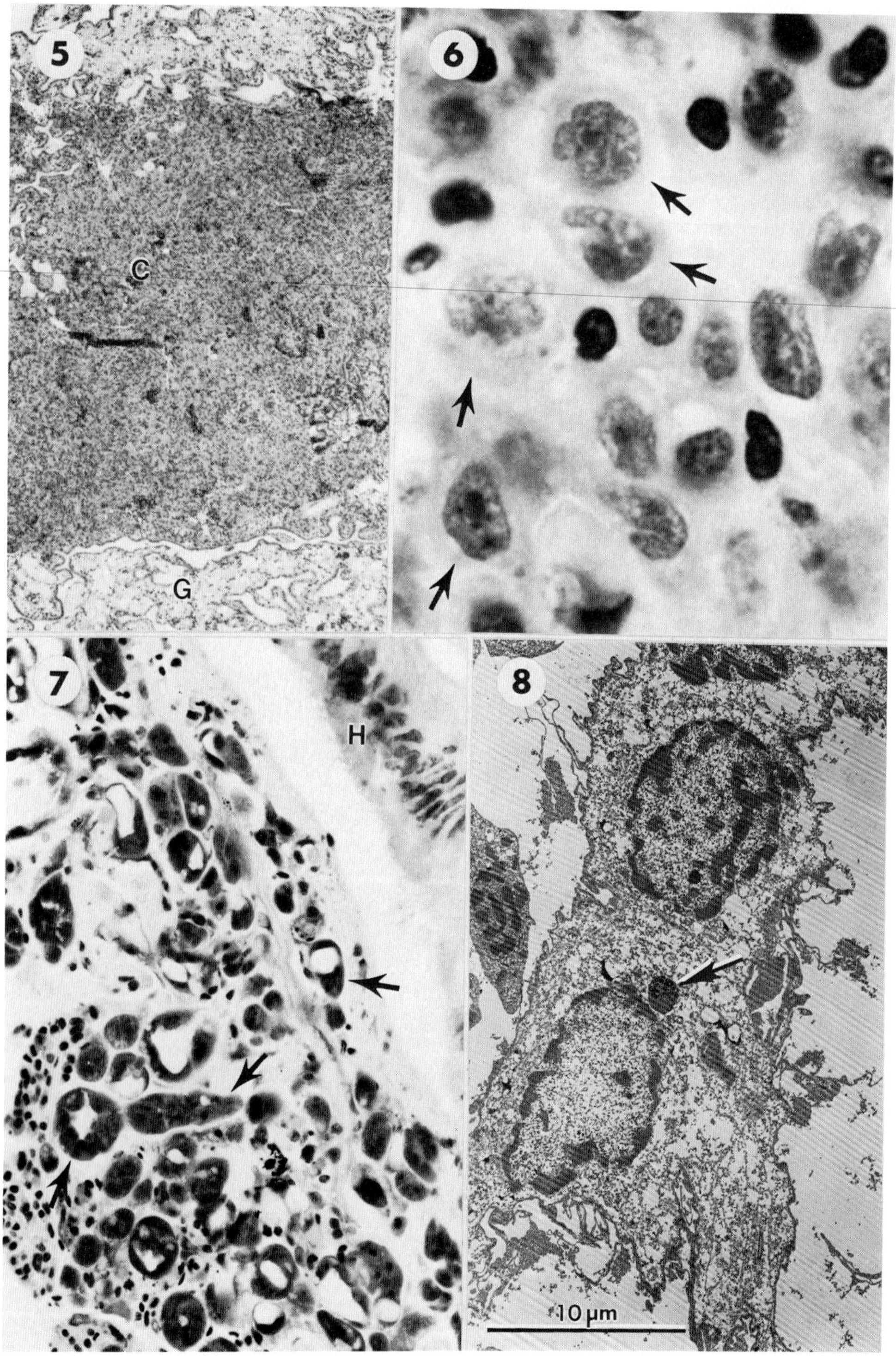
5
C
G
6
7
H
8
10 µm

were closely adherent to all major organs and tissues within the visceral cavity of the host crab including the bases of gills and skeletal muscle within the coxal joints. Rootlets did not appear to extend out into the leg musculature past the coxal joints.

Histological examination of parasitized crab tissues showed parasite rootlets to be tubular structures lined by basophilic vacuolated cells surrounded by an outer refractile cuticle. The rootlets occasionally had a lumen containing proteinaceous material and cytoplasmic cell granules retained the green color visible in fresh tissues (Fig. 3). Parasite rootlets were extremely dendritic infiltrating between skeletal muscle bundles and within the aerolar connective tissues of the urinary bladder, hepatopancreas and adventitia of the pyloric stomach, midgut, rectum, testes, and anterior vas deferens. Rootlets were never observed to invade epithelial boundaries and actually penetrate any of the major organs. However, rootlets did penetrate the neurilemma into the neuropile of the thoracic ganglion and associated nerve tracts (Fig. 4). Affected nervous tissue developed hypertrophy of the neurilemma and some fibrosis accompanied by minor infiltration of granular hemocytes around the invading parasite rootlets. Host cellular recognition of the parasite was generally lacking in other tissues. Testes, vas deferens, and ovaries that still could be found in some of the parasitized crabs contained apparently viable gametes but the organs were atrophied.

Barnacle prevalences have ranged from 0 to as high as 75% in certain king crab populations (Hawkes et al., 1985). In general, barnacle prevalences have been greatest in blue king crabs, with moderate frequencies of infection in golden king crabs, and the lowest numbers of parasitized hosts in red king crab populations (Hawkes et al., 1986b).

Proliferative disorder. Routine necropsy of one male blue king crab from Glacier Bay disclosed two white nodular foci on the surface of the right antennal gland. Histologic examination showed this lesion to be a highly cellular basophilic mass (Fig. 5) primarily composed of a large anaplastic cell type with an irregularly lobed to oval heavily stippled nucleus having one to two eosinophilic nucleoli. Cell cytoplasm was moderate in volume and basophilic (Fig. 6). Mitotic figures were present but not abundant and cells were arranged in whorled patterns with areas of cell necrosis and abundant infiltration by inflammatory hemocytes. The anaplastic cells formed acinar-like structures near the periphery of the lesion that involved both labyrinth and coelomosac portions of the gland. Metastatic focal masses of this large cell type were also found in the testes and myocardium and formed acinar structures within the adventitia and beneath the mucosa of the midgut (Fig. 7). Several other organs contained single cells of this type within hemal spaces. TEM of affected

---

Fig. 5. Proliferative mass of cells (C) within the right antennal gland (G) of a blue king crab from Glacier Bay, Alaska. H & E; 95X.

Fig. 6. Higher magnification of the mass in Fig. 5 showing an anaplastic cell type (arrows) with a large irregular or multilobed nucleus and prominent 1-2 nucleoli. H & E; 2,500X.

Fig. 7. Same type of proliferative cell forming acinar structures (arrows) that have infiltrated the adventitia directly beneath the mucosa of the hindgut (H). H & E; 608X.

Fig. 8. Labyrinth epithelial cells from the lesion area showing an intracytoplasmic inclusion body (arrow) containing virus-like particles. Uranyl acetate & lead citrate; 3,293X; Bar = 10 μm.

antennal gland demonstrated cytoplasmic inclusion bodies within labyrinth epithelial cells that contained microtubular structures associated with virus-like particles, often in paracrystalline arrays (Figs. 8, 9). The particles were about 65 nm in diameter and resembled a birna-type virus. This lesion had neoplastic and malignant characteristics and was diagnosed as an adenocarcinoma of either labyrinth epithelium or hepatopancreas origin. Whether the associated virus was responsible for neoplastic cell transformation could not be determined from the present material.

Intracellular prokaryotic agent. An intracellular Gram-negative prokaryotic rod-shaped organism was associated with holding mortality in two king crabs, one golden and one blue. The organism formed cytoplasmic inclusions causing hypertrophy of host cells in the antennal gland of the blue king crab (Fig. 10) and in the hepatopancreas of the golden king crab. Infected tissues were indurated and friable with granulomatous foci and caseous necrosis. TEM of infected antennal gland from the blue king crab confirmed the intracellular nature of the organism and demonstrated a rickettsia-type morphology (Fig. 11) with oval to rod-shaped organisms measuring about 1.0 to 0.508 μm x 0.262 μm. The ultrastructure of the organism was typically prokaryotic with abundant ribosomes around the periphery of the cytoplasm and two trilaminar limiting membranes that formed frequent rod-like projections or blebs. Organisms within inclusion bodies were surrounded by a finely granular matrix interspersed by parasite cell wall projections (Fig. 11). Only one potential dividing form was observed showing a central zone of constriction suggestive of binary fission. However, the complex arrangement and close fit of organisms within inclusion bodies also may be a result of a cleavage process. Although the organism histologically appears to be the same in both crab species, further characterization would be required for confirmation.

Systemic ciliate infection. After 2 months in captivity, a blue and a golden king crab having external shell injuries died from a systemic infection by a *Paranophrys*-like ciliate protozoan. Grossly visible coagulated hemolymph forming gelatinous nodules and diffusely scattered white foci each representing myriads of the protozoan were presenting signs within the visceral mass of the infected crabs. Generalized tissue displacement by the organism and necrosis were the microscopic features (Fig. 12). The disease may not be a natural occurrence, but rather a secondary result of mechanical injury and stress from confinement as often occurs in Dungeness crabs (Armstrong et al., 1981).

Fig. 9. Composite electron micrograph of virus particles within cytoplasmic inclusions similar to Fig. 8 showing crystalline arrays and associated microtubular structures (arrows) within the viroplasm. Uranyl acetate & lead citrate; 18,800X (Top bar = 1 μm) and 30,000X (Bottom bar = 100 nm).

Fig. 10. Prokaryotic cytoplasmic inclusion bodies within antennal gland cells of a blue king crab from Glacier Bay, Alaska. Rod-shaped organisms (arrows) are apparent. H & E; 2,500X.

Fig. 11. Electron micrograph of the organisms (O) in Fig. 10 showing typical prokaryotic ultrastructure, a portion of the host cell nucleus (N), granular material and cell wall projections or blebs (arrows) within the inclusion body. Uranyl acetate & lead citrate; 100,000X; Bar = 0.1 μm.

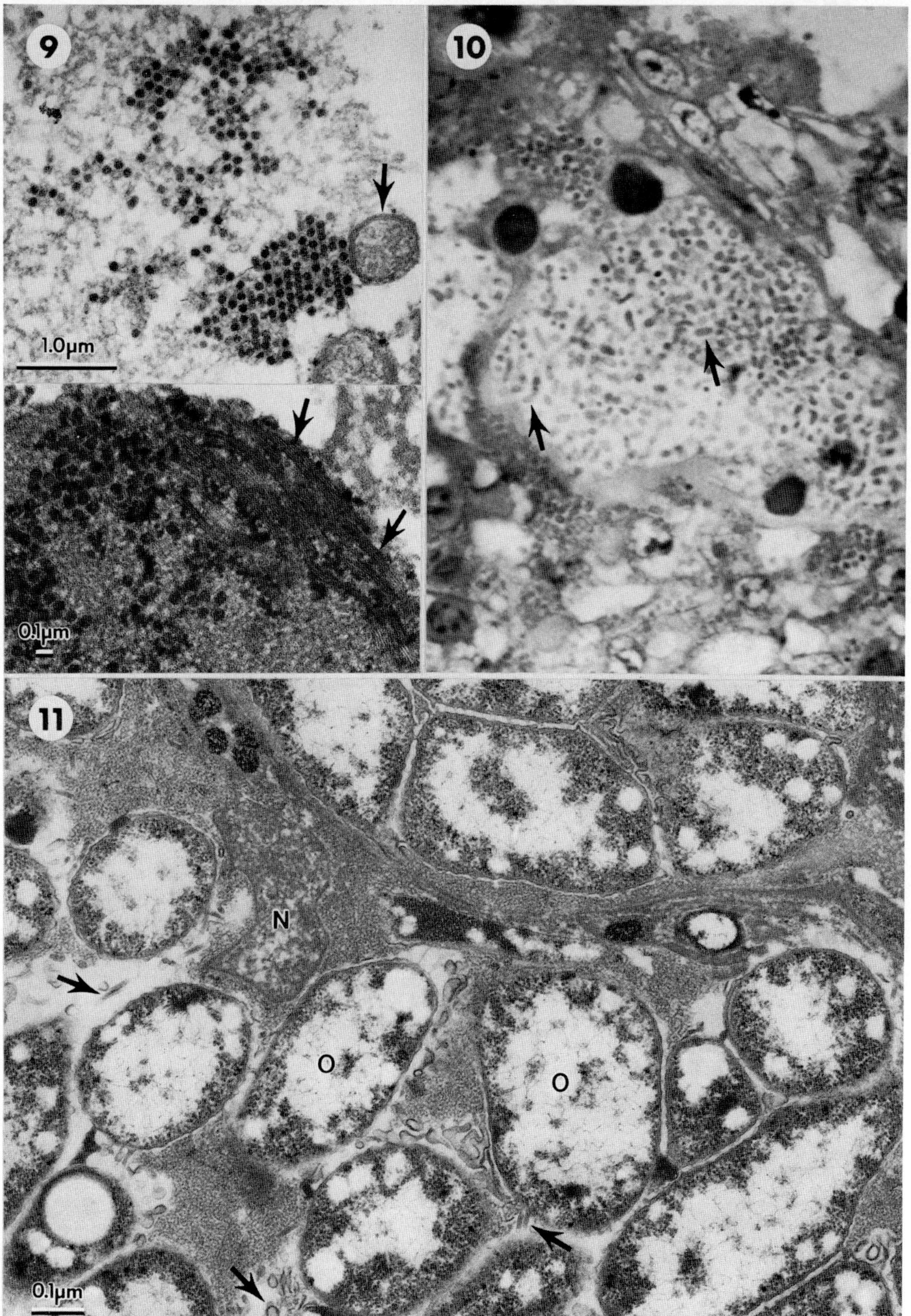
9
1.0μm
0.1μm
10
11
N
O
O
0.1μm

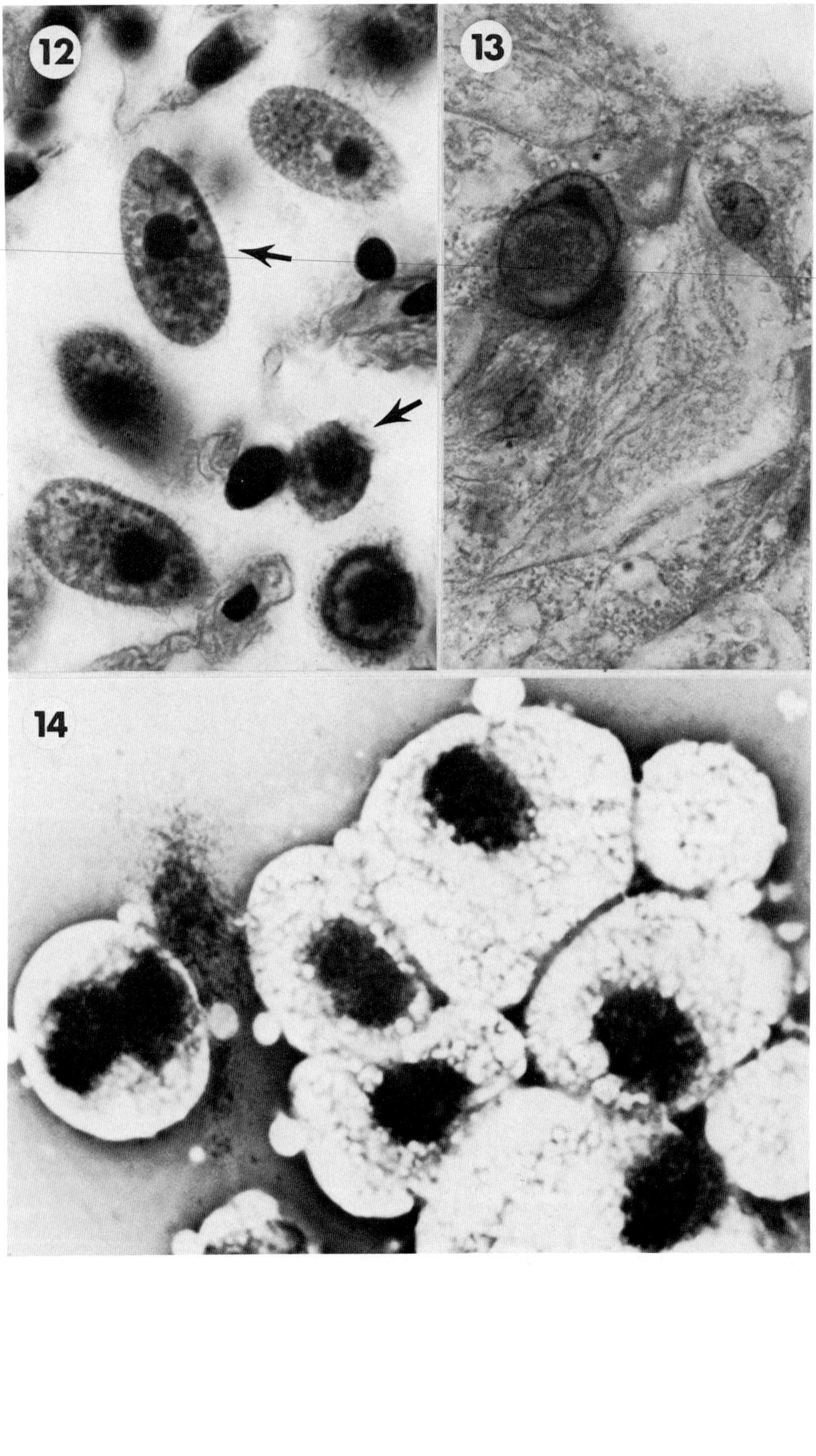
12
13
14

Dungeness crab diseases

Intranuclear inclusions. Four percent of 153 adult Dungeness crabs collected in 1984 and 1985 from Excursion Inlet, Freshwater Bay, and Bridgett Cove, Alaska, had basophilic, sometimes multiple, intranuclear inclusion bodies within the epithelial cells of the urinary bladder, hepatopancreas, and female seminal receptacle. These inclusion bodies caused nuclear hypertrophy, compression and eccentricity of nucleoplasm and the nucleolus, and margination of chromatin. Inclusions in the urinary bladder were somewhat different, appearing less dense and whorled in structure rather than homogeneous and granular (Fig. 13). Whether these inclusion bodies represent one or more virus infections remains for later investigation with TEM. If caused by a virus(es), the infection does not appear to produce significant pathological changes in otherwise healthy crabs.

Tanner crab diseases

Bitter crab disease (Meyers et al., 1987). Blood smears from 1,478 Tanner crabs collected from 12 different commercial fishing areas in northern southeast Alaska demonstrated bitter crab disease in some populations at rates of up to 95%. Bitter crab disease is caused by a *Hematodinium*-like dinoflagellate (Chatton and Poisson, 1931; Newman and Johnson, 1975; MacLean and Ruddell, 1978; Johnson, 1986) infection of the hemolymph occurring in both sexes and all age classes of Tanner crabs (Fig. 14). Infections have resulted in virtually 100% mortality in naturally-infected crabs held in the laboratory and impart a chalky texture and bitter flavor to cooked crab meats making the product unmarketable. The loss in crab meat quality is probably due in part to a substance observed to be exuded from the surface of the vegetative stage of the parasite (Fig. 15). Presenting signs of the disease include an exaggerated pink color of the carapace, milky hemolymph, lethargy, and increased deadloss in commercial holding tanks. Histological examination of infected crab tissues shows extensive infiltration of the parasite throughout all tissues and organs with accompanying degeneration and necrosis.

Experimental and field data suggest the following yearly life cycle for the parasite, which may be somewhat dependent upon sea water temperature: a noninfectious replicating vegetative stage causing most of this chronic disease syndrome persists for 10-11 months from September to the following June or July during which most crab mortality occurs; vegetative stages develop into pre-spore stages and sporulate within 10-14 days; sporulation results in crab death within 24-48 hr and produces only one of two biflagellated types of spores per crab host. The two types of dinospores differ morphologically (Figs. 16, 17) and behaviorally; spores released from decaying crabs in August through October either directly infect other crab hosts or combine outside the crab to form an infectious zygote, either of which

---

Fig. 12. *Paranophrys*-like ciliates (arrows) within the connective tissues beneath the epidermis of a blue king crab held in the laboratory. H & E; 3,800X.

Fig. 13. Intranuclear inclusion body within a bladder cell from a Dungeness crab collected in Southeast Alaska. H & E; 3,800X.

Fig. 14. Vegetative stages of the bitter Tanner crab dinoflagellate within a peripheral hemolymph smear. Note indistinct nuclear outline, foamy cytoplasm, and droplets of nonstaining material exuding from the cell margin. Diff Quik; 3,800X.

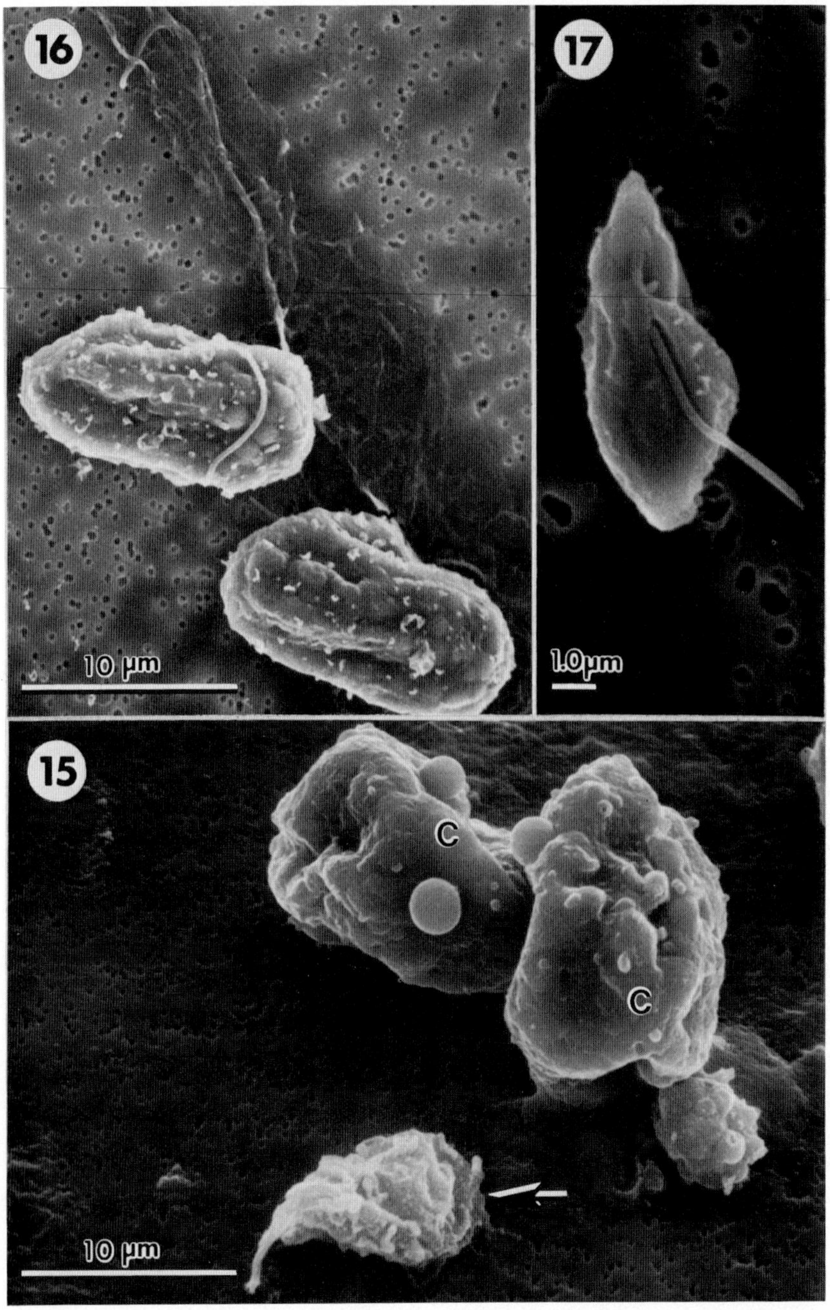

Fig. 15. Scanning electron micrograph of two dinoflagellate vegetative cells (C). Note irregular surface and droplet material exuding from cell pellicle. Host hemocyte (arrow). Gold palladium; 3,000X; Bar = 10 μm.

Fig. 16. Dinoflagellate large spore type exhibiting a warty surface and two flagella (circumferential and whiplash). Gold palladium; 3,000X; Bar = 10 μm.

Fig. 17. Dinoflagellate small spore type showing a smooth surface and flagellar attachment sites. Gold palladium; 6,000X; Bar = 1 μm.

supposedly perpetuates the parasite cycle for another year. Death of the crab host is most likely caused by replacement of hemocytes and loss of immunocompetence with secondary bacterial and protozoal infections, tissue displacement by the parasite resulting in organ dysfunction, and possible toxic effect from organism metabolites such as the exudate observed from the vegetative stages. Bitter crab disease is widespread in northern southeast Alaska, affecting about 2/3 of the commercial harvest areas for *C. bairdi*. Recently, increased deadlosses in the commercial *C. opilio* fishery in the Bering Sea prompted investigation of the cause. Typical bitter crab dinoflagellate vegetative and spore stages were observed in 6-9% of 150 crabs randomly sampled from different areas by the ADF&G. The significance of this disease in the *C. opilio* fishery is currently under investigation.

DISCUSSION

King crab diseases. *Briarosaccus callosus* causes reduced king crab growth (Hawkes et al., 1986a, 1987) as well as castration in red (Boschma and Haynes, 1969), blue (Hawkes et al., 1985), and golden (McMullen and Yoshihara, 1970) king crabs. Additional physiological studies of hemolymph from parasitized king crabs and respective parasites indicated variable host responses, depending upon crab species, which could help explain the more severe effects of parasitism on the blue king crab (Shirley et al., 1986). Regardless of host species, this parasite has considerable potential to negatively impact king crab populations through reduced fecundity and poor growth while parasitized individuals continue to distribute the parasite and compete with healthy crabs for food and space. Commercial fishing exacerbates the problem through harvest of the larger uninfected crabs while leaving the smaller parasitized individuals. A possible management approach to reduce the prevalence of this parasite would be to allow commercial and sport harvest of parasitized king crabs regardless of sex or size since their meats are still marketable.

The carcinoma observed in the single blue king crab is the second such epithelial neoplasm reported in Crustacea (Sparks and Morado, 1987; Lightner and Brock, 1987) and the first in Crustacea to be associated with a virus. Although many different viruses have been described in other crustaceans (Johnson, 1983; Sparks, 1985), only one other virus has been reported from king crabs. Sparks and Morado (1986) observed a herpes-like virus infecting the bladder and antennal gland labyrinthal epithelium of blue king crabs from the Pribilof Islands and the Western Bering Sea. Typical Cowdry type A intranuclear inclusions were produced. Birna-type viruses, such as the one reported here, are not generally associated with tumors in other animal species excepting the eel virus (Berlin) that occurs in association with stomatopapillomas that reach huge proportions in European eels (Pfitzner, 1969; Pfitzner and Schubert, 1969; Peters and Peters, 1977). Whether our king crab virus is indeed oncogenic needs further study.

The intracellular prokaryotic organism observed in the blue king crab is similar in appearance to other rickettsia-like organisms infecting other invertebrates (Sparks, 1985). Only seven other observations of natural rickettsial infections in crustacean hosts are included in the literature: a freshwater amphipod, *Crangonyx* nr. *floridanus* (Federici et al., 1974); a terrestrial isopod, *Armadillidium vulgare* (Vago et al., 1970); a marine crab, *Carcinus mediterraneus* (Bonami and Pappalardo, 1980); blue king crab

(Johnson, 1984); and three penaeid shrimp species, *Panaeus marginatus* (Brock et al., 1986), *P. japonicus*, and *P. merquiensis* (Lightner et al., 1985). Depending upon the species in the first three cases, rickettsial infections involved epidermal and connective tissue cells and organisms were found free in the hemolymph. The king crab and shrimp agents both infected the hepatopancreatic cells. Our rickettsial agent was similar to that reported by Johnson (1984) in a Bering Sea blue king crab regarding size and lack of obvious dividing forms but did not exhibit the different developmental morphologies apparent in her material. The antennal gland was also a different target tissue in our blue king crab samples although the hepatopancreas was similarly infected in our golden king crab. Whether all three agents are similar and infect various tissues, or are different organisms altogether and prefer specific target tissues remains to be determined. The significant pathogenic nature of infection in both of our king crab species was certainly apparent.

The *Paranophrys*-like protozoan is the first report of such an agent in king crabs and given the opportunity is quite pathogenic. Whether the parasite causes disease and mortality in natural populations of king crabs still warrants further investigation.

Dungeness crab diseases. Dungeness crabs have relatively few diseases and no viral infections reported in the literature (Meyers et al., 1985). This is the first report of intranuclear inclusion bodies in Dungeness crab and would be significant from an academic perspective if of viral etiology.

Tanner crab diseases. The bitter crab studies have been the most comprehensive work in the literature regarding parasitic dinoflagellates in decapod crustacea. More importantly, management "around" the disease may be possible if the parasite infection follows a yearly cycle in that an earlier harvest season during the fall would allow crabs to be taken before the disease progresses sufficiently to cause crab death or significant degradation in meat quality. Disease dissemination could be reduced by educating fishermen to recognize and cull infected crabs on the fishing grounds rather than somewhere else miles away where potentially healthy crabs may be exposed to the infected animals. Any crabs culled at the processing plant should be incinerated or buried. Lastly, fishing areas having severely diseased crab populations could be closed to commercial harvest until the situation improves. The occurrence of this disease in *C. opilio* Tanner crabs in the Bering Sea documents a new host species for this parasite and extends the known northern range of its distribution. Which species of Tanner crab, *C. opilio* or *C. bairdi*, is the natural host for this parasite remains debatable. Whether the parasite has spread from *C. bairdi* populations northward also is not certain. Although previous large-scale visual surveys of *C. bairdi* Tanner crabs in the northwestern Gulf of Alaska have been made for *Trichomaris invadens*, a fungal pathogen (Hicks, 1982), infections by the bitter crab dinoflagellate could have been missed without examination of the blood or internal tissues. Extensive disease surveys of *C. opilio* have not been published. The as yet unsubstantiated opinion for the *C. bairdi* populations in Southeast Alaska is that bitter crab disease is spreading. Continued investigations in progress will determine whether this is true.

ACKNOWLEDGEMENTS

We thank Dr. Ralph Elston of Battelle Laboratories in Sequim, Washington, for processing some of the king crab TEM.

LITERATURE CITED

Armstrong, D.A., E.M. Burreson and A.K. Sparks. 1981. A ciliate infection (*Paranophrys* sp.) in laboratory-held Dungeness crabs, *Cancer magister*. J. Invert. Pathol., 37, 201-209.

Bonami, J.R. and R. Pappalardo. 1980. Rickettsial infection in marine crustacea. Experientia, 36, 180-181.

Boschma, H. and E. Haynes. 1969. Occurrence of the rhizocephalan *Briarosaccus callosus* Boschma in the king crab *Paralithodes camtschatica* (Tilesius) in the northeast Pacific Ocean. Crustaceana, 16, 97-98.

Brock, J.A., L.K. Nakagawa, T. Hayashi, S. Teruya and H. Van Campen. 1986. Hepatopancreatic rickettsial infection of the penaeid shrimp, *Penaeus marginatus* (Randall), from Hawaii. J. Fish. Dis., 9, 73-77.

Chatton, E. and R. Poisson. 1931. Sur l'existence, dans le sang des crabes, de peridiniens parasites *Hematodinium perezi* n.g., n sp. (Syndinidae). C.R. Seanc. Soc. Biol., 105, 553-557.

Federici, B.A., E.I. Hazard and D.W. Anthony. 1974. Rickettsia-like organism causing disease in a crangonid amphipod from Florida. Applied Microbiol., 28, 885-886.

Hawkes, C.R., T.R. Meyers and T.C. Shirley. 1985. Parasitism of the blue king crab, *Paralithodes platypus* (Brandt), by the rhizocephalan *Briarosaccus callosus* Boschma. J. Invert. Pathol., 45, 252-253.

Hawkes, C.R., T.R. Meyers and T.C. Shirley. 1986a. Length-weight relationships of blue, *Paralithodes platypus*, and golden, *Lithodes aequispina*, king crabs parasitized by the rhizocephalan, *Briarosaccus callosus* Boschma. Fish. Bull., 84, 327-332.

Hawkes, C.R., T.R. Meyers, T.C. Shirley and T.M. Koeneman. 1986b. Prevalence of the parasitic barnacle *Briarosaccus callosus* on king crabs of southeastern Alaska. Trans. Amer. Fish. Soc., 115, 252-257.

Hawkes, C.R., T.R. Meyers and T.C. Shirley. 1987. Growth of Alaskan blue king crabs, *Paralithodes platypus* (Brandt), parasitized by the rhizocephalan *Briarosaccus callosus* Boschma. Crustaceana, 52, 78-84.

Hicks, D.M. 1982. Abundance and distribution of black mat syndrome on stocks of Tanner crabs, *Chionoecetes bairdi*, in the Northwestern Gulf of Alaska. Proc. Intl. Symp. Chionoecetes, Alaska Sea Grant Rpt. No. 82-10, pp. 565-579.

Johnson, P.T. 1983. Diseases caused by viruses, Rickettsiae, bacteria and fungi. In: "The Biology of Crustacea. Vol. 6, Pathobiology" (A.J. Provenzano, Jr., ed.), pp. 1-78. Academic Press, New York.

Johnson, P.T. 1984. A rickettsia of the blue king crab, *Paralithodes platypus*. J. Invert. Pathol., 44, 112-113.

Johnson, P.T. 1986. Parasites of benthic amphipods: dinoflagellate (Duboscquodinida: Syndinidae). Fish. Bull., 84, 605-615.

Lightner, D.V. and J.A. Brock. 1987. A lymphoma-like neoplasm arising from hematopoietic tissue in the white shrimp, *Penaeus vannamei*, Boone (Crustacea: Decapoda). J. Invert. Pathol., 49, 188-193.

Lightner, D.V., R.M. Redman, R.R. Williams, L.L. Mohoney, J.P.M. Clerx, T.A. Bell and J.A. Brock. 1985. Recent advances in penaeid virus disease investigations. J. World Maricult. Soc., 16, 267-274.

MacLean, S.A. and C.L. Ruddell. 1978. Three new crustacean hosts for the parasitic dinoflagellate *Hematodinium perezi* (Dinoflagellata: Syndinidae). J. Parasitol., 64, 158-160.

McMullen, J.C. and H.T. Yoshihara. 1970. An incidence of parasitism of the deepwater king crab, *Lithodes aequispina*, by the barnacle *Briarosaccus callosus*. J. Fish. Res. Bd. Canada, 27, 818-821.

Meyers, T.R., T.C. Shirley, S.M. Shirley, A.K. Sparks and J.F. Morado. 1985. A review of Dungeness crab diseases in the Pacific Northwest and Alaska. Proc. Symp. Dungeness Crab Biol. Mgmt., Alaska Sea Grant Rpt. No. 85-3, pp. 209-221.

Meyers, T.R., T.M. Koeneman, C. Botelho and S. Short. 1987. Bitter crab disease: a fatal dinoflagellate infection and marketing problem for Alaskan Tanner crabs, *Chionoecetes bairdi.* Dis. Aquat. Organ., 3, 195-216.

Newman, N.W. and C.A. Johnson. 1975. A disease of blue crabs (*Callinectes sapidus*) caused by a parasitic dinoflagellate, *Hematodinium* sp. J. Parasitol., 61, 554-557.

Peters, G. and N. Peters. 1977. Temperature-dependent growth and regression of epidermal tumors in the European eel (*Anguilla anguilla* L.). Ann. N.Y. Acad. Sci., 298, 245-260.

Pfitzner, I. 1969. Zur Atiologie der Blumenkohlkrankheit der Aale. Arch. Fischereiwiss., 20, 24-35.

Pfitzner, I. and G. Schubert. 1969. Ein virus aus dem Blut mit Blumenkohlkrankheit behafteter Aale. Z. Naturforsch., 24B, 790.

Shirley, S.M., T.C. Shirley and T.R. Meyers. 1986. Hemolymph responses of Alaskan king crabs to rhizocephalan parasitism. Can. J. Zool., 64, 1774-1781.

Sparks, A.K. 1985. "Synopsis of Invertebrate Pathology Excluding Insects." Elsevier, New York.

Sparks, A.K. and J.F. Morado. 1986. A herpes-like virus disease in the blue king crab, *Paralithodes platypus*. Dis. Aquat. Organ., 1, 115-122.

Sparks, A.K. and J.F. Morado. 1987. A putative carcinoma-like neoplasm in the hindgut of a red king crab, *Paralithodes camtschatica*. J. Invert. Pathol., 50, 45-52.

Vago, C., G. Meynadier, P. Juchault, J. Legrand, A. Amargier and J. Duthoit. 1970. Une maladie rickettsienne chez les Crustaces Isopodes. C.R. Acad. Sci. Paris, Ser. D, 271, 2061-2063.

DISEASES RELEVANT TO PENAEID MARICULTURE IN TROPICAL AUSTRALIA

Leigh Owens and Sonja Hall-Mendelin

Graduate School of Tropical Veterinary Science,
James Cook University of North Queensland, Australia, 4811

ABSTRACT: In the fledgling Australian prawn mariculture industry, a number of disease problems have arisen. Conditions have also been found in prawns from the wild fishery that could have a serious impact on prawn mariculture. This paper describes those conditions.

Many inclusion bodies have been found and their diagnosis confirmed by electron microscopy. *Plebejus* baculovirus (PBV), hepatopancreatic parvovirus (HPV), and *Chlamydia* were all found in the hepatopancreas of juvenile banana prawns, *Penaeus merguiensis*, from a local estuary. Blue shrimp syndrome was seen in wild caught *Penaeus merguiensis* infected with an 8-spored *Thelohania* species. Microsporidiosis is common in five species of wild prawns in the area of greatest maricultural interest.

In *Penaeus monodon* an actinomyocete-like fungus infection of the hypodermis has caused losses in one prawn farm. Toxic pond, as a result of ill prepared ponds, destroyed one crop causing superficial black death. Cramp tail has been seen when water and air temperatures were high. Muscular degeneration similar to vitamin E and selenium deficiencies of cattle was observed at a farm. The biggest problem to date is the hypersaline conditions before the onset of the wet, which curtails the growth of prawns, and feeding only maintains prawns at their current size.

INTRODUCTION

Tropical Queensland is undergoing a rapid expansion in mariculture with over $60 million already invested and a further $140 million planned. Most projects are concerned with the growing of penaeid shrimp. Disease is always limiting in aquaculture, and recent large scale losses ($500 million) of shrimp in Taiwan were attributed to disease (Rosenberry, 1988). In Australia, there has been considerable debate on the movement of spawners and larvae of fish and shrimp from one geographical location to another as this may spread disease and dilute genetic diversity. Therefore, a study of disease in local maricultural operations and in adjacent wild prawn stocks was undertaken to discover what disease agents occurred locally.

Pathology in Marine Science

adjacent wild prawn stocks was undertaken to discover what disease agents occurred locally.

MATERIALS AND METHODS

In order to collect banana prawns (or shrimp) (*Penaeus merguiensis*), mangrove estuaries were dragged with a beach seine net with 1/2-inch mesh at low tide. Juvenile prawns were caught in Townsville while adult prawns were caught in Ayr. Prawns were kept in glass aquaria and fed commercial prawn food supplemented with fresh squid. Prawns were anesthetized in chilled water, fixed in Davidson's with extra fixative being injected into the hepatopancreas. Prawns from maricultural facilities were fixed in Davidson's, 10% formalin in seawater or frozen and then fixed. Standard histology was performed according to Culling et al. (1985).

Initially, the study was not designed to use the electron microscope. Only after the importance of inclusion bodies was realized did fixing specifically for electron microscopical examination take place. Therefore, paraffin blocks that had shown inclusion bodies were deparaffinized, post-fixed in osmium tetroxide, and embedded in Spurr's resin. Sections of 70-90 nm were cut and stained with uranyl acetate and lead citrate before viewing under the transmission electron microscope (TEM). Semithin epoxy resin sections (1-2 μm) were stained with toluidine blue and fuchsin following the method of Burns and Bretschneider (1981).

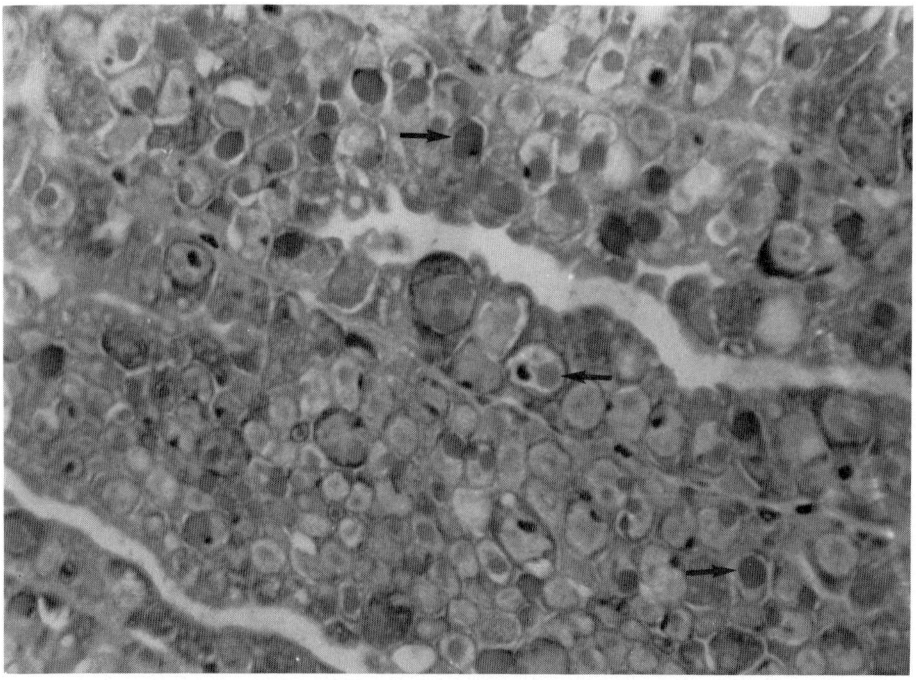

Fig. 1. Numerous intranuclear occlusion bodies (arrows) of *Plebejus* baculovirus in the hepatopancreas of *Penaeus merguiensis*. H & E; 307X.

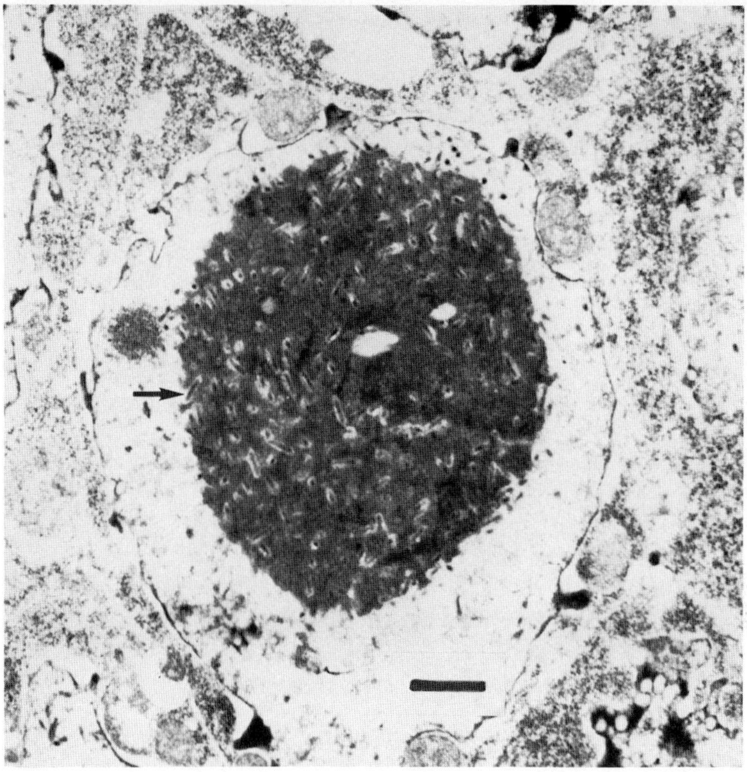

Fig. 2. Electron micrograph of *Plebejus* baculovirus occlusion body showing many virions (arrow) within the matrix of the body. 12,037X. Bar = 0.8μm.

RESULTS

Wild caught penaeids

In juvenile banana prawns, *P. merguiensis*, low levels (4%, Table 1) of round, eosinophilic, intranuclear occlusion bodies were found in the hepatopancreas (Fig. 1). The greatest majority of cells were infected with only one occlusion body. Multiple small occlusion bodies occurred and they appeared to fuse into a single occlusion when larger. Some of the histochemical staining characteristics are listed in Table 2. The most notable were the uniformity in staining with eosin and phloxine and neutral staining with Gram Twort. With electron microscopy, virions were seen in the occlusion bodies (Fig. 2) and had two electron dense layers in the envelope, which suggested the baculovirus was *Plebejus* baculovirus rather than *Monodon* baculovirus. When examining *Plebejus* baculovirus from *P. merguiensis* under the electron microscope, one membrane bound occlusion body showed small particles similar to Reo-like virus particles.

In both adult and juvenile banana prawns, large basophilic, intranuclear inclusions (Fig. 3) consistent with hepatopancreatic parvovirus were found in the cortex layer acini of the hepatopancreas. The odd inclusion body was found in the midgut, associated with a moderate hemocytic infiltration and deposition of

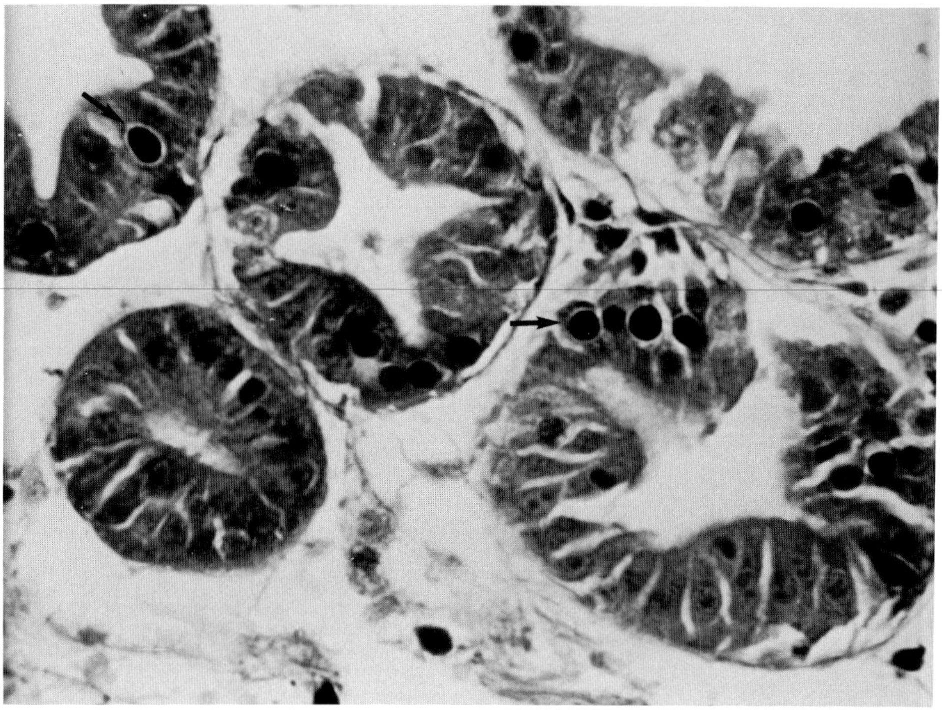

Fig. 3. Intranuclear inclusions of hepatopancreatic parvovirus (HPV) showing densely staining nuclei with clear halos (arrows) in *Penaeus merguiensis*. Macchiavello; 383X.

eosinophilic material in the circular muscular layers. Inclusion bodies were more prevalent in older prawns than the younger prawns (Table 1). In three infected cases, the hepatopancreas was greatly modified with hypertrophic cells containing very large vacuoles filled with masses of eosinophilic material of little structure (Fig. 4). The lumens of the tubules were almost occluded by the hypertrophied cells. One case of a dual infection of parvovirus and *Plebejus* baculovirus was recorded. A further case of a dual infection of parvovirus and *Thelohania duorara* was seen. Histochemically, these inclusions were positive with Macchiavello and phloxine (inclusion body stains), positive with methyl green (DNA), and negative with pyronin (RNA). The method of fixation was unsuitable for Feulgen stain. With both light and electron microscopy, large inclusions with 3-4 electron lucient, intranuclear bodies typical of hepatopancreatic parvovirus were seen. As a result of fixing, resolution under the TEM was too poor to recognize individual virions.

Large cytoplasmic inclusion bodies observed in the hepatopancreas of *P. merguiensis* proved to be *Chlamydia* (Fig. 5). The inclusion bodies were predominately eosinophilic with basophilic spots (tiny nuclei). Staining characteristics included being negative with Macchiavello and Feulgen. While the inclusions were neutral with Gram stain, the tiny nuclei were Gram-negative. The inclusions stained with phloxine and were pink with Giemsa. The inclusion nuclei stained with hemotoxylin and were blue with Giemsa in combination with the above two stains. With semithin sections and TEM, the *Chlamydia* showed the middle stage of nucleoid condensation frequently and the initial body infrequently (Fig. 6).

Most of the adult *P. merguiensis* showed a distinctive navy blue coloration with an underlying milky white flesh. On histopathological examination, an eight-spored *Thelohania* was found infecting only the muscle bundles including those in the center of the abdomen. In electron micrographs, 12-14 polar filaments were observed, suggesting the species may be *Thelohania duorara*. A second microsporidian was found in wild caught adult prawns, *Penaeus latisulcatus, Penaeus longistylus,* and *Penaeus semisulcatus*. In *P. semisulcatus*, the affected areas were raised, tumor-like areas. Tissues infected included the periphery of the muscle bundles, the intestinal wall, subcuticular layers, gonads, gills, heart, nerve cord, and especially the connective tissue. In electron micrographs, 8-9 polar filaments were observed and, coupled with the characteristics of the infected tissues, this suggested the microsporidian might have been *Agmasoma penaei*.

Penaeids in mariculture

Diseases in hatcheries. One hatchery was having mortality problems with larval *P. monodon*. Some post-larvae grew through to day 21. The post-larvae were characterized by large size differences and bright orange hepatopancreases rather than the normal brown coloration. Histopathological examination revealed numerous eosinophilic intranuclear occlusion bodies in the hepatopancreas. As these mostly occurred as single occlusions and stained evenly with eosin, they were considered to result from infection with *Plebejus* baculovirus.

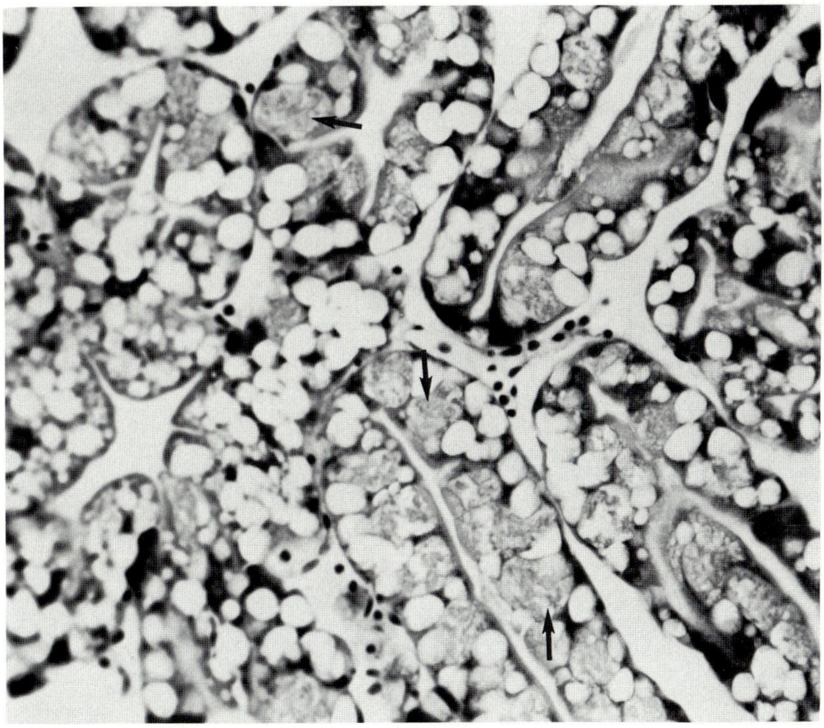

Fig. 4. Hepatopancreas of *Penaeus merguiensis* infected with hepatopancreatic parvovirus showing hypertrophied cells filled with eosinophilic material of little structure (arrows). H & E; 268X.

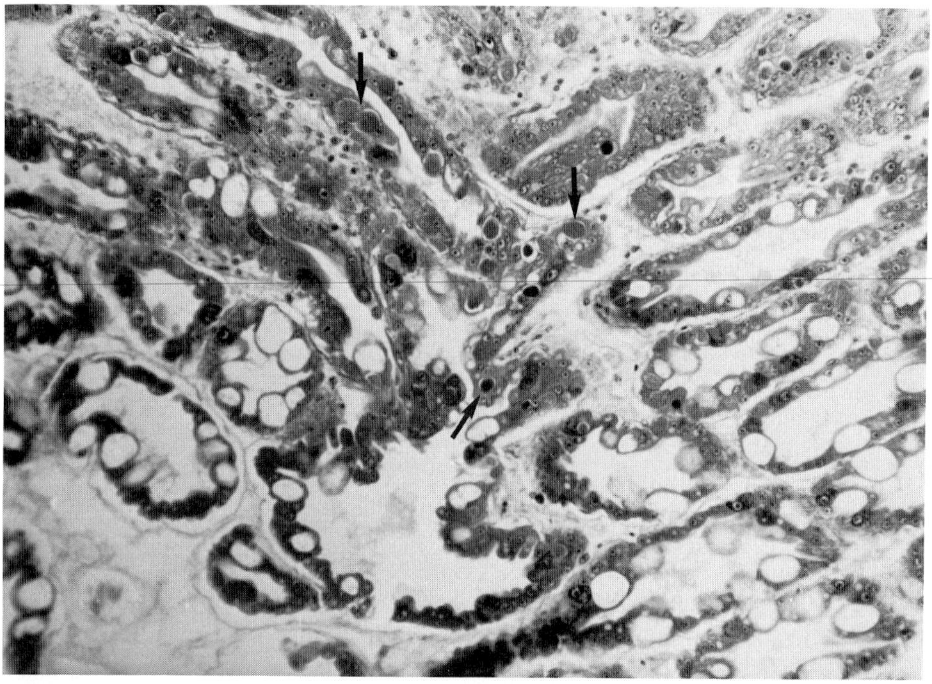

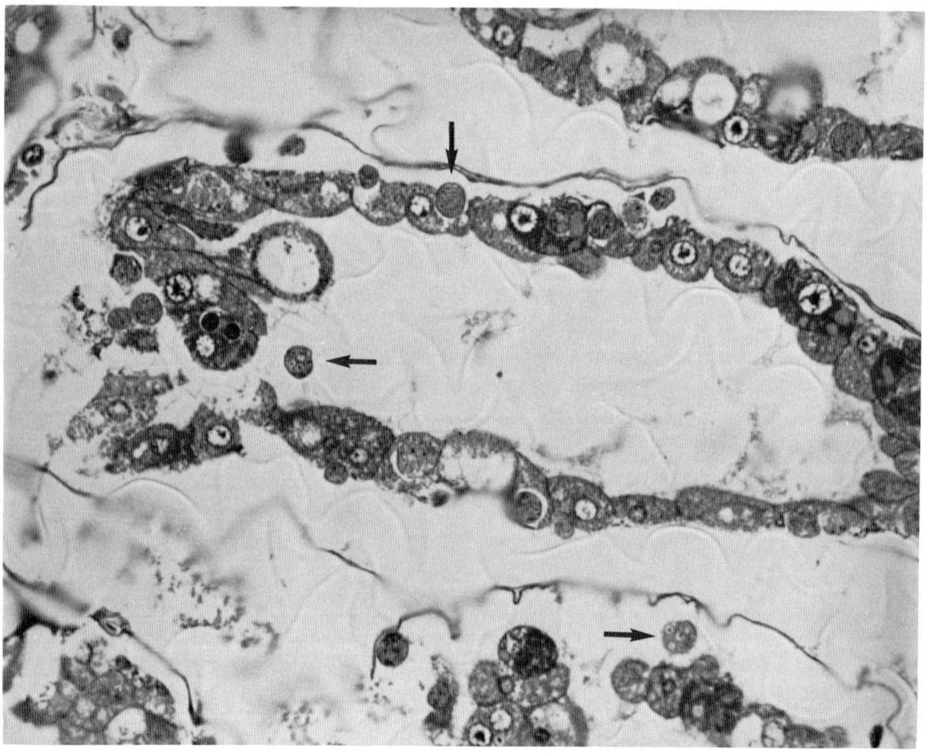

One million 10-day-old post-larvae of *P. merguiensis* were showing increasing fluorescence over a few days in hatchery tanks. A treatment of oxytetracycline at 2 ppm was initiated. Only 45,000 treated post-larvae survived. All nontreated ones died. On postmortem, Gram-negative rods rounded at both ends were found in the tissues of the post-larvae. The causative agent was considered *Pseudomonas* cf. *fluorescens*. Fluorescent larvae have been an intermittent problem in several hatcheries in northern Queensland, particularly in association with spawning slicks of coral reef animals and algal blooms in spring.

Eight- to 11-day-old post-larvae of *P. monodon, P. esculentus*, and *P. plebejus* died over a 2-day period. Histopathological examination showed massive bacterial loads consisting of curved *Vibrio*, rods similar to *Pseudomonas, Diplococcus*, and agglutinating mats of bacteria in the midgut and hepatopancreatic tubules. Rounding of hepatopancreatic cells, presumably caused by bacterial toxins, was a prominent feature. The reason for the crash occurring at this stage of development of the prawns was not determined.

Zoea of *P. monodon* were dying within 12 hr of commencing to feed on a monoculture of *Chaetoceros gracilis*. After initial high ingestion rates, feeding dropped off markedly. Microscopically, zoea were in good condition except for large vacuoles filled with eosinophilic material in hypertrophied hepatopancreatic cells. These vacuoles were very similar to those seen in juvenile *P. merguiensis* infected with parvovirus. In these zoea, no inclusion bodies were apparent. A nutritional or antinutritional factor appeared most likely.

Lagenidium was found to be a problem at one prawn farm by M. Kenway and L. Trott of the Australian Institute of Marine Science. Initially, the problem was with brood stock females and later became a continuous problem with protozoeal and mysis of *P. monodon*. The hyphae were characterized by being aseptate, highly branching, and producing a small discharge tube. No apparent cellular reaction was ellicited by the hyphae. Eggs were also destroyed by invading hyphae. A bath of 10 ppm of Treflan (trifuralin) for spawners and a continuous drip for 24 hr in early larval stages has brought the problem under control.

Diseases in grow-out ponds. An infectious hypodermal necrosis was a chronic problem at one prawn farm. Large *P. monodon* died at the rate of half a dozen per day. Prawns were found listless around the edge of the pond, with greatly expanded gill covers. Prawns remained soft on molting for several days and in aquaria were a sky blue color. Cuticular lesions formed, especially on the last two distal segments of the abdomen, and were invaded by bacteria causing muscle necrosis and eventually, death. Microscopically, the hypodermis contained long strands of necrotic material, brightly eosinophilic with small specs of basophilic material, possibly chromatin. The strands looked like branching, nonseptate fungal hyphae (Fig. 7) and stained weakly with silver methenamine. There was a hemocytic infiltrate in the epidermis, and in extreme cases, melanization of the cuticular lesion. Idiopathic cardiomyopathy with hemocytic infiltration and mild myophagia were also present. Fungi of the genus *Cladosporium* were isolated from diseased prawns but were not believed to be the

---

Fig. 5. Hepatopancreas of *Penaeus merguiensis* showing cytoplasmic inclusion bodies of *Chlamydia* (arrows). H & E; 154X.

Fig. 6. Semithin section (1 μm) of *Penaeus merguiensis* with *Chlamydia*. Some inclusions clearly show the middle stage of development (nucleoid condensation) (arrows). Toluidine blue and Fuchsin; 307X.

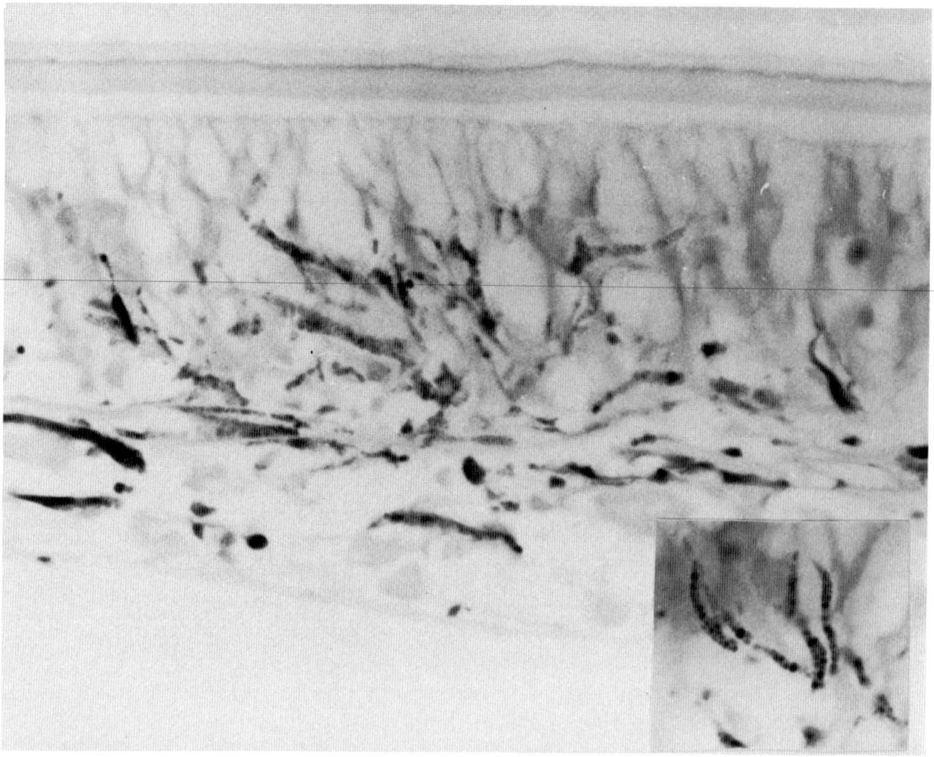

Fig. 7. Fungal-like hyphae growing in the cuticular hypodermis of *Penaeus monodon* associated with sustained low level losses after molting. Gram Twort.; 503X. Insert, higher magnification of hyphal-like material; 760X.

causative agent as they produce septae in culture. The problem did not respond to antiseptic dips (e.g., malachite green, formalin).

A crop of *P. monodon* was lost owing to acidic culture conditions. The ponds were ill-prepared mangrove muds with large organic loads. This resulted in low pH with reducing conditions prevalent at the bottom of the ponds. After a still night with little water change because of a low tide, mass mortality followed. Prawns presented black coloration and highly fouled gills as a result of the reducing conditions and the high organic loads, respectively. Histopathology revealed swollen, edematous gills covered in epicommensal ciliates (many *Rhabdostyla*-like, few *Epistylus*, and rarely *Vorticella*) and mild iron deposition. The fouled gills lead to anoxia under the limited water movement conditions.

Cramp tail has occurred at a number of prawn farms in early summer (November and December) before relief from high temperatures with the onset of summer rains. Better management of ponds has allowed prawns to be harvested before the onset of this condition. Ponds were then restocked with post-larvae, which do not seem so susceptible to this condition and the problem has become less important.

A subclinical focal myopathy was seen in juveniles of *P. monodon*. Small patches of muscle were necrotic, fragmented, and invaded by predominately the granulated

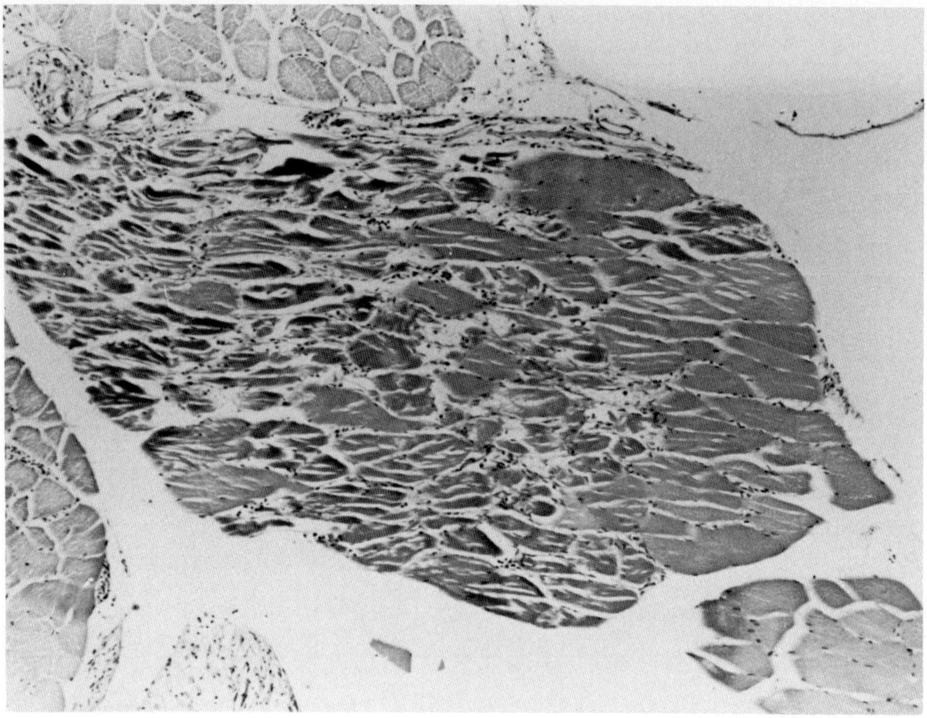

Fig. 8. Subclinical muscular dystrophy in *Penaeus monodon* similar to vitamin E and selenium deficiency. Note muscular fragmentation and necrosis and haemocytic infiltration; 48X.

series of hemocytes (Fig. 8). This picture was very similar to vitamin E and selenium deficiency in ruminants. The farm involved was storing its pelleted food in a hot metal shed, which presumably led to the denaturing of oils and fats.

DISCUSSION

The round occlusion bodies in the hepatopancreas classify the baculovirus as either *Monodon* or *Plebejus* baculovirus. The evenness of staining, the occurrence of predominately one occlusion per cell and the double envelope of the virions suggest that the virus was *Plebejus* baculovirus (Lester et al., 1987). Whether *Monodon* and *Plebejus* are identical baculoviruses but are on different ends of a species cline is undetermined. Finding *Plebejus* baculovirus in widespread Indo-Pacific prawns, *P. merguiensis* and *P. mondon*, as well the Australian endemic prawn, *P. plebejus*, suggests the two baculoviruses producing round occlusions may be the same virus.

Hepatopancreatic parvovirus is certainly widespread in the Indo-west Pacific (Lightner et al., 1985) and Australia, being found in Townsville, Ayr (this report), Mackay (Paynter, pers. comm.), and Moreton Bay (Paynter et al., 1985). Within Australia, hepatopancreatic parvovirus has not been associated with disease but overseas it has caused high accumulative mortalities in mid-juvenile *P. merguiensis* and *P. semisulcatus* (Lightner, 1988a).

Table 1. Prevalence of pathogens in wild caught *Penaeus merguiensis*.

	Juveniles			Adults		
Pathogen	no.	No. Infect.	%	no.	No. Infect.	%
Plebejus Baculovirus	195	8	4	18	0	0
Hepatopancreatic Parvovirus	195	5	3	18	4	22
Chlamydia	195	1	0.5	18	0	0
Thelohania duorara	195	0	0	18	12	67

Table 2. Selected histochemical staining of hepatopancreatic inclusion bodies.

Stain	PBV	HPV	*Chlamydia*
H&E	eosinophilic	basophilic	mostly eosin. basoph. spots
Gram Twort	neutral		neutral with -ve spots
Giemsa			pink with blue spots
Feulgen	-ve	+ve	-ve
Phloxine/ Tartrazine	Phloxophilic	weakly Phloxophilic (no hematoxylin)	Phloxophilic with basophilic spots
Methyl Green/ Pyronin		+ ve MG	+ ve MG
Macchiavello	+ ve with weak halo	+ ve	-ve

Chlamydia have been associated with disease in a concurrent infection with baculoviral midgut gland necrosis in *Penaeus japonicus* (Lightner et al., 1985) and in *P. merguiensis* and *Penaeus indicus* (Chong and Loh, 1984), again in a concurrent infection with hepatopancreatic parvovirus. The single case reported here was apparently free of other infections.

The diagnosis of *Plebejus* baculovirus, hepatopancreatic parvovirus, and *Chlamydia* from wild caught prawns in northern Queensland is of major consequence to the prawning industries. It belays the fears of fishermen that mariculture may spread exotic viruses to new sites where the wild fishery may be at risk. Obviously, the wild fishery is already infected with these pathogens. Furthermore, it has not taken long for maricultural operations to accumulate pathogens from the wild as reported herein.

Bacteria causing problems in prawns were mostly Gram-negative rods that were part of the normal gut flora and became opportunistic pathogens (Lightner, 1985). Bacteria found in northern Queensland were in no way different to those found in other studies. Similarly, the genus *Lagenidium* has been associated with maricultural problems, although the species are regionalized to some extent (Lightner, 1988b). The finding of *Lagenidium* in Australia substantiates the almost prophetic statement of Lightner (1985) in that although new fungal diseases have not emerged, the known geographic range of the "old" fungi has widened. Even though *Cladosporium* has been responsible for disease in octopuses (Polglase et al., 1984) and was recovered in the present study, it is believed the lesions were only superficially fungal-like and another agent was responsible.

The muscle necrosis that was associated with vitamin E and selenium deficiencies, hypodermal necrosis, and acid culture conditions was significantly different to that described as idiopathic muscle necrosis in *Macrobrachium rosenbergii* (Nash et al., 1987). In this study, the myonecrosis was less severe and consequently concentrations of myonuclei and hemocytes were fewer and less intense. Also, the vacuolated "moth-eaten" appearance of muscle fibers described by Nash et al. (1987) was not prominent. However, in the hypodermal necrosis case, the location of the myofibril lesions on the distal abdominal segments suggests activity stress might be involved, which resulted in lactic acidosis and subsequent myofibril degeneration.

In conclusion, the diseases of penaeids of tropical Australia show great affinities with diseases of penaeids in other areas of the Indo-west Pacific. Both *P. merguiensis* and *P. monodon* are widespread throughout the Indo-west Pacific and their diseases appear to be also widespread. Metazoan parasites of *P. merguiensis* similarly show these Indo-west Pacific affinities rather than cosmopolitan affinities (Owens, 1986). The geological isolation of terrestial Australia has not been matched in marine zoogeographical regions for the prawns that have migrated down the east coast of Australia. However, the disease status of the temperate Australian endemic prawns is largely unknown and they may still be relatively free of disease owing to their long (geological) period of isolation. Furthermore, prawns from tropical western Australia appear to have different parasites to eastern Australia (Owens, 1988) and may also have a different disease status to tropical eastern Australia. Consequently, the conservative view that these three Australian prawn stocks are separate entities is probably still warranted.

LITERATURE CITED

Burns, W.A. and A. Bretschneider. 1981. "Thin is in: Plastic Embedding of Tissue for Light Microscopy." Educational Products Division, Am. Soc. Clin. Pathologists, Chicago.

Chong, Y.C. and H. Loh. 1984. Hepatopancreas chlamydial and parvoviral infections of farmed marine prawns in Singapore. Sing. Vet. J., 9, 51-56.

Culling, C.F.A., R.T. Allison and W.T. Barr. 1985. "Cellular Pathology Techniques." 4th ed., Butterworth, London.

Lester, R.J.G., A. Doubrovsky, J.L. Paynter, S.K. Sambhi and J.G. Atherton. 1987. Light and electron microscope evidence of baculovirus infection in the prawn *Penaeus plebejus*. Dis. Aquat. Organ., 3, 217-219.

Lightner, D.V. 1985. A review of the diseases of cultured penaeid shrimps and prawns with emphasis on recent discoveries and developments. Proc. 1st Inter. Con. Culture Penaeid Shrimps/Prawns Iloilo City, Philippines, pp. 79-103.

Lightner, D.V. 1988a. Hepatopancreatic parvo-like virus (HPV) disease of penaeid shrimp. In: "Disease Diagnosis and Control in North American Marine Aquaculture" (C.J. Sindermann and D.V. Lightner, eds.), pp. 30-32. Elsevier, Amsterdam.

Lightner, D.V. 1988b. Larval mycosis of penaeid shrimp. In: "Disease Diagnosis and Control in North American Marine Aquaculture" (C.J. Sindermann and D.V. Lightner, eds.), pp. 58-63. Elsevier, Amsterdam.

Lightner, D.V., R.M. Redman, R.R. Williams, L.L. Mohney, J.P.M. Clerx, T.A. Bell and J.A. Brock. 1985. Recent Advances in penaeid virus disease investigations, J. World Maricult. Soc., 16, 267-274.

Nash, G., S. Chinabut and C. Limsuwan. 1987. Idiopathic muscle necrosis in the freshwater prawn, *Macrobranchium rosenbergii* de Man, cultured in Thailand. J. Fish Dis., 10, 109-120.

Owens, L. 1986. "Parasites as Biological Markers for Banana Prawn (*Penaeus merguiensis* de Man) Stocks in the Gulf of Carpentaria." Ph.D. Thesis, James Cook University of North Queensland.

Owens, L. 1988. Maricultural considerations of the zoogeography of parasites from prawns in tropical Australia. Aust. Mar. Sc. Assoc., Silver Jubilee Cont. Commem. Vol., 193-197.

Paynter, J.L., D.V. Lightner and R.J.G. Lester. 1985. Prawn virus from juvenile *Penaeus esculentus*. In: "Second Australian National Prawn Seminar" (P.C. Rothlisberg, B.J. Hill and D.J. Staples, eds.), pp. 61-64. NPS2, Cleveland, Australia.

Polglase, J.L., N.J. Dix and A.M. Bullock. 1984. Infection of skin wounds in the lesser octopus, *Eledone cirrhosa*, by *Cladosporium sphaerospermum*. Trans. Br. Mycol. Soc., 82, 577-580.

Rosenberry, R. 1988. Crash in Taiwan. Aquac. Digest, 9, 1-2.

EPIDEMIOLOGY/EPIZOOTIOLOGY SECTION: CONCLUDING REMARKS NEED FOR MORE "SECOND-LINE" RESEARCH

P. van Banning

Netherlands Institute for Fishery Investigations, P.O.B. 68,
1970 AB IJmuiden, The Netherlands

Among the different animal production systems that man has developed for food and commercial purposes, in historical terms, marine aquaculture can be considered as a recent and specialized branch. However, within a short period of roughly three decades, marine aquaculture has undergone very rapid development. An effect of this is that pioneer-like situations were, and still are, frequently met, resulting sometimes in unexpected outbreaks of epizootics without good answers for explanation, control, or prevention. Furthermore, marine aquaculture takes place in a complex environment and deals with sensitive organisms from widespread biological groups, e.g., fish, molluscs, crustacea, and algae. As a result of the rapid expansion, together with the large variation of complex biological circumstances and the rapidly increasing economic importance, both a persistent lack and need have been, and still are, felt for adequate knowledge in the field of diagnosis, prevention, and control of diseases and pests.

Research in this field is increasing considerably, resulting in an impressive stream of data on infectious organisms that can cause epizootics and can disturb aquaculture projects. However, if one considers the different topics in this field of research, a situation is seen with a skewed distribution in research direction. Most research is carried out with fish, as these are one of the most important groups in marine aquaculture. However, for the invertebrates, the other important group in marine aquaculture, far more gaps in research programs are present. Most of the problems of invertebrate epizootics are studied within the molluscs. Taking the bivalves as an example of a very important commercial group of molluscs, it is well known today how epizootics can disturb or eliminate culture activities in large areas, resulting in important economic losses. Several countries are, for instance, suffering from protozoan infections in their oyster cultures. Since the late fifties, the culture of the American oyster, *Crassostrea virginica*, has faced serious economical losses as a result of infection by two haplosporidians: *Haplosporidium nelsoni* and *H. costale*. An overview of the studies on these protozoan parasites is, for instance, given by Andrews (1979). Another good example of a disastrous epizootic in oysters is the protozoan *Bonamia ostreae*, introduced by activities of oyster growers in several European countries. This epizootic ruined and eliminated important cultures of the European flat oyster, *Ostrea edulis*, starting in 1980 in France, but also affecting in the next years the oyster culture areas of Spain, The Netherlands, England, and Ireland.

These cases are only two examples out of several epizootics that can affect molluscs.

By analyzing the field of epizootic studies of invertebrates in comprehensive overviews (e.g., Lauckner, 1983), the conclusion can be drawn that most of the topics have been investigated using first-line research, i.e., studies dealing with recognition or detection of the infectious agent, systematic position, prevalence rates, and observed mortalities. Second-line research of invertebrate epizootics is relatively less well represented. Second-line research deals with topics such as defense or immune systems of the invertebrate hosts, agent-host interactions, role of environmental factors, effects of stress conditions on host and infectious organism, detailed knowledge on the complete biological cycle of the infectious organism, and agent establishment. For aquaculture purposes, second-line research is very important because it forms the basic knowledge for control and prevention of epizootics.

Fortunately, an early start has been made into second-line research of invertebrate infections, for example, by Cheng (1987) with his studies on the cellular reaction systems of oysters to infectious agents, and by Miahle et al. (1988) with their studies of immunodiagnosis techniques of the oyster disease bonamiasis.

In the PAMAQ III program, one can see a mirroring of the ratio of problems met in marine aquaculture. The topic of fish is represented by eel, a species that has already been cultured successfully for some time, suggesting a depth of experience. However, it is clearly shown by the presentation that a lot of questions are still open regarding the multifactorial causes that may play a role in the outbreaks of epizootics in eel farming. The presentation is in fact pointing to the need for more second-line research, even in the field of a well-experienced fish farming industry, such as eel farming is today.

The ratio of subjects of this session is clearly heading towards the aquatic invertebrates, underlining that most of the open questions in epizootiology occur in this group of animals. A wide field of infectious agents have been considered, from viruses, bacteria, rickettsiae to protozoan and metazoan parasites, infecting mussel, bay scallop, and crabs. These provide a good cross section of the wide field of problems met in the invertebrate aquaculture of today. However, as discussed above, such topics, although very important, are mainly first-line research. Much more second-line research is required to better understand the complex matter of the factors that contribute to the process of infection. Therefore, I hope additional second-line research will be forthcoming. Only with such comprehensive knowledge can marine aquaculture, especially of invertebrates, develop better prevention and control programs for epizootics. By doing so, several aquaculture projects may be raised out of the challenge and pioneer status into a better manageable and more commercially stabilized position.

LITERATURE CITED

Andrews, J.D. 1979. Oyster diseases in Chesapeake Bay. Mar. Fish. Rev., 41, (1-2), 45-53.

Cheng, T.C. 1987. Some cellular mechanisms governing self and nonself recognition and pathogenicity in vertebrates and invertebrates relative to protistan parasites. Aquaculture, 67, 1-14.

Lauckner, G. 1983. Diseases of molluscs. In: "Diseases of Marine Organisms" (O. Kinne, ed.), pp. 477-983. Biologische Anstalt Helgoland, Federal Republic of Germany.

Miahle, E., E. Bachere, D. Chagot and H. Grizel. 1988. Isolation and purification of the protozoan *Bonamia ostreae* (Pichot et al., 1980), a parasite affecting the flat oyster *Ostrea edulis* L. Aquaculture, 71, 293-299.

NUTRITIONAL PATHOLOGY SECTION

NUTRITIONAL PATHOLOGY: INTRODUCTORY REMARKS

Thomas C. Cheng

Marine Biomedical Research Program and
Department of Anatomy and Cell Biology
Medical University of South Carolina
P.O. Box 12559 (Fort Johnson)
Charleston, South Carolina 29412, U.S.A.

Adequate nutrition is of paramount importance in the maintenance of health in all organisms. The effects of malnutrition are varied. For example, an inadequate amount or the absence of nitrogen compounds can lead to sterility and other pathophysiological effects. My intent is not to summarize the known effects of malnutrition on fish and shellfish maintained in aquaculture ponds. Rather, I wish to point out that a number of diseases can be indirectly associated with inadequate or unwholesome diets.

Immunodeficiency diseases. There is little doubt that the primary method by which all animals defend themselves against infectious diseases is by means of their immune mechanisms, be these innate or acquired, cell-mediated or humoral. In animals maintained on deficient diets, their immunologic defenses usually become compromised and this leads to more conspicuous infections that result in elevated morbidity or mortality. Even normally nonpathogenic or mildly pathogenic invaders may become highly virulent in immunodeficient hosts.

In the case of humoral immunity, retarded ability to synthesize immunoglobulins, which is an energy-expensive process, is associated with deficiencies in dietary nitrogen compounds and utilizable carbohydrates. Likewise, cell-mediated immunity can be compromised by diets deficient in these constituents.

Parasitic diseases. The oral route is probably the most common one for the establishment of parasitic diseases caused by protozoans and metazoans. Such diseases can be associated with diets in one of two ways: (1) inadequately sanitized diets that are contaminated with the infective stages of parasites; or (2) deficient diets that lead to reduced immunocompetence.

Microbial and viral diseases. The introduction of real and potentially pathogenic microbes (i.e., bacteria and fungi), some of which may be toxigenic, and viruses in contaminated feed can be a serious problem in mariculture. As in the case of parasitic diseases, in addition to serving as a vehicle for deleterious microbes and viruses, feed deficient in certain constituents can result in immunodeficiencies that, in turn, favor the manifestation of infectious diseases.

Also, it is now known that for efficient growth, secondary metabolism, differentiation, and survival, microbes require specific quantities of such transition elements as iron, manganese, and zinc. Furthermore, virulence of a microorganism in a specific tissue, organ, or host often depends on the ability of the invader to acquire the necessary amount of a key element. Vertebrates, including fish, and possibly invertebrates, besides recruiting immunologic defenses, can also contribute to the overall defense by altering their metabolism of iron and zinc during microbial invasion. Thus their ability to synthesize metal-binding molecules, which, in turn, is dependent on their diets, is of considerable importance.

Toxicologic diseases. As in the case of infectious disease, the entry of toxins into target organisms is commonly via the oral route. Thus, even if the diet is adequate, the inadvertent introduction of toxins or toxigenic organisms in feed, especially artificial feed, represents a real and serious problem in aquaculture.

Teratologic diseases. In recent years, there have been increasingly frequent reports of abnormal development of fish, especially in their embryonic stages. Although anthropogenic environmental factors have been incriminated as the principal causes, the possibility of deleterious dietary factors associated with the parents should not be overlooked.

Although several categories of diseases, other than nutritional problems in the traditional sense, which are associated with diets have been mentioned briefly, information pertaining to these are based almost exclusively on epizootiology and descriptive pathology. It is my opinion that the time has arrived when studies should be conducted to ascertain the mechanisms responsible for their perpetuation and pathogenesis. It is only when these have been elucidated can we expect to successfully eliminate these diseases.

PATHOGENESIS OF THE EEL IN CULTURE

Jacques Brusle

Laboratoire de Biologie Marine,
Université de Perpignan (66000), France

ABSTRACT: A review of the literature dealing with causative agents responsible for eel pathogenesis in culture shows that, among different levels of diagnosis of pathogenic symptoms, there are very little data on histo- and cytopathology. Ultrastructural analysis of the liver of the European eel has revealed some structural changes in hepatocytes corresponding to a nonpathological response to the diet as well as a true nutritional pathology. Symptoms of such a feeding stress should be considered as secondary etiological effects and as possible inductors of infections by bacteria and viruses.

INTRODUCTION

For several decades, eel production in aquaculture has been widely carried out in Japan, China, and Taiwan. In the late 1970s, production of eel in recirculation systems was introduced on an experimental scale in some European countries. Since then, eel production has expanded rapidly and many production plants have been set up in European countries (France, Italy, U. K., Germany, Denmark, Sweden, etc.).

The eel is a suitable species for culture because of its relative rusticity and various other advantages such as its adaptability to artificial food, tolerance of variations in temperature and salinity, and high stocking densities. However, several limiting factors, for instance, pathology, are encountered by eel culturists. Indeed, it is becoming increasingly apparent that this fish is not as resistant as generally believed, since it is sensitive to various environmental stresses (chemical, physical, and biological) and also to parasites and fungal, bacterial, and viral diseases (Fig. 1) all of which are responsible for lower survival, lower growth, lower commercial value, and for higher costs in preventive and curative treatments.

Disease is inevitable under the present intensive culture conditions prevalent in eel farms, especially on account of the following factors: (1) the temperature for eel production (20-25°C) almost corresponds to the optimum temperature for parasites, and intensity of infection increases as the temperature rises; (2) the high stocking densities (20-50kg/m^2) favor widespread infection because most infectious agents

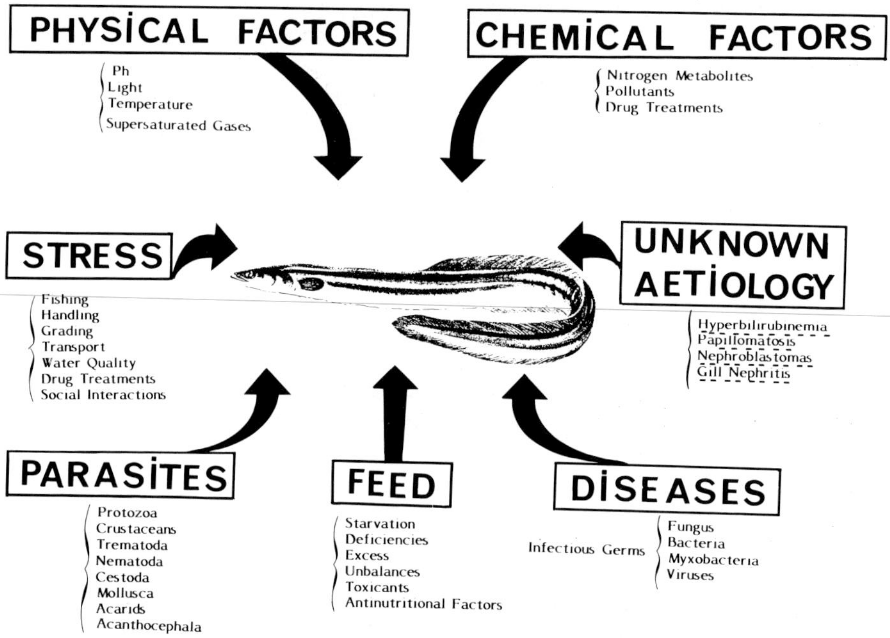

Fig. 1. General pathogenesis of the eel in culture.

multiply rapidly and are easily transferred to the whole population; and (3) the annual introduction of wild elvers from natural waters is necessary. Many parasites (*Myxidium, Pseudodactylogyrus, Anguillicola*) and latent pathogens (fungi, bacteria, and viruses) are thus introduced into the production unit despite the quarantine systems adopted in many countries.

As the importance of eel culture increased, a great deal of research was carried out on the infections that have detrimental effects on farmed eels and considerable financial losses. A lot of scientific papers on the pathology of the eel were therefore published.

The intent of this contribution is firstly to sum up the pathological problems affecting the development of eel culture, which have focused considerable interest on various patterns of pathogenesis. Secondly, it is to show that some reports and assumptions are questionable. This problem arises because the pathogenesis of some infectious agents is still not clearly understood, the knowledge of the factors responsible for the pathogenic and lethal effects of some organisms is still uncertain, and the occurrence of some multifactorial diseases (or "mixed etiology" according to Sindermann, 1983). Finally, it aims to emphasize the need for further investigations in experimental epidemiology and etiology, in histology, and especially the cytology on different target organs.

A REVIEW OF THE PATHOLOGY OF THE EEL

Some books (Rickards, 1978; Matsui, 1980) and papers (about 125) deal with numerous causative agents responsible for eel pathogenesis. They can be classified and summarized as follows: (1) physical and chemical factors of diseases (Fig. 2);

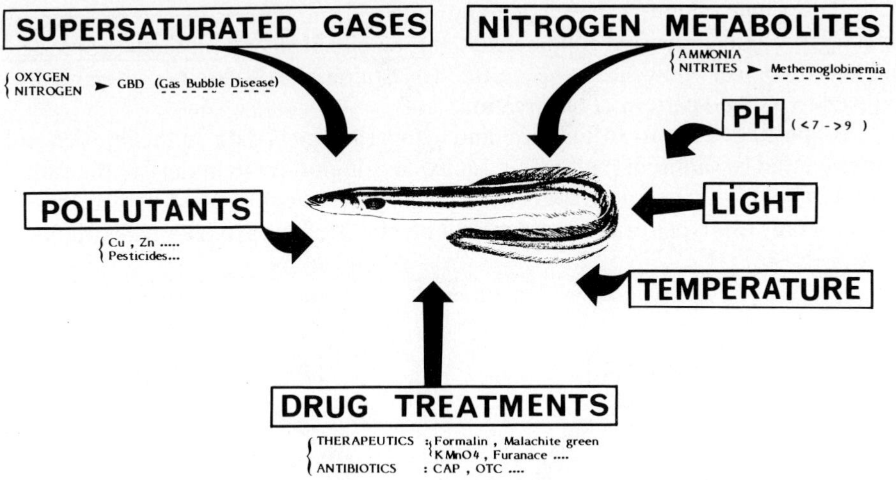

Fig. 2. Physical and chemical factors of diseases.

(2) parasites: Protozoa (Fig. 3), Crustaceans, Trematodes, Nematodes, etc. (Fig. 4); (3) fungal, bacterial, and viral diseases (Fig. 5); (4) stress and stress-mediated diseases (Fig. 6); and (5) nutritional diseases (Fig. 7).

These factors have been the cause of a number of recent epizootics and breakdowns in eel farms, some of which have been described in Japan, China, and Europe (Table 1).

Table 1. Recent Epizootic Breakdowns in Eel Farms.

Countries	Causative Agents	Pathogenicity	Authors
Japan	EEV virus	branchionephritis	Okamoto et al., 1983
id.	*Edwardsiella*	Edwardsiellosis	Aoki et al., 1987
id.	-	microbial diseases	Sano and Fukuda, 1987
China	*Pseudodactylogyrus*	dactylogyrosis	Chan and Wu, 1984
Israel	id.	id.	Bejerano, 1985
Denmark	id.	id.	Buchmann et al., 1987
id.	*Ichthyophthirius*	ichthyophthiriasis	
	Saprolegnia	fungal disease	Mellergaard and
	Aeromonas	red fin disease	Dalsgaard, 1987
	Pseudomonas	red spot disease	
	Dermocystidium	gill disease	
Italy	*Dactylogyrus*	dactylogyrosis	Saroglia et al., 1985
	Ichthyophthirius	ichthyophthiriasis	
Hungary	*Pseudodactylogyrus*	dactylogyrosis	Szekely and Molnar, 1987
Scotland	*Pseudomonas*	red spot disease	Stewart et al., 1983
Portugal	*Ichthyophthirius*	ichthyophthiriasis	Ventura and Paperna, 1985
FRG	AAV2 virus	viral diseases	Moller and Anders, 1983

Pathogenic symptoms have been approached from different levels of diagnosis: gross signs, clinical signs, morphometric signs, biochemical and physiological tests, and microscopical findings. However, the description of a disease has often been reduced to a single pattern of observation.

Regarding the histopathological and cytopathological data on the injuries and lesions caused by different pathogenic agents, it is important to underline the rather small number of detailed descriptions; the available findings on organic disturbances deal with only two types of diseases, environmental stress and nutritional pathology.

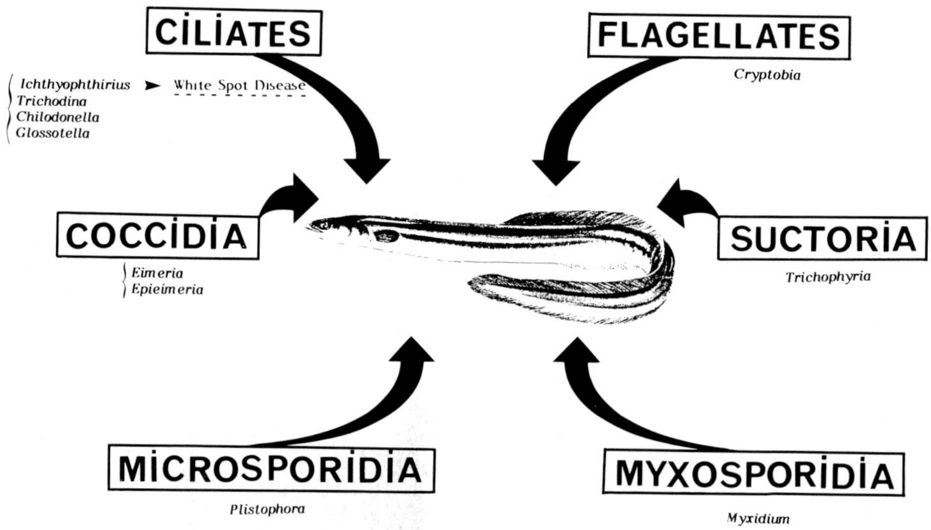

Fig. 3. Parasites : Protozoa.

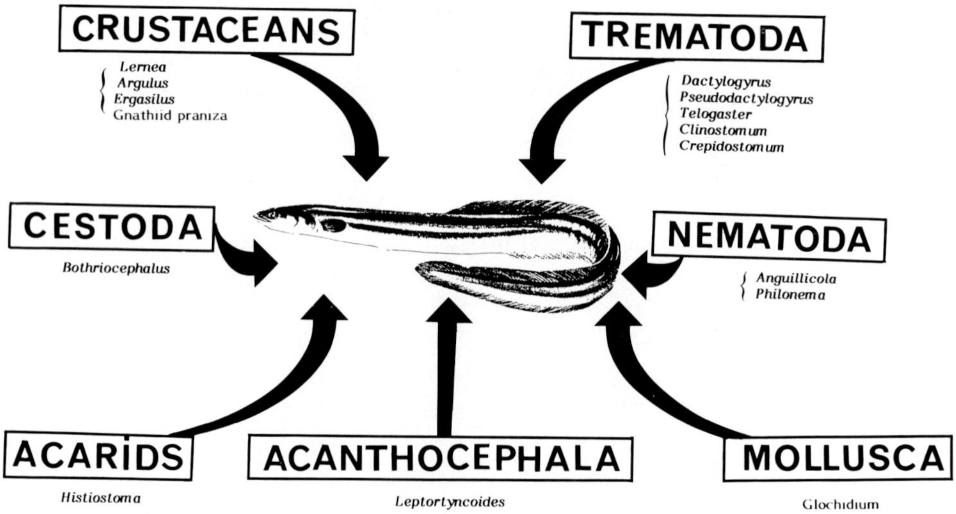

Fig. 4. Parasites.

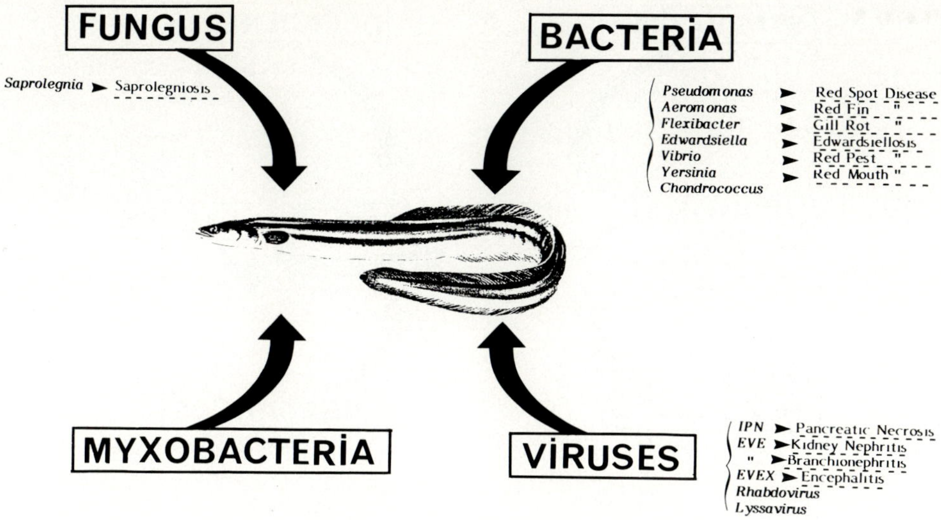

Fig. 5. Fungal, bacterial, and viral diseases.

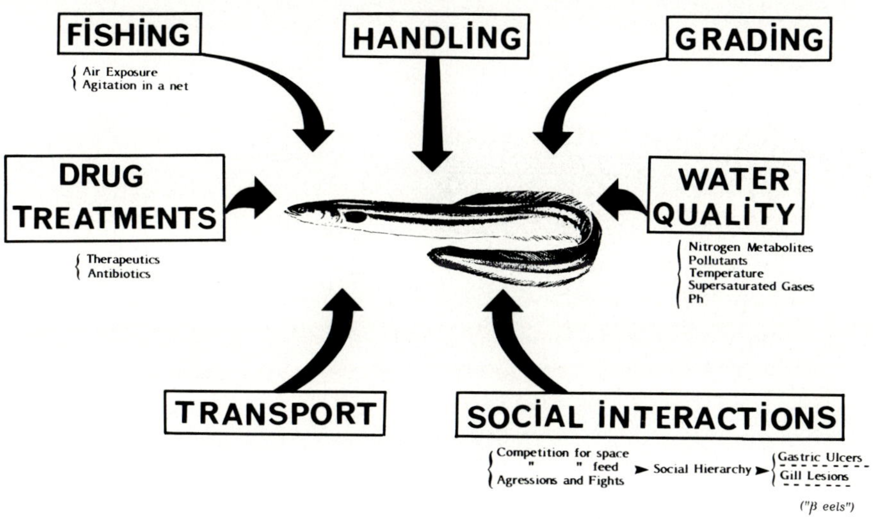

Fig. 6. Stress and stress-mediated diseases.

Environmental stress from social interactions (social stress). The high stocking densities of eels produced permanent confrontations between individuals, intraspecific aggression, and fights. A pronounced social hierarchy was generally established in tanks and pools and the subordinate individuals (β eels) suffered great stress caused by dominant fish (α eels) according Peters et al. (1981) and Peters (1982). The following parameters were demonstrated by Peters et al. (1981) to be good indicators of stress: significant decrease in blood volume, leucocyte count, spleen weight, glycogen content, stomach diameter, and a parallel increase in blood

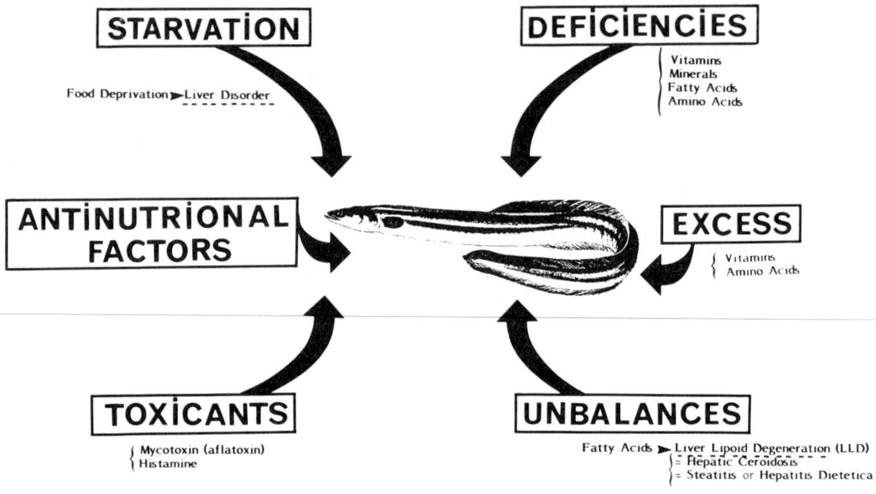

Fig. 7. Nutritional diseases.

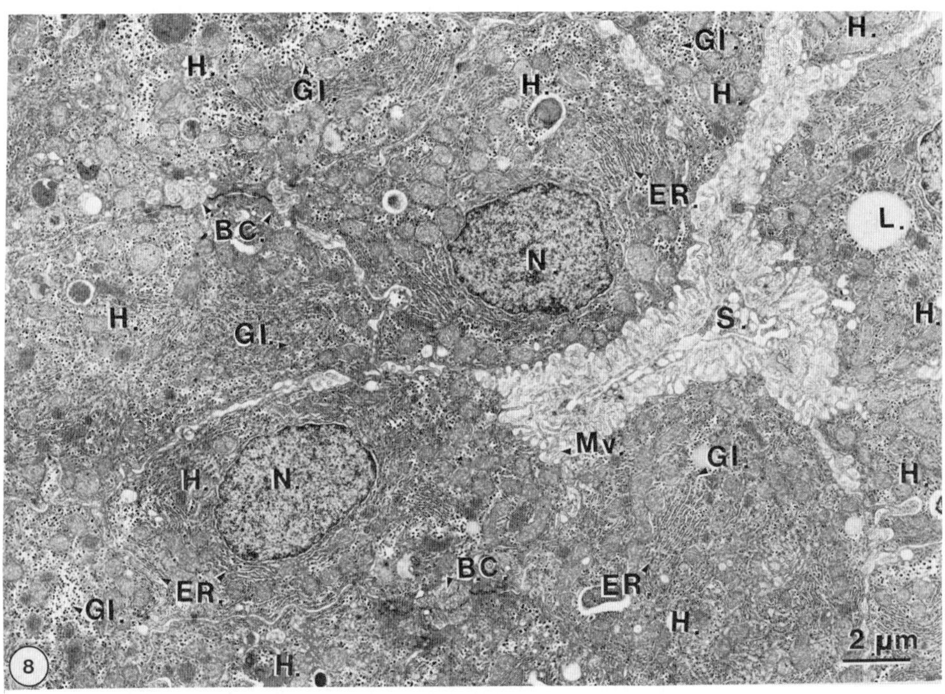

Fig. 8. Control liver (wild eel). Hepatocytes (H) show glycogen storage (G1). Lipid droplets (L) are poorly represented.

Abbreviations for Figs. 8-11. BC, biliary canalicule; CF, collagen fibers; DS, Disse's space; E, erythrocyte; EC, endothelial cell; ER, endoplasmic reticulum; Gl, glycogen granules; H, hepatocyte; L, lipid droplet; Mi, mitochondria; Mv, microvilli; N, nucleus; Ne, nuclear envelope; NU, nucleolus; S, sinusoid.

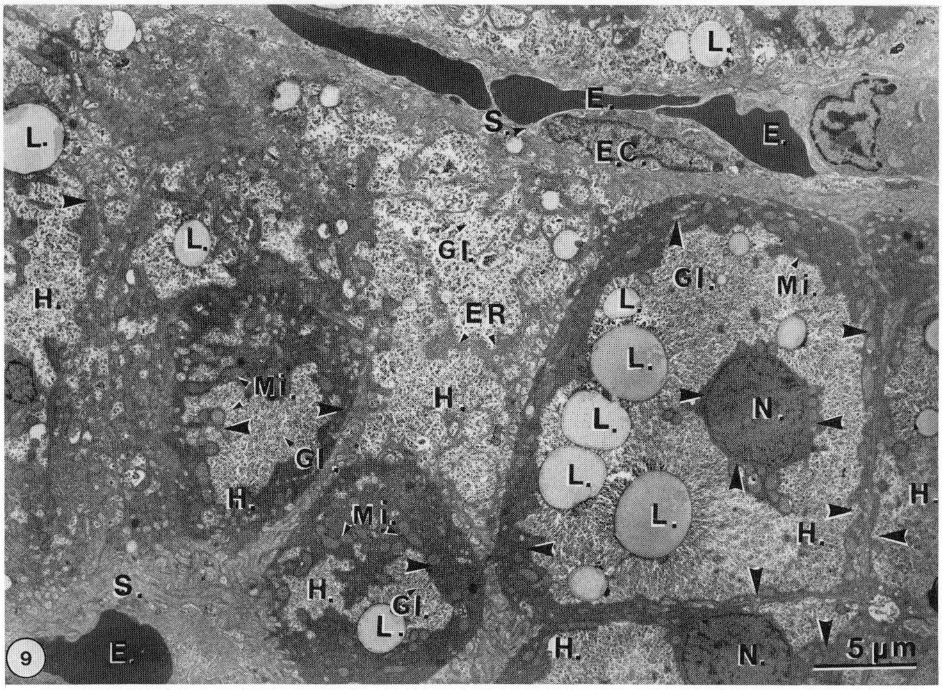

Fig. 9. Eel fed an artificial diet (2 months). A compartmentation of the hepatocytes with mitochondria (Mi) and endoplasmic reticulum (ER) concentrated around the nucleus and in the peripheral area ( ). Lower concentration in glycogen (Gl) but increasing amount of lipid droplets (L)

cortisol, glucose and lactate levels. "Ulcer eels" had crater-like lesions in the gastric epithelium. Moreover, gastric glands were smaller and mucous epithelium showed signs of erosion and degeneration. Stressed eels also showed an increase in the number, enlargement, and dysfunction of the chloride cells in the gill; these became necrotic and displayed a vacuolar degeneration (Peters and Hong, 1985).

Nutritional pathology. A cytological analysis of the liver of the European eel was carried out in our laboratory by Affandi (1986) and by Affandi and Biagianti (1987). The main ultrastructural modifications in the organization of hepatocyte of eels fed on an artificial diet (dough), when compared to wild samples (Fig. 8) were: a compartmentation of the cytoplasm, with both the mitochondria and the rough endoplasmic reticulum (RER) concentrated in the perinuclear region and the peripheral area (Fig. 9); a loss of stored glycogen but, in contrast, a greater accumulation of lipids (in the form of lipid droplets) (Figs. 9, 10); an irregular outline of the nucleus (Fig. 10); a vacuolization of mitochondria with few cristae (Fig. 11); and a deposition of collagen fibers both around the biliary canalicules and the Disse's space (Fig. 11).

These changes in the distribution of RER and mitochondria, together with the accumulation of lysosomes, are probably linked to nutritional imbalance, that is, an excess of lipids in the diet, especially of saturated fatty acids such as oleic, palmitic, and stearic acids. It seems to be simply a nonpathological response to the diet resulting from deviation in the metabolism of fatty acids. However, other structural

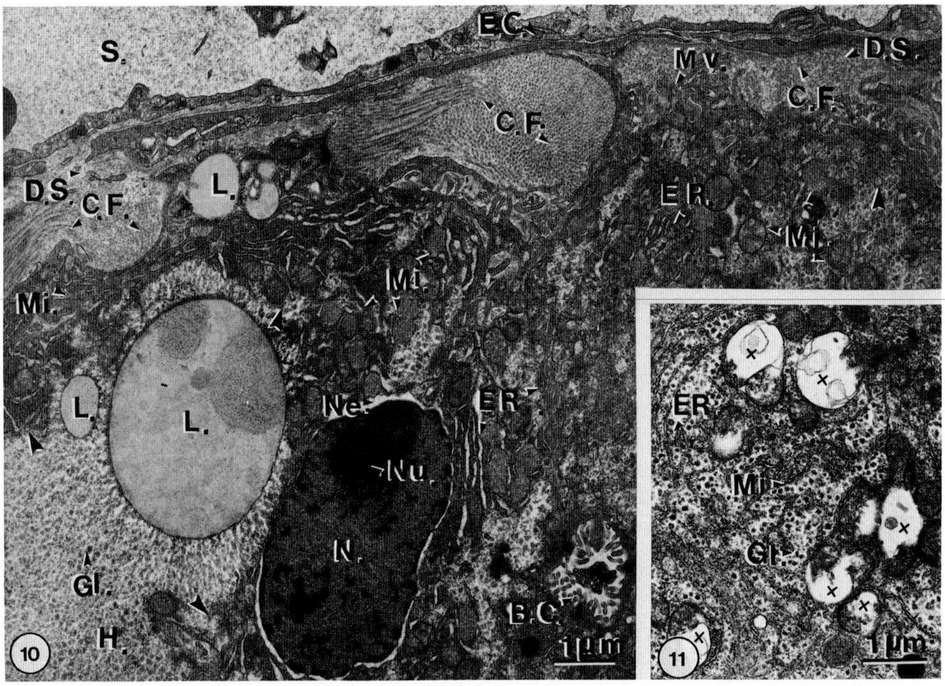

Fig. 10. Collagen fibers (CF) densely deposited in the Disse's space (DS).

Fig. 11. Vacuolization (x) of mitochondria (Mi).

modifications seem to reflect more serious liver disturbances: vacuolization of the mitochondria, irregular nuclear shape, accumulation of collagen fibers in the perinuclear and perisinusoidal areas (such as fibrosis and cholestasis), all of which correspond to a true hepatic dysfunction and nutritional pathology.

SOME CASES OF MULTIFACTORIAL DISEASES (OR MIXED ETIOLOGY)

The majority of observations deal with a single occurrence of diseased eels and each case was correlated with only one precise causative agent. The pathogenic effect of many infectious agents have not been clearly understood, mainly because of a lack of knowledge of the factors responsible for the pathogenic and lethal effects of parasites or infectious germs.

In addition to a small number of precise syndromes revealed in eel farming, it is noted that individual fish in different groups of eels showed a wide variability in susceptibility to infection and, consequently, in the nature and severity of the pathological response to pathogens (Ventura and Paperna, 1985). According to these authors, such intraspecific variability merely reflects differences in the physiological condition of individual eels. Moreover, in most observations, other epidemiological and etiological conditions were not mentioned, nor were environmental factors (e.g., water quality, food value, etc.) taken into account.

Table 2. Infectious diseases associated with stressors in eel culture.

Species	Infectious Agents	Additional Stress	Countries	Authors
Anguilla anguilla	*Saprolegnia* sp.	handling and grading	Denmark	Mellergaard and Dalsgaard, 1987
id.	*Aeromonas hydrophila*	grading and poor water quality	Denmark	id.
id. (elvers)	id.	feeding and non-cleaning tanks	Israel	Levanon et al., 1986
A. japonica	*Flexibacter columnaris*	feeding (formulated feed with white fish meal)	Japan	Sugimoto et al., 1981
A. anguilla (elvers)	*Myxidium giardi*	noxious dietary factors?	Italy	Ghittino, 1978
id.	*Vibrio anguillarum*	chemical pollution (copper)	Norway	Rodsaether et al., 1977
A. japonica	*Edwardsiella tarda* *Pseudomonas anguilliseptica*	id.	Japan	Mushiake et al., 1984
A. anguilla	Viruses	"stress condition"(?)	FRG	Moller and Anders, 1983
A. japonica	*Saprolegnia* sp.	*Aeromonas*	Japan	Egusa and Nishikawa, 1965
A. anguilla	*Myxidium giardi*	high stocking density and stress inherent in intensive culture	U.K.	Copland, 1981
A. japonica	*Chondrococcus columnaris*	mechanical effects of food particles on gill	Japan	Wakabayashi et al., 1970

A few precise cases of synergism between two causative agents, with additional effects, however, have been mentioned in the literature pertaining to eel pathology. These consist of fungal, bacterial, and viral diseases associated with stressors such as poor water quality, handling, and grading or feeding (Table 2).

Findings on species other than the eel (Table 3) strongly support the hypothesis of basic disorder of noninfectious origin; these are being considered as multifactorial disease complexes (Moller, 1985) in which nutritional and metabolic disorders are important factors (Snieszko, 1974; Poppe et al., 1986). The effect of environmental stress on the development of infectious diseases was demonstrated by Snieszko (1974). He demonstrated that these can have a "mixed etiological" basis.

Table 3. Multifactorial diseases in other species.

Species	Pathogenicity	Associated Stressors	Authors
Salmo salar	hemorrhagic syndrome	metabolic disorder vitamin B/selenium deficiency	Poppe et al., 1986
Epinephelus tauvina	bacteriosis (*Pseudomonas*)	pelleted diet of poor quality hepatocellular degeneration	Nash et al., 1987
Seriola quinqueradiata	streptococciosis	thawed sardine and anchovy→vitamin B deficiency	Taniguchi, 1983
id. *Pagrus major*	id.	environmental stress or nutritional disorder	Sano and Fukuda, 1987
Salmo salar	bacterial kidney disease	vitamins and minerals deficiencies in commercial diet	Paterson et al., 1981
Ictalurus punctatus	Edwardsiellosis	vitamin C deficiency	Durve and Lovell, 1982
Salmo gairdneri	lower bacterial (*Yersinia*) immune response	vitamin E deficiency	Blazer and Wolke, 1984
id.	bacteriosis (*Aeromonas*)	social stress	Peters et al., 1988

Several authors have emphasized that good nutrition is essential for resistance against opportunistic pathogens (Taniguchi, 1983; Sano and Fukuda, 1987, in *Seriola*; Patterson et al., 1981, in *Salmo*; Nash et al., 1987 in *Epinephelus*). Such diseases were correlated with a lower resistance of fish as a result of vitamin deficiencies (Blazer and Wolke, 1984) and the relationship between nutrition, immunology, and disease resistance in fish has received considerable attention. One of the most insidious effects of stress in aquaculture is indeed an increased susceptibility to diseases (Walters and Plumb, 1980). The presence of fish pathogens will not result in epizootics unless they are present in overwhelming numbers, or unless there are also unfavorable environmental conditions that have compromised the fish's defense system. Thus, infectious fish diseases are not caused by a single factor but are an outcome of the continuous interaction between the aquatic environment, the fish, and their pathogens.

Therefore, regarding the liver status of the eel investigated by Affandi (1986) and Affandi and Biagianti (1987), although the hepatocytes were not greatly affected by the artificial diet as there was limited cell damage, and even though the symptoms of nutritional pathology are relatively benign when compared with those described in other species (Mosconi-Bac, 1987), it is, nevertheless, likely that such hepatocyte changes correspond to feeding stress. This stress could be sufficient to favor secondary etiological effects and to induce a true syndrome in eel culture (Fig. 12).

Thus, eels do not die because of parasites and pathogens alone but owing to predisposing nutritional effects and also complications involving secondary infections by bacteria and viruses.

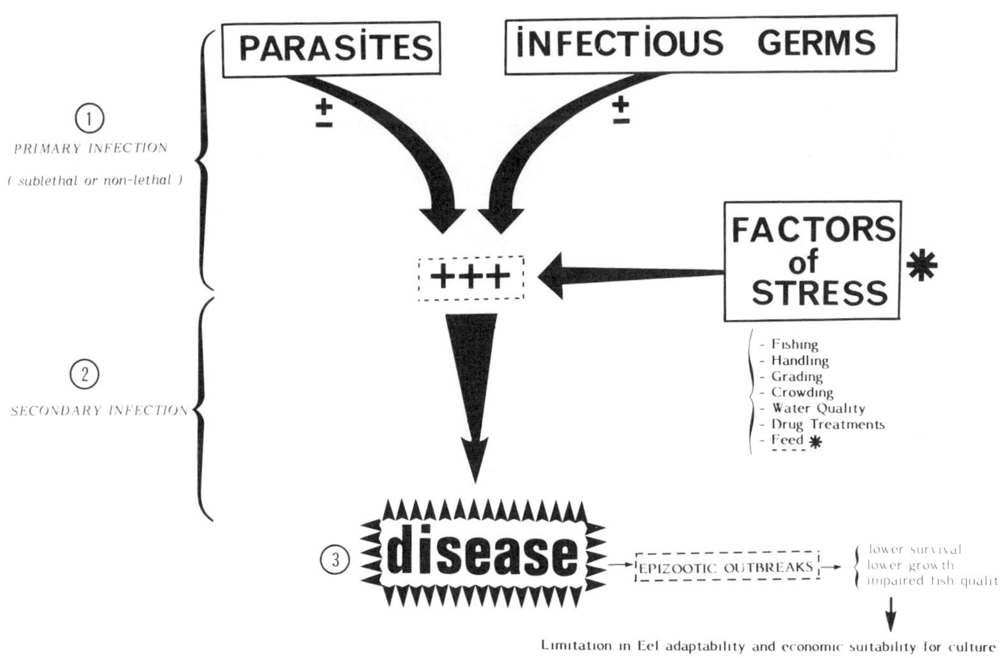

Fig. 12. Pattern of multifactorial diseases.

CONCLUSION

Although growth and survival could continue to be used as accurate measures of health in cultured eels and appear to be satisfactory at a first evaluation, a critical examination at the histocytopathological level appears to be of great interest.

The resolution of pathological problems revealed in eel as in other fish by the complexity of etiological and epidemiological factors has only just started. The malfunction of organs must be assessed at complementary levels: morphological damage, structural changes at the subcellular level, biochemical deviations, and physiological dysfunctions.

The primary task for future research is to clarify the etiology and epidemiology of many syndromes and to focus interest not on a single factor but on the environmental conditions of rearing as a whole. Histological and cytological data must also be monitored to evaluate all the organic disorders of cultured fish.

The preliminary findings on eel feeding draw attention to the nutritional status of the fish as reflected by structural morphology of its liver and the subcellular organization of its hepatocytes. Histological changes in the liver are indeed a good indicator of the quality of the environment in which the eel are reared. The liver seems promising not only for testing the nutritional status of the fish and the nutritional effectiveness of diets, but also for evaluating the general sanitary balance of the fish. Thus, hepatocytes can be used as monitor cells for the diagnosis of nutritional influences and dietary imbalance (Storch et al., 1983; Segner et al., 1987), which may be considered as primary factors of stress responsible for subsequent fish disease.

It would be worthwhile taking into account the following cytological parameters and main cellular changes in the hepatocytes:

Cell Parameters	Pathological Changes
Cellular size	shrinkage
Organelle distribution	intracellular compartmentation
Stored products	losses of glycogen
	accumulation of lipid droplets
Endoplasmic reticulum	disorganization into vesicles
Mitochondria	swelling and shortage of cristae
Lysosomes	increase in number

The increasing urgency of improving the quality of information on pathogenesis of the eel through accurate methodologies for assessing the detrimental effects of the causative agents must be emphasized. The success of the eel culture, at least partly, depends on the solution of such pathological problems.

LITERATURE CITED

Affandi, R. 1986. Etude biologique de l'anguille *Anguilla anguilla* en élevage: intérêt de l'ensilage acide dans l'alimentation. Thèse 3ème cycle, Université de Paris VI.

Affandi, R. and S. Biagianti. 1987. A study of the liver of eels kept in captivity: disturbances induced in hepatocytes by artificial diets. Aquaculture, 67, 226-228.

Aoki, T., T. Sakaguchi and T. Kitao. 1987. Multiple drug-resistant plasmids from *Ewardsiella tarda* in eel culture ponds. Nip. Suisan Gakkaishi, 53, 1821-1825.

Bejerano, F. 1985. Occurrence and control of eel diseases in Israel. Working party on eel, FAO-EIFAC Symp. Perpignan, France.

Blazer, V.S. and R.E. Wolke. 1984. The effects of tocopherol on the immune response and non-specific resistance factors of rainbow trout (*Salmo gairdneri*). Aquaculture, 37, 1-9.

Buchmann, K., S. Mellergaard and M. Køie. 1987. *Pseudodactylogyrus* infections in eel: a review. Dis. Aquat. Organ., 3, 51-57.

Chan, B. and B. Wu. 1984. Studies on the pathogenicity, biology and treatment of *Pseudodactylogyrus*. Acta Zool. Sinica, 30, 173-180.

Copland, J.W. 1981. The occurrence and distribution of *Myxidium giardi* in wild and cultured European eels, *Anguilla anguilla*, in England. J. Fish. Dis., 4, 231-242.

Durve, V.S. and R.T. Lovell. 1982. Vitamin C and disease resistance in channel catfish (*Ictalurus punctatus*). Can. J. Fish. Aquat. Sci., 39, 948- 951.

Egusa S. and T. Nishikawa. 1965. Studies of a primary infectious disease in the so-called fungus disease of eels. Bull. Japan Soc. Scient. Fish., 31, 812-813.

Ghittino, P. 1978. L'Ascite della cieca (*Anguilla anguilla*) d'allevamento da degenerazione lipoidea epatica. Piscicol. Ittiopatol., 13, 97-100.

Levanon, N., B. Motro, D. Levanon and G. Degani. 1986. The dynamics of *Aeromonas hydrophila* in the water of tanks used to nurse elvers of the European eel *Anguilla anguilla*. Bamidgeh, 38, 55-63.

Matsui, I. 1980. "Theory and Practice of Eel Culture." (A.A. Balkema, ed.). Rotterdam.

Mellergaard S. and I. Dalsgaard. 1987. Disease problems in Danish eel farms. Aquaculture, 67, 139-146.

Moller, H. 1985. A critical review on the role of pollution as a cause of fish diseases. In: "Fish and Shellfish Pathology" (A.E. Ellis, ed.), pp. 169-182. Academic Press, London.

Moller, H. and K. Anders. 1983. A compilation of viruses in fish from marine and brackish environments. Int. Coun. Expl. Sea, C.M. 1983/E:49.

Mosconi-Bac, N. 1987. Hepatic disturbances induced by an artificial feed in the Sea Bass (*Dicentrarchus labrax*) during the first year of life. Aquaculture, 67, 93-99.

Mushiake, K., K. Muroga and T. Nakai. 1984. Increased susceptibility of Japanese eel *Anguilla japonica* to *Edwardsiella tarda* and *Pseudomonas anguilliseptica* following exposure to copper. Bull. Japan. Soc. Scient. Fish., 50, 1797-1801.

Nash, G., I.G. Anderson, M. Sharrif and M. Nor Shamsudin. 1987. Bacteriosis associated with epizootic in the giant sea perch, *Lates calcarifer*, and the estuarine grouper, *Epinephelus tauvina*, cage cultured in Malaysia. Aquaculture, 67, 105-111.

Okamoto, N., T. Sano, R.P. Hedrick and J.L. Fryer. 1983. Antigenic relationships of selected strains of infectious pancreatic necrosis virus and European eel virus. J. Fish Dis., 6, 19-25.

Paterson, W.D., S.P. Lall and D. Desautels. 1981. Studies on bacterial kidney disease in Atlantic Salmon (*Salmo salar*) in Canada. Fish Pathol., 15, 283-292.

Peters, G. 1982. The effect of stress on the stomach of the European eel, *Anguilla anguilla* L. J. Fish. Biol., 21, 497-512.

Peters, G. and L.Q. Hong. 1985. Structure and blood electrolyte levels of European eels under stress. In: "Fish and Shellfish Pathology" (A.E. Ellis, ed.), pp. 183-198. Academic Press, London.

Peters, G., H. Delventhal and H. Klinger. 1981. Stress diagnosis for fish in intensive culture systems. Proc. World Symp. on aquaculture in heated effluents and recirculation systems, Stavanger 28-30 May, 1980, Vol. I, 239-248.

Peters, G., M. Faisal, T. Lang and I. Ahmed. 1988. Stress caused by social interaction and its effects on susceptibility to *Aeromonas hydrophila* infection in rainbow trout *Salmo gairdneri*. Dis. Aquat. Organ., 4, 83-89.

Poppe, T.T., T. Hastein, A. Froslie, N. Koppang, and G. Norheim. 1986. Nutritional aspects of haemorrhagic syndrome ("Hitra Disease") in farmed Atlantic salmon *Salmo salar*. Dis. Aquat. Organ., 1, 155-162.

Rickards, W. 1978. "A Diagnostic Manual of Eel Diseases Occuring Under Culture Conditions in Japan." Univ. North Carolina Sea Grant Publ., UNC-SG-78-06, 89 pp.

Rodsaether, M.C., J. Olafsen, J. Raa, K. Myhre and J.B. Steen. 1977. Copper as initiating factor of vibriosis (*Vibrio anguillarum*) in eel (*Anguilla anguilla*). J. Fish. Biol., 10, 17-21.

Sano, T. and H. Fukuda. 1987. Principal microbial disease of mariculture in Japan. Aquaculture, 67, 59-69.

Saroglia, M.G., P. Fantin and G. Arlati. 1985. Eel production in Italy—Problems and Perspectives. Working party on eel, FAO—EIFAC Symp. pp. 17-18. Perpignan, France.

Segner, H., P. Burkhardt, E.M. Avila, V. Storch and J.V. Juario. 1987. Effects of *Chlorella*-feeding on larval milkfish *Chanos chanos*, as evidenced by histological monitoring. Aquaculture, 67, 113-116.

Sindermann, C.J. 1983. An examination of some relationships between pollution and disease. Rapp. P. -v. Reun. Cons. Int. Explor. Mer, 182, 37- 43.

Snieszko, S.F. 1974. The effects of environmental stress on outbreaks of infectious diseases of fishes. J. Fish. Biol., 6, 197-208.

Stewart, D.J., K. Woldermariam, G. Dear and F.M. Mochaba. 1983. An outbreak of "Sekiten-byo" among cultured European eels, *Anguilla anguilla* L. in Scotland, J. Fish. Dis., 6, 75-76.

Storch, V., W. Stahlin and J.V. Juario. 1983. Effect of different diets on the ultrastructure of hepatocytes of *Chanos chanos* fry (Chanidae: Teleostei) an electron microscopic and morphometric analysis. Marine Biol., 74, 101-104.

Sugimoto, N., S. Kashiwagi and T. Matsuda. 1981. Pathogenetic relation between Columnaris disease in cultured eel and the formula feeds. Bull. Jap. Soc. Scient. Fish., 47, 719-725.

Szekely, C. and K. Molnar. 1987. Mebendazole is an efficacious drug against *Pseudodactylogyrosis* in the European eel (*Anguilla anguilla*) J. Appl. Ichtyol., 3, 183-186.

Taniguchi, M. 1983. Effects of the food quality on the appearance of Yellowtail streptococcicosis. Bull. Japan. Soc. Scient. Fish., 49, 363-366.

Ventura, M.T. and I. Paperna. 1985. Histopathology of *Ichthyophthirius multifiliis* infections in fishes. J. Fish. Biol., 27, 185-203.

Wakabayashi, H., K. Kira and S. Egusa. 1970. Studies on Columnaris disease of pond-cultured eels-II—The relation between gill disease and *Chondrococcus columnaris*. Bull Japan. Soc. Scient. Fish., 36, 678-685.

Walters, G.R. and J.A. Plumb. 1980. Environmental stress and bacterial infection in channel catfish, *Ictalurus punctatus*. J. Fish. Biol., 17, 177-185.

HEPATOCYTE DISTURBANCES INDUCED BY ARTIFICIAL FEED IN JUVENILE SEA BASS, *DICENTRARCHUS LABRAX*: ULTRASTRUCTURAL STUDY OF THEIR REVERSIBILITY

N. Mosconi-Bac

Marine Biology Laboratory, University of Perpignan,
Ave. de Villeneuve, F-66025 Perpignan Cedex, France

ABSTRACT: One major anomaly of hepatocytes of juvenile sea bass induced by artificial feed consists of large deposits of lipid droplets that are interpreted as reflecting a deviation in the metabolism of fatty acids. Two phenomena support the interpretation of the reversibility of such hepatocyte disturbances induced by artificial feed. Firstly, the number and size of lipid droplets decreased as did their affinity for osmium in hepatocytes of sea bass returned to natural feed. It would appear that in such cases fatty acids were not stored but metabolized when there were less lipids in the diet. On the other hand, rough endoplasmic reticulum is well developed. This may be associated in some yet undetermined way with the activation of lipid metabolism. In conclusion, it seems that hepatocyte disturbances induced by artificial feed represent a short-term reactive adaptation of the cell. When such a degenerative change is long term, it finally leads to liver necrosis.

INTRODUCTION

Earlier, we studied the effect of artificial feed on the fine structure of livers of juvenile sea bass, *Dicentrarchus labrax*, during their first five months of life. The results of that experiment revealed a precocious and rapid accumulation of lipids in the liver (Mosconi-Bac, 1987). The objective of this study was to ascertain whether hepatocyte disturbances induced by artificial feed are reversible.

MATERIALS AND METHODS

Breeding and sampling. Three-month-old sea bass that had been weaned were fed for 44 days on artificial feed and maintained in two separate tanks. Those in one tank continued to be fed on the same artificial feed, i.e., dry granular feed, whereas those in the second tank were fed on natural feed, i.e., crunched and frozen mullet flesh.

Prior to the beginning of the experiment, 4-month-old fish that had been fed on the artificial diet for 44 days constituted the first sample and the starting point of the experiment. At the time of each sampling, fish fed natural feed were compared to those still fed artificial feed, as indicated in Table 1.

Table 1. Sampling calendar.

	Tank 1	Tank 2
Starting Point Sample No. 1	44 days on artificial feed	
Sample No. 2	52 days on artificial feed	44 days on artificial feed + 8 days on natural feed
Sample No. 3	60 days on artificial feed	44 days on artificial feed + 16 days on natural feed
Sample No. 4	71 days on artificial feed	44 days on artificial feed + 27 days on natural feed
Sample No. 5	88 days on artificial feed	44 days on artificial feed + 44 days on natural feed
Sample No. 6	106 days on artificial feed	44 days on artificial feed + 62 days on natural feed

Electron microscopy. Liver samples were fixed by immersion in 3% glutaraldehyde buffered with 0.12 M cacodylate buffer for 4 hr at 4°C (pH 7.3); $CaCl_2$ was added according to Busson-Mabillot (1971). After post-fixation in 2% osmium tetroxide for 2 hr at 4°C, samples were dehydrated in a graded acetone series and embedded in Epon. Ultrathin sections were stained with uranyl acetate and lead citrate (Reynolds, 1973).

RESULTS

Artificial feed. During the experiment, lipid droplets in hepatocytes were large, numerous, and very osmiophilic. There was a large accumulation of lipid droplets even in some nuclei. In the last sample, after 106 days on artificial feed, lipid droplets

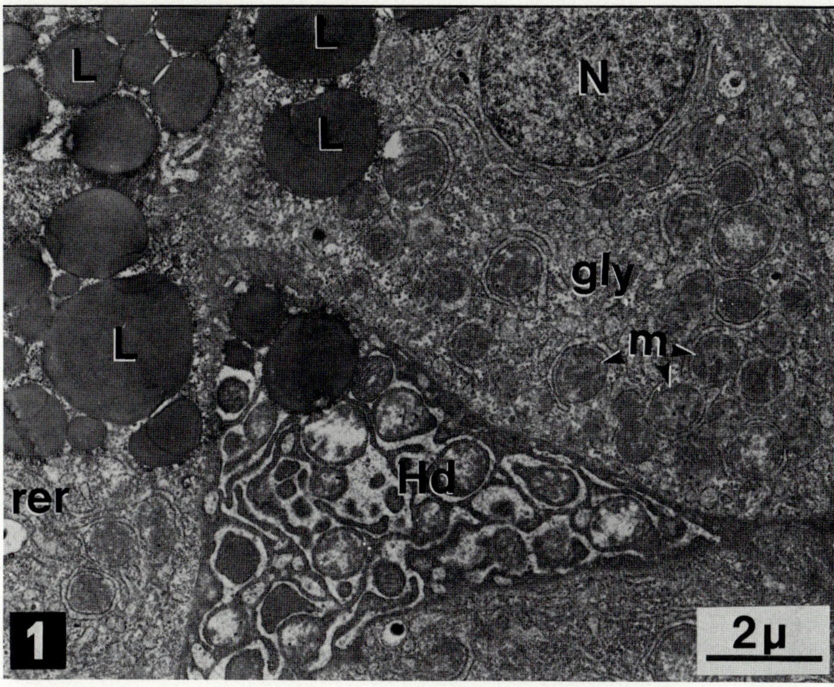

Fig. 1. Sample No. 6. 106 days on artificial feed. gly, glycogen; Hd, Hepatocyte in degenerescence; L, Lipid droplet; m, mitochondrion; N, Nucleus; rer, rough endoplasmic reticulum. 4,000X.

were still numerous and very osmiophilic and a large number of degenerating hepatocytes were present (Fig. 1).

Natural feed. Lipid droplets, large in size and number at the beginning of the experiment, were rapidly vacuolated. Subsequently, they became reduced in size and were of abnormal shapes and became less osmiophilic.

At the same time as the transformation of lipid droplets, well defined rough endoplasmic reticulum became visible.

At the end of the experiment, after 62 days on natural feed, the lipid droplets were few in number and were slightly osmiophilic. The rough endoplasmic reticulum was well developed (Fig. 2).

DISCUSSION AND CONCLUSION

This study confirms that artificial feed is responsible for a conspicious accumulation of lipid droplets in the hepatocytes of juvenile *D. labrax*. These droplets are similar to those in fish hepatocytes fed on a diet too rich in lipids (Storch and Juario, 1983). These investigators also noted an accumulation of lipid droplets in some nuclei of fish hepatocytes.

In our experiment, two phenomena support the interpretation of the reversibility of hepatocyte disturbances induced by artificial feed. Firstly, the number and size of lipid droplets decreased as did their affinity for osmium in hepatocytes of sea bass switched to natural feed. It is noted that Claude (1970) demonstrated that

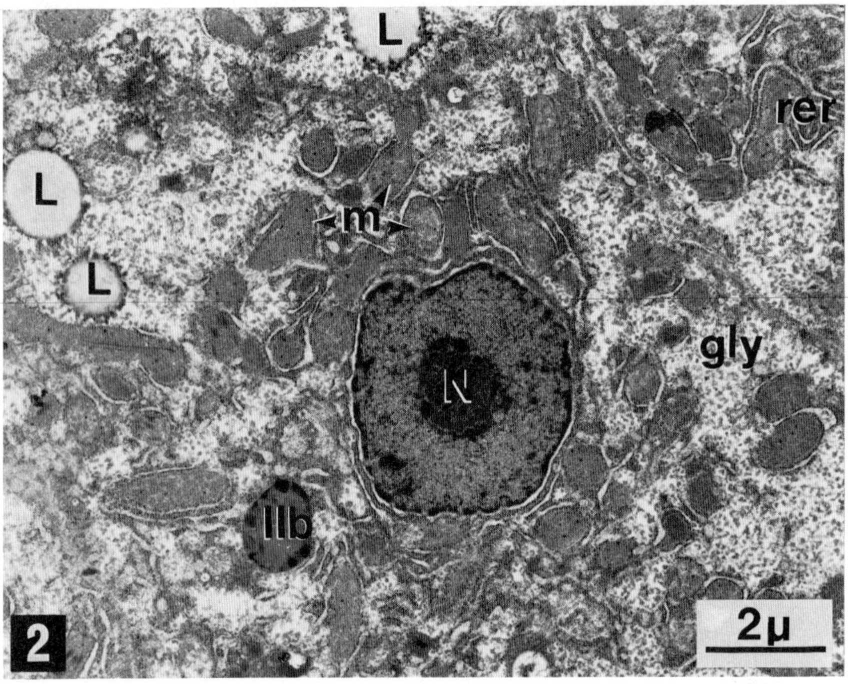

Fig. 2. Sample No. 6. 44 days on artificial feed and 62 days on natural feed. gly, glycogen; L, Lipid droplet; llb, lysosome-like-body; m, mitochondrion; N, Nucleus; rer, rough endoplasmic reticulum. 4,000X.

high osmiophilicity reflects an increased amount of unsaturated fatty acids. Thus, lipid droplets may represent unmetabolized fats. Therefore, as lipid droplets become less osmiophilic and reduced in size, it may mean that fatty acids are not stored but metabolized when there are less lipids in the diet.

On the other hand, rough endoplasmic reticulum presents an important development in hepatocytes of sea bass fed on natural feed. This may reflect elevated lipid metabolism (Stenger, 1970).

Hepatocyte disturbances induced by artificial feed have been interpreted in different ways by different authors. For example, Ghittino (1978) considers these disturbances to be a real nutritional pathology designated as "liver lipoid degeneration." Paperna et al. (1977) consider the fatty liver to be a temporary physiological condition as a result of an unbalanced diet and one that will return to normal when a balanced diet is resumed. Finally, Storch et al. (1984) proposed a compromise between the two interpretations presented. Specifically, they propose that in the short term fatty deposits in hepatocytes reflect a reactive adaptation of the cells to a new metabolic state. In the long term, however, progressive change occurs that finally leads to liver necrosis. We agree with this last interpretation. Thus, it would be interesting to determine how long the condition can persist in hepatocytes before it becomes irreversible.

LITERATURE CITED

Busson-Mabillot, S. 1971. Influence de la fixation chimique sur les ultrastructures. J. Microbiol., 12, 317-347.

Claude, A. 1970. Growth and differentiation of cytoplasmic membranes in the course of lipoprotein granule synthesis in the hepatic cell. J. Cell. Biol., 47, 745-766.

Ghittino, P. 1978. L'ascite de la cieca, *Anguilla anguilla*, d'allevamento da degenerazione lipoida epatica. Riv. Ital. Pisc. Ittiopatol., 13, 97-100.

Mosconi-Bac, N. 1987. Hepatic disturbances induced by artificial feed in the sea bass (*Dicentrarchus labrax*) during their first five months of life. Aquaculture, 67, 93-99.

Paperna, I., A. Colorni, H. Gordin and G.W. Kissil. 1977. Diseases of *Sparus aurata* in marine culture at Eilat. Aquaculture, 10, 195-213.

Reynolds, E.S. 1973. The use of lead citrate at high pH as an electron opaque stain in electron microscopy. J. Biophys. Biochem. Cytol., 17, 208- 212.

Stenger, R.J. 1970. Organelle pathology of the liver: the endoplasmic reticulum. Gastroenterology, 58, 554-574.

Storch, V. and J.V. Juario. 1983. The effect of starvation and subsequent feeding on the hepatocytes of *Chanos chanos* (Forsskal) fingerlings and fry. J. Fish Biol., 23, 95-103.

Storch, V., H. Segner, J.V. Juario and M.N. Duray. 1984. Influence of nutrition on the hepatocytes of *Chanos chanos* (Chanidae : Teleostei). Zool. Anz., 213, 151-160.

NUTRITIONAL PATHOLOGY: CONCLUDING REMARKS

Jacques Brusle

Laboratoire de Biologie Marine, Université de Perpignan, France

Everybody agrees that "good" nutrition is essential for disease resistance against opportunistic pathogens; the relationship between nutrition and immunology in fish has received considerable attention. It is becoming more evident that the mere presence of fish pathogens, unless present in overwhelming numbers, will not result in epizootics unless unfavorable environmental conditions, including food and feeding, also exist, that have compromised the fish's defense system.

Because food is one of the most critical factors that must be understood for a successful fish culture, and in view of the present state of knowledge, new patterns of research should be developed in which chemists, pharmacologists, physiologists, anatomists, cell biologists, and pathologists collaborate in the following:

(1) The use, not only of routine histological techniques, but also of fine cytological studies as a supplementary source of information to biochemical and physiological parameters and to statistical data on rearing.

(2) The determination of the effectiveness, sensitivity, and practicability of employing cytological methods in order to establish a standardization of the protocols and of the definition of pathological characters as morphologic proof, using a precise vocabulary in pathological descriptions.

(3) The development of experimental controls in laboratory conditions, taking into account the epidemiological and etiological factors, including "mixed etiology."

(4) The expression of results in quantitative terms and acquisition of more objective data through statistical values (surface area, volume, number and size of cell structures), because disease is not always associated with new different structures and functions but rather with quantitative alterations of existing structures and functions.

(5) An account on the sanitary history of the fish, differences between "facts" and "artefacts", and the potential reversibility of lesions.

Pathology in Marine Science

IMMUNOLOGY SECTION

THE PRESENT STATE OF IMMUNOLOGICAL RESEARCH IN MARINE AQUACULTURE

Katsuyoshi Mori

National Research Institute of Aquaculture,
Fisheries Agency, Nansei, Mie, 516-01 Japan

Renewal of interest in the immunity of marine invertebrates. Marine animals were the first to be studied regarding defense mechanisms in invertebrates. The works deserving special mention in this area are those pertaining to hemolysins, etc., in the horseshoe crab (Noguchi, 1903), the observation of phagocytosis by starfish larvae and *Daphnia* (Metchnikoff, 1884), and the experiment on autochthonous transplantation in pecten (Drew and de Morgan, 1910). The work of Noguchi predates that of Glaser (1918), one of the oldest reports on the humoral defense reaction of land insects, by 15 years. Needless to say, the observation carried out by Metchnikoff has been regarded as the origin of the theory on cellular immunity based on phagocytosis. Metchnikoff was deeply interested in Darwin's theory of evolution and suggested that phagocytes play a significant role in the defense mechanisms of all evolutionary stages of animals from unicellular animals to mammals.

The above mentioned facts indicate that marine invertebrates served fruitfully as experimentally materials for phylogenetic studies on the defense mechanisms in the Animal Kingdom from the late 19th century to the beginning of 20th. From then until the middle of the 1960s, however, these invertebrates were scarcely spotlighted in this field.

As described in the preface of "Immunity in Invertebrates" edited by Dr. M. Brehélin, the renewal of interest in the subject over the past few years is now well established, and invertebrates are no longer the poor relatives in immunological research. The main reasons for the renewal of interest are as follows (Brehélin, 1986): (1) there is hope of finding defense mechanisms in invertebrates that are new or of using phylogenetic information to explain phenomena in vertebrates that are still unaccounted for; and (2) the development of intensive mariculture requires better knowledge of the phenomena that govern whether crustaceans and molluscs resist or are killed by potentially pathogenic microorganisms.

Characteristics of the defense mechanisms in marine invertebrates and fish. In order that aquatic animals may maintain physiological homeostasis and structural characteristics and complete their growth and reproduction, coping with changes in the external environment and internal factors, it is necessary that various internal

Pathology in Marine Science

mechanisms regulating their physiological functions work smoothly. Among these mechanisms, the defense system occupies an important position along with the nervous and endocrine systems. Unnecessary or waste cells and substances are inactivated and removed from the body by means of the defense system.

The defense system in fish as well as in other vertebrates includes immunity in the narrow sense as one of its constituents. In invertebrates such as molluscs, crustaceans, and echinoderms, however, it has been reported that no immune phenomena based on the clonal selection theory in which the division and differentiation of antigen-specific clonal lymphocytes and the synthesis of antibody and sensitized lymphocytes are regarded as important. Migratory phagocytes, instead, take the initiative in supporting the defense system of invertebrates in which some humoral factors, such as lectins and lysozymes, actively participate. Such factors play less significant roles in the defense systems of higher vertebrates such as birds and mammals. Some of these factors are inducible not only in vertebrates but also in invertebrates after immunologic challenge. Enhancement of the baseline killing power in invertebrates for bacteria occurs very rapidly and is followed by no secondary response, whereas the acquired immune response of mammals usually entails a latent period of over 1 week and is accompanied by a typical secondary response.

Recently, a molecular approach to the elucidation of the roles of various humoral factors involved in pro-phenoloxidase activation (Wago, 1986) and coagulogen-mediated coagulation cascade (Iwanaga et al., 1986) is being made in relation to crustacean defense mechanisms. It is very likely that echinoderms possess a component similar to factor B, which functions in the alternative pathway of complement activation and C_3 receptor (D'Cruz et al., 1986). It has been suggested that the major histocompatibility complex-like substance is responsible for the determination of colony specificity in protochordate ascidians (Yokozawa et al., 1987).

As stated above, invertebrate defense mechanisms are considerably diverse and complex and have become elaborated during evolution, although immunoglobulins have not been found in any invertebrate. In the Cyclostomata, the lowest vertebrate animals, rosette formation and the occurrence of B lymphocytes in the hematopoietic organs has been demonstrated, together with the production of antibody; specifically, immunoglobulin M. No thymus or thymus-like organ has yet been observed. The classic complement pathway is absent or ineffective, while the alternative pathway functions well. Hence, it is presumed that in cyclostomes the role of antibody in defense mechanisms is not very important, whereas the role of the phagocytic system facilitated by the alternative pathway is very important (Fujii and Katagiri, 1986).

Vital reactions to a variety of nonself materials, governed by molecular mechanisms, have been analyzed in aquatic invertebrates and fish resulting in a partial understanding of the specificity and universality of each defense factor. The importance of the phagocytic system has consequently come to be realized. Phagocytes are capable of a broad range of nonself discrimination and active phagocytosis, contributing greatly to the maintenance of physiological homeostasis in the individual.

The defense mechanisms of the animals that we are studying in the area of aquaculture cover a wide range of constituent elements from various factors that traditional immunology almost never deals with to that part of immunity in the narrow sense that comprises the core of the subject.

Very new rationale for applying immunological research to marine aquaculture. As stated by Cheng (1983), the rationales for studying invertebrate immune mechanisms have been assigned a less prominent position in the broad spectrum of immunologic investigations, although they are scientifically valid. In part, no doubt, this is a reflection of anthropocentrism and, in part, because of the small number of investigators who have elected to concentrate on invertebrates. However, he emphasized that in recent years a new rationale has evolved, applying what has been discovered about cellular immunity in invertebrates, especially marine bivalve molluscs, to an urgent and practical problem, i.e., monitoring pollution in the marine environment. In view of this, it seems significant and timely to find a simple and precise method that can make the best use of cellular defense and associated factors as a practical means of assessing the physiological activity of marine bivalves not only in their natural habitats but also in hanging culture areas (Mori et al., 1980a,b, 1984).

Even though the failure of many pathological investigations to recognize specific pathogens tends to strengthen our "physiological stress theory," which has been reviewed by Sindermann (1979), it cannot be denied that bivalves are vulnerable to infections or diseases that possibly could lead to their death when physiologically inactive ones happen to encounter a bacterial or parasitic pathogen at its strong stage of pathogenicity. However, if bivalves are in their elevated stage of physiological activity or the virulence of the pathogen they encounter is weak even in the low stage of their physiological activity, the possibility that they suffer from infectious disease will decrease even in an infective environment.

In brief, the question is the relationship between the power of vital resistance of bivalves and the strength of virulence of causative organisms. Infection does not always lead to disease. It is necessary, therefore, to investigate seasonally and chronologically the relationship between the defensive capacity of bivalves and the pathogenicity of pathogens in the principal bivalve culture areas of the world. In order to make such important investigations feasible, it is preferable to establish molluscan tissue culture cells that contain a certain quantity of known defense factor and retain a fixed defensive capacity. At the same time, it is also desirable to culture the pathogen that contains a certain quantity of known antigen and show a fixed level of pathogenicity. If such tissue culture cells and pathogen are provided as the internationally common indicator cells and organism for test, it will become possible to carry out both fundamental and practical studies of molluscan mass mortalities on a worldwide scale (Mori, 1987).

In a certain marine environment, there should be bivalve molluscan species or race best fitted to it from the viewpoints of both vital resistance to disease and growth rate. For the purpose of selecting such a species or race, it is advisable to determine quantitatively the total defensive capacity of each bivalve mollusc in its culture area. Solutions such as how to modify the culture technique in reference to the total defensive capacity are also necessary. In addition, since even the same species or race of bivalves may possibly show different defensive capacities in different environments, it is hoped that cooperation among countries will become closer so that comparisons of the same species or race of bivalves from different culture areas of the world can be made. Such comparative studies, involving geneticists, would provide much useful information on the possibility of introducing a new species or race of bivalve mollusc from abroad.

Problems related to molluscan immunology. There are still many problems related to molluscan immunology awaiting solution. For example, the site where hemocytes are produced in bivalve molluscs still remains undetermined, although the generally accepted belief is that these cells arise from differentiation of connective tissue cells (Cheng, 1983). This has yet to be confirmed experimentally. Hemocytes produce humoral factors, but very little is known about the functions, biochemistry, and production sites of these products (Sminia and van der Knaap, 1987).

Recently, we separated and purified a lectin from the serum of the pearl oyster, *Pinctada fucata martensii,* by a combination of affinity chromatography on Sepharose 4B coupled with bovine submaxillary gland mucin, anion exchange chromatography on Mono Q, and gel filtration of Superose 6 (Suzuki and Mori, 1988). The purified lectin was shown to be homogeneous by polyacrylamide electrophoresis and rechromatography on Mono Q. It was approximately 440,000 in molecular weight, being composed of identical subunits with a molecular weight of approximately 20,000. D-galactose and N-acetylgalactosamine gave a 50% inhibition of agglutination of horse erythrocytes by it at 0.3 and 1.2 mM, respectively. The antibody obtained from rabbit immunized with it was monospecific to it as judged by the hemagglutination blocking test, immunoelectrophoresis, and immunoblotting. By using the purified lectin and the antibody prepared against it, we are now studying the physiological functions of the serum lectin of the pearl oyster as a part of our research on the self-nonself discrimination system of various bivalve molluscs.

In conclusion, we believe that as an applied interdisciplinary science, the field of immunology in the area of marine aquaculture should proceed either toward molecular biology, with regard to the immune responses, or toward disease control by performing artificial manipulation of immune responses.

LITERATURE CITED

Brehélin, M. 1986. Preface. In: "Immunity in Invertebrates" (M. Brehélin, ed.), pp. v-vi. Springer-Verlag, Berlin.

Cheng, T.C. 1983. Triggering of immunologic defense mechanisms of molluscan shellfish by biotic and abiotic challenge and its applications. Mar. Technol. Soc. J., 17, 18-25.

D'Cruz, O., R.A. Good and N.K. Day. 1986. Musekitsuidôbutsu no menekigenshô [Immunity in invertebrates]. In: "The Iwanami Immunology Series 4" (R.A. Good, ed.), pp. 1-57. Iwanami Shoten, Tokyo.

Drew, G.H. and W. de Morgan. 1910. The origin and formation of fibrous tissue produced as a reaction to injury in *Pecten maximum*, as a type of the Lamellibranchiata. Quart. J. Microsc. Sci., 55, 595-610.

Fujii, T. and C. Katagiri. 1986. Yatsumeunagirui no menekiôtôkei [Immune response in lampreys]. In: "The Iwanami Immunology Series 4" (R.A. Good, ed.), pp. 77-96. Iwanami Shoten, Tokyo.

Glaser, R.W. 1918. On the existence of immunity principles in insects. Psyche, 25, 39-46.

Iwanaga, S., T. Morita, T. Miyata, T. Nakamura and J. Aketagawa. 1986. The hemolymph coagulation system in invertebrate animals. J. Protein Chem., 5, 255-268.

Metchnikoff, E. 1984. Uber eine Sprosspilzkrankheit der Daphnien. Beitrag zur Lehre uber den Kamph der Phagocyten gegen Krankheitserreger. Virchow Arch. Pathol. Anat. Physiol. Klin. Med., 96, 177-195.

Mori, K. 1987. Oyster self-resistance mechanisms to diseases. In: "Aquaculture: Shellfish Culture Development and Management" (International Seminar held at La Rochelle, France, 1985), pp. 127-159. IFREMER, Brest, France.

Mori, K., S. Itsukaichi, K. Murayama and T. Nomura. 1980a. Activities of agglutinin and bactericidin in oyster tissues. Bull. Jap. Soc. Sci. Fish., 46, 1385-1389.

Mori, K., Y. Tone, T. Suzuki, K. Kasahara and T. Nomura. 1980b. Defense mechanisms of molluscs-I. Bactericidal and agglutinin activities in the scallop tissues. Bull. Jap. Soc. Sci. Fish., 46, 717-722.

Mori, K., K. Murayama, N. Kanno, M. Nakamura, E. Ohira, Y. Kato and T. Numura. 1984. Occurrence and characterization of the defense-factors in the Japanese oyster *Crassostrea gigas*. Tohoku J. Agr. Res., 35, 55-68.

Noguchi, H. 1903. A study of immunization – haemolysins, agglutinins, precipitins and coagulins in cold-blooded animals. Zentralbl. Bakt. Abt. Orig., 33, 352-362.

Sindermann, C.J. 1979. Oyster mortalities and their control. In: "Advances in Aquaculture" (FAO Technical Conference on Aquaculture, Kyoto, 1976), (T.V.R. Pillay and Wm. A. Dill, eds.), pp. 349-361. Fishing News Books Ltd., Farnham, Surrey, England.

Sminia, T. and W.P.W. van der Knaap. 1987. Cells and molecules in molluscan immunology. Dev. Comp. Immunol., 11, 17-28.

Suzuki, T. and K. Mori. 1988. A galactose-specific lectin from the hemolymph of the pearl oyster, *Pinctada fucata martensii*. Comp. Biochem. Physiol., in press.

Wago, H. 1986. Host defense mechanisms of invertebrates from the phylogenetic viewpoints. Protein, Nucl. Acid Enzy., 31, 886-903.

Yokozawa, H., K. Azumi and S. Ishii. 1987. Gensakudôbutsuhoyarui no seitaibôgyoinshi [Defense-factors in ascidians]. Seikagaku, 59, 564. (Abstract).

INFESTATION BY THE PENNELID COPEPOD *PHRIXOCEPHALUS CINCINNATUS* MODULATES CELL MEDIATED IMMUNE RESPONSES IN THE PACIFIC ARROWTOOTH FLOUNDER, *ATHERESTHES STOMIAS*

M. Faisal[1*], P.S. Perkins[2], and E.L. Cooper[2]

[1]Department of Avian and Aquatic Animal Medicine,
Faculty of Veterinary Medicine, University of Alexandria, Egypt
[2]Department of Anatomy and Cell Biology, School for Medicine,
University of California, Los Angeles, California 90024, USA

ABSTRACT: Leucocytes were obtained from the anterior kidney, spleen, and heart blood of the Pacific arrowtooth flounder. A clear redistribution of cytotoxic cells was observed after infection with *Phrixocephalus cincinnatus*. While anterior kidney leucocytes of control, noninfested fish possessed the greatest cytotoxicity, blood leucocytes in infested fish showed a significant elevation ($P < 0.001$). Infestation with *P. cincinnatus* also evoked alterations in mitogenic responses. In unilateral infestation, there was a significant increase in mitogenic activity of blood leucocytes to all mitogens. However, splenic and anterior kidney leucocytes showed suppressed mitogenic activity to PHA and Con A. Moreover, in hyperinfested fish (>2 copepods/fish) an overall suppression of all immune parameters except for LPS was observed. Our information reveals that aspects of cell mediated immunity are affected by parasitic infestations.

INTRODUCTION

Infestation of the Pacific arrowtooth flounder, *Atheresthes stomias*, by the pennelid copepod, *Phrixocephalus cincinnatus,* is common in the Pacific coastal areas of the United States and Canada (Kabata, 1984). The free-swimming juvenile female parasite completely enters the eye of the host. Once inside, it moves through the pupillary opening and courses along the lens perimeter until contacting the retina. A

*Present address: Virginia Institute of Marine Science, School of Marine Science, College of William and Mary, Gloucester Point, Virginia, 23062, USA.

developing holdfast system pierces the choroid layer resulting in the formation of small blood lacunae. Simultaneously, the posterior end of the parasite elongates and eventually ruptures through the eye. Once established, *P. cincinnatus* feeds on the host's blood (Kabata 1967, 1969). Although infestation by *P. cincinnatus* retards host growth and inflicts extensive damage to its eyes, which results in total blindness (Kabata and Forrester, 1974), no host mortalities have been recorded.

The effect of temporary and long term parasitism on the host immune system has been the subject of several recent studies (Damian, 1987; Morrison, 1987; Pavia, 1987). Infection by certain parasites apparently modulates host immune responses (Graves et al., 1985; Leid, 1987) and exacerbates the pathogenesis of parasitic diseases (Holmes, 1987). Because *P. cincinnatus* makes long-term, non-lethal contact with its host's blood, it provides a unique model to examine the modulatory role of parasitism in two innate cell-mediated immune responses: (1) non-specific cytotoxicity and (2) proliferative responses to the mitogens phytohemagglutinin, concanavalin A, and lipopolysaccharides derived from *Escherichia coli*.

MATERIALS AND METHODS

Fish. A total of 59 specimens of *Atheresthes stomias* were collected from shallow water off the California coast at Santa Barbara during the months of February and March, 1988. Fish with an average standard length of 17.2 cm ($\pm$3.1 cm) were selected for the study. The fish were transported alive to the laboratory where they were maintained for 1 week in a marine aquarium (Instant Ocean) at 15°C, and 11 mg O_2/liter. After anesthetization (200 mg/liter MS 222, Sandoz, Switzerland), the body cavity was opened, heparinized blood was obtained by cardiac puncture, and the spleen and anterior kidney were removed. The eyes were carefully opened and examined for the presence of developmental stages. Uninfected fish and specimens harboring only mature parasites were used.

Medium. All media and salt solutions were adjusted to marine teleost isotonicity by adding NaCl to 0.2 M as described for marine fish by Hinuma et al. (1980) as well as McKinney and Schmale (1988). Cell cultures were performed in a complete medium (CM) that consisted of RPM1-1640 (Gibco) supplemented with 10% heat inactivated fetal calf serum (FCS) and antibiotics (100 units penicillin, 100 μg streptomycin/ml). For cell washing, Hank's balanced salt solution (HBSS) was used.

Cell preparation. Heart blood was diluted three times with HBSS, overlaid on Histopaque 1077 (Sigma), centrifuged for 20 min at 450*g* (room temperature) and washed three times with HBSS. The anterior kidney and splenic tissues were gently pressed through 80-gauge mesh stainless steel sieves. The resulting cell suspension was further overlaid on Histopaque 1077, centrifuged, and cells at the interface were washed three times with HBSS and finally suspended in CM. Cell counts and cytospin preparations were performed by standard techniques.

Assessment of nonspecific cytotoxicity. Cytotoxicity was assessed by the ^{51}Chromium-release assay using the murine cell line YAC-1 (Kiessling et al., 1975) as the target. Chromation of the target cells was achieved by incubating 5 x 10^6 cells with

100 μCi of ^{51}Cr-Na_2CrO_4 for 60 min at 37°C. Following three washes, the target cells were diluted to 10^5 cells/ml in medium, and delivered (100 μl) to each of the wells of a 96-well, round bottom Linbro-plate (Flow Laboratories, McLean, Virginia). Effector cells were added in a volume of 100 μl at a ratio of 100:1, 50:1, 25:1 and 10:1. The plates were centrifuged (200*g*, 5 min, at room temperature), and incubated for 4 hr at 28°C. The incubation was terminated by centrifuging as before. One hundred μl of the supernatant fluid were removed from each well for counting in a Beckman gamma counter. Percent cytotoxicity was computed as:

$$100 \times \frac{\text{(experimental CPM - spontaneous CPM)}}{\text{(maximum release - spontaneous CPM)}}$$

and the data transformed into lytic units at 40% lysis (LU_{40}) per 10^6 cells with the aid of the computerized nonlinear equation program described by Pross et al. (1981).

Conjugate formation. Effector cells (1 x 10^4) were incubated with 1 x 10^5 YAC-1 target cells in 1 ml of CM in 12x75 mm glass tubes, sedimented by centrifugation at 200*g* for 5 min and incubated for 1 hr at 4°C. Pellets were resuspended, and cytocentrifuged smears were stained with Giemsa. Percentage of conjugate forming cells (CFCs) was determined by counting 200 leucocytes (bound and free) in triplicate samples.

Assessment of lymphocyte proliferation. Phytohemagglutinin (PHA, Difco Laboratories, Detroit, Michigan) (2 mg/ml), concanavalin A (Con A, Pharmacia, Sweden) (1 mg/ml), and lipopolysaccharide from *E. coli* (LPS, Difco) (4 mg/ml) were reconstituted as stock solutions in Hank's balanced salt solution (HBSS, Gibco). Isolated leucocytes(10^5/200 μl/well) were incubated in triplicate in 96-well plates (Flow) in the presence of mitogen (1 μg/ml PHA; 2.5 μg/ml Con A; 100 μg/ml LPS) for 4 days at 28°C. On day 3, the cultures were supplemented with 1 μCi/200 μl of ^3H-thymidine (5 Ci/mM) for 24 hr. The cells were harvested onto glass fiber filters, and the radioactivity assessed in a Beckman liquid scintillation counter, using Beckman Ready-Safe (Beckman, Fullerton, California) as scintillant.

Statistical analysis. Mean and standard deviation were calculated. The statistical analysis was performed using the Student t test.

RESULTS

Diagnosis. *P. cincinnatus* was identified according to the morphological characteristics described by Kabata (1967, 1969). We selected 27 fish with a single parasite in one eye (unilateral, UI), 17 specimens with a bilateral infestation (BI) harboring multiple parasites in both eyes, and 15 uninfected controls. In all infested fishes, the parasite was alive and mature as judged by the emergence of the genital segment from the eye. Furthermore, the eyes did not contain other developmental stages of *P. cincinnatus*.

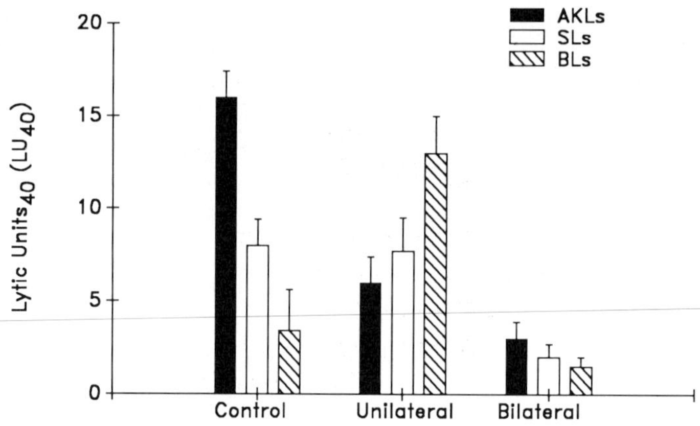

Fig. 1. Cytotoxic activity of *A. stomias* leucocytes in response to *P. cincinnatus* infestation. Data are expressed as means ± standard deviation. Abbreviation used: AKLs = anterior kidney leucocytes, SLs: splenic leucocytes, BLs: blood leucocytes.

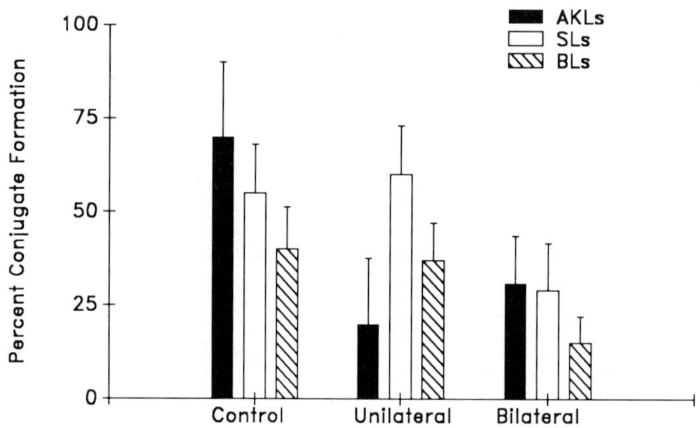

Fig. 2. Percentage of conjugate forming cells. Data are expressed as means ± standard deviation. Abbreviations as in Fig. 1.

Nonspecific cytotoxicity. YAC-1 cells were used as targets for *A. stomias* effectors. In noninfested flounders, the anterior kidney leucocytes (AKLs) exhibited the highest level of cytotoxicity (16 ± 1.4 LU_{40}), followed by spleen (8 ± 1.4 LU_{40}), and the blood (3.4 ± 2.2 LU_{40}). This pattern was altered in all infested fishes. Flounders with unilateral infestations displayed a suppressed activity in AKLs (6 ± 1.4 LU_{40}); however, blood lytic activity was markedly augmented (13 ± 2 LU_{40}) and spleen cytotoxic activity was unaffected (Fig. 1). These alterations in immunoreactivity are statistically significant ($P < 0.001$). As shown in Figure 2, the augmented cytotoxic activity of blood was not associated with a parallel increase in the percentage of conjugate forming cells. Fish with bilateral infestations exhibited significant ($P < 0.001$) suppression in all tissues.

Mitogenic response. Leucocytes from *A. stomias* responded to mitogen stimulation *in vitro*. Splenic leucocytes (SL) from control fish exhibited a marked proliferative response to stimulation by all three mitogens (Fig. 3). Anterior kidney leucocytes responded to a lesser extent to PHA and Con A, and responded poorly to LPS. Similarly, blood leucocytes exhibited a low level of responsiveness to mitogenic stimulation (Fig. 3).

Infestation with *P. cincinnatus* evoked alterations in e mitogen response patterns. In unilateral infestations, there was a significant ($P < 0.001$) increase in mitogenic activity by blood leucocytes to all mitogens, especially PHA. When exposed to Con A, SL and AKL activity was significantly suppressed ($P < 0.05$ and $P < 0.02$, respectively), but AKL response to LPS was slightly augmented ($P < 0.05$). Bilateral infestations were associated with a general suppression in response to PHA or Con A. However, exposure to LPS caused an augmentated response by AKLs and BLs ($P < 0.01$, $P < 0.001$, respectively).

DISCUSSION

Infestation by the copepod parasite, *P. cincinnatus* has a modulatory effect on host cell mediated immune parameters. This modulation was particularly evident in unilateral infestations. In this case, there was a redistribution of cytotoxic activity of host leucocytes such that blood leucocyte activity was greatly enhanced. Similar observations were made in channel catfish, *Ictalurus punctatus,* infected with the ciliate parasite, *Ichthyophthirius multifiliis* (Graves et al., 1985), and in the gilthead seabream, *Sparus auratus*, infected with lymphocystis virus (Faisal, 1989).

The redistribution of nonspecific cytotoxicity in response to parasitic invasion may be a defense mechanism by which the host attempts to counteract the pathogen during early stages of infection prior to its establishment in the host. Nonspecific cytotoxic cells are known to lyse cells expressing nonself antigens without prior exposure (Lanier and Phillips, 1986). The augmented blood NCC was surprisingly not associated with a parallel increase in the number of conjugate forming cells thereby excluding the possibility that the AKLs have migrated to the blood. Alternatively, the augmentation may be explained by an increase in the recycling capacity of cytotoxic cells (Graves et al., 1985). This pathway requires further investigation.

The immunocompetent cells from unilaterally infested *A. stomias* exhibited heterogeneity in their proliferative responses to mitogens. Spleen leucocytes and BLs, but not AKLs, displayed an augmented response to PHA, a known stimulator of T-cell-like leucocytes in fish (Caspi et al., 1984). In addition, AKLs and SLs, but not BLs, were suppressed when exposed to Con A, another T-cell stimulator (Sizemore et al., 1984); however, blood leucocyte activity was augmented in the presence of LPS, a stimulator of macrophages and B cells (Kaatari and Irwin, 1985).

The data from bilaterally infested fish reflects an overall suppression of immune function. Despite the absence of information on mortality induced by *P. cincinnatus*, it is likely that bilateral infestation is lethal. Besides total blindness and inability to capture prey (Kabata and Forrester, 1974), such fish are more vulnerable to invasion by other pathogens than unilateral or uninfested fishes. Sindermann (1987), as well as Dobson and May (1987), published excellent reviews concerning the impact of parasitic infection on marine fish populations. Our observed modulation of immune

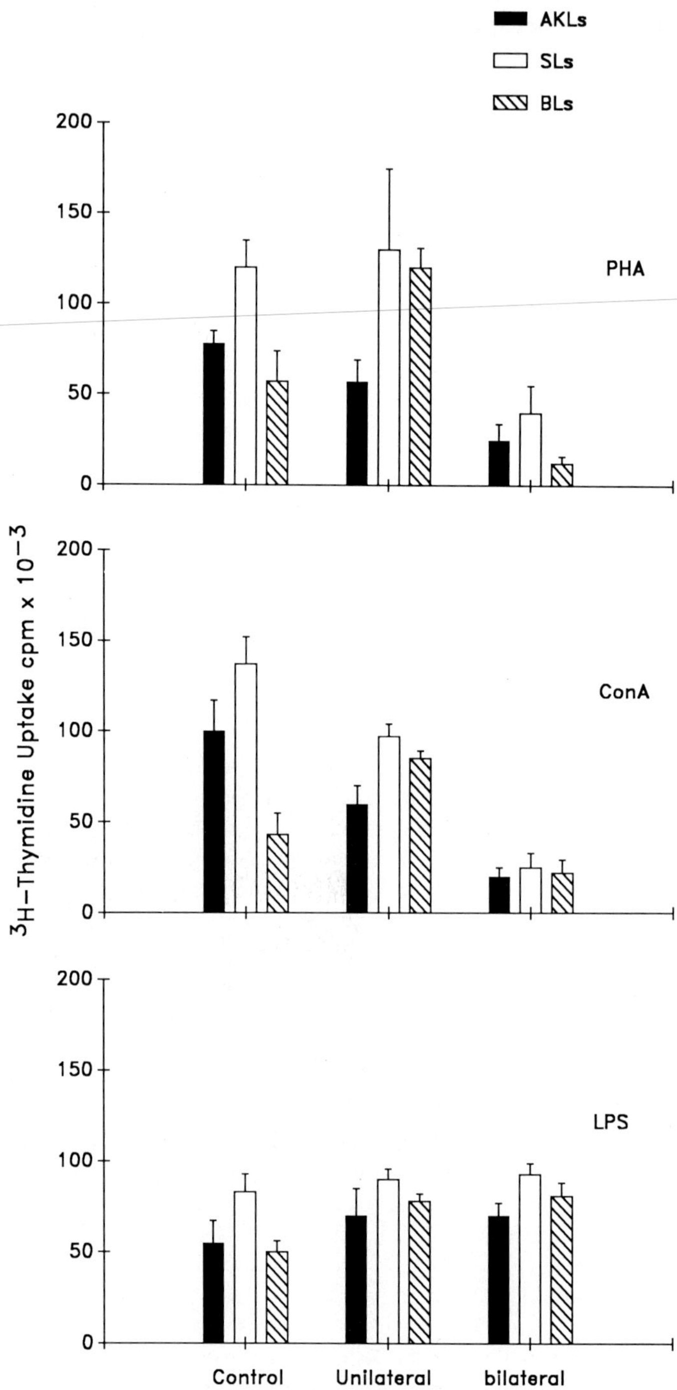

Figure 3. Proliferative responses to mitogens as measured by ^{3}H-thymidine uptake. Data are expressed as means ± standard deviation. Abbreviations as in Fig. 1.

parameters by parasitism expands this previous information and constitutes a new promising array of research questions in host-parasite relationships.

In summary, our data provides evidence for the presence of different leucocyte subsets in the hemopoeitic tissues of *A. stomias*, and that these subsets react differentially in response to prolonged parasitic infestation. Since recompartmentalization of different leucocyte subsets is a common phenomenon in mammals exposed to various pathogens, these observations are important from the phylogenetic standpoint. The exact nature of the reacting leucocyte subset(s) that responds to parasitic invasion deserves further clarification.

ACKNOWLEDGEMENTS

Dr. M. Faisal would like to thank the Fulbright Commission, Cairo, Egypt, for financial support.

LITERATURE CITED

Caspi, R.R., R. Shahrabani, T. Kehati-Dan and R.R. Avtalion. 1984. Heterogeneity of mitogen responsive lymphocytes in carp (*Cyprinus carpio*). Dev. Comp. Immunol., 8, 61-70.

Damian, R.T. 1987. The exploitation of host immune responses by parasites. J. Parasitol., 73, 3-13.

Dobson, A.P. and R.M. May. 1987. The effects of parasites on fish populations: theoretical aspects. Int. J. Parasitol., 17, 363-370.

Faisal, M. 1989. Lymphocystis disease alters innate cell-mediated immune responses in the gilthead seabream *Sparus auratus* L. In: "Viruses of Lower Vertebrates" (W. Ahne, ed.). Springer Verlag, Stuttgart, in press.

Graves, S.S., D.L. Evans and D.L. Dawe. 1985. Mobilization and activation of nonspecific cytotoxic cells (NCC) in the channel catfish (*Ictalurus punctatus*) infected with *Ichthyophthirius multifiliis*. Com. Immunol. Infect. Dis., 8, 43-51.

Hinuma, S., T. Abo, K. Kumagai and M. Hata. 1980. The potent activity of freshwater fish kidney cells in cell killing. I. Characterization and species distribution of cytotoxicity. Dev. Comp. Immunol., 4, 653-666.

Holmes, P.H. 1987. Pathophysiology of parasitic infections. Parasitology, 94 (Suppl.), 529-551.

Kaatari, S.L. and Irwin, M.J. 1985. Salmonid spleen and anterior kidney harbor populations of lymphocytes with different B cell repertoires. Dev. Comp. Immunol., 9, 433-444.

Kabata, Z. 1967. Morphology of *Phrixocephalus cincinnatus* Wilson (1908) (Copepod: Lernaeoceridae): J. Fish. Res. Bd. Canad., 24, 515-526.

Kabata, Z. 1969. *Phrixocephalus cincinnatus* Wilson 1908 (Copepoda: Lernqeoceridae): morphology, metamorphosis and host-parasite relationship. J. Fish. Res. Bd. Canad., 26, 921-934.

Kabata, Z. 1984. Diseases caused by metazoans: crustaceans. In: "Diseases of Marine Animals" IV (1), Introduction, Pisces (O. Kinne, ed.), pp. 321-381. Biologische Anstalt Helgoland, Hamburg.

Kabata, Z. and C.R. Forrester. 1974. *Atheresthes stomias* (Jordaqn and Gilbert 1880) (Pisces:Pleuronectiformes) and its eye parasite *Phrixocephalus cincinnatus* Wilson 1908 (Copepoda:Lernaeoceridae) in Canadian Pacific waters. J. Fish. Res. Bd. Canad., 31, 1589-1595.

Kiessling, R., E. Klein and H. Wigzell. 1975. Cytotoxic cells with specificity for mouse Moloney leukemia cells. Specificity and distribution according to genotype. Eur. J. Immunol., 5, 112-118.

Lanier, L.L. and J.H. Phillips. 1986. Evidence for three types of human cytotoxic lymphocyte. Immunology Today, 7, 132-134.

Leid, R.W., C.M. Suguet and L. Tanigoshi. 1987. Parasite defense mechanisms for evasion of host attack: a review. Vet. Parasitol., 25, 147-162.

McKinney, E.C. and M.C. Schmale. 1988. Proliferative and cytotoxic responses of healthy bicolor damselfish. Dev. Comp. Immunol., 12, 207-212.

Morrison, W.I. 1987. Host effector mechanism against parasites. Vet. Parasitol., 25, 163-176.

Pavia, C.S. 1987. Thymocyte-dependent immunity to toxoplasmosis in the normal and immunocompromised guinea-pig host. Parasite Immunol., 9, 205-218.

Pross, H.F., M.G. Baines, P. Rubin, P. Schragge and M.S. Patterson. 1981. Spontaneous human lymphocyte-mediated cytotoxicity against tumor target cells. IX. The quantitation of natural killer cell activity. J. Clin. Immunol., 1, 51-63.

Sindermann, C.J. 1987. Effects of parasites on fish populations: practical considerations. Int. J. Parasitol., 17, 371-382.

ULTRASTRUCTURE OF PHAGOCYTOSIS BY HEMOCYTES OF THE AMERICAN OYSTER

Gertrude W. Hinsch and Michele Hunte

Department of Biology, University of South Florida, Tampa, Florida 33620, USA

ABSTRACT: The electron microscopical study of oyster hemocytes has shown that electron dense particles within the cytoplasm are extruded from the cells. At the same time, latex beads are internalized and are found in membrane-bound vesicles within the cells. The vesicles appear to be primary phagosomes. Beads phagocytosed in funnel-shaped pseudopodium appear to change shape during the process. The primary phagosomes appear to be capable of fusing as seen by the presence of several beads within a single membrane.

INTRODUCTION

Phagocytosis by molluscan hemocytes is the major line of defense against invading microorganisms and foreign materials. The number of hemocytes identified by various authors varies considerably and appears in part to depend on the technique utilized in isolating them. Rifkin et al. (1969) described a single type of hemocyte, the hyalinocyte. Feng et al. (1971) described three types of hemocytes, Cheng et al. (1980) identified five subpopulations of hemocytes on the basis of separation on discontinuous gradient differential centrifugation and Hawkins and Howse (1982) identified four types of cells. The latter grouped three types into a granular or slightly granular type and the fourth group as granular. Renwrantz et al. (1979) classified hemocytes as hyalinocytes and three subpopulations of single hemocyte type within a developmental sequence. More recently, Bachere et al. (1988) classified eight fractions of hemocytes based on their diameters when separated by density gradient centrifugation and counterflow centrifugal elutriation.

Hemocytes adhere and become bound to surfaces. These form filopodia, which play an important role in phagocytosis (Bang, 1961). Foley and Cheng (1975) identified the granulocytes as the hemocytes involved in phagocytosis. They exposed hemocytes of *Crassostrea virginica* to bacteria and evaluated the role of the filopodia in phagocytosis. More recently, Fisher and Newell (1986) and Fisher and Tamplin (1988) have studied the effects of salinity while Cheng (1988a,b) tested various concentrations of metallic ions on hemocytes. Fisher (1988) reported on the binding of *Bonamia ostreae* and latex beads by hemocytes of oysters while Fisher (1988) and Fisher and Tamplin (1988) have described the adherence of beads and presumed their internalization.

Phagocytosis of polystyrene beads and the dynamics thereof have not been studied. Thus, the present work was undertaken to study phagocytosis of beads *in vitro* by hemocytes of *C. virginica* and to determine whether the polystyrene beads are internalized by means of scanning and transmission electron microscopy.

MATERIALS AND METHODS

Oysters, *C. virginica*, were obtained from a site in the upper Old Tampa Bay area and were maintained in a circulating artificial sea water system of 10-21°C with a salinity of 24 ‰. The hemocytes aspirated from the pericardial cavity were fixed for electron microscopy immediately upon collection or were permitted to settle onto a lawn of polystyrene beads either in culture flasks or on coverslips. After allowing time for phagocytosis, the hemocytes exposed to beads were released from the flask surface by trypsinization and fixed in 4% paraformaldehyde and 5% buffered gluteraldehyde, rinsed in buffer, and post-fixed with OsO_4. They were pelleted and the pellet dehydrated in a graded alcohol series and embedded in LR White resin. Sections were cut with a diamond knife, mounted on copper grids, stained with uranyl acetate and lead citrate, and viewed with an Hitachi EM 500 at 50kV.

Free hemocytes as well as those that were maintained on the coverslips were fixed as above, dehydrated in a graded acetone series, critical point dried, coated with gold-paladium, and viewed with a Zeiss Novascan.

RESULTS

Hemocytes collected from the pericardial cavity are round and exhibit few projections from their cell surface (Fig. 1a). Upon contact with the coverslip surface, they commence flattening out and produce numerous filopodia, some with knob-like endings (Fig. 1b). Those hemocytes exposed to polystrene beads can be seen to be in contact with beads as well as to appear to contain one to several beads within the cell (Fig. 2a,b).

Most of the hemocytes collected and studied appeared to be granulocytes with varying numbers of lysosomes within the cytoplasm (Fig. 3). Polystrene beads (Fig. 4a) were found in contact with the plasmalemma of the hemocytes (Fig. 4b) as well as in the process of being internalized (Fig. 4c). Several of the hemocytes had varying numbers of fully internalized polystrene beads within a membrane. The beads within the cytoplasm of the hemocytes present a reticular profile and are surrounded by a membrane. Some hemocytes with one or two internalized beads also exhibited lysosomes (Fig. 5a). Hemocytes with large numbers of beads, however, lacked such organelles. The primary phagosomes that contain the beads may fuse with each other so that several beads come to be surrounded by the same membrane (Fig. 5b).

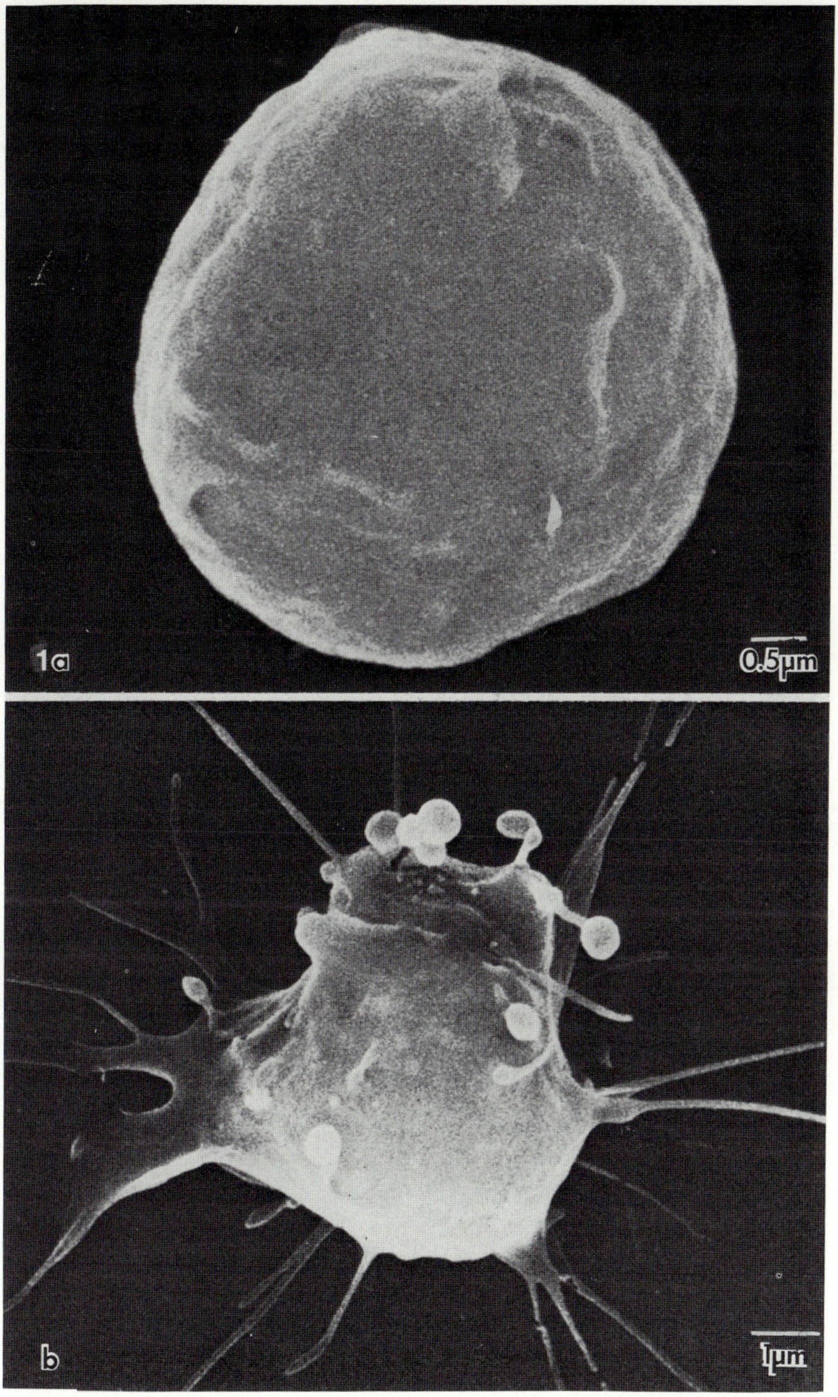

Fig. 1a. Scanning micrograph of a hemocyte taken from the pericardial cavity. b. Scanning micrograph of a hemocyte that has been allowed to spread out and attach to the surface of a coverslip. Note the bulblike tips of the filopodia.

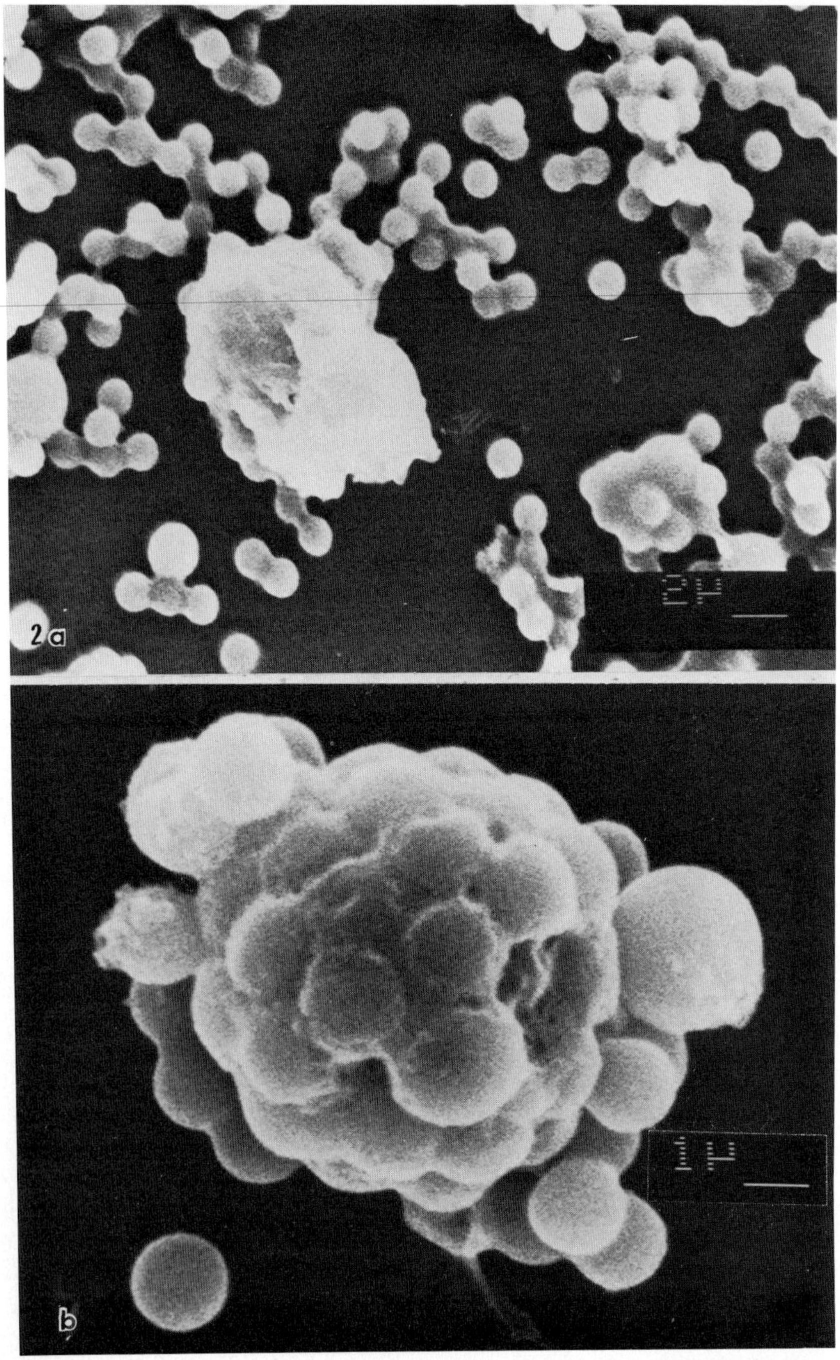

Fig. 2a. Scanning micrograph of hemocytes that have been placed on a coverslip covered with polystyrene beads. b. Scanning micrograph of hemocyte that has internalized several polystyrene beads.

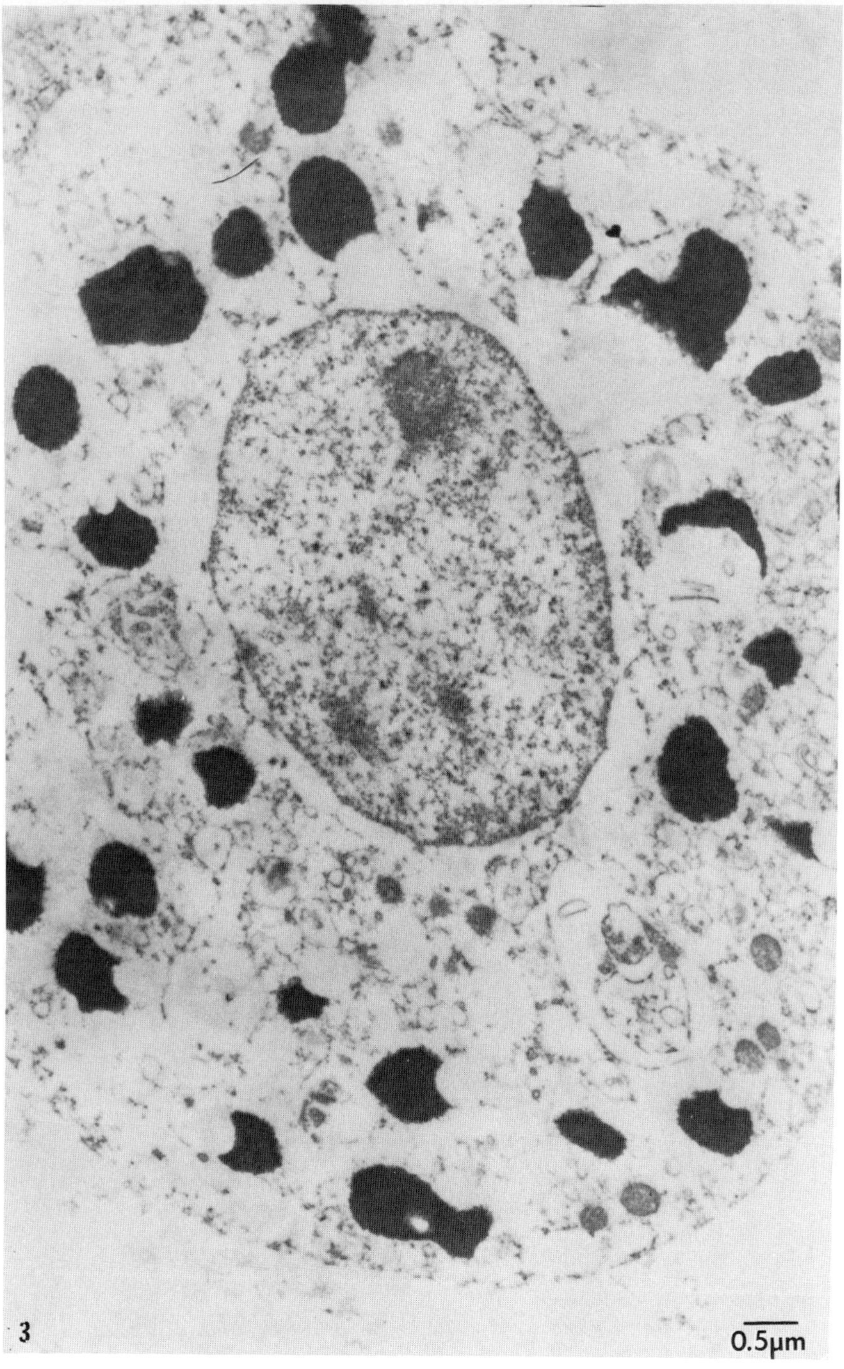

Fig. 3. Transmission micrograph of a hemocyte taken from the pericardial cavity. Note the numerous electron dense granules distributed throughout the cytoplasm.

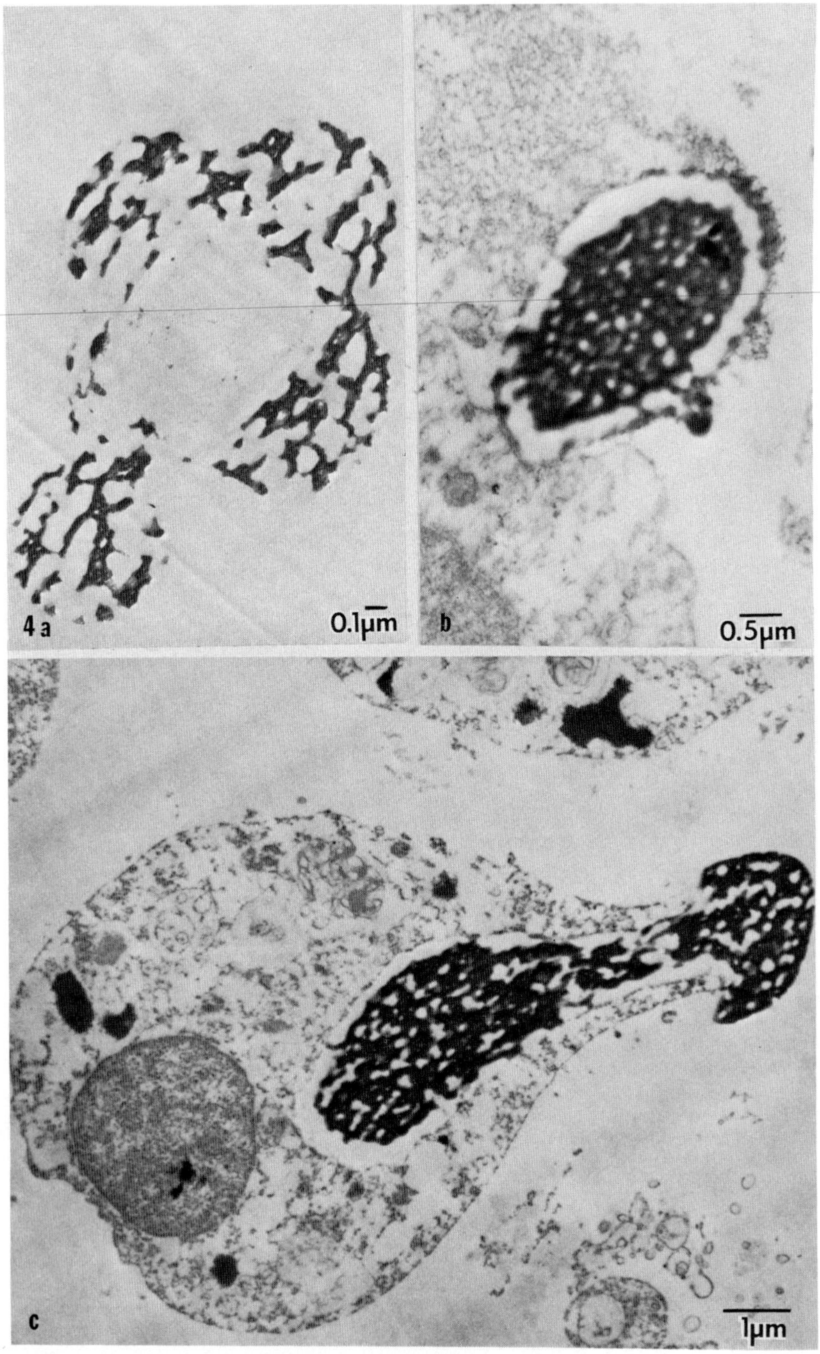

Fig. 4a. Sections through several polystyrene beads. b. A polystyrene bead in contact with the surface of a hemocyte. c. Thin section of a funnel-like pseudopod around a polystyrene bead in the process of internalization. Note the distortion of the bead. Note also that few electron dense granules persist in the cytoplasm of this hemocyte.

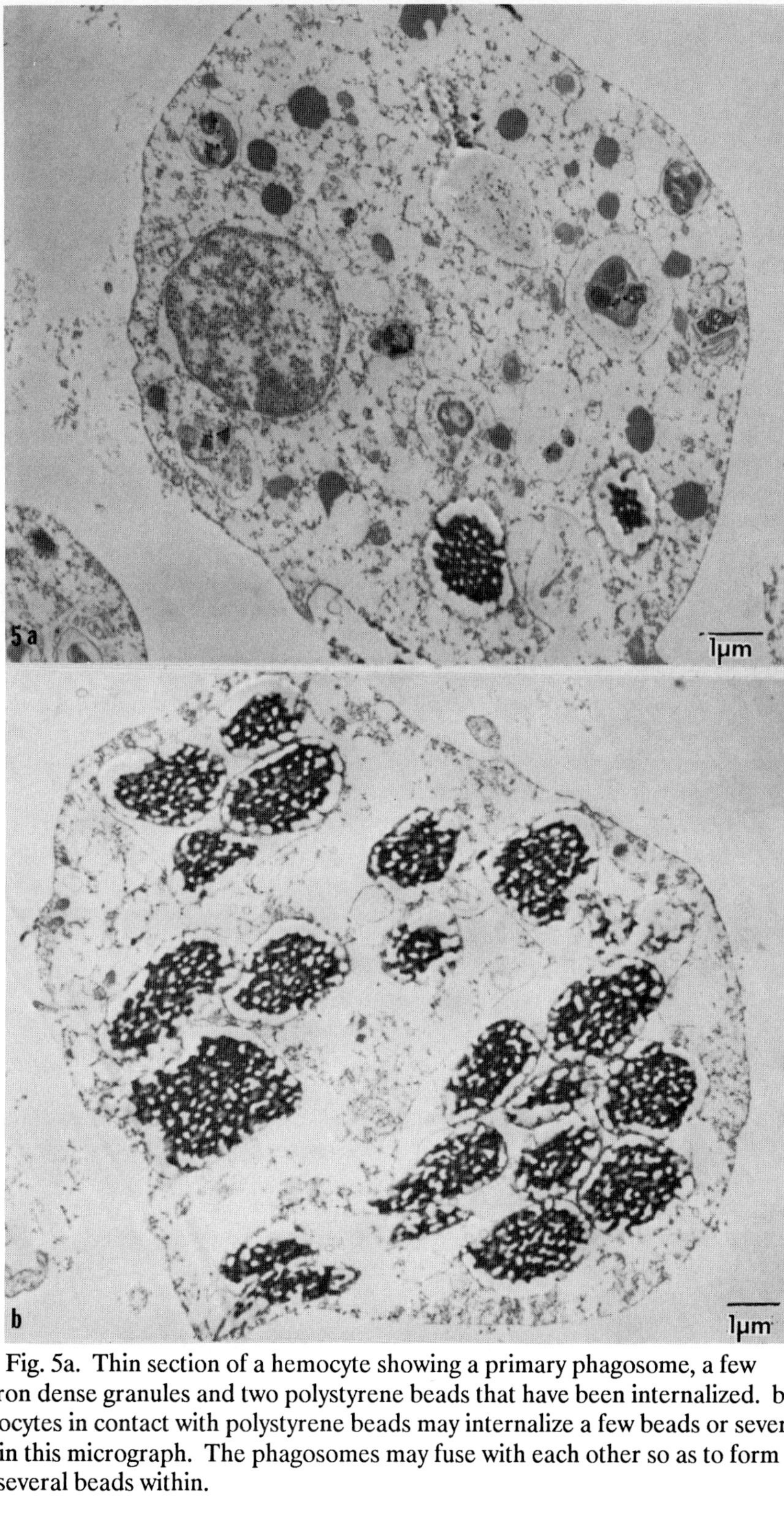

Fig. 5a. Thin section of a hemocyte showing a primary phagosome, a few electron dense granules and two polystyrene beads that have been internalized. b. Hemocytes in contact with polystyrene beads may internalize a few beads or several as seen in this micrograph. The phagosomes may fuse with each other so as to form one with several beads within.

DISCUSSION

Fisher (1988) reported the binding of latex beads to hemocytes of *Crassostrea gigas* and *Ostrea edulis*. In the present study, we have reported that not only do such beads bind to hemocytes but that they become internalized by phagocytosis. Cells may contain a single bead or as many as 22 beads within phagosomes. Each internalized bead within a phagosome is surrounded by a membrane. Phagosomes may fuse with each other and thus contain several beads.

Prior to and during the time of phagocytosis of the beads, the electron dense granules appear to be discharged to the outside of the cell with the end result that they are absent from the cells containing several beads. We have been unable to ascertain whether or not the contents are discharged directly to the environment or whether they are released from the surface surrounded by intact membranes as has been reported in *Mercenaria mercenaria* (Mohandas et al., 1985). Cheng (1975) considered degranulation, which occurs with phagocytosis, to be associated with the release of lysozyme enzymes. Leippe and Renwrantz (1988) reported that the release of molecules from hemocytes occurs when the hemocytes come into contact with a foreign surface. One must consider whether or not degranulation of oyster hemocytes is a necessary prerequisite for phagocytosis.

The disposition of the membranes surrounding these dense aggregates during degranulation is of interest when one considers the quantity of membrane that must be generated to surround the beads upon their phagocytosis. Oyster hemocytes that contain numbers of beads lack any concentration of cytoplasmic organelles. If one considers the quantity of membrane necessary to surround the beads internalized in the hemocytes, one would expect to see quantities of organelles usually associated with the synthesis of membrane precursors. Lacking these organelles, one must postulate the source of the membranes that surround the beads. One possible mechanism by which the membrane needed to surround the beads might be acquired would be from a recycling of the membranes retained after degranulation, which could become the wall of the phagosomes.

Cheng (1975) reported that primary degradation of bacteria occurred in the primary phagosome followed by transference to a secondary phagosome for further degradation. We have been unable to identify secondary phagosomes in this study.

C. virginica hemocytes have been reported to internalize rat erythrocytes (Renwrantz et al. 1979) or lectin-treated gastropod hemocytes (Schoenberg and Cheng, 1980) through a single funnel-shaped pseudopodium. Such pseudopodia have been seen incorporating beads into the hemocytes of the oysters in this study as well. Such incorporation implies involvement of the cytoskeleton. Indeed, Alvarez et al. (1989) showed inhibition of phagocytosis by cytochalasin B. The means by which this incorporation occurs is unknown. The pseudopodia are apparent only if the funnel is extended in the plane of the substrate (Schoenberg and Cheng, 1980). Renwrantz et al. (1979) and Schoenberg and Cheng (1980) believe that direct contact between the hemocyte and object to be incorporated is unnecessary. Renwrantz et al. (1979) suggested that direct chemotactic stimulus might be involved. In the case of the polystyrene beads, we cannot assume such a chemotactic mechanism.

ACKNOWLEDGEMENTS

This work is a result of research sponsored by NOAA, Office of Sea Grant, Department of Commerce, under Grant Number NA86AA-D-SG068. The U.S. Government is authorized to produce and distribute reprints for governmental purposes notwithstanding any copyright that may appear hereon.

This article was developed under the auspices of the Florida Sea Grant College Program grant no. IR-88-2.

LITERATURE CITED

Alvarez, M.R., F.E. Friedl, J.S. Johnson and G.W. Hinsch. 1989. Factors affecting *in vitro* phagocytosis by oyster hemocytes. J. Invert. Pathol., in press.

Bachere, E., D. Chagot and H. Grizel. 1988. Separation of *Crassostrea gigas* hemocytes by density gradient centrifugation and counterflow centrifugal elutriation. Devel. Comp. Immunol., 12, 549-559.

Bang, F.B. 1961. Reaction to injury in the oyster (*Crassostrea virginica*). Biol. Bull., 121, 57-68.

Cheng, T.C. 1975. Functional morphology and biochemistry of molluscan phagocytes. Ann. N.Y. Acad. Sci., 266, 343-379.

Cheng, T.C. 1988a. *In vivo* effects of heavy metals on cellular defense mechnisms of *Crassostrea virginica*: Total and differential cell counts. J. Invert. Pathol., 51, 207-214.

Cheng, T.C. 1988b. *In vivo* effects of heavy metals on cellular defense mechanisms of *Crassostrea virginica*: Phagocytic and endocytotic indices. J. Invert. Pathol., 51, 215-220.

Cheng, T.C., J.W. Huang, H.Karadogan, L.R. Renwrantz and T.P. Yoshino. 1980. Separation of oyster hemocytes by density gradient centrifugation and identification of their surface receptors. J. Invert. Pathol., 36, 35-40.

Feng, S.Y., J.S. Feng, C.N. Burke and L.H. Khairallah. 1971. Light and electron microscopy of the leucocytes of *Crassostrea virginica* (Mollusca: Pelecypoda). Z. Zellforsch., 120, 222-245.

Fisher, W.S. 1988. *In vitro* binding of parasites (*Bonamia ostreae*) and latex particles by hemocytes of susceptible and insusceptible oysters. Dev. Comp. Immunol., 12, 43-53.

Fisher, W.S. and R.I. Newell. 1986. Salinity effects on the activity of granular hemocytes of American oysters, *Crassostrea virginica*. Biol. Bull., 170, 122-134.

Fisher, W.S. and M. Tamplin. 1988. Environmental influence on activities and foreign-particle binding by hemocytes of American oysters, *Crassostrea virginica*. Can. J. Fish. Aquat. Sci., 45, 1309-1315.

Foley, D.A. and T.C. Cheng. 1975. A quantitative study of phagocytosis by hemolymph cells of the Pelecypods *Crassostrea virginica* and *Mercenaria mercenaria*. J. Invert. Pathol., 25, 189-197.

Hawkins, W.E. and H.D. Howse. 1982. Ultrastructure of cardiac hemocytes and related cells in the oyster (*Crassostrea virginica*). Trans. Am. Micros. Soc., 101, 241-252.

Leippe, M. and L. Renwrantz. 1988. Release of cytotoxic and agglutinating molecules by *Mytilus* hemocytes. Dev. Comp. Immunol., 12, 297-308.

Mohandas, A., T.C. Cheng and J.B. Cheng. 1985. Mechanism of lysosomal enzyme release from *Mercenaria mercenaria* granulocytes: a scanning electron microscopy study. J. Invert. Pathol., 46, 189-197.

Renwrantz, L.R., T. Yoshino, T. Cheng and K. Auld. 1979. Size determination of leucocytes from the American oyster, *Crassostrea virginica* and description of a phagocytosis mechanism. Zool. J. Physiol., 83, 1-12.

Rifkin, E., T.C. Cheng and H.R. Hohl. 1969. An electron microscope study of the constitutents of encapsulating cysts in *Crassostrea virginica* formed in response to *Tylocephalum* metacestodes. J. Pathol., 14, 211-226.

Schoenberg, D.A. and T.C. Cheng. 1980. Phagocytic funnel-like pseudopodia in lectin-treated gastropod hemocytes. J. Invert. Pathol., 36, 141-143.

CYTOMETRIC STUDIES ON *MERCENARIA* HEMOCYTES[1]

Frank E. Friedl and Marvin R. Alvarez

Department of Biology, University of South Florida, Tampa, Florida 33620, USA

ABSTRACT: Hemocytes of two species of *Mercenaria, M. campechiensis* and *M. mercenaria*, were compared relative to size, total cellular protein, and nuclear DNA content by employing flow cytometry. No significant differences were found in these characteristics. Cytometric profiles of nuclear DNA appear superposable for *Mercenaria* and easily distinguished from *Crassostrea, Lymnaea,* and the guinea fowl, *Numida.*

INTRODUCTION

Hemolymph of bivalve molluscs contains large numbers of suspended multifunctional cells. They generally are considered to be major factors in cellular defense mechanisms of these animals, performing such diverse functions as phagocytosis of nonself materials, encapsulation of foreign bodies, and nacrezation (Stauber, 1950; Tripp, 1960; Mohandas et al., 1985). Their ability to digest foreign cellular materials with an accompanying synthesis of glycogen (Cheng and Cali, 1974), their mobility, and their assumed association with digestive organs as tissue residents, also suggest intimate relationships with intracellular digestion and other nutritional processes. Although the extent has not been established, it seems inescapable that they play primary roles in bio-accumulation and depuration.

---

[1]This article was developed under the auspices of the Florida Sea Grant College Program with support from the National Oceanic and Atmospheric Administration, Office of Sea Grant, U.S. Department of Commerce, Grant Number NA86AA-D-SG068.

Because hemocytes are so functionally significant, particularly in their abilities to recognize self from nonself, their cytological characterization as part of a primitive, but basic, immune system assumes the same importance as similar studies in higher animals. Such information can lead to an understanding of their hemopoietic origins and ontogeny, and perhaps to an element of control or optimization of cellular immunity in bivalves of commercial importance.

Hemolymph cells of the hard clam, *Mercenaria mercenaria*, have been reported to be of three types: granulocytes, fibrocytes, and hyalinocytes (Foley and Cheng, 1974); however, subsequent electron microscopical investigation by these workers led them to conclude that only two types actually existed, since fibrocytes appeared to be "degranulated" granulocytes (Cheng and Foley, 1975).

We feel that it is additionally important to know the characteristics of these cells in their rounded, suspended forms, as similar as possible to the way they actually exist in the hemolymph of the clam. Accordingly, we have investigated some cytometric parameters of hemocytes and their nuclei from two kinds of *Mercenaria* (a close Florida relative or counterpart of the northern quahog, *M. mercenaria*, and the southern quahog or hard clam, *M. campechiensis*) using flow cytometric techniques. These procedures have the advantage of permitting the analysis of thousands of cells per minute, thus enabling large samples of a cell population to be evaluated. Specifically, we were interested in investigating size, total protein content (as a correlate of size), and nuclear deoxyribonucleic acid (DNA) content (as related to cell division and polyploidy).

MATERIALS AND METHODS

Experimental animals. *Mercenaria campechiensis*, the southern quahog or hard clam, was obtained from a site near the mouth of Tampa Bay on the Gulf coast of Florida. Specimens of *Mercenaria mercenaria* were collected on the east coast of Florida, near Melbourne[2]. *M. mercenaria* is better known and has superior storage qualities after harvesting, a shortcoming that apparently limits the commercial use of *M. campechiensis* (Menzel, 1968a). The animals were maintained in a circulating, filtered artificial sea water system at a temperature of 22°C and a salinity of 26 ‰.

Cell sources. After drilling a hole with a carbide-tipped bit, cells were removed by aspirating pericardial hemolymph into a plastic syringe with a 16-gauge needle. The cells were processed immediately. Both species typically yielded 1.5×10^6 cells/ml of hemolymph by hemocytometer count.

---

[2]On the Florida east coast *Mercenaria mercenaria* and *M. campechiensis* can be found in the same regions and can hybridize (Menzel, 1968a). However, according to Arnold and Bert (1988, pers. commun.), the *Mercenaria* from the Florida east coast that were selected for our use in this investigation are phylogenetically much more closely related to the northern *M. mercenaria* than to *M. campechiensis* from the west Florida coast. Because of this we refer to the east coast form as *Mercenaria mercenaria* throughout this article.

Size analyses. For size determinations, cells in hemolymph were immediately diluted 1:1 with 3.7% formaldehyde or 2.5% glutaraldehyde in saline. Several methods of preparation were tested to ascertain that sizes were not unduly influenced by handling procedures (see also, Friedl et al., 1988, on oysters). Hemocyte sizes were determined electronically using a Coulter counter and Channelizer (Model C-256) with appropriate standardization.

Total cellular protein. To determine total protein, fixed, pelleted cells were suspended in a dansyl chloride solution (1 mg/ml in acetone), and subsequently repelleted and suspended in water. As necessary, such preparations were passed through a 26-gauge needle to remove clumps, and microscopically examined for quality. Dansylated cell fluorescence signals were detected and measured using a single parameter flow cytometer equipped with a microcomputer-based pulse height analyzer. Chemical analyses of protein in centrifuged samples of known numbers of cells were made using the method of Lowry et al. (1951).

Nuclear DNA content. Nuclei were isolated to determine DNA content. Freshly aspirated cells were allowed to settle and attach to the bottoms of 25-ml glass beakers (20 min). The fluid was then withdrawn, and the cells washed with 0.54 M NaCl. This was decanted, and a nuclear isolation medium containing the detergent NP-40 and the DNA specific fluorochrome DAPI (4′,6-diamidino-2-phenylindole) was added (Thornthwaite et al., 1980). This treatment simultaneously isolates and stains nuclei. The cell layer was scraped off, and the nuclear suspension diluted to give a concentration suitable for analysis. Clumped nuclei were separated by pulling the sample through a 26-gauge needle, and the quality of the preparation ascertained by direct observation with an epi-illuminated fluorescence microscope.

Flow cytometric analyses of nuclear DNA content as indicated by DAPI fluorescence were done using a single parameter instrument with a microcomputer-based pulse height analyzer.

Display of cytometric data. Data analysis was done using a utility program developed in our laboratory to supplement commercially available software. Flow cytometric and cell size frequency distribution histograms are displayed as point plots of frequency vs. channel number (signal intensity) with curves illustrating general tendencies generated by smoothing these data using the "Lowess" procedure (Robust Locally Weighted Regression) of Cleveland and McGill (1985).

RESULTS

Characteristics of *M. campechiensis* hemocytes.

Size. Table 1 shows the results of several experiments where electronic size was determined following various preparative methods. Using these data, the mean median diameter for *M. campechiensis* hemocytes was found to be 6.86 μm. The methods of preparation employed do not appear to introduce important artifactual size differences. Figure 1 shows the size distribution found for 2,396 glutaraldehyde-fixed cells. The distribution is unimodal at channel 43, with the median hemocyte size of 6.73 μm found in channel 54. For reference, the size in channel 256 would be 10.91 μm.

Table 1. Electronic size measurements of *Mercenaria* hemocytes (*M. campechiensis*).

Median Cell Diameter (in μm)	Cell Preparation Method
6.548	3.7% Formaldehyde/hemolymph 1:1
6.392	same
6.769	same
6.733	2.5% Glutaraldehyde/hemolymph 1:1
7.870	Unfixed (EDTA-Saline/hemolymph 1:1)
6.860	Mean Median Cell Diameter from Above Data

Protein content. Figure 2 shows a distribution of fluorescence signals from dansylated hemocytes. The distribution strongly resembles that of Figure 1, which might be expected if protein were uniformly distributed and content proportional to size and fluorescence intensity. The chemical measurement of protein from 12

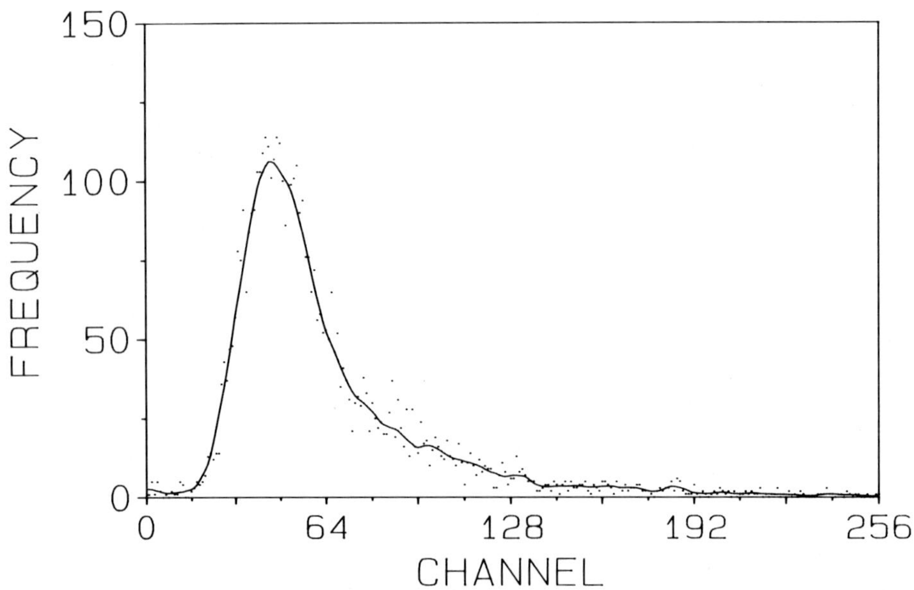

Fig. 1. Hemocyte size analysis of cells from *Mercenaria campechiensis*. Glutaraldehyde fixed. Median cell diameter estimated at 6.73 μm in channel 54. Size in channel 256 is 10.91 μm. Data smoothed for plotting curve.

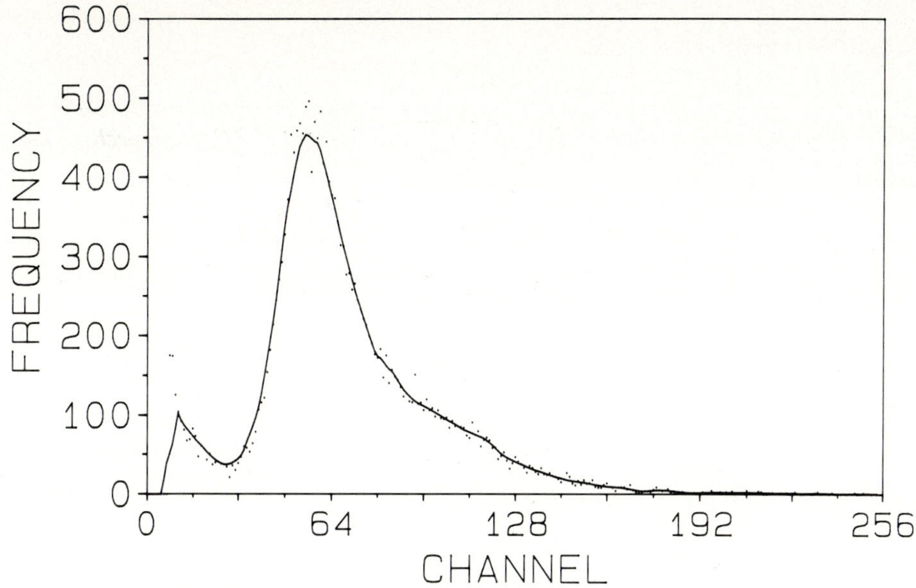

Fig. 2. Total protein analysis of hemocytes from *Mercenaria campechiensis*. Formaldehyde fixed and reacted with dansyl chloride. Fluorescence intensity is proportional to channel number with data smoothed for plotting curve.

hemocytometer-counted cell samples indicated an average concentration of 103 picograms protein per cell, presumably that contained by hemocytes near the modal channel in the frequency distribution.

Nuclear DNA. A smoothed flow histogram of DNA from isolated nuclei is shown in Figure 3. The distribution of 15,918 signals is unimodal at channel 38 and has a coefficient of variation (c.v.) of 5.6%. We interpret the single peak to represent nuclei from cells in the GO/G1 phase of the cell cycle. We see no evidence of signals from S or G2 + M phases nor is polyploidy apparent.

Characteristics of *M. mercenaria* hemocytes.

Size. A representative distribution of electronically determined size for *M. mercenaria* is shown in Figure 4. The distribution of 1,798 hemocytes is unimodal at channel 42 with a median diameter of 6.94 μm. An average of two such determinations was 7.44 μm.

Protein content. The distribution of fluorescence signals from dansylated hemocytes shown in Figure 5 is unimodal and, as for *M. campechiensis*, similar to the size distribution (Fig. 4). The average chemically-measured protein concentration for three visually counted samples was 112 picograms per cell.

Nuclear DNA. Figure 6 shows a smoothed frequency distribution of DNA fluorescence from 18,462 nuclei. The distribution is unimodal at channel 53 with a c.v. of 3.4%. We interpret the peak to represent the GO/G1 phase of the cell cycle and any signals to its left to be from broken nuclei and other debris. As for *M. campechiensis*, there is no convincing evidence for S and G2 + M phases or polyploidy.

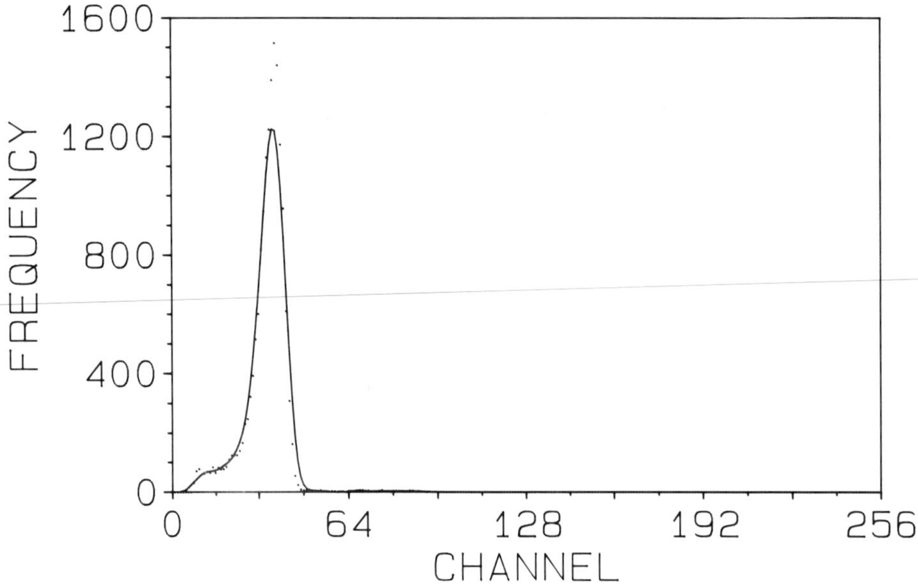

Fig. 3. Nuclear DNA analysis of hemocytes from *Mercenaria campechiensis*. Attached cells reacted with nuclear isolation medium and fluorochromed with DAPI (see text). Mode in channel 38 has a coefficient of variation of 5.6%. Fluorescence intensity is proportional to channel number with data smoothed for plotting curve.

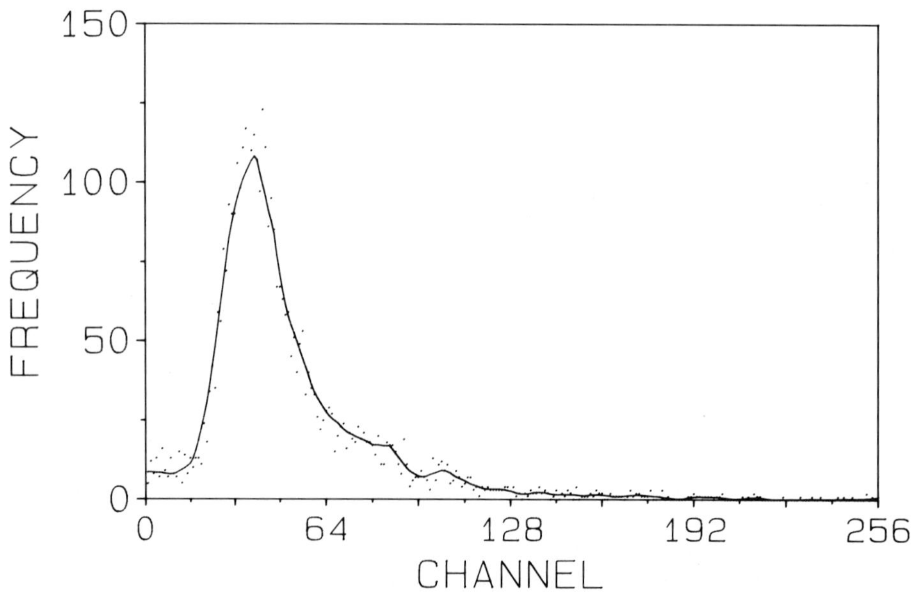

Fig. 4. Hemocyte size analysis of cells from *Mercenaria mercenaria*. Formaldehyde fixed. Median cell diameter estimated at 6.94 μm in channel 42. Smoothed as in Fig. 1.

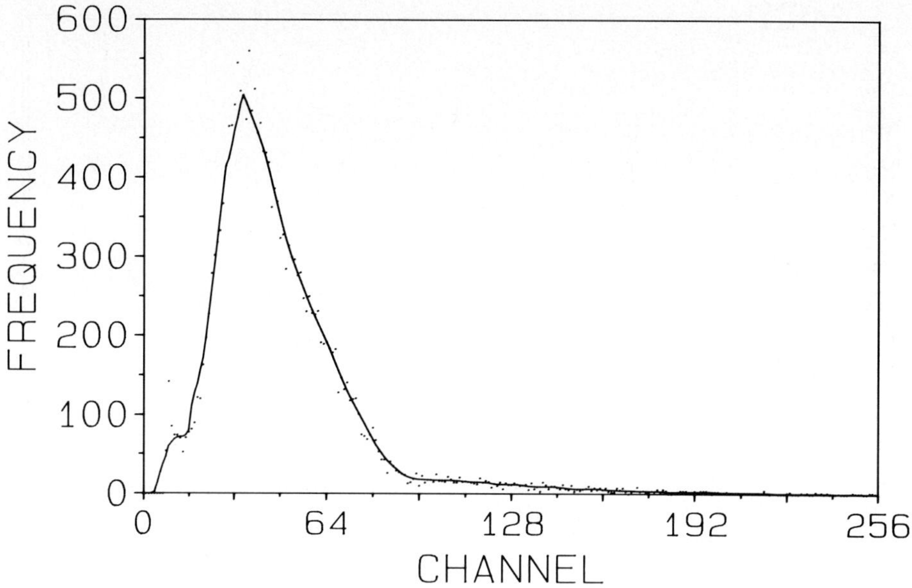

Fig. 5. Total protein analysis of hemocytes from *Mercenaria mercenaria*. Formaldehyde fixed and reacted with dansyl chloride. Displayed as in Fig. 2.

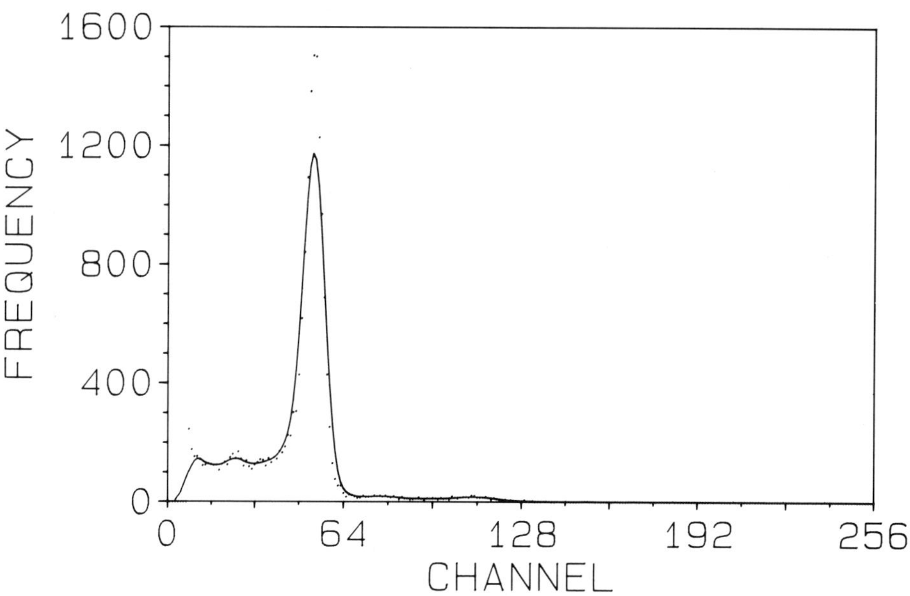

Fig. 6. Nuclear DNA analysis of hemocytes from *Mercenaria mercenaria*. Attached cells reacted with nuclear isolation medium and fluorochromed with DAPI (see text). Mode is in channel 53 with a coefficient of variation of 3.4%. Signals outside of the modal peak are interpreted as nuclear debris and aggregates. Displayed as in Fig. 3.

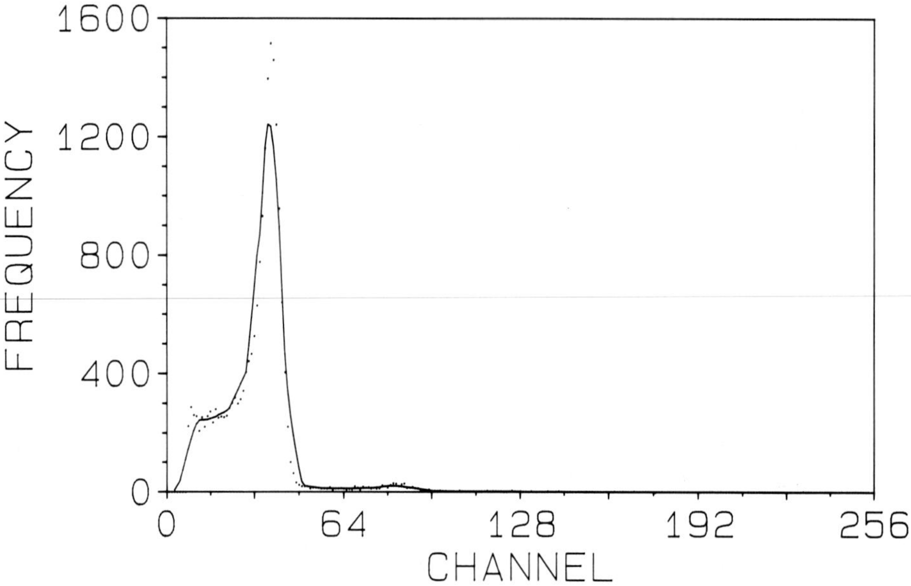

Fig. 7. Nuclear DNA histogram data from *Mercenaria campechiensis* and *Mercenaria mercenaria* combined. The mode is in channel 38 and the coefficient of variation is 4.8%. Signals outside of the modal peak are interpreted as nuclear debris and aggregates and are most likely from the *M. mercenaria* preparation. The fluorescence signals from nuclei of both species appear to contribute to a single modal peak with a coefficient of variation similar to measurements on one species. Displayed as in Fig. 3.

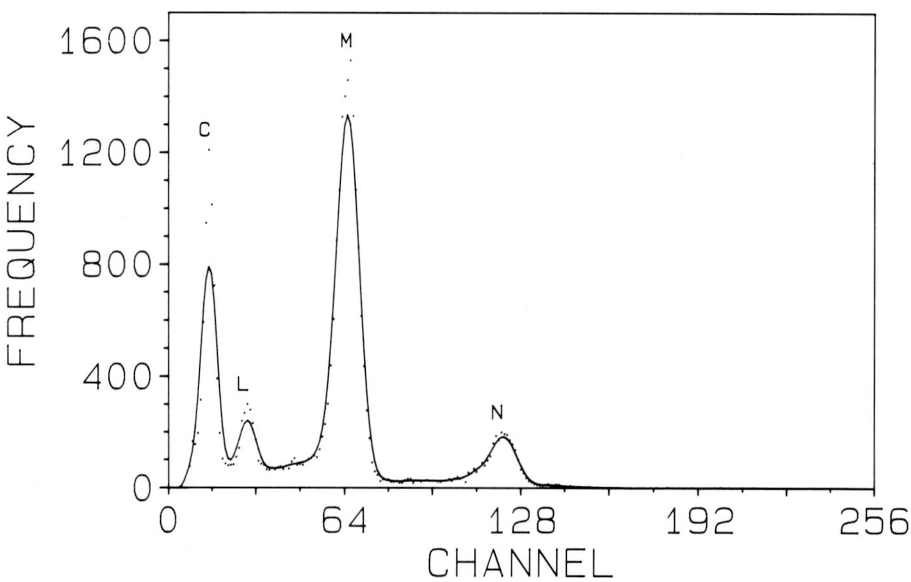

Fig. 8. Nuclear DNA histogram data combined from flow cytometric analyses of *Crassostrea virginica* (C), *Lymnaea stagnalis* (L), *Mercenaria campechiensis* (M), and the Guinea Fowl, *Numida* sp. (N). Displayed as in Fig. 3.

Interspecies comparisons.

Relative nuclear DNA content of the two clams. The distribution of fluorescence of isolated nuclei from *M. campechiensis* recorded over a similar preparation of *M. mercenaria* nuclei is shown in Figure 7. It is unimodal at channel 38 with a c.v. of 4.84% (N = 19,820). The signals from the two species appear to be directly superposable, thus indicating very similar nuclear concentrations of DNA.

Nuclear DNA content of *Mercenaria* hemocytes compared to that of other species. Hemocyte nuclei isolated from two bivalves (*Crassostrea virginica*, 2n = 20, and *Mercenaria campechiensis*, 2n = 38), a gastropod (*Lymnaea stagnalis*, 2n = 36), and a guinea hen (*Numida* sp. erythrocyte nuclei, 2n = 78) were run through the flow cytometer to identify peak positions, and then the records combined at identical instrument settings to give a composite histogram. Such a record is shown in Figure 8. It can be seen that nuclear DNA signals from all four species are easily separable. In all cases the signals most likely represent nuclei of cells in the GO/G1 phase of the cell cycle.

DISCUSSION AND CONCLUSIONS

Recognizing the difficulties associated with characterizing invertebrate amoebocytes, it was felt that additional information might be obtained from cytometric studies of isolated nuclei and of cells in their rounded condition as similar as possible in form to those found suspended in hemolymph. Flow cytometric techniques can give such information and provide quantitative data to assist in the interpretation of homogeneity or diversity of cell populations within a species. Such data can also be of comparative value and provide cytological information useful in identifying taxonomic categories. With this in mind, we have studied pericardial hemocytes of two kinds of *Mercenaria* using flow techniques that permit measurements on large numbers of cells to assist in quantifying various cellular characteristics. Specifically, we have measured size, total protein, and nuclear DNA concentration, characteristics that can discriminate between cell types and provide information on mitotic activity and interspecific content of genetic material.

Size. On the basis of electronic size distributions of hemocytes, there is no evidence of discrete size classes in either species. This has been true of selected samples prepared in various ways to avoid fixation artifacts. In all cases the distributions have been unimodal with average median diameters of about 7 μm for both *M. campechiensis* and *M. mercenaria*.

Total protein content. Total protein content of dansylated cells distributed itself unimodally in a pattern similar to that found for size. This might be expected since uniformly distributed protein content should be proportional to size. It thus provides supporting evidence that there are no discontinuous size categories in the hemocyte cell populations we sampled. Chemically determined reference values for protein indicated similar concentrations for both species (103 and 112 picograms/cell for *M. campechiensis* and *M. mercenaria*, respectively).

Nuclear DNA. The distributions of DNA concentrations from isolated nuclei of attached hemocytes showed single peaks with coefficients of variation not exceeding 5.6% in the measurements figured. For both species we interpret this peak to represent nuclei from cells in the G0/G1 phase of the cell cycle. Signals representing S and G2 + M phases of the cycle or polyploidy do not appear to be present . It is our opinion that any slight fluorescence in channels higher than the mode represent aggregations of debris or doublets. Some fluorescence was seen in channels lower than the mode in preparations from *M. mercenaria*. Since the nuclei were isolated from attached cells as a purification step, it is possible that these signals represent nuclear debris. Perhaps the nuclei of this species are more fragile than those of the other.

We feel the data indicate that the hemocytes we measured were from nondividing populations of euploid cells, and are representative of those suspended in the hemolymph of *Mercenaria*. It is not clear what nuclear behavior might result if such cells were additionally "activated," but one might infer that a nonproliferating, circulating population such as this originated in tissue-residing primordia. That hemocytes of bivalves arise from connective tissue cell differentiation has been stated as a "generally accepted belief" (Cheng, 1981; Cheng, 1984).

Mercenaria **nuclei compared.** When *M. campechiensis* and *M. mercenaria* nuclei were measured for DNA content, and the records were combined, it was found that one peak was obtained with a coefficient of variation similar to separate peaks for each species. Nuclear DNA histograms for these two species thus appear to be directly superposable. It would appear that their nuclear DNA contents are very similar and flow cytometric measurements of this parameter will not easily discriminate between them.

Mercenaria **nuclei compared with other genera.** It was of interest to see whether the nuclear DNA content of *Mercenaria* sp. resembled that of other molluscs. When hemocyte nuclear DNA histograms from *C. virginica* (the eastern oyster, 2n = 20), *L. stagnalis* (a pulmonate snail, 2n = 36), *M. campechiensis* (2n = 38), and guinea fowl erythrocyte nuclei (*Numida*, 2n = 78) were combined, fluorescence peaks were easily demonstrable for all genera. This experiment demonstrates the selectivity of the method and suggests that it may be of potential cytogenetic value in certain types of taxonomic and ecological research. It also suggests a considerable variability of DNA content in and between molluscan groups[3].

We have tried to choose certain cytometric descriptors of *Mercenaria* hemocytes that would assist in their cytological characterization, contribute to an understanding of their ontogeny, and be of possible comparative value in discriminating descriptively and functionally between the northern and southern species of this clam.

---

[3]The information on chromosome numbers was obtained from the following sources: *Crassostrea*, Menzel (1968a,b); *Lymnaea*, Patterson and Burch (1978); *Mercenaria*, Menzel and Menzel (1965); *Numida*, Makino (1951).

It is our contention that it is essential to study certain morphological parameters on rounded, unattached cells, as close as possible to their condition in hemolymph. Appropriate preparative methods combined with flow techniques can accomplish this and provide quantitative data on large samples of a cell population. These methods are well established for vertebrate cell biology, but are only emergent for lower animals. Certain parameters, such as size and protein fluorochroming, would clearly separate certain categories of leucocytes in vertebrate blood. Additionally, these patterns would be expected to be reproduceable from sample to sample, within and among individuals of a species.

In our studies of *M. campechiensis* and Florida *M. mercenaria*, we can find no separate classes or heterogeneity in electronically measured cell size within or between species. The same can be said of total protein content, a supporting parameter obtained by a different method. The hemocytes of these two species not only are similar in these characteristics, but appear to be a continuously variable population of polymorphic and multifunctional cells. Using nuclear DNA as an indicator, it appears that hemolymph cells are nondividing, and in the GO/G1 phase of the cell cycle. There is no evidence for S and G2+M phases in our preparations. It is suggestive that these hemocytes, although polymorphic and polyfunctional, are a single population at an ontogenetic end-point, with hemopoiesis taking place outside of the circulating hemolymph (a review of several molluscan hemocyte ontogenies can be found in Cheng, 1984).

As indicated above, hemocytes from the two species of *Mercenaria* were found to be similar in distributions of size and protein content. As one might expect, nuclear DNA content of the two species was seen to be very similar as indicated by the apparent superposibility of DNA fluorescence signals. Nonetheless, *Mercenaria* hemocyte DNA content can be easily separated from that of the eastern oyster (*C. virginica*), and the pulmonate snail (*L. stagnalis*), a feature indicative of potential research value of such cytometric techniques in other areas of invertebrate zoology.

ACKNOWLEDGEMENTS

The authors wish to acknowledge the excellent technical assistance of Ms. Jacqueline Johnson in this research. They are also grateful to Mr. William Arnold of the State of Florida Marine Research Institute, St. Petersburg, for providing the Florida east coast *Mercenaria*.

LITERATURE CITED

Arnold, W. and T. Bert. 1988. Personal Communication, State of Florida Marine Research Institute, St. Petersburg, Florida.

Cheng, T. 1981. Bivalves. In: "Invertebrate Blood Cells" (N. Ratcliffe and A. Rowley, eds.), Vol. 1, pp. 233-300. Academic Press, New York.

Cheng, T. 1984. A classification of molluscan hemocytes based on functional evidences. In: "Comparative Pathobiology. Invertebrate Blood: Cells and Serum Factors" (T. Cheng, ed.), Vol. 6, pp. 111-146. Plenum Press, New York.

Cheng, T. and A. Cali. 1974. An electron microscope study of the fate of bacteria phagocytized by granulocytes of *Crassostrea virginica*. Contemp. Top. Immunobiol., 4, 25-35.

Cheng, T. and D. Foley. 1975. Hemolymph cells of the bivalve mollusc *Mercenaria mercenaria*: an electron microscopical study. J. Invert. Pathol., 26, 341-351.

Cleveland, W. and R. McGill. 1985. Graphical perception and graphical methods for analyzing scientific data. Science, 229, 828-833.

Foley, D.A. and T. Cheng. 1974. Morphology, hematologic parameters, and behavior of hemolymph cells of the quahaug clam, *Mercenaria mercenaria*. Biol. Bull., 146, 343-356.

Friedl, F., M. Alvarez, J. Johnson and H. Gratzner. 1988. Cytometric investigations on hemocytes of the American oyster, *Crassostrea virginica*. Tissue and Cell, 20, 933-939.

Lowry, O., A. Rosebrough, A. Farr and R. Randall. 1951. Protein measurement with the Folin phenol reagent. J. Biol. Chem., 193, 265-275.

Makino, S. 1951. "An Atlas of the Chromosome Numbers in Animals." 2nd ed. Iowa State College Press, Ames, Iowa.

Menzel, R.W. 1968a. Cytotaxonomy of species of clams (*Mercenaria*) and oysters (*Crassostrea*). In: "Marine Biological Association of India, Proceedings of the Symposium on Mollusca," Part I, pp. 75-84.

Menzel, R.W. 1968b. Chromosome number in nine families of marine pelecypod molluscs. Nautilus, 82, 45-50.

Menzel, R.W. and M. Menzel. 1965. Chromosomes of two species of quahog clams and their hybrids. Biol. Bull., 129, 181-188.

Mohandas, A., T.C. Cheng and J.B. Cheng. 1985. Mechanism of lysosomal release from *Mercenaria mercenaria* granulocytes: A scanning electron microscope study. J. Invert. Pathol., 46, 189-197.

Patterson, C. and J. Burch. 1978. Chromosomes of pulmonate molluscs. In: "Pulmonates: Systematics, Evolution, and Ecology" (V. Fretter and J. Peake, eds.), Vol. 2A, pp. 171-218. Academic Press, New York.

Stauber, L. 1950. The fate of india ink injected intracardially into the oyster, *Ostrea virginica* Gmelin. Biol. Bull., 98, 227-241.

Thornthwaite, J., E. Sugarbaker and W. Temple. 1980. Preparation of tissues for DNA flow cytometric analysis. Cytometry, 1, 229-237.

Tripp, M. 1960. Mechanisms of removal of injected microorganisms from the American oyster, *Crassostrea virginica* (Gmelin). Biol. Bull., 119, 273-282.

FACTORS AFFECTING *IN VITRO* PHAGOCYTOSIS BY HEMOCYTES OF THE AMERICAN OYSTER[1]

Marvin R. Alvarez and Frank E. Friedl

Department of Biology, University of South Florida, Tampa, Florida 33620, USA

ABSTRACT: Proposed herein is a new method for assaying phagocytosis of polystyrene beads by hemocytes of *Crassostrea virginica* involving flow cytometry. In addition, it has been demonstrated that low temperature and cytochalasin B inhibit internalization. Anaerobic conditions did not inhibit bead phagocytosis.

INTRODUCTION

Phagocytic cells of the hemolymph appear to be the primary defense of molluscs against foreign particulates entering the tissues (Robohm, 1984 for review). A thorough understanding of the dynamics of the phagocytic process in these cells has practical applications for disease control in oyster fishery and aquaculture and also provides an interesting view into the functioning of a primitive cellular immune system, which differs in a number of ways from the better understood mammalian macrophage.

The hemocytes of oysters can be easily aspirated in suspension in hemolymph and their avidity for foreign particulates evaluated by direct microscopical examination of the cells. Using this approach, a number of investigations have been conducted on the effects of environmental factors such as temperature (Feng, 1965; Feng and Feng, 1974; Foley and Cheng, 1975), salinity (Fisher and Newell, 1986; Fisher et al., 1987; Fisher and Tamplin, 1988), and concentrations of metallic ions

---

[1]This article was developed under the auspices of the Florida Sea Grant College Program with support from the National Oceanic and Atmospheric Administration, Office of Sea Grant, U.S. Department of Commerce, Grant Number NA86AA-D-SG068.

(Cheng, 1988a,b) on hemocyte phagocytosis. Intrinsic factors, such as substrate utilization and energy requirements of the hemocytes, have also been examined (Cheng, 1976).

The intent of the present study was to develop a standardized quantitative assay system that could be generally applied as a test for the effects of environmental factors on the cellular immune response of the American oyster, *Crassotrea virginica*. In this procedure, fluorescent polystyrene beads were used as the test particles. The beads were immobilized on a surface as a uniform monolayer whose density ensured that the hemocytes would contact one or more beads immediately upon settling out of the overlayering hemolymph. Use of this bead to cell ratio insures that the amount of phagocytosis measured is independent of cell motility, which can often be influenced by the test conditions. After a prescribed period of time, the cells were released from the beads by gentle trypsinization, fixed, and separated from unbound beads through a discontinuous density gradient. The amount of bead uptake that occurred was determined by flow cytometry and fluorescence microscopy.

We report in this communication the application of this assay to the determination of rate of phagocytosis as influenced by temperature and oxygen availability. These studies illustrate the potential of this standardized, quantitative procedure as a generalized method for evaluating the effects of environmental stress on the oyster's immune response.

MATERIALS AND METHODS

Specimens of *Crassostrea virginica* were obtained from a site in upper Old Tampa Bay (27°53′20″ north latitude, 82°38′13″ west longitude) and maintained at 21°C in a circulating artificial sea water system with salinity at 24 ‰. The oysters were used for experiments within 30 days of placement in the sea water system.

Standardized assay system. Plastic tissue culture flasks (Costar, 25 cm^2) were internally coated with freshly drawn, cell-free hemolymph and allowed to stand at room temperature for 20 min, providing an attachment surface for the test particles. Fluorescent polystyrene microspheres (beads) of 2.02 μm mean diameter (Polysciences) were dispersed by repeated flowing through a 26-gauge needle into a syringe containing Millipore® filtered (0.45 μm) sea water. Volumes of the bead suspension were placed in the flasks and centrifuged gently onto the bottom surface to give a uniform monolayer of beads with a density of $2.2 \times 10^4/mm^2$. The beads remained attached to the flask surface after pouring off the suspending sea water.

Hemolymph was removed from the pericardial cavity through a hole drilled in the right valve using a 26-gauge needle fitted to a 1-ml syringe. The cell concentration was determined and a volume added to the bead monolayer to give a 2:1 bead:cell ratio. The fluid level over the bead monolayer was approximately 1 mm deep. In the time course experiments, the amount of time the hemolymph remained in contact with the beads varied from 15 to 120 min. In all other experiments, the contact time was 60 min. Attached cells were released by trypsinization (0.4% trypsin in Hank's balanced salt solution for 5 min) and fixed immediately for 30 min. Cells used for fluorescence microscopy and flow cytometry were fixed in 3.7% formaldehyde in 0.5 M NaCl. For electron microscopy, the cells were fixed in a mixture of 5% buffered glutaraldehyde and 4% paraformaldehyde. The fixed cells were centrifuged at 250*g* and the pellet resuspended in 1 ml of fixative. This suspension was layered over a discontinuous sucrose gradient. The cells were allowed to settle through the gradient overnight at 4°C and the top layer, containing suspended unbound beads, was discarded. The cells were pelleted by centrifugation and resuspended in fixative. This procedure eliminated unattached beads and cell surface-attached beads as evidenced both by fluorescence and electron microscopy.

The number of cells with one or more phagocytized beads was scored by fluorescence microscopy. One thousand cells were scored for each experiment, the data were expressed as percent cells with one or more phagocytized beads. Experiments were repeated at least twice.

Flow cytometry. Flow analysis of the number of beads phagocytized per hemocyte was done with a flow cytometer constructed after the design by Shapiro (1985) interfaced to a microcomputer-based pulse height analyzer. The instrument consistently gave coefficients of variation of 2% or less with 4′, 6-diamidino-2-phenylindole HCL (DAPI)-stained bird erythrocytes (Alvarez and Stone, 1988). Cell injection rates were adjusted to give flow rates of 200 cells/sec through the measurement point.

Data analysis was done using a utility program developed in our laboratory for graphic and numerical analyses. Various statistical parameters can be obtained and channel frequencies can be integrated to give cell numbers in selected histogram regions.

Oxygen availability study. The effects of oxygen concentration on the rate of phagocytosis were determined by carrying out the phagocytosis assays under atmospheres of different gas composition. Nitrogen or helium passed through a solution of alkaline pyrogallol, were bubbled through hemolymph (diluted 1:1 with 0.54 M NaCl to obtain approximately 10^6 cells/ml) for 60 min on ice. The culture flask was purged of oxygen with the experimental gas for 30 min prior to addition of the cells. The oxygen concentration inside the flask was monitored during the experiment and maintained at zero throughout the incubation period by passing a low pressure stream of the test gas through the flask. The assay system was maintained at room temperature.

Inhibition of phagocytosis with cytochalasin B. Actin polymerization was inhibited by treatment of undiluted hemocyte suspensions with 10 μg/ml and 25 μg/ml final concentrations of cytochalasin B. The cells were exposed to these inhibitors for 60 min at room temperature (23°C) after adding the cells to the bead monolayer. The number of cells containing one or more beads were counted by fluorescence microscopy. These experiments were repeated twice.

RESULTS

Transmission electron micrographs of hemocytes fixed after 1 hr incubation with beads showed numerous beads contained within single-membrane bound vesicals presumed to be phagosomes (Fig. 1a). These micrographs show that the beads were

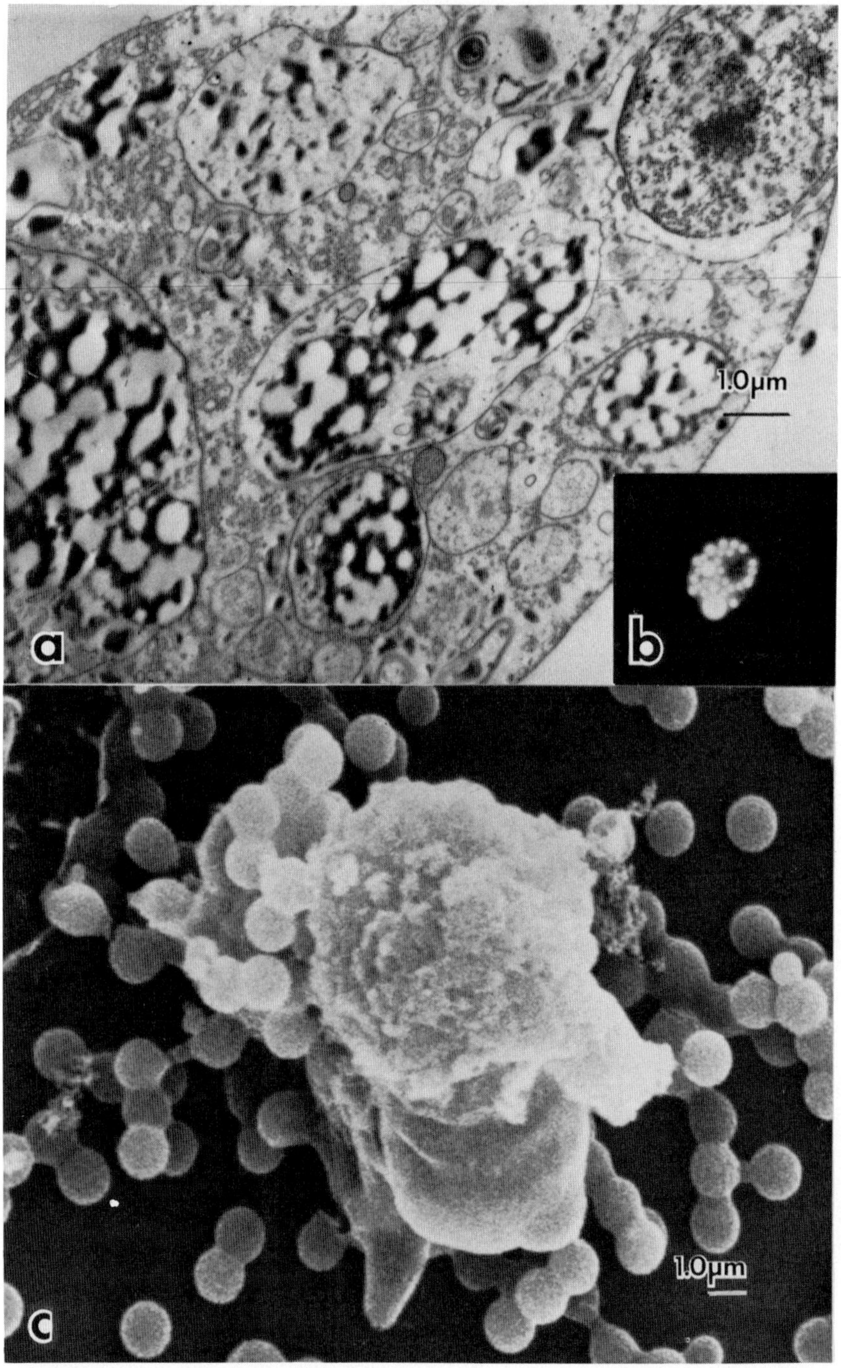

Fig. 1. Section through hemocyte containing several beads in phagosomes (1a). Similar cell viewed by fluorescence microscopy (1b). Scanning electron micrograph of hemocyte settled on surface of beads (1c). The uniformity of the bead layer was disturbed by preparative procedure.

in fact internalized under the experimental conditions of the assay system. Figure 1b shows a similar cell as viewed by fluorescence microscopy. Figure 1c is a scanning electron micrograph of a hemocyte situated on the bead layer. The cell is simultaneously in contact with numerous beads.

Figure 2 shows the number of beads taken up per cell as a function of time. The data are presented as frequency distribution histograms, with the number of cells containing one or more beads presented on the ordinate and the number of beads/cell as channel number on the abscissa. The numeral over each peak represents the number of beads within the cells comprising that peak. Individual histograms represent exposure times of 15, 45, 60, and 120 min.

The most frequent number of beads found per cell was one, the frequency subsequently decreasing with number of beads/cell. As many as four distinct peaks are visible at 15 min exposure and up to six can be seen at the longer exposure periods. However, no distinct differences in the relative sizes of the peaks are seen regardless of exposure time, i.e., the number of beads/cell was not a function of exposure time. Since the cells were in contact with several beads immediately upon settling on the bead monolayer, it is evident that they are capable of multiple uptake within 15 min after addition to the flask.

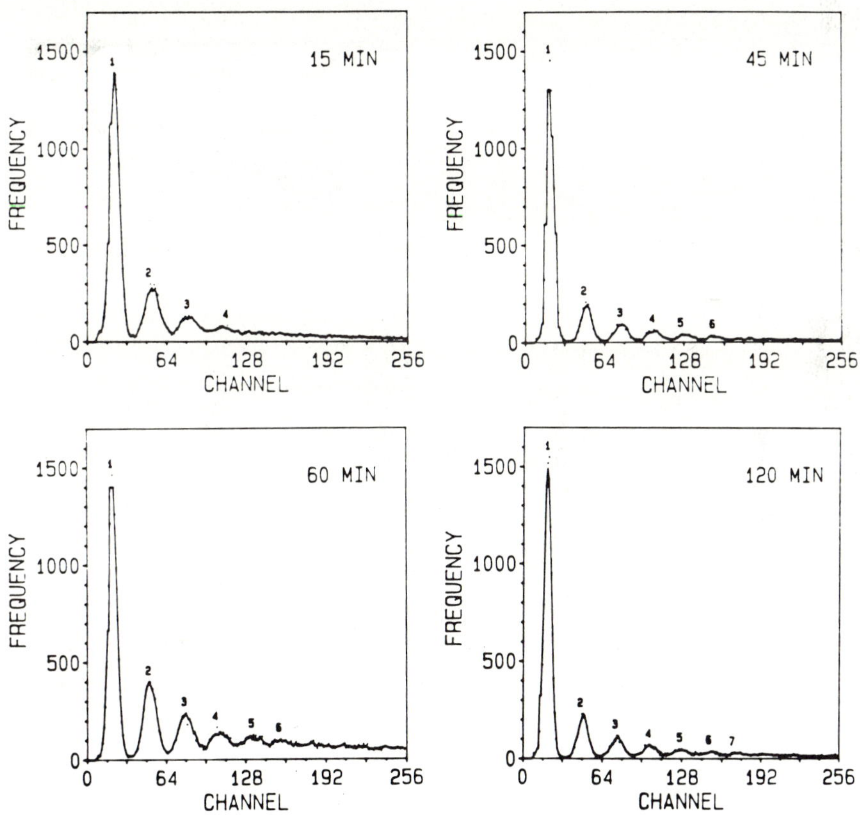

Fig. 2. Flow cytometric profiles of number of beads per cell taken up over time by oyster hemocytes. Histograms smoothed for plotting. Numbers over the peaks represent the number of beads within the cells under the peak.

The effect of temperature on hemocyte phagocytosis is shown in Figure 3 as percent cells containing beads. In the temperature range between 8° and 37°C, the percentage of cells containing beads ranged from 50-78%. At 4°C, approximately 10% of hemocytes contained beads. At 0°C about 8% of the cells contained beads.

The number of beads/cell was measured by flow cytometry in hemocyte populations exposed to beads at various temperatures (Fig 4.) At least four peaks are evident from cells exposed at 0°C indicating between 1 and 4 beads/cell. A similar number of peaks were discernable at 4°, 12°, and 23°C and the relative heights of the peaks remained about the same regardless of the temperature at which phagocytosis took place. Thus, while the number of hemocytes that internalize beads is affected at the extremes of the oyster's biological temperature range, the number of particles phagocytosed by a given cell is relatively independent of temperature within this range.

The results described above suggested that low temperatures might affect cell attachment. Thus, cells that did attach could phagocytose several beads with which they came into contact, but, since fewer cells attach, the number of cells with internalized beads might be lower. This would account for the same number of beads/cell for each temperature but lower percent cells with beads at low temperatures. To test this, cells suspended in hemolymph were added over the bead monolayer and incubated for 1 hr at 0°, 4°, and 10°C. The cell concentration in each

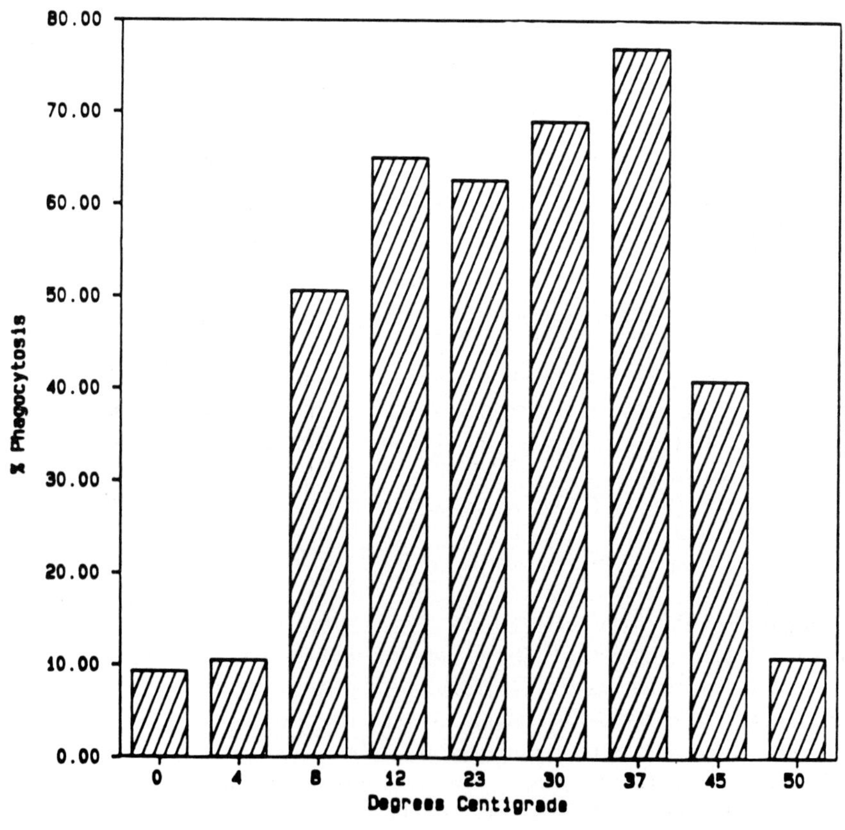

Fig. 3. Percent hemocytes containing one or more beads as a function of temperature.

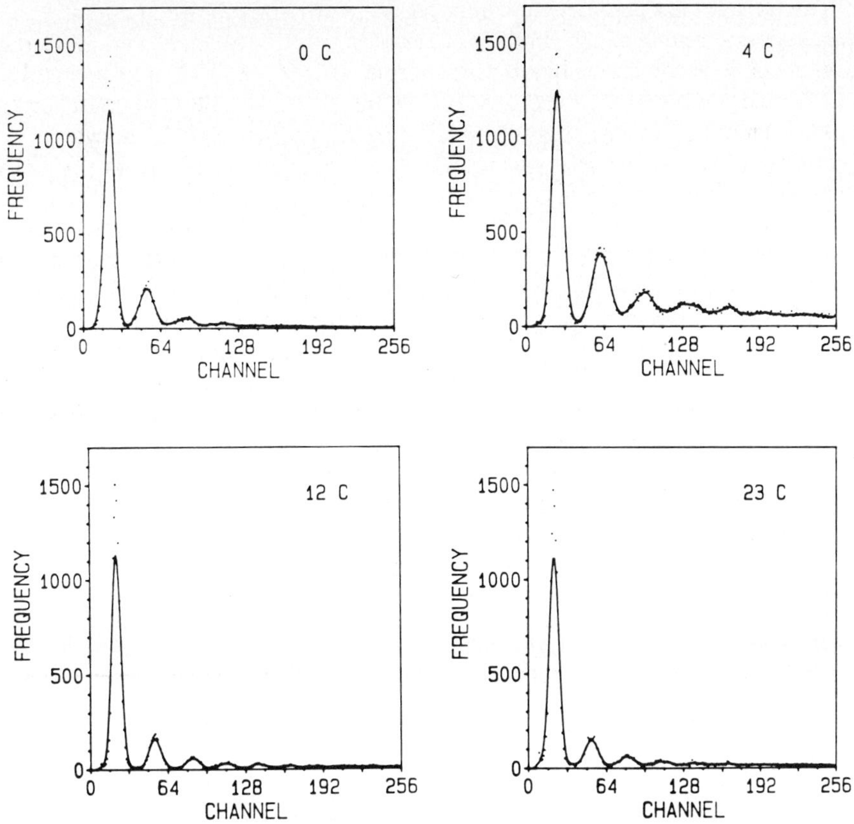

Fig. 4. Flow cytometric profiles of number of beads per cell taken up as a function of temperature by hemocytes of *C. virginica*. Histograms smoothed for plotting.

assay was determined prior to addition of the hemolymph to the flasks. After 1 hr, the hemolymph was decanted without trypsinization and the concentration of cells in suspension determined. At 0°C, 89.7% of the hemocytes remained unattached; at 4°C, 37.5% were unattached; and at 10°C, 22.6% of 7 the cells were unattached, these data show that cell attachment decreases sharply at low temperatures and is probably the major cause of reduced percent phagocytosis. These data are shown in Table 1.

The effect of oxygen availability on bead uptake was tested by constructing completely anaerobic environments using helium or nitrogen atmospheres during incubation of the cells with beads (Fig. 5). The results are summarized in Table 2. The mean values for helium and nitrogen atmospheres do not differ significantly from each other nor from those in air ($P = 0.05$), showing that hemocytes maintained in completely anaerobic atmospheres for 1.5 hr are able to continue particle phagocytosis at the same time as in an aerobic environment.

The effect of inhibition of actin polymerization on the hemocytes' ability to phagocytize beads was tested by adding cytochalasin B to cells in hemolymph and incubating the cells with beads for 1 hr at room temperature. At a cytochalasin B concentration of 10 μg/ml, 12% of the cells took up beads. At 25 μg/ml, only 0.01%

Table 1. Effect of low temperatures on attachment of hemocytes to microsphere monolayers.

Temperature (°C)	Suspended Hemocytes (cells/ml) Before Incubation	After Incubation	Percent Unattached Cells
0	$1.55x10^6$	$1.39x10^6$	89.7
4	$2.90x10^6$	$1.09x10^6$	37.5
10	$3.05x10^6$	$0.69x10^6$	22.6

Table 2. Effect of air and oxygen-free atmospheres on phagocytosis.

Experiment Number	Atmosphere Gas	Percent Cells with Internalized Microspheres
1	Air	63.2
2	Air	63.6
3	Air	49.6
4	Air	51.8
5	Air	57.6
		Mean 57.1 ± 6.4
6	N_2	68.2
7	N_2	66.5
8	N_2	65.8
9	N_2	57.8
10	N_2	64.8
		Mean 64.6 ± 4.0
11	He	60.8
12	He	56.8
13	He	47.3
14	He	53.2
		Mean 54.5 ± 5.7

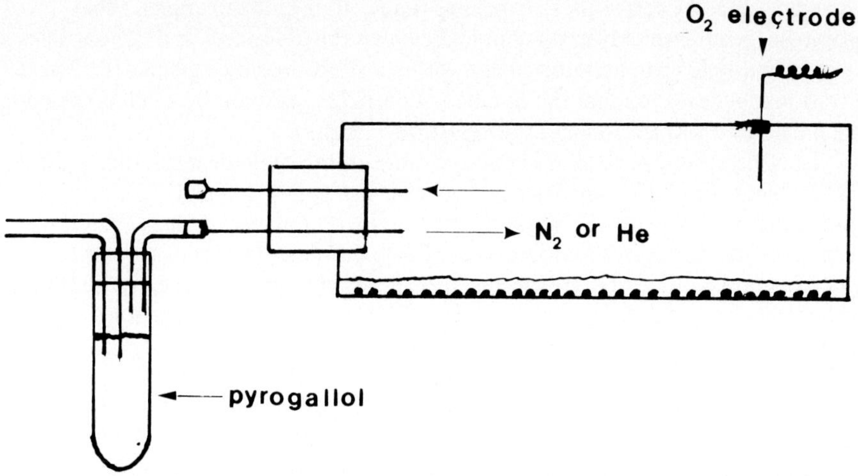

Fig. 5. Diagram of apparatus used for creating anaerobic environment during exposure of hemocytes to beads.

of the hemocytes contained beads. These data show that cytochalasin B is a powerful inhibitor of bead phagocytosis by oyster hemocytes.

DISCUSSION AND CONCLUSIONS

The purpose of the studies presented here was to develop a standardized, quantitative assay system and to use the procedure to test the effects of temperature and oxygen availability on hemocyte particle uptake. In order to achieve this goal, it was necessary to rule out the possibility that the end point being measured by fluorescence microscopy and flow cytometry might include surface-attached as well as internalized beads. In the assay system described, the fluorescent test particles were immobilized as a uniform monolayer of a density such that a cell setting onto the monolayer would be in contact with several beads simultaneously. This high bead density was chosen to rule out the possibility that the test conditions might affect cell motility. If that were the case, the amount of phagocytosis measured would be due to the effect of the test conditions on both cell motility and phagocytic avidity.

Examination of slides prepared from the various assays by either phase contrast or fluorescence microscopy showed no sign of free beads nor cell surface attached beads. Examination of the cells by electron microscopy showed clearly that the beads were internalized within membrane bound vesicals. Treatment of hemocytes with cytochalasin B, which is known in other cell systems to inhibit actin polymerization and phagocytosis (Stossel and Cohn, 1976), almost completely inhibited particle uptake. Because of the high particle to cell ratio used, it is unlikely that the observed inhibition by cytochalasin B was due to reduced cell motility. When considered together, the above observations convince us that the uptake measured in our experiments was indeed phagocytic internalization of the test particles.

Flow analysis of hemocyte populations exposed to the test particles for 15, 45, 60, and 120 min showed that the distribution of the number of particles contained per cell was the same at each of these exposure times. It, therefore, appears that phagocytosis occurs quickly upon contact between the hemocyte and the particles and that several particles can be internalized within a short time by a given cell. These observations suggest then that the number of particles taken up by a cell is dependent on the particle concentration and the cell surface area.

The effect of temperatures in the low range of the biological tolerance of the oyster has been studied by various experimental strategies. Feng (1965), for example, examined the pinocytic uptake of rhodamine-labeled human gamma globulin by oyster hemocytes at various temperatures. Feng and Feng (1974) studied the phagocytic uptake of avian erythrocytes by injecting the cells into the circulation of oysters maintained at different temperatures. These authors noted that low temperatures both reduced the uptake of materials into the cell and slowed intracellular digestion. Foley and Cheng (1975) exposed hemocytes of *Crassotrea virginica* and *Mercenaria merceneria* to various bacterial species, evaluating the association of the microbial cells with the hemocytes microscopically. These authors found that in every case low temperatures decreased the extent of association.

Our data also show a reduction of particle uptake at low temperatures. At 0°C, less than 10% of the hemocytes contained fluorescent particles, while at 8°C over 50% of the cells had internalized beads. Thus we see a four-fold increase in uptake in less than a 10 degree temperature range. This observation suggests to us that inhibition of particle uptake at lower temperatures in our system may be due to more than just inhibition of the metabolic process that drives phagocytosis.

Earlier studies with snail hemocytes (Robohm, 1984) showed that low temperatures prevented these cells from attaching to glass surfaces. We found that while at 10°C approximately 78% of the hemocytes had attached to the bead monolayer; at 0°C, nearly 90% of the hemocytes remained unattached. Furthermore, our flow analysis showed that approximately the same number of particles were internalized per cell at all of the test temperatures. It, therefore, appears that those cells that do attach at low temperatures are as capable of particle uptake as those in the higher temperature range. These observations indicate to us that the primary inhibitory effect of low temperature is prevention of cell-surface attachment of the particles.

It has been previously shown that hemocytes of *M. mercenaria*, unlike mammalian macrophages, do not show increased oxygen uptake during phagocytosis. In our studies, particle uptake was quantified in cells exposed to test particles in completely anaerobic environments. No significant difference in percent phagocytosis was found between cells in aerobic or anaerobic conditions. These results, even though arrived at by a different experimental approach, are consistent with those of Cheng (1976) who found no inhibition of phagocytosis by KCN in *Mercenaria* hemocytes, which utilize both glucose and glycogen and produce lactate. It therefore appears the hemocytes of these two mollusc species are capable of carrying out phagocytic uptake in the absence of oxygen and that the process is driven by an anaerobic metabolism.

LITERATURE CITED

Alvarez, M.R. and D.J. Stone. 1988. Hypoploidy and hyperplasia in the developing brain exposed to alcohol *in utero*. Teratology, 37, 233-238.

Cheng, T.C. 1976. Aspects of substrate utilization and energy requirement during molluscan phagocytosis. J. Invert. Pathol., 27, 263-268.

Cheng, T.C. 1988a. *In vivo* effects of heavy metals on cellular defense mechanisms of *Crassostrea virginicia*: Total and differential cell counts. J. Invert. Pathol., 51, 207-214.

Cheng, T.C. 1988b. *In vivo* effects of heavy metals on cellular defense mechanisms of *Crassostrea virginica:* Phagocytic and endocytic indices. J. Invert. Pathol., 52, 215-220.

Feng, S.Y. 1965. Pinocytosis of proteins by oyster leucocytes. Biol. Bull., 129, 95-105.

Feng, S.Y. and J.S. Feng. 1974. The effect of temperature on cellular reactions of *Crassostrea virginica* to the injection of avian erythrocytes. J. Invert. Pathol., 23, 22-37.

Fisher, W.S. and R.I.E. Newell. 1986. Salinity effects on the granular hemocytes of the American oyster, *Crassostrea virginica*. Biol. Bull., 170, 122-134.

Fisher, W.S. and M. Tamplin. 1988. Environmental influence on activities and foreign-particle binding by hemocytes of American oysters, *Crassostrea virginica*. Can. J. Fish Aquat. Sci., 45, 1309-1315.

Fisher, W.S., N. Auffret and G. Balouet. 1987. Acclimation of European flat oyster (*Ostrea edulis*) hemocytes to acute salinity and temperature changes. Aquaculture, 67, 179-190.

Foley, D.A. and T.C. Cheng. 1975. A quantitative study of phagocytosis by hemolymph cells of the pelecypods *Crassostrea virginica* and *Mercenaria mercenaria*. J. Invert. Pathol., 25, 189-197.

Robohm, R.A. 1984. *In vitro* phagocytosis by molluscan hemocytes: A survey and critique of methods. In: "Comparative Pathobiology" (T.C. Cheng, ed.), Vol. 6, pp. 147-172. Plenum Press, New York.

Shapiro, H.M. 1985. "Practical Flow Cytometry." Alan R. Liss, New York.

Stossel, T.P. and Z.A. Cohn. 1976. Phagocytosis. In: "Methods in Immunology and Immunochemistry," (C.A. Williams and M.W. Chase, eds.), pp. 261-301. Academic Press, New York.

EFFECTS OF *IN VIVO* EXPOSURE OF *CRASSOSTREA VIRGINICA* TO HEAVY METALS ON HEMOCYTE VIABILITY AND ACTIVITY LEVELS OF LYSOSOMAL ENZYMES[1]

Thomas C. Cheng

Marine Biomedical Research Program and Department of Anatomy and Cell Biology, Medical University of South Carolina, Charleston, South Carolina 29412, USA

ABSTRACT: Immunity in bivalves is comprised of cell-mediated and humoral mechanisms. Since marine bivalves, which are not only ectotherms but also possess open circulatory systems, are continuously exposed to fluctuating environmental factors and additives, a series of studies has been conducted to ascertain if *in vivo* exposure of the American oyster, *Crassostrea virginica*, to 1 ppm of Cu^{2+} or Cd^{2+} would influence several aspects of its cell-mediated and humoral immunity. Reported herein are the findings that Cu^{2+} is more cidal to oyster hemocytes than Cd^{2+}, Cd^{2+} inhibits the release of lysozyme from granulocytes into serum, and Cu^{2+} inhibits the release of acid phosphatase from cells into serum. Since the viability of oyster hemocytes is directly correlated with their ability to phagocytose invading agents and since the release of lysosomal hydrolases from cells into serum represents one aspect of humoral defense in antigenically challenged molluscs, it may be concluded that *in vivo* exposure to 1 ppm of Cu^{2+} or Cd^{2+} can reduce the immunologic competence of *C. virginica*.

INTRODUCTION

Immunity in molluscs can be conveniently categorized as being either cell-mediated or humoral although there is little doubt that the two are interrelated. Within the realm of cell-mediated immunity, three types are generally recognized: phagocytosis, encapsulation, and nacrezation. Among these, phagocytosis is the most

[1]This research was supported by a contract (DE-AS09-83ER60132) from the U.S. Department of Energy.

common. In fact, as Cheng and Rifkin (1970) have suggested, encapsulation represents aborted attempts at phagocytosis since the nonself body is too large to be internalized. In a more recent study involving tracing cellular reactions in the gastropod *Biomphalaria glabrata* implanted with allografts, Jourdane and Cheng (1987) have reported that encapsulation occurs in two phases. (1) During the initial phase, granulocytes adhere to the graft, become hypertrophied, and elongate. This results in chemical alteration of the graft surface. (2) At the commencement of the second phase, the altered graft emits a signal(s) that recruits additional molluscan granulocytes.

What is known about humoral immunity in molluscs has been reviewed by Fries (1984) and more recently by Cheng and Combes (1989). Among known molecules associated with humoral immunity are the lysosomal enzymes. Specifically, it is known that when molluscs are antigenically challenged, there is hypersynthesis of lysosomal hydrolases within their granulocytes. Furthermore, these enzymes are released into serum where they have been demonstrated to have antimicrobial and antiparasitic functions (see Cheng, 1978, 1979, 1983, for reviews).

In view of the above and knowing that pelecypods and gastropods are ectotherms that possess open circulatory systems, the question has been asked as to whether fluctuations in and additives to the immediate environment of these animals, especially the aquatic species, will influence the effectiveness of their immune mechanisms. What is being reported herein represents the results of studies designed to ascertain the effects of *in vivo* exposure of the American oyster, *Crassostrea virginica*, to two heavy metals, copper and cadmium, on the viability of its hemocytes. Also being reported are the effects of *in vivo* exposure of *C. virginica* to the same two metals on the levels of serum lysozyme and cellular and serum acid phosphatase activities.

MATERIALS AND METHODS

Oysters. The specimens of *C. virginica* employed in this study were collected from Bull's Bay, South Carolina, during June and July, 1984. Each oyster measured 8-12 cm in shell length.

After being brought into the laboratory, all oysters were depurated in a metal-free, continuous flow, artificial sea water system (20 ‰ salinity, 25°C) for at least 6 weeks. Weekly checks of soft tissues of representative specimens for Cu^{2+} and Cd^{2+} contents by following the method provided by the Environmental Protection Agency (1980) for preparation of tissues, followed by analysis of metal contents in ash, involving use of a Perkin-Elmer (Model 306) flame atomic absorption spectrophotometer, revealed that the Cu^{2+} content decreased from 4.9 ± 3.2 ppm to < 0.06 ppm and the Cd^{2+} content decreased from 0.5 ± 0.8 ppm to < 0.04 ppm after 6 weeks of depuration. During the depuration period, the oysters were fed two to three times weekly with the alga *Thalassiosira pseudonana* that had been cultured in metal-free artificial sea water and F-2 medium (Guillard, 1973). At each feeding period, $\sim 10 \times 10^6$ algal cells suspended in 0.1 liter of sea water were added to each 5000-liter tank containing 100 oysters. All oysters appeared healthy and were pumping.

Exposure of oysters. After depuration, the oysters included in the experiments were divided into five groups, 48 specimens per group. The first group was separated into three subgroups. The 16 oysters that comprised each subgroup were placed in a polyethylene tank containing 94.6 liters of metal-free artificial sea water (20 ‰ salinity, 22°-23°C). To each of the three tanks was added sufficient Cu^{2+} (as $CuSO_4 \cdot 5H_2O$) so that the Cu^{2+} concentration was 1 ppm.

The second group of oysters was similarly divided into three subgroups. To each of the three tanks was added sufficient Cd^{2+} (as $CdCl_2$) so that the Cd^{2+} concentration was 1 ppm.

The third group of 48 oysters was likewise divided into three subgroups that were placed in identical tanks; however, to each tank was added sufficient Cu^{2+} +EDTA (ethylenediametetraacetic acid) at an equal molar concentration as Cu^{2+} to give a concentration of 1 ppm. The fourth group of oysters was also divided between three tanks to each of which was added sufficient Cd^{2+} +EDTA at an equal molar concentration as Cd^{2+} to give a concentration of 1 ppm.

The fifth group was identically divided into three subgroups and placed in three tanks. No metal or metal chelate was added to these control tanks.

The concentrations of Cu^{2+}, Cu^{2+} +EDTA, Cd^{2+}, and Cd^{2+} +EDTA were checked every other day and if required, were brought up to 1 ppm.

The concentration of 1 ppm was selected because it had been determined earlier that *in vitro* exposure of *C. virginica* hemocytes to this concentration of Cu^{2+} and Cd^{2+} did not alter their phagocytic activity (Cheng and Sullivan, 1984). Consequently, I wished to determine whether *in vivo* exposure of oyster hemocytes to the same concentration of these two heavy metals would result in the same or altered results.

The groups of oysters exposed to Cu^{2+} +EDTA or Cd^{2+} +EDTA were included to determine whether the effects, if any, detectable in the groups exposed to Cu^{2+} or Cd^{2+} were due to these heavy metals. Specifically, it is well known that EDTA is an effective chelator of Cu^{2+} and Cd^{2+} (Williams, 1977) and its being chelated to these metals should negate or reduce the possible effects of the metals alone.

During the exposure periods, all oysters were fed by adding $2x10^4$ *T. pseudonana* cells to each tank every other day.

Collection of hemolymph. A 2- to 3-ml hemolymph sample was intracardially collected from four oysters from each subgroup by use of a sterile 10-gauge hypodermic needle and a 5-ml syringe that had been precooled at 4°C at four time intervals, 24 hr, 72 hr, 1 week, and 2 weeks, after initiation of exposure to sea water containing Cu^{2+}, Cd^{2+}, Cu^{2+} +EDTA, Cd^{2+} +EDTA, or non-metal-containing sea water.

Viability test. To determine the viability of hemocytes obtained from the experimental and control subgroups of oysters at the four time intervals, the trypan blue exclusion test (Phillips, 1973) was performed. Specifically, 25 µl of fresh whole hemolymph was pipetted onto a dichromate-cleaned, paraffin-ringed glass slide. To this preparation was added 25 µl of 1% trypan blue in the basic salt solution of Tripp et al. (1966). Each slide was then incubated at room temperature in a humidified chamber for 10 min after which they were examined microscopically. The number of viable granulocytes, i.e., those that did not take up trypan blue, per 100 cells was recorded. The test was performed on the four samples individually of each subgroup at each time interval.

Lysozyme activity. The method used for determining serum lysozyme activity levels was that of Shugar (1952) as modified by Rodrick and Cheng (1974). As whole hemolymph samples were collected at each time interval, the cells were separated from the serum by centrifugation at 7200*g* for 10 min at -5°C in a Sorvall RC-2B refrigerated centrifuge. Subsequently, enzyme assays were carried out on the sera involving the use of 0.1 M glycylglycine as the buffer at pH 5.5 and 0.2 mg/ml of dried *Micrococcus lysodeikticus* cells (Sigma).

Lysozyme activities, derived from a standard curve, were recorded as ΔOD_{540}/min/mg protein and subsequently converted to specific activity units by multiplying the ΔOD_{540} value by 35.03 (Cheng, 1989). The determinations were made on the four samples of each subgroup of oysters individually at each time interval.

Acid phosphatase activity. Acid phosphatase activity levels were determined in both the cells and 4000*g* serum of individual oysters of each subgroup at each of the time intervals by monitoring the hydrolysis of sodium p-nitrophenylphosphate at 405 μm. The protocol followed to prepare hemocytes for determination of cellular acid phosphatase levels is that of Cheng and Downs (1988) except that separation of hemocytes into subpopulations was not carried out. All determinations were carried out at 25°C and pH 4.8 (Fishman and Lerner, 1953) and the specific activities were recorded as milliunits (mU)/mg protein. The assays were made on the four samples of each subgroup individually at each time interval.

Protein concentrations. The total protein concentrations of 4000*g* supernatant (serum), except in the case of hemolymph samples used for serum lysozyme determinations where 7200*g* supernatant were employed, and pellets (cells) of representative aliquots of hemolymph of each oyster belonging to each subgroup collected at the four time intervals were determined by the method of Lowry et al. (1951). Crystalline bovine serum albumin (Sigma) was used as the standard.

Statistical analyses. The data resulting from all of the tests at the four exposure periods were compared in all possible combinations by using the Bonferroni test for multiple comparisons (Neter et al., 1983).

RESULTS

Hemocyte viability. The results of the hemocyte viability tests on all five subgroups of oysters at the four time intervals are tabulated in Table 1.

Lysozyme levels. The serum lysozyme specific activity levels in *C. virginica* of all five subgroups at the four time intervals are tabulated in Table 2.

Acid phosphatase levels. The levels of acid phosphatase activity associated with both cells and sera of the five subgroups of oysters bled at the four time intervals were determined. The values of serum acid phosphatase are tabulated in Table 3.

Table 1. Number of viable *Crassostrea virginica* hemocytes per 100 cells at four time intervals of exposure to 1 ppm of Cu^{2+}, Cu^{2+} +EDTA, Cd^{2+}, Cd^{2+} +EDTA, and sea water. The figures represent means ± standard errors (n = 4/reported value). Values designated by an asterisk are significantly different from the corresponding controls. Superscript numbers indicate where different treatments are significantly different.

Treatment	Period of exposure			
	24 hr	72 hr	1 week	2 weeks
Cu^{2+}	98.33 ± 0.72	97.00 ± 0.58	$96.29 \pm 0.47^{**1}$	$97.36 \pm 0.47^{*}$
Cu^{2+} +EDTA	99.27 ± 0.28	96.80 ± 0.46	98.60 ± 0.29	98.40 ± 0.50
Cd^{2+}	$96.60 \pm 0.70^{*2}$	97.85 ± 0.63	$99.70 \pm 0.15^{1,2}$	99.15 ± 0.21
CD^{2+} +EDTA	97.93 ± 0.62	$95.97 \pm 1.36^{*3}$	$99.76 \pm 0.11^{1,3}$	99.33 ± 0.24^{3}
Sea water (controls)	99.17 ± 0.20	98.41 ± 0.36	98.98 ± 0.20	98.95 ± 0.21

*$p<0.05$; **$p<0.01$

[1]The result of exposure to Cu^{2+} for 1 week is significantly less than those resulting from exposure to Cd^{2+} ($p<0.01$) and Cd^{2+} +EDTA ($p<0.01$) for 1 week.

[2]The result of exposure to Cd^{2+} for 24 hr is significantly less ($p<0.05$) from that resulting from exposure to Cd^{2+} for 1 week.

[3]The result of exposure to Cd^{2+} +EDTA for 72 hr is significantly less ($p<0.01$) from those resulting from exposure to Cd^{2+} +EDTA for 1 and 2 weeks.

Table 2. Serum lysozyme specific activity levels in *Crassostrea virginica* at four time intervals of exposure to 1 ppm of Cu^{2+}, Cu^{2+} +EDTA, Cd^{2+}, Cd^{2+} +EDTA, and sea water. The figures represent specific activities ± standard errors (n = 4/reported value). Value designated by an asterisk is significantly different from the corresponding control. Superscript numbers indicate where different treatments are significantly different.

Treatment	Period of exposure			
	24 hr	72 hr	1 week	2 weeks
Cu^{2+}	5.00 ± 0.35[1]	4.53 ± 0.47	3.50 ± 0.48	3.27 ± 0.39[4]
Cu^{2+} +EDTA	4.80 ± 0.39	5.11 ± 0.41[2,4]	4.67 ± 0.28	4.89 ± 0.44
Cd^{2+}	3.50 ± 0.41	3.77 ± 0.52	2.83 ± 0.23*[1,2,3]	3.57 ± 0.19
CD^{2+} +EDTA	3.93 ± 0.43	3.70 ± 0.59	3.03 ± 0.24[2,3]	2.96 ± 0.18[2,3]
Sea water (controls)	5.05 ± 0.33[3,4]	4.93 ± 0.34[3]	3.76 ± 0.26	3.76 ± 0.22

*$p < 0.05$

[1]The result of exposure to Cu^{2+} for 24 hr is significantly higher ($p < 0.05$) than that resulting from exposure to Cd^{2+} for 1 week.

[2]The result of exposure to Cu^{2+} +EDTA for 72 hr is significantly higher than those resulting from exposure to Cd^{2+} for 1 week ($p < 0.05$), Cd^{2+} +EDTA for 1 week ($p < 0.05$) and 2 weeks ($p < 0.05$).

[3]The results of exposure to Cd^{2+} for 1 week ($p < 0.01$, 0.05, respectively), Cd^{2+} +EDTA for 1 week ($p < 0.01$, 0.05, respectively) and 2 weeks ($p < 0.01$, 0.05, respectively) are significantly lower than those resulting from exposure to sea water for 24 hr and 72 hr.

[4]The result of exposure to sea water for 24 hr is significantly higher than that resulting from exposure to Cu^{2+} for 2 weeks ($p < 0.05$).

Table 3. Serum acid phosphatase activity levels in *Crassostrea virginica* at four time intervals of exposure to 1 ppm of Cu^{2+}, Cu^{2+} +EDTA, Cd^{2+}, Cd^{2+} +EDTA, and sea water. The figures represent mU/mg protein ± standard errors (n = 4/reported value). Values designated by an asterisk are significantly different from the corresponding controls. Superscript numbers indicate where different treatments are significantly different.

Treatment	Period of exposure			
	24 hr	72 hr	1 week	2 weeks
Cu^{2+}	0.02 ± 0.003	0.01 ± 0.003*[1,2,4]	0.01 ± 0.004*[1,2]	0.02 ± 0.004
Cu^{2+} +EDTA	0.03 ± 0.005	0.02 ± 0.004[1]	0.01 ± 0.001*[1,3,4,5]	0.03 ± 0.0006
Cd^{2+}	0.03 ± 0.003	0.04 ± 0.005[1]	0.02 ± 0.004[1]	0.02 ± 0.003[1]
CD^{2+} +EDTA	0.02 ± 0.003	0.04 ± 0.005[2,3]	0.03 ± 0.004[3]	0.01 ± 0.002[1,2,4,6]
Sea water (controls)	0.03 ± 0.002[5]	0.04 ± 0.006[5,6]	0.04 ± 0.006[4,6]	0.02 ± 0.002

* $p < 0.01$

[1]The result of exposure to Cd^{2+} for 72 hr is significantly higher than those resulting from exposure to Cu^{2+} for 72 hr ($p < 0.01$) and 1 week ($p < 0.01$), Cu^{2+} +EDTA for 72 hr and 1 week ($p < 0.05$, 0.01, respectively), Cd^{2+} for 1 and 2 weeks ($p < 0.05$, 0.05, respectively), and Cd^{2+} +EDTA for 2 weeks ($p < 0.01$).

[2]The result of exposure to Cd^{2+} +EDTA for 72 hr is significantly higher than those resulting from exposure to Cu^{2+} for 72 hr ($p < 0.01$) and 1 week ($p < 0.01$) and Cd^{2+} +EDTA for 2 weeks ($p < 0.01$).

[3]The results of exposure to Cd^{2+} +EDTA for 72 hr and 1 week are significantly higher than that resulting from exposure to Cu^{2+} +EDTA for 1 week ($p < 0.01$, 0.01, respectively).

[4]The result of exposure to sea water for 1 week is significantly higher than those resulting from exposure to Cu^{2+} for 72 hr ($p < 0.05$), Cu^{2+} +EDTA for 1 week ($p < 0.01$), and Cd^{2+} +EDTA for 2 weeks ($p < 0.05$).

[5]The result of exposure to Cu^{2+} +EDTA for 1 week is significantly lower than those resulting from exposure to sea water for 24 hr ($p < 0.05$) and 72 hr ($p < 0.01$).

[6]The result of exposure to Cd^{2+} +EDTA for 2 weeks is significantly lower than those resulting from exposure to sea water for 72 hr ($p < 0.01$) and 1 week ($p < 0.05$).

Statistical analyses of the cellular acid phosphatase activity levels in the five subgroups of oysters bled at the four time intervals in all possible combinations (data not shown) revealed that none of the differences was significant.

DISCUSSION AND CONCLUSIONS

A number of factors must be considered when interpreting the data presented herein. It is well known that because of the occurrence of open circulatory systems in aquatic pelecypods (and gastropods) and the occurrence of diapedesis and influx of water, the volume of whole hemolymph fluctuates significantly. As a result, the number of hemocytes, the volume of serum, and the serum protein concentration also fluctuate. Consequently, the minor inconsistencies reflected in the data presented in Tables 1, 2, and 3 are to be expected. In spite of these, there are some conspicuous trends. These are presented below.

Hemocyte viability. The results presented pertaining to hemocyte viability indicate that *in vivo* exposure to 1 ppm of Cu^{2+}, Cd^{2+}, Cu^{2+} +EDTA, and Cd^{2+} +EDTA have significantly different effects. Initially (24 hr), only exposure to Cd^{2+} has a significant toxic effect on hemocytes. After 72 hr, only exposure to Cd^{2+} +EDTA significantly reduces the number of viable cells; after 1 and 2 weeks, only those cells that had been exposed to Cu^{2+} are significantly reduced in number. It is noted that in all instances where significant differences exist, exposure to sea water alone results in less death of hemocytes than exposure to 1 ppm of Cu^{2+}, Cd^{2+}, Cu^{2+} +EDTA, or Cd^{2+} +EDTA.

The conspicuous trends apparent in the data presented in Table 1 are as follows. (1) It appears that 1 ppm of Cd^{2+} is less cidal to *C. virginica* hemocytes than the same concentration of Cu^{2+}. (2) Exposure to 1 ppm of Cd^{2+} +EDTA is less deleterious than 1 ppm of Cu^{2+} after 1 week of exposure. This is not surprising since the binding constant (log K_o) of EDTA to Cd^{2+} is 16.5 (Williams, 1977). Consequently, as it has been shown that 1 ppm of Cd^{2+} alone is less deleterious than Cu^{2+}, it is reasonable to expect that 1 ppm of Cd^{2+} +EDTA would be less toxic. (3) Cu^{2+} +EDTA appears to be less toxic than Cu^{2+}. Again, this is not surprising since the binding constant (log K_o) of EDTA for Cu^{2+} is 18.7 (Williams, 1977), which is greater than for Cd^{2+}.

From the data presented, it may be concluded that 1 ppm of Cu^{2+} is more cidal to *C. virginica* hemocytes when exposed *in vivo* than the same concentration of Cu^{2+} +EDTA, Cd^{2+}, or Cd^{2+} +EDTA.

Lysozyme levels. The assays of serum lysozyme levels in all five subgroups of oysters at the four time periods have revealed that exposure to both Cu^{2+} and Cu^{2+} +EDTA does not alter the levels of this hydrolase when compared to the corresponding controls (Table 2); however, the serum lysozyme levels in oysters that had been exposed to Cu^{2+} and Cu^{2+} +EDTA are significantly higher than in oysters that had been exposed to Cd^{2+} and Cd^{2+} +EDTA at several time intervals (Table 2). Furthermore, the serum lysozyme levels in oysters that had been exposed to Cd^{2+} and Cd^{2+} +EDTA are significantly lower than those that had only been exposed to sea water (controls) at several time periods (Table 2). Thus, it may be concluded that a trend exists wherein *in vivo* exposure to 1 ppm of Cd^{2+} causes a reduction in lysozyme release from hemocytes into serum while exposure to Cu^{2+} does not quantitatively alter the release of cellular lysozyme into serum.

Acid phosphatase levels. Contrary to the results pertaining to lysozyme, analysis of the data relative to acid phosphatase has revealed that *in vivo* exposure to Cd^{2+} results in significantly higher levels of this hydrolase in the serum than as a result of exposure to Cu^{2+} at several time periods (Table 3). Moreover, when compared to sea water, exposure to Cu^{2+} and Cu^{2+} + EDTA causes significant reductions in the levels of serum acid phosphatase at 72 hr and 1 week. Also, exposure to Cd^{2+} or Cd^{2+} + EDTA does not cause significant fluctuations in the levels of serum acid phosphatase when compared to the controls (Table 3).

From the above, it may be concluded that exposure to Cu^{2+} appears to inhibit the release of acid phosphatase into serum while exposure to Cd^{2+} does not appear to alter the release of acid phosphatase from hemocytes into serum.

Analysis of the data relative to intracellular acid phosphatase has revealed that there were no significant fluctuations in the levels of this enzyme in any of the categories of oysters.

The influence of environmental additives and fluctuations on the immunocompetence of aquatic gastropods and bivalves has only recently attracted the attention of comparative immunologists. Crowding is known to alter the ability of molluscs to defend themselves against invaders. For example, Combes and Cheng (1986) have reported that crowding results in a significantly higher death rate of the gastropod *Biomphalaria glabrata* when exposed to a strain of *Aeromonas hydrophila* isolated from moribund snails.

Relative to the influence of environmental factors, Fisher and Newell (1986) have reported that salinity affects hemocyte locomotion in *C. virginica.* Specifically, they have shown that the locomotion of oyster hemocytes decreases with acute rises in salinity (from 6-8 ‰ to 27-32 ‰). Since hemocyte locomotion is directly correlated with the cell-parasite attraction phase of phagocytosis, the implication that may be drawn is that acute rises in salinity could influence cell-mediated immunity in marine bivalves.

In addition to the above, Fisher et al. (1987) have demonstrated that an acute decrease in salinity (from 30 ‰ to 15 ‰) decreases adherence of latex beads to hemocytes of *Ostrea edulis* at 4 hr and 8 hr post change. Similarly, these investigators have shown that an increase in temperature (from 20°C to 25°C and from 15°C to 20°C and 25°C) decreases adherence of latex beads to hemocytes of *O. edulis* at 4 hr and 8 hr post change. Also, Fisher and Tamplin (1988) have reported that an acute rise in salinity (from 6 ‰ to 30 ‰) increases binding of latex beads and also increases the time required for *C. virginica* hemocytes to spread. Since both the spreading and binding ability of *C. virginica* hemocytes are attributes associated with their ability to phagocytose nonself particles, it may be concluded that alterations in ambient temperature and salinity may affect cell-mediated immunity in *O. edulis* and *C. virginica.*

More recently, Cheng (1988a) has reported that *in vivo* exposure of *C. virginica* to 1 ppm of Cu^{2+} reduces the percentage of hyalinocytes after 1 and 2 weeks of exposure and *in vivo* exposure to 1 ppm of Cd^{2+} elevates the percentage of hyalinocytes after 1 and 2 weeks of exposure. Since Foley and Cheng (1975) have demonstrated that granulocytes of *C. virginica* (and also *Mercenaria mercenaria*) are significantly more phagocytic than hyalinocytes, it would appear that *in vivo* exposure to 1 ppm of Cd^{2+} may compromise cell-mediated immunity in *C. virginica.*

In addition to the above, Cheng (1988b) has reported that *in vivo* exposure of *C. virginica* retards phagocytosis but stimulates endocytosis of *Escherichia coli*; however, *in vivo* exposure of the same oyster to 1 ppm of Cd^{2+} suppresses endocytosis but stimulates phagocytosis of *E. coli.* Thus, in view of the interpretations presented herein and from what is known, it may be concluded that *in vivo* exposures to Cu^{2+} and Cd^{2+} have different effects. It is now also apparent that *in vivo* and *in vitro* exposures of *C. virginica* hemocytes to Cu^{2+} and Cd^{2+} have different effects. This conclusion is based on the comparison of the results of Cheng (1988b) pertaining to *in vivo* exposure to those of Cheng and Sullivan (1984) relative to *in vitro* exposure of oyster hemocytes to 1 ppm of Cu^{2+} or Cd^{2+}. In the earlier study, no differences were detected in the ability of *C. virginica* granulocytes to phagocytose polystyrene beads.

In conclusion, it is becoming increasingly more apparent that in the case of aquatic ectothermic animals with open circulatory systems, such as certain pelecypods and gastropods, environmental factors, including fluctuations in temperature and salinity and additives can compromise both their cell-mediated and humoral immune mechanisms.

ACKNOWLEDGEMENTS

I wish to thank Caroline S. Keller for her capable technical assistance and Dr. Hurshell H. Hunt, Department of Biometry, Medical University of South Carolina, for his expert assistance in statistical analysis.

LITERATURE CITED

Cheng, T.C. 1978. The role of lysosomal hydrolases in molluscan cellular response to immunologic challenge. Comp. Pathobiol., 4, 59-71.

Cheng, T.C. 1979. The role of hemocytic hydrolases in the defense of molluscs against invading parasites. Haliotis, 8, 193-209.

Cheng, T.C. 1983. The role of lysosomes in molluscan inflammation. Am. Zool., 23, 129-144.

Cheng, T.C. 1988a. *In vivo* effects of heavy metals on cellular defense mechanisms of *Crassostrea virginica*: total and differential cell counts. J. Invert. Pathol., 51, 207-214.

Cheng, T.C. 1988b. *In vivo* effects of heavy mmetals on cellular defense mechanisms of *Crassostrea virginica*: phagocytic and endocytotic indices. J. Invert. Pathol., 51, 215-220.

Cheng, T.C. 1989. Immunodeficiency diseases in marine molluscs: measurement of some parameters. J. Aquat. Anim. Health, in press.

Cheng, T.C. and C. Combes. 1989. Influence of environmental factors on the invasion of molluscs by parasites: with special reference to Europe. In: "Biological Invasions in Europe and the Mediterranean Basin" (F. di Castri and A.J. Hansen, eds.), in press. Dr. W. Junk, Fed. Rep. Germany.

Cheng, T.C. and J.C.U. Downs. 1988. Intracellular acid phosphatase and lysozyme levels in subpopulations of oyster, *Crassostrea virginica*, hemocytes. J. Invert. Pathol., 52, 163-167.

Cheng, T.C. and E. Rifkin. 1970. Cellular reactions in marine molluscs in response to helminth parasitism. In: "Diseases of Fish and Shellfish," pp. 443-496. Am. Fisher. Soc. Symp. Vol. 5, Washington, D.C.

Cheng, T.C. and J.T. Sullivan. 1984. Effects of heavy metals on phagocytosis by molluscan hemocytes. Mar. Environ. Res., 14, 305-315.

Combes, C. and T.C. Cheng. 1986. Control of biomedically important molluscs. Arch. Inst. Pasteur Algérie, 55, 153-193.

Environmental Protection Agency (EPA). 1980. "Interim Methods for the Sampling and Analysis of Priority Pollutants in Sediments and Fish Tissue." U.S. EPA, Cincinnati, Ohio.

Fisher, W.S. and R.I.E. Newell. 1986. Salinity effects on the activity of granular hemocytes of American oyster, *Crassostrea virginica*. Biol. Bull., 170, 122-134.

Fisher, W.S. and M. Tamplin. 1988. Environmental influence on activities and foreign-particle binding by hemocytes of American oysters, *Crassostrea virginica*. Canad. J. Fisher. Aquat. Sci., 45, 1309-1315.

Fisher, W.S., M. Auffret and G. Balouet. 1987. Response of European flat oyster (*Ostrea edulis*) hemocytes to acute salinity and temperature changes. Aquaculture, 67, 179-190.

Fishman, W.H. and F. Lerner. 1953. A method for estimating the serum acid phosphatase of prostatic origin. J. Biol. Chem., 200, 89-97.

Foley, D.A. and T.C. Cheng. 1975. A quantitative study of phagocytosis by hemolymph cells of the pelecypods *Crassostrea virginica* and *Mercenaria mercenaria*. J. Invert. Pathol., 25, 189-197.

Fries, C.R. 1984. Protein hemolymph factors and their roles in invertebrate defense: a review. Comp. Pathobiol., 6, 49-109.

Guillard, R.L. 1973. Methods for microflagellates and nannoplankton. In: "Handbook of Phycological Methods – Culture Methods and Growth Measurements," (J.R. Stein, ed.), pp. 69-85. Cambridge Univ. Press, London.

Jourdane, J. and T.C. Cheng. 1987. The two-phase recognition process of allografts in a Brazilian strain of *Biomphalaria glabrata*. J. Invert. Pathol., 49, 145-158.

Lowry, O.H., N.J. Rosebrough, A.L. Farr and R.I. Randall. 1951. Protein measurement with Folin phenol reagent. J. Biol. Chem., 193, 265-275.

Neter, J., W. Wasserman and M.H. Kitner. 1983. "Applied Linear Regression Models." R.D. Irwin, Homewood, Illinois.

Phillips, H.J. 1973. Dye exclusion tests for cell viability. In: "Tissue Culture Methods and Applications" (P.F. Fruse, Jr. and M.K. Patterson, Jr., eds.), pp. 406-408. Academic Press, New York.

Rodrick, G.E. and T.C. Cheng. 1974. Activities of selected hemolymph enzymes in *Biomphalaria glabrata* (Mollusca). J. Invert. Pathol., 24, 374- 375.

Shugar, D. 1952. The measurement of lysozyme activity and the ultraviolet inactivation of lysozyme. Biochim. Biophy. Acta, 8, 302-309.

Tripp, M.R., A. Bisignani and M.T. Kenny. 1966. Oyster amoebocytes *in vitro*. J. Invert. Pathol., 8, 137-140.

Williams, R.J.P. 1977. Calcium chemistry and its relation to protein binding. In: "Calcium-Binding Proteins and Calcium Function" (R.H. Wasserman, R.A. Corradino, E. Carafoli, R.H. Kretsinger, D.H. MacLennan and F.L. Siegel, eds.), pp. 3-12. North-Holland, New York.

H_2O_2 PRODUCTION BY HEMOCYTES AS A METHOD OF EVALUATING THE DEFENSIVE CAPACITY OF SCALLOPS UNDER CULTURE

Katsuyoshi Mori[1], Motoichi Nakamura[2], and Tadashi Nomura[2]

[1]National Research Institute of Aquaculture, Fisheries Agency, Nansei, Mie, 516-01 Japan
[2]Faculty of Agriculture, Tohoku University, Sendai, 980 Japan

ABSTRACT: Hydrogen peroxide production by hemocytes of the scallop, *Patinopecten yessoensis*, in various sown culture areas in Hokkaido, Japan, was ascertained. As a result, it has been demonstrated that the healthiest scallops produced the highest level of H_2O_2. Thus, it is proposed that the measurement of H_2O_2 production by scallop hemocytes could be employed as an indicator of the health of the animal maintained in mariculture.

INTRODUCTION

Physiological activity in marine bivalve molluscs has been investigated in relation to their energy metabolism mainly during sexual maturation and spawning since this is regarded as one of the most fundamental pieces of information for analyzing the causes of unusual mass mortalities among molluscs under culture (Mori et al., 1965; Mori, 1979). The environmental conditions in and around bivalve molluscan culture areas in Japan are not always favorable for their culture and, in addition, there are many problems related to cultural techniques awaiting solution, strongly suggesting the possibility that unusual mass mortalities of marine bivalve molluscs may reoccur frequently in some principal culture areas. Hence, it will become increasingly necessary to forecast the occurrence of such mass mortalities through daily management, which requires much accurate information on the physiological activity of the animals under culture, to take preventive measures against mortalities.

In the past we have assessed the physiological activity in such important bivalve molluscs as the oyster, *Crassostrea gigas*, and the scallop, *Patinopecten yessoensis*, by analyzing the results on the respiratory metabolism of tissues, ciliary activity of gill, glycogen content of whole soft body, and mortality rate (Mori, 1975, 1979). Now a simpler and more precise method makes use of aspects of defense mechanisms as a

Pathology in Marine Science

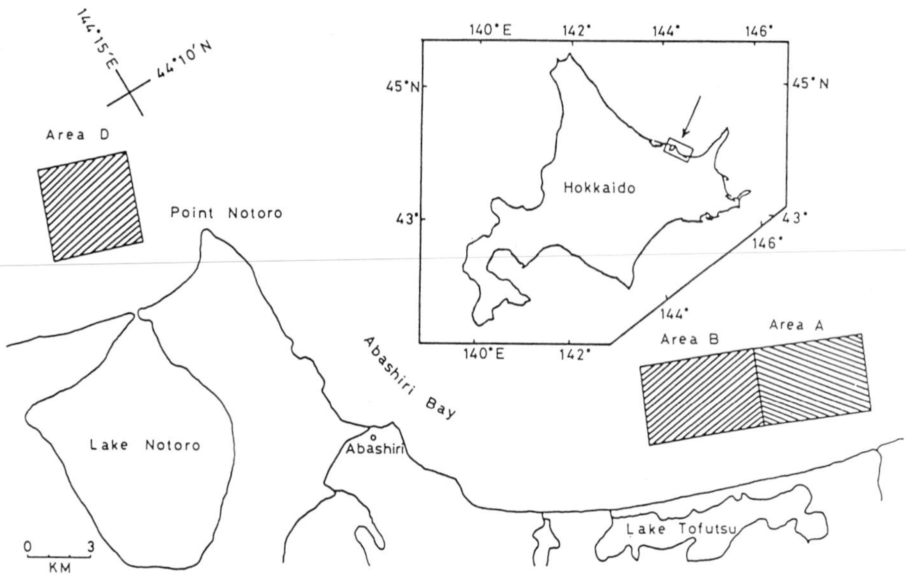

Fig. 1. Location of sampling areas (A, B, and D) for scallops in Abashiri waters, Hokkaido, Japan.

practical means of evaluating physiological activity, including natural agglutinin(s) in oysters (Mori et al., 1980a, 1984; Mori, 1987). Thus, a new dimension has been added to the existing significance of invertebrate immunologic defense reactions.

Our previous study (Nakamura et al., 1985) demonstrated the presence of an H_2O_2-generating system in scallop hemocytes. The purpose of this paper is to ascertain whether the capacity for H_2O_2 production by hemocytes can be practically used as an indicator for the total defensive capacity, i.e., the physiological activities of bivalve molluscs under culture.

H_2O_2 production by scallop hemocytes. The scallops, *P. yessoensis*, were collected from the sowing cultures in Abashiri waters, Hokkaido, Japan (Fig. 1). They were transferred to the laboratory at Tohoku University, Sendai, by airplane and maintained in a recirculating sea water system at $12 \pm 2^{\circ}C$ until use.

Homovanillic acid (HVA, 4-hydroxy-3-methoxyphenylacetic acid), horseradish peroxidase (HRP, type II), and disodium EDTA were purchased from Sigma Chemical Co., USA. Other reagents were of analytical reagent grade.

HEPES (N-2-hydroxyethylpiperazine-N′-2-ethanesulfonic acid) buffered saline (0.55 M NaCl, 0.01 g/liter phenol red, and 20 mM HEPES, pH 8.0), 3% (w/v) disodium EDTA in HEPES buffered saline, HEPES buffered artificial sea water (HASW, 460 mM NaCl, 9.4 mM KCl, 10.8 mM $CaCl_2$, 21.3 mM $MgCl_2 \cdot 6H_2O$, 20.7 mM $MgSO_4 \cdot 7H_2O$, 0.01 g/liter phenol red, and 20 mM HEPES, pH 8.0), H_2O_2-reaction medium (HASW) containing 0.5 mM HVA, 2.0 mM glucose, 2.0 mM NaN_3, and 2 nM HRP), and 0.4 M glycine-NaOH buffer, pH 12.0, containing 25 mM EDTA were employed for collection of hemocytes and H_2O_2 assay.

To obtain hemocytes, 0.9 ml of hemolymph was drawn from the ventricle using a disposable syringe containing 0.1 ml of 3% EDTA solution, fitted with a 24-gauge needle. Subsequently, the hemolymph sample was kept at 4°C until use. The number of cells was determined with a Thoma hemocytometer by use of a phase contrast microscope.

Quantitative assay of H_2O_2 was carried out according to the methods of Takeshige et al. (1979) and Ruch et al. (1983) with a slight modification. Figure 2 shows the principle of the detection of H_2O_2 by peroxidase-mediated oxidation of reduced HVA.

A hemolymph sample (0.1 ml) was added to 0.8 ml of the H_2O_2-reaction medium in a sterile glass tube (10 x 75 mm), and this cell suspension was preincubated for 10 min at 12°C. After the addition of bacteria (0.1 ml), the incubation was continued for 30 min at 12°C. Then, 0.15 ml of 0.4 M glycine-NaOH buffer was added to stop the oxidative reaction of HVA, followed by centrifugation at 1,000*g* for 10 min. The increase of fluorescence in the cell-free supernatant fluid was measured with a Hitachi 204 fluorospectrophotometer (excitation at 315 nm and emission at 425 nm).

The control experiments were performed with the following systems: (1) reaction medium + hemocytes + saline (without bacteria), (2) reaction medium without HRP + hemocytes + bacteria, and (3) reaction medium without HVA + hemocytes + bacteria.

Micrococcus luteus IFO 3333 were cultured for 20 hr in trypticase soy broth and autoclaved for 25 min at 121°C. The resulting heat-killed bacteria were washed with HEPES-buffered saline four times by centrifugation at 10,000*g* for 10 min and suspended in the same solution at the concentration of 10 mg wet weight/ml.

A 10 mM stock solution of H_2O_2 in distilled water was stored at 4°C in the dark. The actual concentration of this solution was determined from its adsorption at 230 nm, using an extinction coefficient of 81 M^{-1} cm^{-1} (Homan-Müller et al., 1975). Hydrogen peroxide standard solutions were prepared by diluting the stock solution with HASW just before use.

The data are presented as mean ± standard error (SE). The statistical significance was evaluated on the basis of the Fisher's randomization test.

Scallop hemocytes produced some amount of H_2O_2 even at the resting state (Table 1, Assay 1), indicating that the production of H_2O_2 by hemocytes is one of the physiological or normal functions in the scallop. A significant stimulation of the

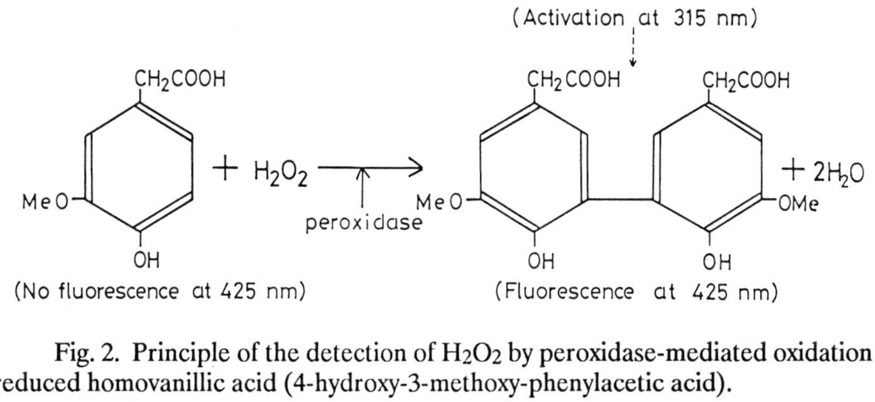

Fig. 2. Principle of the detection of H_2O_2 by peroxidase-mediated oxidation of reduced homovanillic acid (4-hydroxy-3-methoxy-phenylacetic acid).

production was observed when the addition of bacteria was made on the assay system (Assay 2). No production of H_2O_2, however, was detected in the assay solution without HVA (Assay 3) or HRP (Assay 4), showing that the H_2O_2 assay used here is based on the oxidation of HVA mediated by HRP.

Table 1. H_2O_2 production by scallop hemocytes: conditions of the assay.

Assay No.	Composition	H_2O_2 production[a]
1	Reaction medium[b] + hemocytes (10^6)	0.45 ± 0.06
2	Reaction medium + hemocytes (10^6) + bacteria[c] (1.0 mg wet wt)	0.77 ± 0.06
3	Reaction medium without HVA + hemocytes (10^6) + bacteria (1.0 mg wet wt)	0.00 ± 0.00
4	Reaction medium without HRP + hemocytes (10^6) + bacteria (1.0 mg wet wt)	0.00 ± 0.00

[a]nmoles H_2O_2/ml/30 min ± SE (1 and 2: on 5 samples; 3 and 4: on 3 samples). Temperature during measurements = 12°C.

[b]Reaction medium: modified Herbst's artificial sea water containing 2 nM horseradish peroxidase (HRP), 0.5 mM homovanillic acid (HVA), 2 mM NaN_3 and 2 mM glucose.

[c]*Micrococcus luteus* (IFO 3333)

1 vs. 2: $p < 0.05$ (t test)

H_2O_2 production by hemocytes as a method of evaluating defensive capacity of scallops under culture. In Abashiri Bay (Fig. 1), scallop sowing-culture began in 1978. In 1980, however, high mortalities and markedly slow growth of sown scallops were observed in the Abashiri A area; the seed scallops sown in 1978 suffered from mass mortalities involving more than 96% (as of September, 1980) of the population. Hence, we have studied the growth and the survival of scallops in Abashiri sowing-culture grounds from 1982 to 1984. In Table 2 is tabulated a part of the results obtained in this study.

Usually the total wet weight of sown scallop at 3 years of age in Funka Bay (Hokkaido), the Okhotsk Sea coast, and the Abashiri A area are 250g, 170g, and 80g, respectively. The scallops in the Abashiri A area grow up to only 1/3 the weight of those in Funka Bay. It is known that scallops grow well in the Abashiri D and in other grounds located in the western part of the Okhotsk Sea coast. In the Abashiri B area, they grow as poorly as in the Abashiri A area.

The scallops collected from the Abashiri A area (Table 2, Expt. 3) are a part of the animals surviving the catastrophic mortalities. Hence, it is presumed that the physiological activity of these animals is very high. In actuality, the capacity for H_2O_2 production by hemocytes of these survivors was much higher than that of other samples taken at the same time (Expts. 1 and 2).

Lower production of H_2O_2 was found in the scallops deliberately exposed to high water temperatures (18° - 20°C) overnight before H_2O_2 assay (Expts. 4 and 5). The natural water temperatures in their habitats were falling to almost 0°C during their sampling.

Hydrogen peroxide production by hemocytes of the scallops from the sowing cultures in the Abashiri D area decreased with an increase in water temperature from May to July (Expts. 6, 8, 11); this decrease in H_2O_2 production partially coincides with such seasonal changes in physiological activity of the scallops as has been reported (Mori, 1975; Mori et al., 1980b).

Special field experiments were carried out during the winter and spring seasons, including the drifting ice period by using underwater cage enclosures or corrals (Fig. 3), 5x5x1.5 meters, which were constructed from stretched nylon mesh attached to a frame of galvanized pipe. A mesh skirt 60-cm deep with galvanized chain at its outer perimeter extended outwards from each corral. Four corrals were deployed on natural substrates at a depth of 30 meters in the Abashiri A area (Fig. 1). The speed of water current inside the corral varied from 76.0 to 89.5% of that outside it, depending upon the location of measurement. The average speed was 82.0%. Placing and sampling of the experimental animals were conducted by professional divers. The scallop density in the corral was 3 individuals/meter^2.

In Abashiri waters, the octopus (*Paractopos dofleini dofleini*) and the starfish (*Asterias amurensis* and *Distolasterias nippon*) are the main predators of scallops in

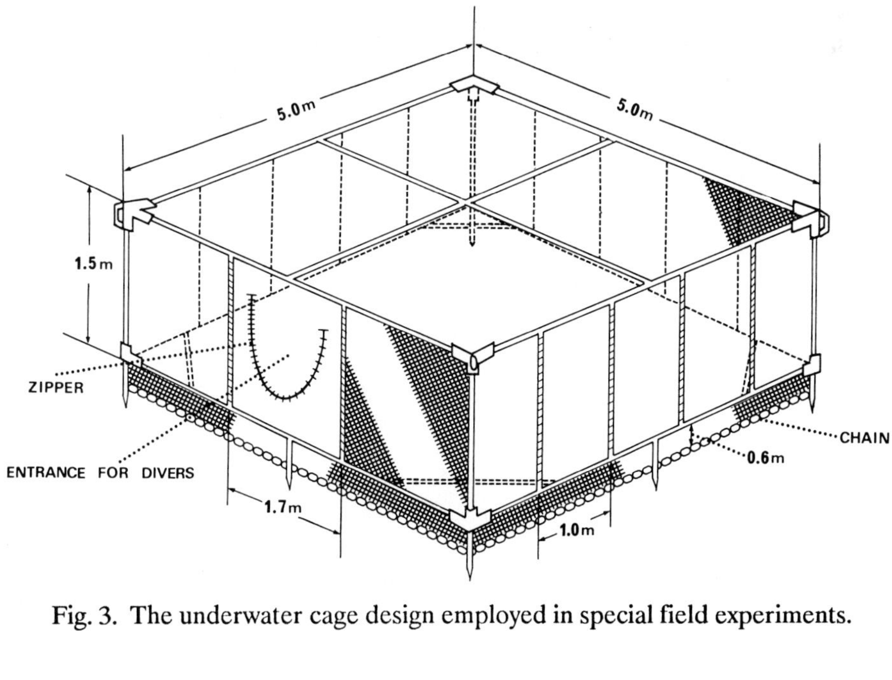

Fig. 3. The underwater cage design employed in special field experiments.

Table 2. H_2O_2 production by hemocytes of the scallops from the sowing cultures in Abashiri, Hokkaido, Japan.

Expt. No.	Culture area	Sampling time	Age (mos.)	H_2O_2 production[a] ± SE		Fold increase[b]
				Resting	Phagocytosing	
1	D (6)	Dec. 1982	31	1.24 ± 0.24	2.04 ± 0.26	1.65
2	B (5)	Dec. 1982	43	0.94 ± 0.24	2.13 ± 0.43	2.27
3	A (5)	Dec. 1982	67	2.43 ± 0.53	3.15 ± 0.50	1.30
4	A (5)[c,d]	Jan. 1983	32	0.60 ± 0.17	1.24 ± 0.24	2.07
5	D (6)[d]	Jan. 1983	32	0.54 ± 0.21	1.21 ± 0.21	2.24
6	D (12)	May 1983	36	2.92 ± 0.48	4.25 ± 0.57	1.46
7	B (12)	May 1983	48	1.38 ± 0.32	3.00 ± 0.70	2.17
8	D (23)	June 1983	37	2.39 ± 0.23	4.07 ± 0.38	1.70
9	A (12)[e]	June 1983	37	2.19 ± 0.23	4.76 ± 0.62	2.17
10	A (12)[c]	June 1983	37	1.57 ± 0.18	3.12 ± 0.41	1.99
11	D (12)	July 1983	38	1.98 ± 0.44	3.33 ± 0.51	1.68
12	B (12)	July 1983	50	1.44 ± 0.23	2.14 ± 0.27	1.49

Hemocytes ($0.2 - 3.3 \times 10^6$) were added to 0.8 ml of the reaction medium and preincubated for 10 min. Then, further incubation was carried out for 30 min at 12°C. Figure in parentheses shows the number of scallops examined. Resting: added no bacteria; phagocytosing: added 1 mg (wet wt.) of bacteria (*M. luteus*).

[a]nmoles H_2O_2/ml/1.0×10^6 cells/30 min.

[b](Phagocytosing)/(Resting).

[c]Native population.

[d]Exposed to high water temperature (18 - 20°C) overnight.

[e]The Abashiri-D scallops reared for 6 months in the underwater cage enclosure on the natural bottom substrate at 30 m in the Abashiri A area. 3 vs. 1 and 2, 4 and 5 vs. 1, 6 vs. 11 and 9 vs. 10: $p < 0.05$ (t test).

sowing cultures. Hence, the above mentioned corral is considered one of the best structures for the growth and survival of the scallops under culture since it guaranteed their natural habitat with no predation by starfish, etc.

The field experiments consisted of three different systems. In the first, the corrals contained no predators inside (Table 2, Expt. 9). No predators from outside the corrals were observed to enter a corral. In the second, they contained starfish such as *A. amurensis* and *D. nippon*, which were placed inside deliberately. In the third, the experimental scallops were caught from outside the corrals (Expt. 10). The mean final mortality of scallops reared for 6 months in duplicate corrals containing no predators was approximately 5.5%. No surviving scallops, however, were observed in duplicate corrals containing starfish 6 months after the start of the experiment.

Table 2 shows that there was a significant difference in H_2O_2 production by hemocytes between the Abashiri-D scallops reared for 6 months in the corral deployed in the Abashiri A area (Expt. 9) and the scallops native to the same area (Expt. 10); this difference suggests that the level of H_2O_2 production by hemocytes is influenced by the life history of scallops.

Since the increase is the ratio of H_2O_2 production at the phagocytosing state to H_2O_2 production at the resting state, lower H_2O_2 production at the resting state usually leads to higher increases. When H_2O_2 production (nmoles H_2O_2/ml/$1x10^6$ cells/30 min) at the resting state is less than 1, the increase goes up to over 2. The increase of the scallops reared for 6 months in the corral was more than twofold in spite of high H_2O_2 production (over 2) at the resting state, reflecting that extremely high H_2O_2 production was found in the phagocytosing state (Expt. 9). This means that the scallops grown in the underwater cage enclosure, namely under the most ideal living conditions, have a high degree of cellular defense capacity. A combination of the H_2O_2 production level at the resting state of 2.0 nmoles H_2O_2/ml/$1.0x10^6$ cells/30 min or more and the increase level of twofold or above can be assumed to serve as a practical index of a high degree of cellular defense; that is, the very good condition of scallop health.

The above results reveal that the capacity for H_2O_2 production by hemocytes is available for practical use as an indicator for the defensive capacity of scallops under culture.

LITERATURE CITED

Homan-Müller, J.W.T., R. S. Weening and D. Roos. 1975. Production of hydrogen peroxide by phagocytizing human granulocytes. J. Lab. Clin. Med., 85, 198-207.

Mori, K. 1975. Seasonal variation in physiological activity of scallops under culture in the coastal waters of Sanriku district, Japan, and a physiological approach to a possible cause of their mass mortality. Bull. Mar. Biol. Stn. Asamushi Tohoku Univ., 15, 59-79.

Mori, K. 1979. Effects of artificial eutrophication on the metabolism of the Japanese oyster *Crassostrea gigas*. Mar. Biol., 53, 361-369.

Mori, K. 1987. Oyster self-resistance mechanisms to diseases. In: "Aquaculture: Shellfish Culture Development and Management" (International Seminar held at La Rochelle, France, 1985), pp. 127-159. IFREMER, Brest, France.

Mori, K., T. Imai, K. Toyoshima and I. Usuki. 1965. Studies on the mass mortality of the oyster in Matsushima Bay. IV. Changes in the physiological activity and the glycogen content of the oyster during the stages of sexual maturation and spawning. Bull. Tohoku Reg. Fish. Res. Lab., 25, 49-63.

Mori, K., S. Itsukaichi, K. Murayama and T. Nomura. 1980a. Activities of agglutinin and bactericidin in oyster tissues. Bull. Jap. Soc. Sci. Fish., 46, 1385-1389.

Mori, K., Y. Tone, T. Suzuki, K. Kasahara and T. Nomura. 1980b. Defense mechanisms of molluscs-I. Bactericidal and agglutinin activities in the scallop tissues. Bull. Jap. Soc. Soc. Fish., 46, 717-722.

Mori, K., K. Murayama, N. Kanno, M. Nakamura, E. Ohira, Y. Kato and T. Nomura. 1984. Occurrence and characterization of the defense factors in the Japanese oyster *Crassostrea gigas*. Tohoku J. Agr. Res., 35, 55-68.

Nakamura, M., K. Mori, S. Inooka and T. Nomura. 1985. *In vitro* production of hydrogen peroxide by the amoebocytes of the scallop, *Patinopecten yessoensis* (Jay). Dev. Comp. Immunol., 9, 407-417.

Ruch, W., P.H. Cooper and M. Baggiolini. 1983. Assay of H_2O_2 production by macrophages and neutrophils with homovanillic acid and horseradish peroxidase. J. Immunol. Methods, 63, 347-357.

Takeshige, K., T. Matsumoto, R. Shibata and S. Minakami. 1979. Simple and rapid method for the diagnosis of chronic granulomatous disease, measuring hydrogen peroxide and superoxide anions released from leukocytes in whole blood. Clin. Chim. Acta, 92, 329-335.

BLUE CRAB HEMOCYTES: A PRELIMINARY STUDY

Jerome F. La Peyre and Fu Lin E. Chu

Virginia Institute of Marine Science, School of Marine Science,
The College of William and Mary, Gloucester Point, Virginia 23062, USA

ABSTRACT: A preliminary study on hemocyte activity of the blue crab, *Callinectes sapidus*, revealed that hemocytes are very labile *in vitro*. Adhesion and phagocytosis of horse red blood cells by crab hemocytes was low. After separation into three subpopulations of hemocytes, using continuous Percoll gradients, viability was low. The problems associated with the preparation of monolayers of blue crab hemocytes for immunological assays are discussed.

INTRODUCTION

Three broad classes of crustacean hemocytes are generally recognized: hyaline cells, semigranular cells, and granulocytes (Bauchau, 1981). It is believed that these cells form a continuous differentiation series, with classification being arbitrarily based on the number and size of granules in the cytoplasm of these cells. Hyaline cells have no or few small granules, semigranular cells have a moderate number of granules, and granulocytes have many granules and some of these are large. The functions of the various types of inclusion bodies for the most part are unknown (Bodammer, 1978). Recently, Hose et al. (1987) classified crustacean (shrimp) hemocytes into four types comprising two distinct hemocyte lines.

Crustacean hemocytes are multifunctional. They are involved in hemolymph coagulation, wound repair, phagocytosis, capsule and nodule formation, transport and synthesis of carbohydrates, molting, and vitellogenesis (Bauchau, 1981). There are functional differences between the three morphological types (hyaline cells, semigranular cells, and granulocytes) of hemocytes. The prophenoloxidase system, which plays an important role in the host cellular immunity (e.g., cell communication, opsonization, microbial killing), is located in the secretory vesicles of semigranular cells and granulocytes in most species (Söderhäll and Smith, 1986). The hyaline cells appear to be the most phagocytic (Bauchau, 1981). As for other decapods, blue crab hemocytes can be distinguished as hyaline cells, semigranular cells, and granulocytes, even though agranular hyaline cells are not always found (Bodammer, 1978).

Hemocyte activities of several decapods have been studied and phagocytosis of various foreign particles by hemocytes has been reported, but little information is available on the hemocyte activities and internal defense of the blue crab, *Callinectes sapidus*. Correspondence between the cell types and functions need to be investigated. This paper reports the preliminary results of the first stage of our study on blue crab hemocytes, namely evaluation and modification of existing methodology for cell separation and for measuring cellular activities of *C. sapidus*.

MATERIALS AND METHODS

Blue crabs were captured in traps at the Virginia Institute of Marine Science's pier on the York River (salinity was about 20 ‰). Hemolymph samples (1 ml) were taken from each adult crab that appeared active and healthy. Hemolymph was drawn with a sterile syringe equipped with a 19-gauge needle from the infrabranchial sinus at the base of the fifth pereopod, after swabbing the exoskeleton with 70% ethanol. To prevent coagulation, the syringe was filled with blue crab saline (9 ml) (Table 1) containing cysteine (Sigma, free base, 20 mg/ml) as an anticoagulant (McKay and Jenkins, 1970). Blue crab saline was made to resemble the ionic composition of *C. sapidus* hemolymph for crabs collected in 20 ‰ (Colvocoresses et al., 1974). More than 90% of hemocytes were found viable 6 hr after collection as determined by the trypan blue (0.2%) exclusion method. Total and differential cell counts, hyaline cells, semi-granular cells, and granulocytes, were determined with a hemacytometer.

Table 1. Composition of blue crab saline.

Constituents	g/l
NaCl	16.947
Na_2So_4	2.700
$MgCl_2 \cdot 6H_2O$	2.541
$CaCl_2$	1.685
KCl	0.522
$NaHCO_3$	0.618
Dextrose	0.550
HEPES	2.380
H_2O	1.0 Liter
Adjusted to pH 7.8 with 1 N NaOH	

Two milliliters of hemocyte suspension (1×10^6 cells) in blue crab saline was added to a 3.5-cm Petri dish containing a glass coverslip (16 x 16 mm). Hemocytes were allowed to settle for 15 min.

Horse red blood cells (HRBC) were washed twice with blue crab saline and resuspended at 1×10^9 cells/ml. Foreign particle adherence was measured by the addition of 100 μl of HRBCs (1×10^9 cells/ml) to the crab hemocyte monolayer and incubated at room temperature (22°C) for 1.5 hr. After incubation, the monolayers were rinsed with blue crab saline, fixed in methanol (5 min), and stained with Hemal 1 (12 sec) and Hemal 2 (22 sec) solutions (Hemal stain Co.). Percent of adhesive hemocytes was determined by counting 500 hemocytes with a microscope. Phagocytosis of HRBCs by hemocytes was also quantified using a similar method; the only difference was that nonphagocytosed HRBCs were lysed by rinsing the monolayer with 0.2% NaCl after incubation.

The efficacy of using continuous Percoll gradients to separate blue crab hemocytes was assessed as follows. Continuous gradients of 50% Percoll (Pharmacia) in 24 ‰ NaCl were formed by centrifugation in round bottom polycarbonate tubes in an angle-head rotor (34°) at 25,000*g* for 20 min (highspeed centrifuge Sorvall 30, Beckman). One milliliter of hemocyte suspension (5×10^7 cells) was introduced below the surface of the gradient and centrifuged in a swing out rotor at 400*g* for 20 min at 20°C (refrigerated centrifuge CR411, Jouan). The percentage of each cell type in each band was measured as well as cell viability in Percoll. Each hemocyte fraction in Percoll was then diluted by a factor of five, centrifuged at 60*g* for 10 min, and resuspended in blue crab saline. Cell viability of each hemocyte subpopulation in blue crab saline was determined as described earlier.

RESULTS

C. sapidus has an average concentration of circulating hemocytes of 5.8×10^7 cells/ml. Hyaline cells represented less than 5% of the hemocyte population. Roundish cells with no or few granules and a large centrally located nucleus were classified as hyaline cells. Ovoid cells with non-refractile granules were classified as semi-granular cells. Granulocytes have typical refractile granules. Semi-granular cells and granulocytes accounted for 75% and 20%, respectively, of the hemocyte population when counted immediately after bleeding. A majority of the hemocytes gradually changed from a round or ovoid shape to a spindle shape with apparent agranular cytoplasm. Thus, it made the distinction between spindle shaped hyaline cells and semigranular cells very difficult.

Separation yielded two distinct cell bands (1 and 3) and one diffuse band (2) (Fig. 1). Hemocyte viability in Percoll was higher for band 1 (71%) and band 2 (94%) than for band 3 (41%) (Table 2). Hemocyte viability in blue crab saline after washing, was low for all cells in band 1 (28%), band 2 (7%), and band 3 (13%). All cells in band 1 were spindle-shaped so that the ratio of hyaline cells to semigranular cells in this band could not be estimated. Band cell purity was based on the ratio of refractile cells (granulocytes) or non-refractile cells (hyaline and semigranular cells) to the total number of cells times 100.

Hemocytes were bright and compact in blue crab saline. Upon contact with the glass coverslips, non-refractile cells (hyaline or semigranular cells) started to extend spike-like extensions and extensive cell spreading followed. Refractile cells (granulocytes) did not spread until about 30 min, followed by degranulation and vacuolation. Viability of blue crab hemocyte attached to the monolayer after 30 min

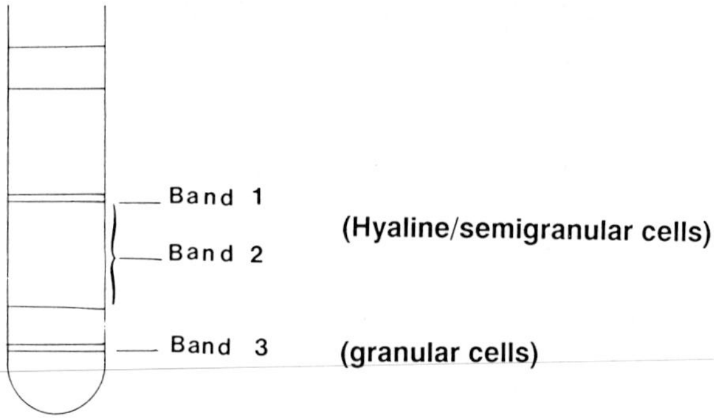

Figure 1. Cell bands after separation with continuous Percoll gradient.

ranged from 100% to 62%. After 1.5 hr, the percent of viable cells ranged from 83% to 42%.

Adhesion of HRBC to blue crab hemocyte was found to be low (<5%). Similarly, phagocytosis of HRBC was very low (<1%).

Table 2. Percentage viability in Percoll gradient after hemocyte separation and after resuspension in blue crab saline.

	% Viability		% Purity
	In Percoll	In saline	
Band 1 (Hyaline/Semigranular cells)	71% ± 6.8	28% ± 14.5	98% ± 1.0
Band 2 (Hyaline/Semigranular cells)	41% ± 19.1	7% ± 12.1	96% ± 2.7
Band 3 (Granular cells)	94% ± 9.2	13% ± 9.0	100

(1) Percent purity for both band 1 and 2 is calculated by taking the ratio of the hyaline and semi-granulocytes to the total number of cells times 100.

(2) Cells from four different crabs were separated. For each crab two Percol gradients were loaded with the hemocyte suspension.

DISCUSSION AND CONCLUSIONS

Viability of blue crab hemocytes on monolayer observed in our study (62% ± 17.0) was not as high as those reported for other crustaceans (> 95 %) (McKay and Jenkins, 1970; Paterson and Stewart, 1974). It is possible to separate blue crab hemocytes into three cell bands, but the viability of hemocytes was low after being washed with blue crab saline (Table 2). The cause of this may be due to the deficiency of certain amino acids in the blue crab saline. For example McKay and Jenkins (1970) used a crayfish saline supplemented with dialyzed crayfish plasma. Paterson and Stewart (1974) added MEM essential amino acids plus glutamine and MEM vitamins to their lobster saline.

It has also been shown that the prophenoloxidase (ProPo) system located in the secretory vesicles of semigranular and granular cells, once released from the cells and activated, has lytic and degranulating properties towards the hemocytes (Smith and Söderhäll, 1983). Any stimulus (e.g., fungal or bacterial cell wall components) or disturbances that may be responsible for releasing ProPo molecules need to be controlled. A number of changes may thus improve blue crab hemocyte viability during cell separation and monolayer preparation. These include the addition of essential amino acids to the blue crab saline and the use of endotoxin-free solutions and equipment such as that used in insect studies to improve cell stability (Mead et al., 1986). Additionally, for cell separation, the following modifications may help: (1) decreasing the pH of the medium, which may increase plasma membrane stability (Bell et al., 1971); (2) lowering the temperature during cell manipulations; and (3) switching to the citrate-EDTA anticoagulant buffer (Smith and Söderhäll, 1983).

Phagocytosis of foreign particles by decapod hemocytes is generally low (0.2 - 9.8% for untreated and opsonized sheep red blood cells) (McKay and Jenkins, 1970; Paterson and Stewart, 1974; Smith and Ratcliffe, 1978) but adherence (27.7 -29.6%) is usually much higher than our finding in blue crabs. Efficient phagocytosis usually depends on the specificity of serum opsonins (McKay and Jenkins, 1970; Paterson and Stewart, 1974).

Phagocytosis includes the processes of recognition, adherence, ingestion, destruction, and disposal. Low percent HRBC adherence may indicate that the ability of blue crab hemocyte to recognize and adhere to HRBC is poor. Consequently, the number of HRBC ingested by hemocytes is very low. As mentioned earlier, the ProPo system may have been released in our preparation and may be responsible for the variation of hemocyte viability on the monolayers and the low percent of phagocytosis. Depletion of the granules from semigranular cells of crustaceans following activation by the ProPo system appears to abolish their phagocytic potential (Söderhäll and Smith, 1986).

There appears to be considerable variation in hemocytic responses between decapod species. These variations may be real, but may be generated in part from the differing and controversial hemocyte classification schemes (morphological and functional). More cytophysiological studies on crustacean hemocytes are evidently needed. How endogenous and exogenous factors affect hemograms and cell functions needs to be determined. Separation of hemocytes by density differences may reflect differences in some but not all functions. As expressed by Bodammer (1978), the potential functional complexity of hemocyte is attested to by the various types of inclusion bodies in crustacean hemocytes. The multifunctionality of the hemocyte must be kept in mind.

The complete understanding of hemocyte monolayer characteristics and reproducibility of measurements are prerequisites for the use, ultimately, of the assay system for the future study of various pathological and immunological parameters.

ACKNOWLEDGEMENTS

Contribution No. 1515 from the Virginia Institute of Marine Science, The College of William and Mary. We wish to thank J. Walker for typing this manuscript. Mr. J. E. Warinner and Dr. B. J. Barber kindly reviewed the manuscript.

LITERATURE CITED

Bauchau, A.G. 1981. Crustaceans. In: "Invertebrate blood cells," Vol. 2 (N.A. Ratcliffe and A.F. Rowley, eds.), pp. 386-420. Academic Press, New York.

Bell, M.L., M.H. Lazarus, A.H. Herman, R.H. Egdahl and A.M. Rutenbury. 1971. pH dependent changes in cell membrane stability. Proc. Soc. Exp. Biol. Med., 136, 298-299.

Bodammer, J.E. 1978. Cytological observations on the blood and hemopoietic tissue in the crab, *Callinectes sapidus*. I. The fine structure of hemocytes from intermoult animals. Cell. Tissue Res., 187, 79-86.

Colvocoresses, J.A., M.P. Lynch and K.L. Webb. 1974. Variations in serum constituents of the blue crab, *Callinectes sapidus*: Major cations. Comp. Biochem. Physiol., 49A, 787-803.

Hose, J.A., G.G., Martin, V.A., Nguyen, J., Lucas and T. Rosenstein. 1987. Cytochemical features of shrimp hemocytes. Biol. Bull., 173, 178-187.

McKay, D. and C.R. Jenkins. 1970. Immunity in the invertebrates: The role of serum factors in phagocytosis of erythrocytes by haemocytes of crayfish (*Parachaeraps bicarinatus*). Aust. J. Exp. Biol. Med. Sci., 48, 130-170.

Mead, G.P., N.A. Ratcliffe and L.R. Renwratz. 1986. The separation of insect haemocyte types on Percoll gradients: Methodology and problems. J. Insect. Physiol., 32, 167-177.

Paterson, W.D. and J.E. Stewart. 1974. In vitro phagocytosis by haemocytes of the American lobster (*Homarus americanus*). J. Fish. Res. Bd. Canad., 31, 1051-1056.

Smith, V.J. and N.A. Ratcliffe. 1978. Host defence reactions of the shore crab *Carcinus maenas* (L) *in vitro*. J. Mar. Biol. Assoc. UK., 58, 367-379.

Smith, V.J. and K. Söderhäll. 1983. Induction of degranulation and lysis of haemocytes of the freshwater crayfish *Astacus astacus* by components of the prophenoloxidase activating system *in vitro*. Cell. Tiss. Res., 233, 295- 303.

Söderhäll, K. and V.J. Smith. 1986. The prophenoloxidase activating system: The biochemistry of its activation and role in arthropod cellular immunity, with special reference to crustaceans. In: "Immunity in Invertebrates" (M. Brehelin, ed.), pp. 208-233. Springer-Verlag, Berlin, Heidelberg, Germany.